Sheaf Theory through Examples

Sheaf Theory through Examples

Daniel Rosiak

The MIT Press
Cambridge, Massachusetts
London, England

The open access edition of this book was made possible by generous funding from Arcadia—a charitable fund of Lisbet Rausing and Peter Baldwin.

The MIT Press would like to thank the anonymous peer reviewers who provided comments on drafts of this book. The generous work of academic experts is essential for establishing the authority and quality of our publications. We acknowledge with gratitude the contributions of these otherwise uncredited readers.

This book was set in LaTeX by the author. Printed and bound in the United States of America.

Library of Congress Cataloging-in-Publication Data

Names: Rosiak, Daniel, author.
Title: Sheaf theory through examples / Daniel Rosiak.
Description: Cambridge, Massachusetts : The MIT Press, [2022] | Includes bibliographical references and index.
Identifiers: LCCN 2021058949 | ISBN 9780262542159 (paperback)
Subjects: LCSH: Sheaf theory.
Classification: LCC QA612.36 .R67 2022 | DDC 514/.224–dc23/eng20220521
LC record available at https://lccn.loc.gov/2021058949

To Fernando Zalamea, who guided my early foray into sheaves, and whose encouragement and example helped my love of category theory flourish.

Contents

Acknowledgments

I didn't set out to write a book on sheaves. In 2017, as I was wrapping up my PhD in philosophy, I received a Fulbright grant to work with Fernando Zalamea in Bogotá, the ostensible aim of which was to support the completion of a brief supplementary chapter giving a high-level overview of certain philosophical elements of interest in category theory and sheaf theory. While I had already been doing research in category theory for some years, as I came to devote more attention to sheaf theory in its own right that year I became entirely engrossed and found myself thinking about sheaves all day every day. By the end of that year, I found that I had produced over 200 pages of notes on sheaves with novel examples and reflections. A number of mathematicians with whom I shared these early notes found them stimulating and encouraged me to circulate them. While it was never my intention to write a book, many readers of those early notes encouraged me to develop them into something more. Over the next four years, as my research led me further into the land of sheaves, and guided by feedback from new readers, I continued to add new examples and content here and there, and at some point it became evident that these were no longer notes. It had grown into something close to the present book.

Over the years, I have tried to fill out the story in a number of ways. But on account of its somewhat accidental genesis, the book remains necessarily incomplete and my selection of examples and points of focus is certainly partial. On the other hand, after years of new additions and revisions, I do hope some of the initial excitement of those early notes still shines through.

There are many people whose feedback over the years has greatly improved this work, and many people from whom I have learned an immense amount along the way. Above all, I would like to express my profound gratitude to Fernando Zalamea, without whose support and encouragement, from the earliest stages, this book would never have come to be. Among the colleagues who offered encouragement and/or valuable feedback on various versions of this manuscript, I would like to thank Colin McLarty, Emily Riehl, Michael Robinson, David Spivak, Walter Tholen, and Philipp Zahn. I am especially grateful to Justin Curry and Alan Hylton, whose detailed comments, suggestions for substantial improvements to the organization of the manuscript, stimulating discussions, and ongoing encouragement helped me get the book over the finish line. Let me also express my indebtedness to William Lawvere, whose work has been a well of inspiration for me over the years; without his particular wisdom, I am sure that I would never have come to aspire to the new knowledge of reality that is thinking categorically. I would also like to thank the anonymous reviewers from MIT Press for their helpful feedback and corrections on an earlier version.

I am very grateful to the contributors to the nLab, for spending their time putting together many impressively rich reflections on category theory, sheaves, and even some philosophy. In an age when it is so easy to despair of the mountains of superficial content online, and so easy to find unrewarding ways of spending one's time looking at a screen, thank you for creating a corner of the internet with something profound and rewarding.

Finally, I would like to thank the editors at MIT Press, Jermey Matthews, Haley Biermann, and Marcy Ross; as well as Yasuyo Iguchi who designed the cover; Michael Durnin who copyedited the book; and Jim Mitchell and Theresa Carcaldi who managed production of the book. The book is much improved from what it was many versions ago. Thanks also to the team for helping with various LaTeX issues that came up, and for saving me the headaches. I am especially grateful to MIT Press for helping to make the book available in an open access version, allowing it to reach more readers.

And of course, my family and friends: for all their patience, love, and support. Without Michelle, I am certain that this book wouldn't have seen the light: thank you for encouraging me whenever I wanted to give it up and for reminding me of the point of it all.

Introduction

An Invitation

In many cases, events and objects are given to observation as extended through time and space, and so the data about these is local and distributed in some fashion. For now, we can think of this situation in terms of the data being indexed by, or attached to, given delimited regions or domains of some sensors. In a very general and rough way, by *local* we typically understand that something is being compared to what is around or nearby it; this is as opposed to the *global*, generally understood to mean compared to everything or across an entire domain of interest. Satisfying a property at a local level does not necessarily entail that the same will obtain at the global level. In saying that the data is local, we just mean that it holds throughout, or is defined for, a certain limited region; that is, its validity is restricted to a prescribed region or partial domain or reference context. We also use this language of locality to describe a way of evaluating a property or data ascribed to a part or point of its extended domain in terms of what that property or data looks like viewed from its immediate surroundings—that is, whenever it holds somewhere, it should also hold *nearby*.

We collect temperature and pressure readings and thus form a notion of ranges of possible temperatures and pressures over certain geographical regions; we record the fluctuating stockpile of products in a factory over certain business cycles; we accumulate observations or images of certain patches of the sky or the earth; we gather testimonies or accounts about particular events understood to have unfolded over a certain region of space-time; we build up a collection of test results concerning various parts of the human body; we amass collections of memories or recordings of our distinct interpretations of a certain piece of music; we develop observations about which ethical and legal principles or laws are respected throughout a given region or network of human actors; we form a concept of our kitchen table via various observations and encounters, assigning certain attributes to those regions of space-time delimiting our various encounters with the table, where we expect that the ascribed properties or attributes are present throughout the entirety of a region of their extension. Even if certain phenomena are not intrinsically local, frequently their measurement or the method employed in data collection may still be local.

But even the least scrupulous person does not merely accumulate or amass local or partial data points. From an early age, we try to understand the various modes of *connections* and *cooperations* between the data, to patch these partial pieces together into a larger whole whenever possible, to resolve inconsistencies among the various pieces, to go on to build coherent and more global visions out of what may have been given to us only in pieces. As informed citizens or as scientists, we look at the data given to us on arctic sea-ice melting rates, on temperature changes in certain regions, on concentrations of greenhouse gases at various latitudes and various ocean depths, and so on, and we build a more global vision of the changes to our entire planet on the basis of the connections and feedbacks between these various data. As investigators of a crime, we must piece together a complete and consistent account of the events from the partial accounts of various witnesses. As doctors, we must infer a diagnosis and a plan of action from the various individual test results concerning the parts of a patient's body. We take our many observations concerning the behavior of certain networks of human actors and try to form global ethical guidelines or principles to guide us in further encounters.

Yet sometimes information is simply not local in nature. Roughly, one might think of such nonlocality in terms of how, as perceivers, certain attributes of a space may appear to us in a particular way but then cease to manifest themselves in such a way over subparts of that space, in which case one cannot really think of the perception as being built up from local pieces. For a different example: in the game of Scrabble™, one considers the assignment of letters, one by one, to the individual squares in a lattice of squares, with the aim of building words out of such assignments. One might thus suspect that we have something like a "local assignment" of data (letters in the alphabet) to an underlying space (15×15 grid of squares). Yet this assignment of letters to squares to form words is not really local in nature, since, while we do assign letters one by one to the grid of squares, the smallest unit of the game is really a *legal word*, but not all subwords or parts of words are themselves words, and so a given word (data assignment) over some larger region of the board may cease to be a word (possible data assignment) when we restrict attention to a subregion.

Even when information is local, there are many instances where we cannot synthesize our partial perspectives into a more global perspective or conclusion. As investigators, we might fail to form a coherent version of events because the testimonies of the witnesses cannot be made to agree with what other data or evidence tells us regarding certain key events. As musicians, we might fail to produce a compelling performance of a piece because we have yet to figure out how to take what is best in each of our trial interpretations of certain sections or parts of the entire score and splice them together into a coherent single performance or recording of the entire piece. A doctor who receives conflicting information from certain test results, or testimony from the patient that conflicts with the test results, will have difficulty making a diagnosis. In explaining the game of rock-paper-scissors to children, we tell them that rock beats scissors, scissors beats paper, and paper beats rock, but we cannot tell the child how to win *all the time*, that is, we cannot answer their pleas to provide them with a global recipe for winning this game.

For distinct reasons, differing in the gravity of the obstacle they represent, we cannot always lift what is local or partial up to a global value assignment or solution. A problem may have a number of viable and interesting local solutions but still fail to have even a single global solution. When we do not have the "full story," we might make faulty inferences. Ethicists might struggle with the fact that it is not always obvious how to pass from the instantiations or particular variations of a seemingly locally valid prescription, valid or binding for a subset of a network of agents, to a more global principle, valid for a larger network. In the case of the doctor attempting to make a diagnosis out of conflicting data, it may simply be a matter of either collecting more data, or perhaps resolving certain inconsistencies in the given test results by ignoring certain data in deference to other data. Other times, as in the case of rock-paper-scissors, there is simply nothing to be done to overcome the failed passage from the given local ranking functions to a global ranking function, for the latter simply does not exist. The intellectually honest person will eventually want to know if their failure to lift the local to the global is due to the inherent particularity or contextuality of the phenomena being observed or whether it is simply a matter of their own inabilities to reconcile inconsistencies or repair discrepancies in data-collecting methods so as to patch together a more global vision out of these parts.

Sheaf theory is the roughly seventy-year-old collection of concepts and tools designed by mathematicians to tame and precisely comprehend problems with a structure like the sorts of situations introduced above. I hope the reader will have noticed a pattern in the various situations just described. We produce or collect assignments of data indexed to certain regions where, whenever data is assigned to a particular region, we expect it to be applicable throughout the entirety of that region. In most cases, these observations or data assignments come already distributed in some way over the given network formed by the regions; but if not, they may become so over time, as we accumulate and compare more local or partial observations. In certain cases, together with the given value assignments and a natural way of decomposing the underlying space, revealing the relations between the regions themselves, there may emerge correspondingly natural ways of restricting assignments of data along the subregions of given regions. In such cases, in this movement of decomposition of the space and restriction of the data assigned to the space, the glue or system of translations binding the various data together, permitting some sort of transit between the partial data items, becomes explicit. In this way, an internal consistency among the parts may emerge, enabling the controlled gluing or binding together of the local data into an integrated whole that now specifies a solution or system of assignments over a larger region embracing all of those subregions. Such structures of coherence emerging among the partial patches of local data, once explicitly acknowledged and developed, may enable a unique global observation or solution, that is, an observation that no longer refers merely to yet another local region but now extends over and embraces all the regions at once. As such, it may even enable predictions concerning missing data or at least enable principled comparisons between various given groups of data.

Sheaves provide us with a powerful tool for precisely modeling and working with the sort of local-global passages indicated above. Whenever such a local-global passage is possible, the resulting global observations make transparent the forces of coherence between the local data points by exhibiting to us the principled connections and translation formulas between the partial information, making explicit the glue by which such partial and distinct clumps of data can be fused together, and highlighting the qualities of the distribution of data. And once in this framework, we may even go on to consider systematic passages or translations between distinct such systems of local-to-global data.

On the other hand, when faced with *obstructions* to such a local-global passage, we typically revise our basic assumptions, or perhaps the entire structure of our data, or maybe just our manner of assigning the data to our regions. We are usually motivated to do this in order to allow precisely such a global passage to come into view. When we can satisfy ourselves that nothing can be done to overcome these obstructions, we examine what the failure in this instance to pass from such local observations to the global can tell us about the phenomena at hand. *Sheaf cohomology* is a tool used for capturing and revealing precisely obstructions of this sort.

The very natural distinction between local and global, hinted at above, in fact posits a large class of problems involving relations between the local and global. For instance, given an overall domain of interest, or space, if we consider some part of that, when is it possible, just through knowledge of such portions, to deduce knowledge about the whole domain of interest? Perhaps unsurprisingly, the antagonism between the local and the global found its initial articulation within the frameworks of geometry and topology (the study of space), where there is a very natural account of locality or what it means for something to hold locally. One of the virtues of sheaves and associated techniques (like sheaf cohomology) is to have allowed for an appreciation of how this local-global dialectic is still more universal and reaches beyond its initial appearance in the context of topology and geometry.

The main purpose of this book is to provide an inviting and (I hope) gentle introduction to sheaf theory, where the emphasis is on explicit constructions and applications, using a wealth of examples from many different contexts. Sheaf theory is typically presented as a highly specialized and advanced tool, belonging mostly to algebraic topology and algebraic geometry (the historical homes of sheaves), and sheaves accordingly have acquired a somewhat intimidating reputation. Even when the presentation is uncharacteristically accessible, emphasis is typically placed on abstract results, and it is left to the reader's imagination (or "exercises") to consider some of the things they might be used for or some of the places where they can be found. This book's primary aim is to dispel some of this fear, to demonstrate that sheaves can be found all over (and not just in highly specialized areas of advanced math), and to give a wider audience of readers a more inviting tour of sheaves.

Especially over the last few years, the interest in sheaves among less specialized groups of people appears to be growing immensely, but whenever I spoke to newcomers about sheaves, they often expressed that the existing literature was either too specialized or too forbidding. This book accordingly also aims to fill a gap in the existing literature, which for the most part tends to either focus exclusively on a particular use of sheaves or assumes a formidable preexisting background and high tolerance for abstraction. I do not share the view that applications or concrete constructions are mere corollaries of theorems, or that examples are mere illustrations with no power to inform deeper conceptual advances. I am not sure if I would go as far as to endorse Vladimir Arnold's idea that "the content of a mathematical theory is never larger than the set of examples that are thoroughly understood," but I do believe that one barrier to the wider recognition of the immense power of sheaf theory lies in the tendency to present much of it as if it were a forbiddingly abstruse or specialized tool, or as belonging mainly to one area of math. One thing this book aims to show is that it is no such thing. Moreover, well-chosen examples are not only useful, both pedagogically and psychologically, in helping newcomers get a better handle on the abstract concepts and advance forward with more confidence but they can even jostle experts out of the rut of the "same old examples" and present interesting challenges both to our fundamental intuitions of the underlying concepts and to preconceptions we might have about the true scope of applicability of those concepts.

Before outlining the contents of the book and discussing some of its unique features, the next section offers a more explicit, but still naive, glimpse into the *idea* of a sheaf via a toy construction, with the aim of better establishing intuitions about the underlying sheaf idea.

A First Pass at the Idea of a Sheaf

Suppose we have some region, which, for the moment, we can represent very naively and abstractly as

We are less interested in the "space itself" and more in how the space serves as a site where various things *take place*. In other words, we think of this region as really just an abstract domain supporting various *happenings*, where such happenings carry information for appropriate sensors or measuring instruments (in a very generalized sense), so that interrogating the space becomes a matter of asking the sensors about what is happening on the space.[1] For instance, the region might be the site of some happenings that supply *visual information*, so that as a sensor monitors the happenings over a region (or some part of it), it collects specifically visual information about whatever is going on in the area of its purview:

1. The description of sheaves as "measuring instruments" or the "meter sticks" associated to a space that we are invoking—so that the set of all sheaves on a given space supply one with an arsenal of all the meter sticks measuring it, yielding "a kind of superstructure of measurement"—ultimately comes from Grothendieck, who was largely responsible for many of the key ideas and results in the early development of sheaf theory. In speaking of (another early sheaf theorist) Jean Leray's work in the 1940s, Grothendieck said this:

> The essential novelty in his ideas was that of the (Abelian) sheaf over a space, to which Leray associated a corresponding collection of cohomology groups (called "sheaf coefficients"). It is as if the good old standard "cohomological metric" which had been used up to then to "measure" a space, had suddenly multiplied into an unimaginably large number of new "meter sticks" of every shape, size and form imaginable, each intimately adapted to the space in question, each supplying us with very precise information which it alone can provide. This was the dominant concept involved in the profound transformation of our approach to spaces of every sort, and unquestionably one of the most important mathematical ideas of the 20th century. (Grothendieck 1986, promenade 12)

Then the sheaves on a given space will incorporate

> all that is most essential about that space . . . in all respects a lawful procedure (replacing consideration of the space by consideration of the sheaves on the space), because it turns out that one can 'reconstitute,' in all respects, the topological space by means of the associated 'category of sheaves' (or 'arsenal' of measuring instruments). . . . [H]enceforth one can drop the initial space. . . . [W]hat really counts in a topological space is neither its 'points' nor its subsets of points, nor the proximity relations between them; rather, it is the *sheaves on that space, and the category that they produce.* (Grothendieck 1986, promenade 13).

The reader for whom this is overwhelming should press on and rest assured that we will have a lot more to say about all this later in the book, and the notions and results alluded to in the above will be motivated and discussed in detail.

The related "sensor" perspective has been developed more recently, to great effect, in the work of Robert Ghrist, Michael Robinson, and Justin Curry, for example, Curry (2014, chap. 10).

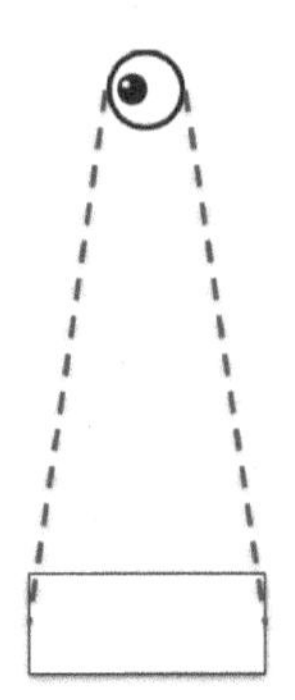

There might then be another sensor, taking in visual information about another region or part of some overall space, offering another "point of view" on another part of the space; and it may be that the underlying regions monitored by the two sensors overlap in part:

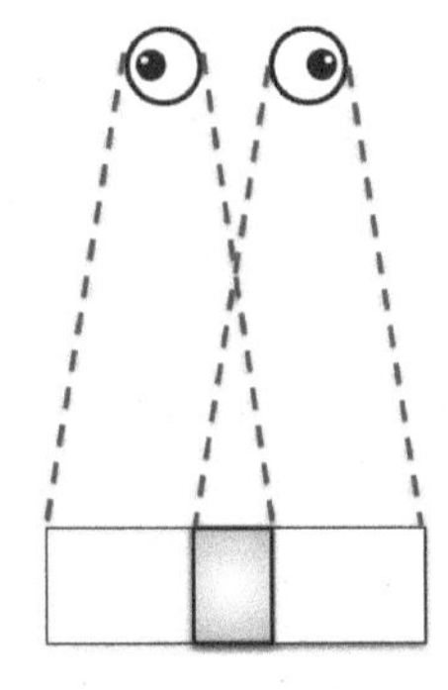

Since we are ultimately interested in the informative happenings attached to the space, we want to see how the distinct perspectives on what is happening throughout the space are themselves related; to this end, a very natural thing to do is to ask how the data collected by such neighboring sensors are related. Specifically, it is very natural to ask whether and how the perspectives are *compatible* on such overlapping subregions, whenever there are such overlaps between the underlying regions over which they, individually, collect data.

A little more explicitly: if we assume the first sensor collects visual data about its region (call it U_1), we may imagine, for concreteness, that the particular sort of data available to the sensor consists of sketches, say, of characters or letters (so that the underlying region acts as some sort of generalized sketchpad or drawing board):

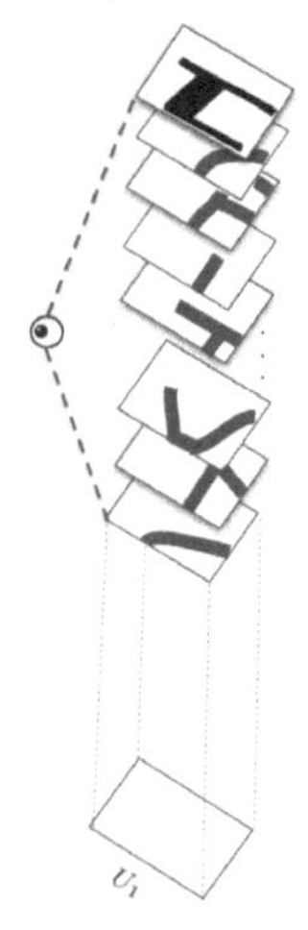

While not really necessary, the sensor might even be supposed to be equipped to process the information it collects, translating such visual inputs into reasonable guesses about which possible capital letter or character the partial sketch is supposed to represent. In any event, attempting to relate the two points of views by considering their compatibility on the region where their two surveyed regions overlap, we are really thinking about first making a selection from each of the collections of data assigned to the individual sensors:

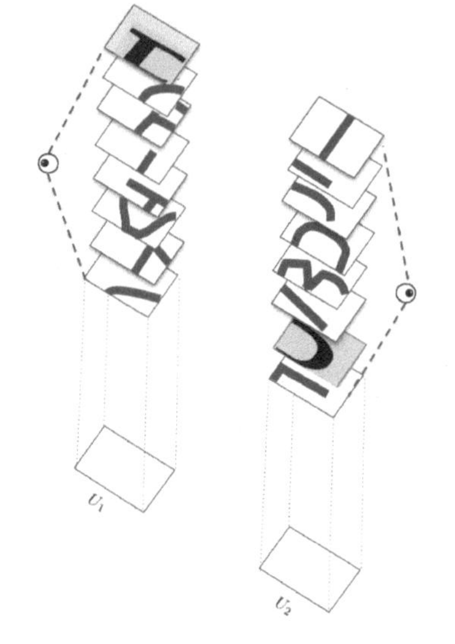

Corresponding to how the underlying regions are naturally related by a relation of inclusion, the compatibility question, undertaken at the level of the selections (highlighted in gray above) from the collections of all informative happenings on the respective regions, will involve looking at whether those data items "match" (or can otherwise be made "compatible") when we restrict attention to that region where the individual regions monitored by the separate sensors overlap:

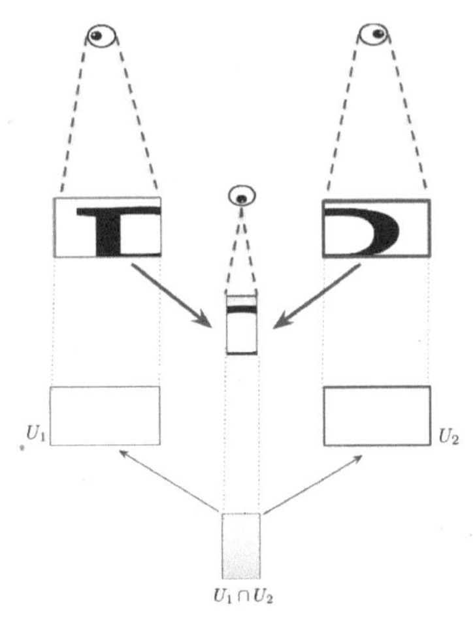

If the given selection from what they individually "see" does match on the overlap, then, corresponding to how the regions U_1 and U_2 may be joined together to form a larger region,

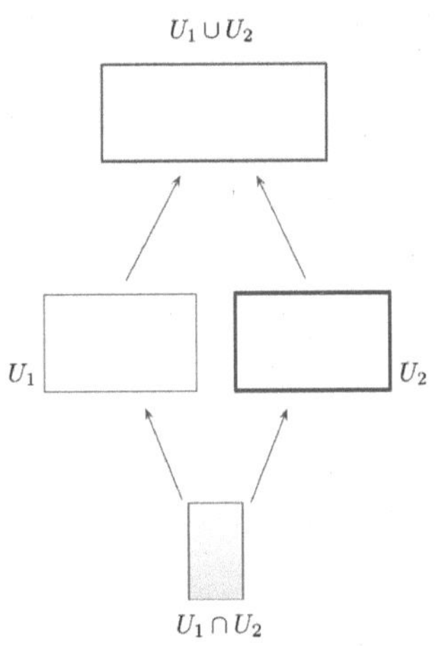

at the level of the data on the happenings over the regions, we can pull this data back into an item of data given now over the entire space $U_1 \cup U_2$, with the condition that we expect that restricting this new, more comprehensive, perspective back down to the original individual regions U_1 and U_2 will give us back whatever the two individual sensors originally saw for themselves:

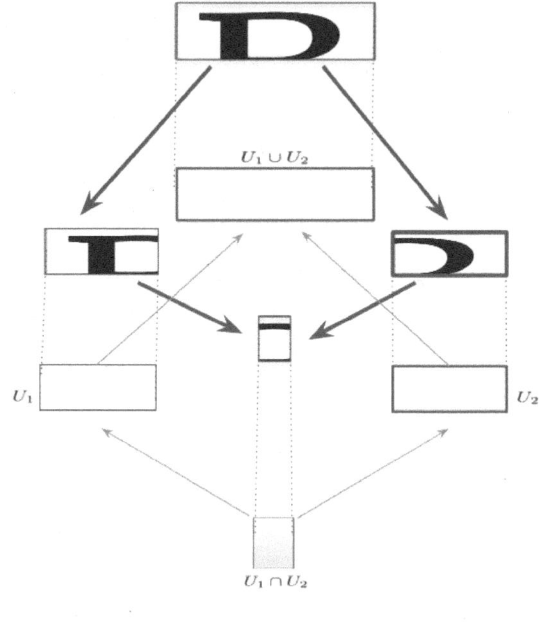

In other words, given some selection from what sensor 1 sees as happening in its region U_1 and from what sensor 2 sees as happening in its region U_2, provided their "story" agrees about what is happening on the overlapping region $U_1 \cap U_2$, then we can paste their individual visions into a single and more global vision or story about what is happening on the overall region $U_1 \cup U_2$—and we expect that this story ultimately "comes from" the individual stories of each sensor, in the sense that restricting the "global story" to region U_1, for instance, will recover exactly what sensor 1 already saw on its own.

Another way to look at this is as follows: while the sensor on the left, when left to its own devices, will believe that it may be seeing a part of any of the letters $\{B, E, F, P, R\}$, checking this assignment's compatibility with the sensor on the right amounts to constraining what the left sensor believes by what the sensor on the right knows, in particular that it cannot be seeing an E or an F. Symmetrically, the sensor on the right will have its own "beliefs" that might, in the matching with the left sensor, be constrained by whatever the left sensor "knows." In matching the two sensors along their overlap, and patching their perspectives together into a single, more collective perspective now given over a larger region (the union of their two regions), we are letting what each sensor individually knows constrain and be constrained by what the other knows.

In this way, as we cover more and more of a "space" (or, alternatively, as we decompose a given space into more and more pieces), we can perform such compatibility checks at the level of the data of the happenings on the site (our collection of regions covering a given space) and then glue together, piece by piece, the partial perspectives represented by each sensor's local data collection into more and more embracing or global perspectives. More concretely, continuing with our present example, suppose there are two additional regions, covering now some southwest and southeast regions, respectively, so that, altogether, the four regions cover some region (represented by the main square), where we have left implicit the obvious intersections ($U_1 \cap U_2$, $U_3 \cap U_4$, $U_1 \cap U_3$, etc.):

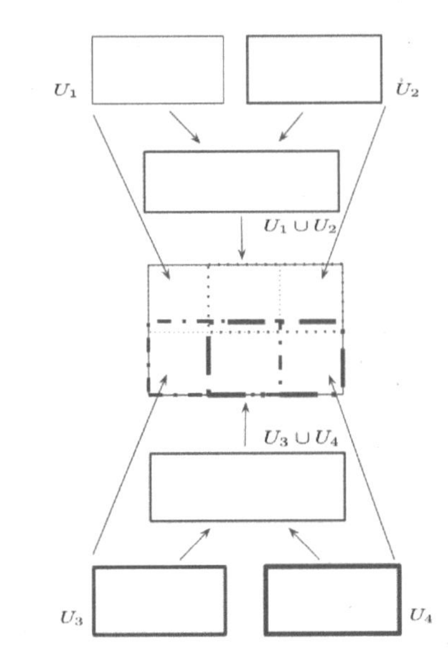

With the four regions U_1, U_2, U_3, and U_4, to each of which there corresponds a particular sensor, we have got the entire region $U = U_1 \cup U_2 \cup U_3 \cup U_4$ "covered." Part of what this means is that, were you to invite *another* sensor to observe the happenings on some further portion of the space, in an important sense this extra sensor would be superfluous—since, together, the four regions monitored by the four individual sensors already have the overall region covered.

For concreteness, suppose we have the following further selections of data from the data collected by each of these new (southwest and southeast) sensors, so that altogether, having performed the various compatibility checks (left implicit), the resulting system of points of view on our site can be represented as follows:

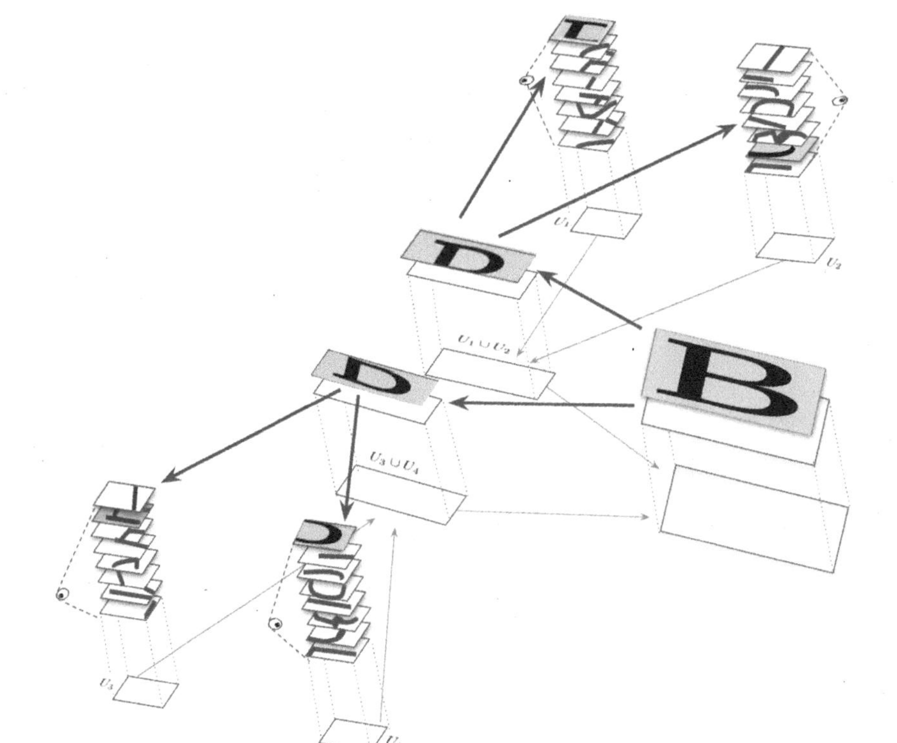

This system of mutually compatible local data assignments or "measurements" of the happenings on the space—where the various data assignments are, piece by piece, constrained by one another, and thereby patched together to supply an assignment over the *entire* space covered by the individual regions—is, in essence, what constitutes our *sheaf*. The idea is that the data assignments are being "tied together" in a natural way

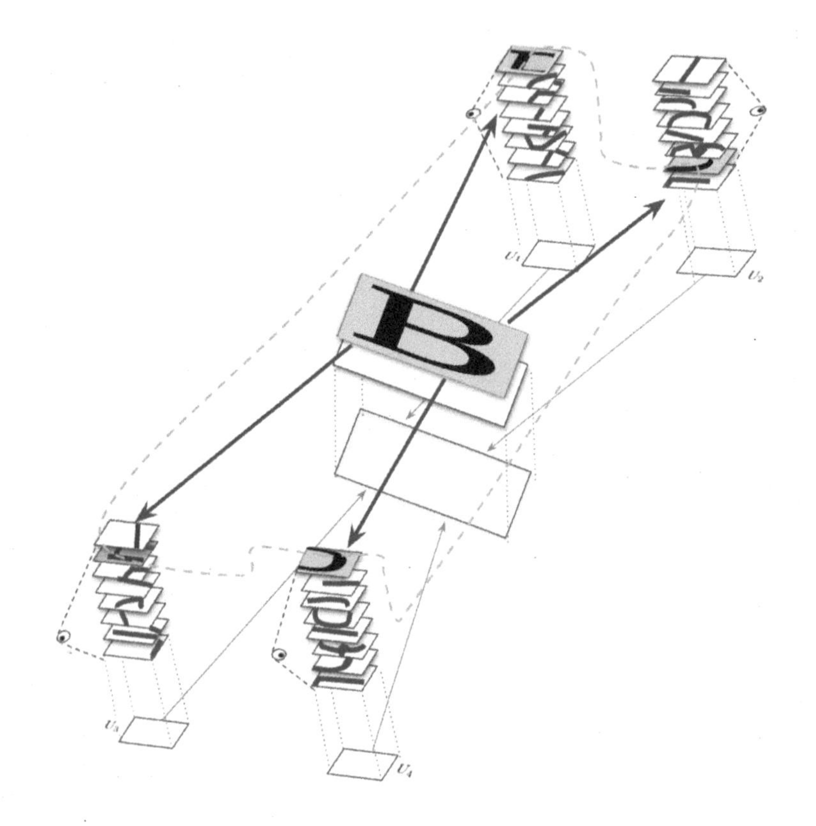

where this last picture is meant to serve as motivation or clarification regarding the agricultural terminology of "sheaf":

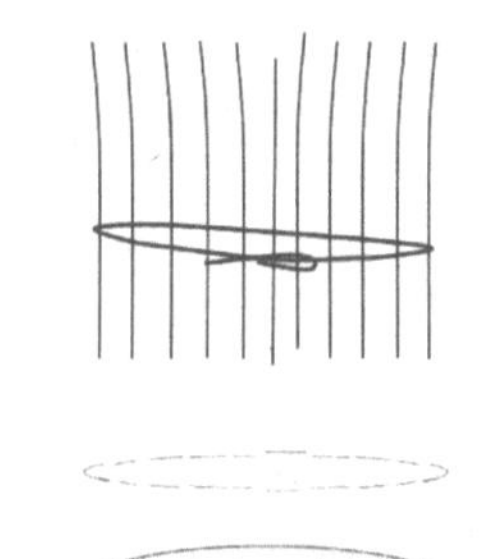

Here one thinks of various regions as the parcels of an overall space covered by those pieces, the collection of which then serves as a site where certain happenings are held to take place, and the abstract sensors capturing local snapshots or measurements of all that is going on in each parcel are then regarded as being collected together into "stalks" of data, regarded as sitting over (or growing out of) the various parts of the ground space to which they are attached. A selection of a particular snapshot made from each of the individual stalks (collections of snapshots) amounts to a cross-section and the process of restriction (along intersecting regions) and collation (along unions of regions) of these sections captures how the various stalks of data are bound together.

To sum up, then: the first characteristic feature of this construction is that some information is received or assigned *locally*, so that the records or observations made by each of the individual sensors are understood as being *about*, or indexed to, the entirety of some limited region, so that whenever something holds or applies at a point of that region, it will hold nearby as well. Next, since together the collection of regions monitored by the individual sensors may be seen as *collectively covering* some overall region, we can check that the individual sensors that cover regions that have some overlap can "communicate" their observations to one another, and a natural expectation is that, however different their records are on the nonoverlapping region, there should be some sort of *compatibility* or *agreement* or *mutual constraining* of the data recorded by the sensors over their shared, overlapping region; accordingly, we ask that each such pair of sensors covering overlapping regions "check in" with one another. Finally, whenever such compatibility can be established, we expect that we can bind together the information supplied by each sensor, and regard them as patching together into a *single sensor supplying data over the union* of the underlying (and partially overlapping) individual regions, in such a way that were we to restrict that single sensor back down to one of the original regions, we would recover exactly the partial data reported by the original sensor assigned to that individual region.

While most of the more fascinating and conspicuous examples of such a construction come from pure and applied math, something very much like the sheaf construction appears to be operative in so many areas of everyday life. For instance, related to the toy example discussed above, even the way our binocular vision system works appears to involve something like the collation of images into a single image along overlapping regions whenever there is agreement (from the input to each separate eye).[2] More generally, image and face recognition appears to operate, in a single brain (where clusters of neurons play the role of individual sensors), in something like the patchwork "sum of parts" way described above. Moving beyond the individual, collective knowledge itself appears to operate in a fundamentally similar way: a society's store of knowledge consists of a vast patchwork built up of partial records and data items referring to delimited (possibly overlapping) domains of interest, each of which data items can be (and often are!) checked for compatibility whenever they involve data that refers to, or makes claims about, the same underlying domain.

The very simple and naive presentation given to it above admittedly runs the risk of downplaying the power and scope of this construction; it would be difficult to overstate just how powerful the underlying idea of a sheaf is. An upshot of the previous illustration, though, is that while sheaves are often regarded as highly abstract and specialized constructions whose power derives from their sophistication, the truth is that the underlying idea is so ubiquitous, so "right before our eyes," that one might even be impressed that it

2. That visual information processing itself seems to fundamentally involve some sort of sheaf-like process appears even more acutely in other species, such as certain insects, like the dragonfly, whose compound eyes contain up to 30,000 facets, each facet within the eye pointing in a slightly different direction and taking in light emanating from only one particular direction, resulting in a mosaic of partially overlapping images that are then integrated in the insect brain.

was finally named explicitly so that substantial efforts could be made to refine our ideas of it. In this context, one is reminded of the old joke about the fish, where an older fish swims up to two younger fish, and greets them, “Morning, how’s the water?” After swimming along for some time, one of the younger fishes turns to the other and says,

> “What the hell is water?”

In this same spirit, Grothendieck would highlight precisely this “simplicity” of the fundamental idea behind sheaves (and, more generally, toposes):

> As even with the idea of sheaves (due to Leray), or that of schemes, as with all grand ideas that overthrow the established vision of things, the idea of the topos had everything one could hope to cause a disturbance, primarily through its “self-evident” naturalness, through its simplicity (at the limit naive, simple-minded, “infantile”)—through that special quality which so often makes us cry out: “Oh, that’s all there is to it!” in a tone mixing betrayal with envy, that innuendo of the “extravagant,” the “frivolous,” that one reserves for all things that are unsettling by their unforeseen simplicity, causing us to recall, perhaps, the long buried days of our infancy. (Grothendieck 1986, promenade 13)

Outline of Contents

The rest of the book is structured as follows. The first three chapters, together with the sixth and seventh chapters, are dedicated to exposition of the most important category theoretic concepts, tools, and results needed for the development of sheaves. Category theory is indispensable to the presentation and understanding of the notions of sheaf theory. While in the last decade there have appeared a number of accessible introductions to category theory,[3] feedback from readers of earlier drafts of this book convinced me that the best approach to an introduction to sheaves that aims to reach a much wider audience than usual would need to be as self-contained as possible. In these chapters, all the necessary categorical fundamentals are accordingly motivated and developed. The emphasis here, as elsewhere in the book, is on explicit constructions and creative examples. For instance, the concept of an *adjunction*, and key abstract properties of such things, is introduced and developed first through an extended example involving “dilating” and “eroding” an image, then through the development of “possibility” and “necessity” modalities applied to modeling the consideration of attributes of a person applied to them *qua* the different “hats” they wear in life, and then applied to graphs of traveling routes.

Chapter 1 introduces categories, some important examples of categories, and some of what one can do with categories.

Chapter 2 develops functors and presheaves in considerable depth. It discusses four main perspectives on presheaves, works through some notable examples of each of them, and develops some useful ways of understanding such constructions more generally. This is done both for its own sake and in order to build up to the following chapters,

3. The general reader without much, or any, background in category theory is especially encouraged to have a look at the engaging and highly accessible Spivak (2014). Readers with more prior mathematical experience may find Riehl (2016) a compelling introduction, displaying as it does the ubiquity of categorical constructions throughout many areas of mathematics. Lawvere and Rosebrugh (2003) are also highly recommended, especially for those readers content to be challenged to work many things out for themselves through thought-provoking exercises, often giving one the feeling of rediscovering things for oneself.

especially chapter 5, dedicated to the initial development of the sheaf concept. Natural transformations are also introduced in this chapter.

Chapter 3 covers universal properties and some important universal constructions.

All that is needed to offer a definition of a sheaf is the notion of a presheaf (covered in chapter 2) and some basic notions from topology, such as that of a cover. With the aim of exposing the reader to the sheaf notion sooner rather than later, chapter 4 covers the requisite notions from general topology, and raises some more philosophical questions that are taken up in later parts of the book (including the appendix).

Chapter 5 introduces sheaves (on topological spaces) and some key sheaf concepts and results through some initial examples. Throughout this chapter, some of the vital conceptual aspects of sheaves in the context of topological spaces are motivated, teased out, and illustrated through the various examples.

Chapter 6 is dedicated to the Yoneda results—perhaps the most important idea in category theory—and the associated Yoneda philosophy.

Chapter 7 returns to, and completes, the treatment of categorical foundations for sheaves, by covering adjunctions. As usual, the key features of this construction are teased out through a variety of examples and worked-out constructions.

Chapter 8 returns to sheaves and covers some more involved results, rooted in historically significant examples. This chapter also includes a section on what is *not* a sheaf, or when and how the sheaf construction fails, as well as an important case where the notions of sheaf and presheaf coincide.

Chapter 9 is dedicated to a "hands on" introduction to sheaf cohomology. The centerpiece of this chapter is an explicit construction, with worked-out computations, involving sheaves on complexes. There is also a brief look at *cosheaves* and an interesting example relating sheaves and cosheaves.

Chapter 10 revisits and revises a number of earlier concepts, and develops sheaves from the more general perspective of Grothendieck toposes. The important notions are motivated and developed through a variety of examples.

We move through various layers of abstraction, from sheaves on a site (with a Grothendieck "topology") to elementary toposes, the topic of chapter 11. The later sections of chapter 11 are devoted to illustrations, through concrete examples, of some slightly more advanced topos-theoretical notions. The book concludes with an abridged presentation of some special topics, including a brief glimpse into *cohesive toposes*. There are many other directions the book could have taken at this point, and more advanced sheaf-theoretical topics that might have been considered, but in the interest of space, attention has been confined in that final section to the special topic of cohesive toposes.

Finally, there is an appendix, dedicated to exploring in greater depth the open philosophical questions raised in chapter 4 on general topology and the concept of space, doing so by building on some of the constructions introduced in chapter 7's treatment of adjunctions.

Remarks on Distinct Features of This Book

This book has three notable features that may deserve brief discussion:

1. an emphasis on pictures;
2. an emphasis on detailed worked-out examples from different areas of application; and

3. an emphasis on ideas.

Regarding the first of these: a colleague once told me that they had read an entire book on sheaf theory, but it was not until years later, after they saw a simple and evocative picture drawn of a certain sheaf, that they finally felt like they understood what sheaves were about. I suspect that this person is by no means alone in their experience. If this is really a fair description of the experience of some newcomers to sheaves, you could say that, at least as far as sheaves are concerned, a picture is worth not a thousand words but many thousands of words! Inspired by this experience, I have tried to include, throughout the book, a great many pictures.

The second feature of the book is that it takes part in the burgeoning area of *applied category theory*, and as such aims to expand the repertoire of examples of sheaves, beyond those that have already had great impact within mathematics. As in any area of life, there can be a kind of "groupthink" that takes over an academic niche, and examples are usually the first things to suffer the negative consequences of this common phenomenon—for instance, many standard texts on sheaves start with the constant sheaf and then are satisfied to mention a handful of other standard examples and well-established uses within mathematics, before pressing on with abstract results. Especially in recent years, there has been something of a push against this, with a number of exciting new applications of sheaves to topological data analysis,[4] to sensor networks,[5] to opinion dynamics (including selective opinion modulation and lying) on social networks,[6] to target tracking,[7] to dynamical systems and behavior types,[8] to name just a few. This book has been greatly inspired by such efforts.

Regarding the third feature of this book: throughout each chapter, I occasionally pause for a few pages to highlight, in a more philosophical fashion (in what I call "Philosophical Passes"), some of the important conceptual features to have emerged from the preceding technical developments. The overall aim of the "Philosophical Pass" sections is to periodically step back from the technical details and examine the contributions of sheaf theory and category theory to the broader development of ideas. These sections may provide some needed rest for the reader, letting the brain productively switch modes for some time, and giving them something to think about beyond the formal details. A lot of category theory, and the sheaf theory built on it, is deeply philosophical, in the sense that it speaks to, and further probes, questions and ideas that have fascinated human beings for millennia, going to the heart of some of the most lasting and knotty questions concerning, for instance:

- What is an object (and can we give an entirely relational account of objects, that is, display an object in terms of all its relations)?
- What is universality?
- What is negation?
- What fundamental notions are codified by our concept of space?

4. As in Curry (2014).
5. As in Robinson (2016b).
6. As in Hansen and Ghrist (2020).
7. As in the work of Robert Ghrist.
8. As in Schultz and Spivak (2017) and Schultz, Spivak, and Vasilakopoulou (2016).

While a number of other issues will be discussed, some of the main philosophical issues that will be explored in the book engage a few decisive dialectics, notably that of the

- local-global,
- continuous-discrete,
- particular-universal, and
- object-relation.

The struggle to articulate the peculiar relations and antagonisms between each of the members of such pairs has been ongoing for centuries, and while mathematics has advanced inquiry into these matters more than any other discipline, it remains the case that there is a great history to investigating such dialectics, and they are not the sole property of mathematics. Occasionally stepping back to ground specialized treatments of these matters in the broader discussion is useful not only for reminding us of some of the stakes of our formalism, but also for connecting the activity of mathematics back to the longer history and future of inquiry, as human beings, into such fundamental questions.

A word about philosophy.[9] Specialization has manifold benefits, and even if it didn't, it seems to be the price we must pay, as beings with very limited resources, for doing something well. At times though—especially during times like ours, an age of increased specialization—the incentive structures for engaging with something outside one's specialization and subspecializations can deteriorate. Whether the thick boundaries of the adult's specialized world have barely been felt by them, or because they still have the luxury of not being overly concerned with the pursuit of excellence, children are good at refreshingly disrespecting the adult's divided world. As we grow out of being a child, those boundaries become more and more real for us, yet most of us (even the hardened specialist) do not really entirely outgrow or utterly forget that state of the child, nor do we ever come to fully believe in the reality of those boundaries. And even if we tell ourselves that we do, the child seems to return, however faintly or mischievously, in unexpected ways. We find ourselves wondering if such a thing as humor can be defined within music in a purely musical way, or if certain growth patterns found throughout nature can tell us something about the impulse movements of financial markets. Through a mixture of curiosity, a drive to unify and organize, or sometimes just a stupid whim, we retain something of this impulse to *take concepts beyond where we are told they belong*. Such inquiries can only be vague and tentative at first, and there is always a risk they will not lead anywhere. Over time, certain inquiries mature and start to appear a little differently to us: we find ourselves seriously considering if there is life on other planets, or how we can get machines to learn complicated behaviors purely using reinforcements built into the environment, the way so many animals do. If we look carefully, even in those more established questions we can still recognize that same childlike impulse to disregard the myriad cues that exert pressure on the questioner to leave concepts where they belong: "'Life' is a *here* thing!" "Learning is something only carbon-based beings can do!"

9. No part of this book rests on the remarks made in the next three paragraphs. They are provided for context, and were prompted by questions I received from separate mathematician colleagues curious about how I, as a professional philosopher, understood "philosophy" and its relation to category theory.

When it works, taking concepts beyond the confines of their native setting can have the effect of attaining greater generality. This impulse to attain greater generality—which is born out of taking concepts beyond where they belong—is the minimal working sense of "philosophical" that I intend in the present context. In this sense, philosophy is something that we all do and that does not at all belong to "the philosopher." And while it is perhaps one of the greatest beneficiaries of the advantages of specialization, mathematics fundamentally shares this same strong drive toward the general—which may in part account for why, throughout the centuries, there has been a great deal of interaction between the disciplines of philosophy and mathematics, even to the point that for much of history it would have been difficult to draw a sharp line between the two. This intimate bond becomes especially evident with category theory. One could argue that, at least in large part, philosophy (in a more traditional sense) has evolved as the informal study of universality (and universal phenomena). One could argue that category theory is the formal study of formal universality. As such, it is no surprise that there appear to be a number of especially strong connections between the matters pursued by category theorists and those of philosophers.

I happen to believe that many of the staple questions that were originally the provenance of the philosopher will eventually be handled with the care they deserve once they are adequately framed as problems within category theory, and that in the near future every major philosophical dialectic—universal-particular, continuous-discrete, global-local, quality-quantity—and even less obvious problems, such as those of "personal identity," will be handed over to, and considerably enriched by, the category theorist. In the other direction, a variety of basic elements of category theory appear to raise philosophical questions of their own, and certain more advanced developments (such as with cohesive toposes, discussed in chapter 11) seem almost inherently philosophical, and poised to attack a number of the traditional philosophical problems. But we are probably at least 100 years away from a world in which one can adequately realize that category theory is everything philosophy ever strove to be, and let that long, rich, and frustrating tradition take on a new form. In the meantime, one of the aims of this book is to encourage those from each camp to engage with the other—and the "Philosophical Pass" sections are opportunities to step back from the formal details, gather our thoughts, relate the mathematical concepts to broader or tangential conceptual developments, and occasionally engage in a little pushing of the formal concepts beyond where they belong.

I would encourage all readers to pursue the philosophical sections of this book—though they are set off in boxes to mark them off from the rest of the text so that the more narrowly focused reader can easily find their way around them should they insist on reading only the mathematics. To encourage the more strictly mathematical reader to engage with those sections, though, I will just add that it seems that nearly all great mathematicians of the past have let themselves be provoked by, and at times have even engaged with, the philosophical dimensions of their work.

Finally, while emphasis on concrete examples from unexpected areas beyond the confines of pure mathematics is already unusual enough for a text on sheaf theory, and while engagement with philosophical dimensions of the mathematics is itself atypical for a primarily mathematical text, the reader might be even more surprised to find these two things paired together. In response to this reaction, let me bastardize the philosopher Kant and say this: knowledge of examples and applications without a sense of the general ideas these exemplify and are powered by is *blind*, while knowledge of general ideas without a familiarity with all sorts of examples and applications is *empty*. Various philosophers of the past, like Aristotle and Spinoza, have set as an ideal for the most demanding and adequate kind of knowledge one that can look past the apparent immediacy of the universal and the particular, taken on their own, and instead achieve a more unified understanding of the subtle mediations between our knowledge of the universal and of the particular. Moreover, it has been my experience that often the only way to really grasp the most general, and to appreciate the various needs to keep pushing things in the direction of the more general, is to sink as deeply as possible into certain particular problems. In a peculiar slogan: often what is furthest (most general) can be most readily approached through closer consideration of what is nearest (least general). In this connection, I believe that the ideal mathematician would represent some sort of fusion between the Grothendieckian impulse toward extreme abstraction and general ideas, on the one hand, and the intimate exploration and care for particulars embodied by the likes of a Ramanujan, on the other. While I would not pretend to achieve anything remotely close to this fusion for myself, I do believe that it is a noble ideal to strive for, and the atypical pairing of engagement with general ideas and respect for examples found in this book has been influenced by that belief.

What the Book Is (and Requirements of the Reader)

I should add a word about what this book aims to be and who it is for. One reviewer of an earlier version characterized the book's most significant contribution as

> providing an accessible sheaf theory book filled with fun examples, with a broad philosophical bent.

I think this is a very clear statement of what I have wanted to achieve with this book. There also happens to be a great gap between the few accessible books on the basic category theory (and other prerequisites) needed to develop sheaves and any currently published book on sheaf theory. Anyone who would find a bridge over that chasm useful, or who would be engaged by a sheaf theory book that meets the above description, will likely find this book valuable.

Realistically, though, anyone who would find their way to this book will likely have some prior mathematical training and interests. The primary audience of this book should include open-minded mathematicians, scientists and engineers with some broader mathematical interests, and mathematically inclined philosophers. Because of the distinctness of these three groups, I highlight, at the beginning of each chapter, the mathematical and philosophical goals and topics explored. As for those with interests of the practical sort: there are a number of examples, constructions, and discussions that should be of interest; however, there may be certain sections (appealing to those with more abstract aspirations or those with a philosophical bent) that might be of less interest to such a reader. Such readers might try dipping their toes into those sections and skimming on first reading, focusing most of their attention on the examples.

As for general requirements of the reader, I have tried to make this book as self-contained as possible and minimize the prerequisites in order to extend the reach as far as possible to nonexperts. I thus assume only some basic familiarity with set theory and mathematical reasoning—all other concepts needed for the formulation and understanding of sheaves, including the basics of category theory, topology, and anything else, are motivated and introduced in this book.

In the end, I have tried to write the book I wish I had when I was first learning sheaf theory. There are some outstanding books on sheaf theory—notably Mac Lane and Moerdijk's *Sheaves in Geometry and Logic*—but such texts can be rather demanding on the beginner, assume a great deal of mathematical maturity, and generally appeal to a rather expert and self-sufficient audience. In this book, I have tried to assume a great deal less than such texts, to engage a broader audience, and generally adopt a more gentle approach.

What the Book Is Not

As one might already imagine, given its unique aims and approach, this book is not meant to be a standard textbook for experts learning about sheaf theory as it is usually taught in one of its specialized contexts, such as algebraic geometry. An expert reader who has certain expectations about what this book should be, based on standard specialized references on sheaves, will surely have those expectations violated.

In this connection, this book deliberately minimizes treatment of applications to problems in algebraic geometry, one of the historical homes of sheaf theory. This was intentional—in part since these applications require a level of mathematical maturity which this book tries not to assume of the reader, in part because there are already many references devoted to sheaves in algebraic geometry. Beyond this, the omission is also somewhat philosophical. Tom Leinster wrote, in 2010, a blog post entitled "Sheaves Do Not Belong to Algebraic Geometry":

> They are, of course, very *useful* in algebraic geometry (as is the equals sign). Also, human beings discovered them while developing algebraic geometry, which is why many of them still make the association. But. . . sheaves are an inevitable consequence of general ideas that have nothing to do with algebraic geometry.

This is a perspective I share, and I have accordingly sought to avoid including applications to algebraic geometry, with the aim of redistributing the somewhat disproportionate control algebraic geometers have taken over these (demonstrably more general and far-reaching) ideas.

This book is also not meant to be a complete reference. This is part of a trade-off one must make when attempting to appeal to, and sustain the interest of, a wider audience of nonexperts. There are a number of additional topics I would have loved to cover, and further examples I would have loved to include, yet doing so with the aim of completeness could have easily made this book extend to over 1,000 pages. It seemed to me more desirable to welcome more newcomers to sheaves with a book of a more manageable size.

1 Categories

In which we meet categories, explore some notable examples, consider distinctions of size, formulate some of the important ways of constructing new categories from old, and reflect on an alternative definition.

The language of category theory is indispensable to the presentation and understanding of the notions of sheaf theory. Chapters 1–3, together with chapters 6 and 7, will motivate and develop all the category theory needed in this book, emphasizing constructions and perspectives that will take center stage in the development of sheaves.

1.1 Categorical Preliminaries

Fundamentally, the specification of a category involves two main tasks: establishing some *data* or series of givens, and then ensuring that this data conforms to two simple axioms or laws. To define, or verify that one has, a category, one should first make sure the right data is present. This first main step of establishing the data of a category really involves doing four things.

First of all, it means identifying a collection of *objects*. Especially when one is assembling a category out of already established mathematical materials, these objects will typically already go by other names, like vertices, sets, vector spaces, topological spaces, types, various algebras or structured sets, and so on.

Second, one must assemble or specify a collection of *morphisms* (also called *arrows* or *maps*), each with two objects associated to it, namely a dedicated "source" object and "target" object. Fundamentally, a morphism is just some principled way of establishing connections between the objects. We depict morphisms with arrows—for example, a morphism f with source object A and target object B is represented diagrammatically by

$$A \xrightarrow{f} B.$$

Again, when dealing with already established structures, morphisms will usually already have names, like directed edges, functions, linear transformations, continuous maps, terms, homomorphisms or structure-preserving maps, and so on. Many of the categories one meets in practice have sets with some structure or supplementary furnishings attached to them for objects and (the corresponding) maps or functions between the underlying sets for morphisms (where these "respect the structure"), so this is a good model to keep in mind at the outset.

Third, and perhaps most important, one must specify an appropriate notion of *composition* for the morphisms, where for the moment this can be thought of in terms of specifying an operation that enables us to form a "composite morphism" that goes directly from object A to object C whenever there is a morphism from A to some B such that this morphism can be juxtaposed with another morphism that lands in C (in particular, whenever the "source" of this second morphism is the same B that was the "target" of the first morphism). In other words, given any pair of morphisms

$$A \xrightarrow{f} B_1 \qquad\qquad B_2 \xrightarrow{g} C$$

such that the target of the first morphism f is in fact the same object as the source of the second morphism g (i.e., $B_1 = B_2$)—in which case, they are *compatible for composition*—then there exists a specified way of combining these mappings to get a resulting morphism $g \circ f : A \to C$, called the *composite* (of f and g). We use the notation $\circ$ to denote composition, and it is read as "following" (or "after"), so that with $g \circ f$, for instance, we would have "g following f." In other words, g gets applied after f—as such, one might parse this by reading right-to-left: first apply f, then run g on the result. As a very simple example, when dealing with sets of numbers equipped with structural mappings corresponding to addition and multiplication, if f is the function defined by $f(x) = x^2$ and g is the function defined by $g(x) = x + 3$, then we know what the composite of f and g must be, namely $(g \circ f)(x) = g(f(x)) = x^2 + 3$.

We shall see that this composition operation in fact already determines the fourth data item of a category: that for each object, there is assigned a unique *identity* morphism that starts out from that object and returns to itself.

$$A \xrightarrow{\mathrm{id}_A} A$$

If you think for the moment of the model supplied by those commonly encountered categories that have structured sets for objects and structure-preserving maps for morphisms, then you might think of the identity morphism as the "trivial" action, the one that trivially preserves the structure by effectively "doing nothing." These four constituents—objects, morphisms, composite morphisms, and identity morphisms—supply us with the data of the (candidate for a) category.

Next, one must show that the data given above conforms to two very natural laws or axioms governing compositions. The first axiom concerns the behavior of the identity morphisms under composition. Consider how, for an arbitrary morphism $A \xrightarrow{f} B$, such a morphism will automatically be compatible for composition with the identity morphisms (on either end). The first axiom effectively says that morphisms that are *supposed* to do nothing (i.e., the designated identity morphisms) *really* do nothing when composed with other morphisms (i.e., really act as identities, or the "trivial" action, with respect to the morphisms with which they can be composed). More specifically, it stipulates that if we have a morphism f (as above) from source object A to target object B, then first applying the identity morphism on A and then traveling along the morphism f should be the same thing as "just" traveling along the morphism f; and the same goes for applying the morphism f straightaway and following this with the identity morphism on the target object B. In short, it is required that the identity morphisms do not do anything to change other morphisms with which they may be composed, in the precise sense that $f \circ \mathrm{id}_A = f = \mathrm{id}_B \circ f$. Observe that this identity axiom is effectively a condition on composition, namely that composing any morphism with the identity morphisms (on either side) is equal to the original morphism.

Second, suppose you have a string of morphisms

$$A \xrightarrow{f} B \xrightarrow{g} C \xrightarrow{h} D,$$

which are compatible for composition, by construction. Given such morphisms, a variety of distinct composites can be formed, yielding (in principle) distinct paths from A to D. For instance, while you could of course get from A to D by stepping through each of the individual morphisms as above, you might instead form the composite map $h \circ g$ and then use this, after taking f to B, to end up in D, as in

$$A \xrightarrow{f} B \xrightarrow{h \circ g} D.$$

Similarly, you could form the composite map $g \circ f$ and use this to get from A to C, and then take h to D, as in

$$A \xrightarrow{g \circ f} C \xrightarrow{h} D.$$

The second axiom just says that if you have a string of morphisms f, g, and h as above, then it should make no difference whether you choose first to go directly from A to C (using the composite map $g \circ f$ that we have by virtue of the third step in the data construction) followed by the map from C to D, or if you go from A to B followed by a direct map from B to D (using the composite map $h \circ g$). This axiom effectively says that composition is associative, in the sense that the outermost arrows of the following diagram are equal:

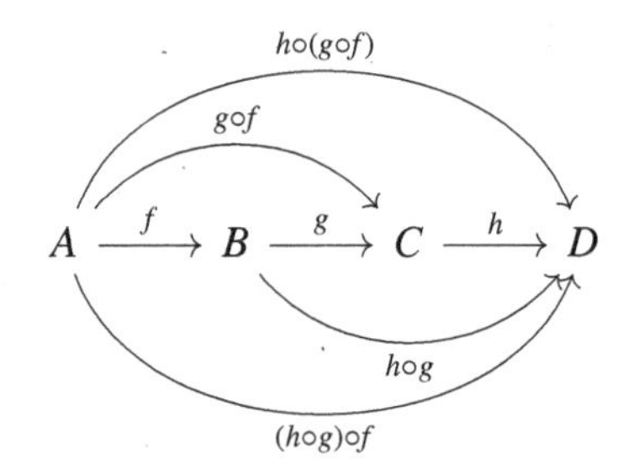

An entity that has all the data specified above, data that in turn conforms to the two laws described in the preceding two paragraphs, is a category. The mostly informal description given in the preceding paragraphs is given more formally in the following definition.

Definition 1 A *category* **C** consists of the following data:[10]

- A collection of *objects* $A, B, C, \ldots$;[11]

10. Throughout this book, categories are generally designated with bold font. However, sometimes we may use script font instead, especially when dealing with things like orders (discussed below), where each individual order is already a category. Context should always make it clear what category we are working with, so this should not be a problem.

11. The reader who finds themselves worried about what is meant by this apparently somewhat vague word "collection"—which, for now, you may take to mean "set" more or less (though "class" would be better, if that means something to you)—should be reassured that we will address what is going on here in section 1.3.

- A collection of *morphisms* $f, g, h, \ldots$, where each morphism has designated source object and target object—so that, for instance, $f: A \to B$ signifies that f is a morphism with source A and target B;[12]
- To each object A in the collection of objects is assigned a designated morphism in the collection of morphisms from A to A (i.e., where source and target are both the object A), denoted by id_A (or 1_A), called the *identity morphism* on A;
- For any pair of morphisms f, g such that the target of f is equal to the source of g—as in $A \xrightarrow{f} B$ and $B \xrightarrow{g} C$—there exists a *composite morphism* $A \xrightarrow{g \circ f} C$, with source equal to the source of f and target equal to the target of g.

This data gives us a category provided it further satisfies the following two axioms:

- *Associativity* (of composition): for any composable triple of morphisms f, g, h, as in $A \xrightarrow{f} B \xrightarrow{g} C \xrightarrow{h} D$, we have $h \circ (g \circ f) = (h \circ g) \circ f$.

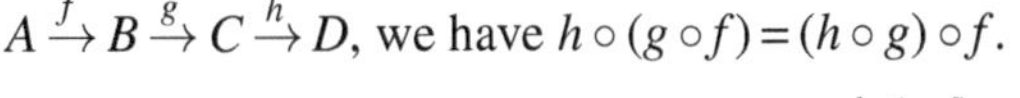

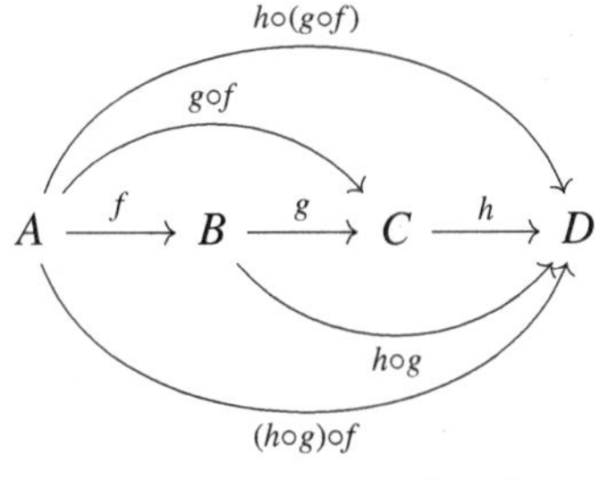

- *Identity*: for any morphism $f: A \to B$, the composites formed by composing f with the identity morphism (on either side, i.e., on source or on target) are equal to f itself—that is, we have $f \circ \mathrm{id}_A = f$ and $\mathrm{id}_B \circ f = f$.

Before turning to examples of categories, let us make a brief observation. The associativity of composites may be expressed with what is called a *commutative diagram*, which is why a diagram appeared in the associativity axiom of the definition. While we will have much more to say about diagrams in later chapters, and be more precise about all this, for now observe how we have been displaying the various objects of a category, together with their morphisms and morphism composition, in the form of *diagrams*, where these are effectively directed graphs with morphisms as directed edges and objects as the names of the implied vertices or nodes attached to such edges. With the display of a morphism f with source A and target B as

$$A \xrightarrow{f} B,$$

we already have an example of a basic diagram. Now suppose we have a diagram involving three morphisms, as in

12. The term "morphism" comes from *homomorphism*, which is how one refers in abstract algebra to a structure-preserving map between two algebraic structures of the same type (such as groups or rings). The morphisms of a category are also commonly referred to as "arrows" or "maps."

I should also note that while, in this definition, we are using $A, B, C, \ldots$ to range over objects, and $f, g, h, \ldots$ to range over morphisms, this convention will not always be respected. For instance, sometimes we will use $a, b, c, \ldots$, or some other natural names, for objects, and similarly reserve other appropriate notation for the arrows. These notation choices mostly just follow what is customary in the special topic being treated categorically, and this should not cause confusion.

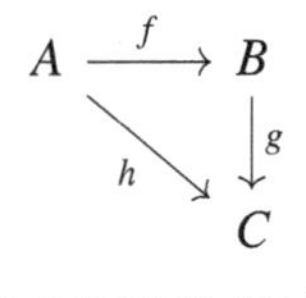

Provided we are indeed working with a category, we know that the composable morphisms $f: A \to B$ and $g: B \to C$ will support the existence of the composite $g \circ f$ morphism, leaving us with a parallel pair of morphisms from A to C, as in

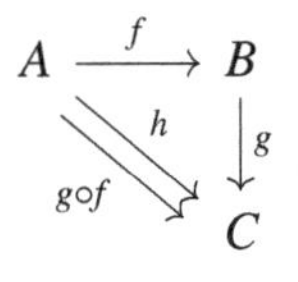

In principle, the two morphisms of such a parallel pair need not be the same—but when they are, so that $h = g \circ f$, we then say that this diagram (the triangle) commutes, and display it with the diagram

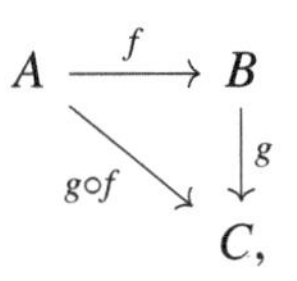

where occasionally a checkmark ✓ is drawn in the center if one wants to really stress that a diagram commutes. Other than "degenerate" diagrams involving identities, commutative triangles give us the most basic and paradigmatic instance of commutativity in diagrammatic form. But commutative diagrams one meets in the wild can have many more moving parts. In general, we will say that a diagram *commutes*, or is *commutative*, provided all directed paths with the same start and endpoints compose to give the same morphism—that is, a commutative diagram visualizes equalities between composite morphisms. As such, commutative diagrams are sometimes said to be for the category theorist what equations are for the algebraist. Commutative diagrams, and diagrammatic reasoning in general, are the main tool of the category theorist. They are not merely visual, intuitively spatial aids to the understanding of formal facts, but they even supply new proof techniques—especially that of the "diagram chase," which involves establishing a property of a particular morphism by tracking the components of a commutative diagram. A lot more will be said about diagrams throughout the book. For now, let us consider some examples of categories.

Example 2 **Set** is a category, where this consists of sets for objects and functions (with specified domain and codomain) for morphisms. Composition is given by the usual function composition (which is moreover associative), and identity morphisms are exactly what you imagine (where the identity functions moreover behave as the "units" for composition).

Example 3 (*Order Categories*) A *relation* between sets X and Y is just a subset $R \subseteq X \times Y$, so that a *binary relation* on X is a subset $R \subseteq X \times X$. It is customary to use infix notation for binary relations and use the symbol $\leq_X$ (or just $\leq$, if the carrier set is understood) for the relation on a set X, so that, for instance, one writes $a \leq b$ for $(a, b) \in R$.

We can then define a *preorder* as a set with a binary relation (call it $\leq$) that further satisfies the properties of being *reflexive* and *transitive*. In other words, it is a pair $(X, \leq_X)$ where we have

- $x \leq x$ for all $x \in X$ (reflexivity); and
- if $x \leq y$ and $y \leq z$, then $x \leq z$ (transitivity).

Then a *partially ordered set* (*poset* for short) is a preorder that is additionally *antisymmetric*, where this means that having $x \leq y$ and $y \leq x$ implies that $x = y$.

It is often useful to represent a given poset (or preorder) with a diagram. For instance, suppose we have an order-structure on the set $X = \{a, b, c, d\}$ given by $a \leq c, b \leq c, b \leq d$, together with the obvious identity (reflexivity) $x \leq x$ for all $x \in X$, which we will leave implicit. For a reason that will be better appreciated in a moment, the data of this poset may be displayed by the diagram:

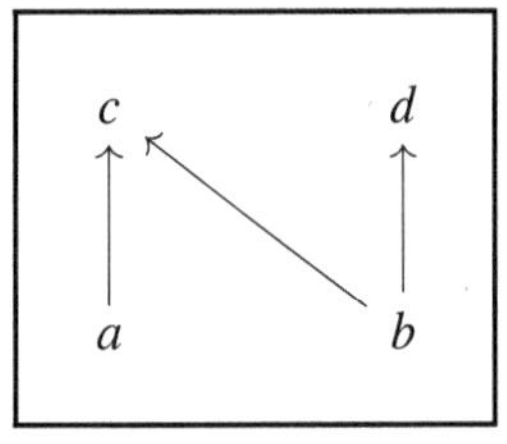

Preorders (posets) can themselves be related to one another, and the right notion here is that of a monotone (or order-preserving) map.

Definition 4 A *monotone* (*order-preserving*) map between preorders (or posets) $(X, \leq_X)$ and $(Y, \leq_Y)$ is a function $f : X \to Y$ on the carrier sets satisfying that for all elements $a, b \in X$,

$$\text{if } a \leq_X b, \text{ then } f(a) \leq_Y f(b).$$

There is then a category **Pre**, the category having preorders for objects and order-preserving functions for morphisms. **Pos** is another category, one having posets for objects and order-preserving functions for morphisms.[13] Each identity arrow will just be the corresponding identity function, regarded as a monotone map. One can verify that for two monotone maps $X \xrightarrow{f} Y$ and $Y \xrightarrow{g} Z$ between orders, the function composition $g \circ f$ is also monotone.

While we are already talking about orders, let us introduce a few other order-theoretic notions, allowing us to further expand our repertoire of order categories. Suppose we further add the property to a poset that for all $a, b \in X$, either $a \leq b$ or $b \leq a$, that is, any two objects are *comparable*. Adding this condition gives us what are called *linear orders* (or, sometimes, *total orders*). If we take finite nonempty linear orders as our objects and monotone (order-preserving) functions between such linear orders as our morphisms, we get another category: **FLin**, the category of finite nonempty linear orders. For a simple example of such an order, take a natural number $n \in \mathbb{N}$, and consider the standard linear order $[n] = (\{0, 1, \ldots, n\}, \leqslant)$, where every such finite linear order may be represented by

$$\overset{0}{\bullet} \xrightarrow{\leq} \overset{1}{\bullet} \xrightarrow{\leq} \overset{2}{\bullet} \xrightarrow{\leq} \overset{3}{\bullet} \xrightarrow{\leq} \cdots \xrightarrow{\leq} \overset{n}{\bullet}$$

13. As can be seen from the few examples given thus far, it is common to let the *objects* determine the name of a category in question. While this is an entirely sensible practice, it is worth noting that it is at odds with the general spirit or philosophy of category theory, which gives priority to the morphisms (or at least demands that objects be considered together with their morphisms), a matter that is explored further in section 1.5.

When we take such finite linear orders $[n]$ for our objects (one for each $n \in \mathbb{N}$) and the order-preserving maps between these for our morphisms—defined, for each pair of objects $[m], [n]$, as all the functions $f : \{0, 1, \ldots, m\} \to \{0, 1, \ldots, n\}$ such that for every pair of elements $i, j \in \{0, 1, \ldots, m\}$, if $i \leqslant j$, then $f(i) \leqslant f(j)$—we get what is called the *simplicial category* (or *simplex category*), typically denoted by $\boldsymbol{\Delta}$. As it turns out, these two categories can be shown to be equivalent, but we postpone making this precise until chapter 2, once we have the resources to do so. For now, this observation may justify thinking of the apparently enlarged category of finite nonempty linear orders in terms of the more manageable objects of the sort depicted above.

For a final example of this sort, we can also consider what is called a *cyclic order*. Just as the names imply, while a linear order is effectively an arrangement of elements along a line (where this has a prescribed direction), a cyclic order can be regarded as an arrangement of elements on a circle (where there is again a direction, such as clockwise). Observe that for elements arranged on a line with a specified direction, it makes sense to say things like "a comes before (or after) b"; however, for elements arranged in a circle with a direction (say, clockwise), the same sort of thing cannot be said—for instance, it doesn't make sense to say that "a is more (or less) clockwise than b." This observation motivates the definition of a cyclic order: it is given not in terms of a binary relation, but more naturally as a ternary relation $[a, b, c]$ (read "after a, one arrives at b before c") on a set. The months of the year form such a cyclic order, where we have, for instance, [*March*, *September*, *February*] but not [*March*, *February*, *September*]. More formally, a cyclic order on a set is a ternary relation that satisfies the following properties (effectively ternary versions of the properties characterizing a linear order):

1. cyclicity: if $[a, b, c]$, then $[b, c, a]$;
2. asymmetry: if $[a, b, c]$, then not $[c, b, a]$;
3. transitivity: if $[a, b, c]$ and $[a, c, d]$, then $[a, b, d]$;
4. totality: if a, b, and c are distinct, then we have either $[a, b, c]$ or $[c, b, a]$.

The notion of a monotone (order-preserving) function between linear orders has its counterpart for cyclic orders in the following: given the cyclic orders $(X, []_X)$ and $(Y, []_Y)$, we say that a function $f : X \to Y$ is *monotone* provided it preserves the relation for any elements that have pairwise distinct images under f, that is, if given $[a, b, c]$ and the images $f(a), f(b)$, and $f(c)$ are pairwise distinct, then $[f(a), f(b), f(c)]$. This data ultimately gives us a category, one that has for objects all finite nonempty cyclically ordered sets, and for morphisms the monotone maps (in the above sense).

Similar to how we were able to generate a further category of interest by confining attention to the standard linear orders $[n]$, it is also useful to consider the standard cyclic orders, where a cyclic order on a finite set with n elements can be pictured as an (evenly spaced) arrangement of the set on an n-hour clockface.

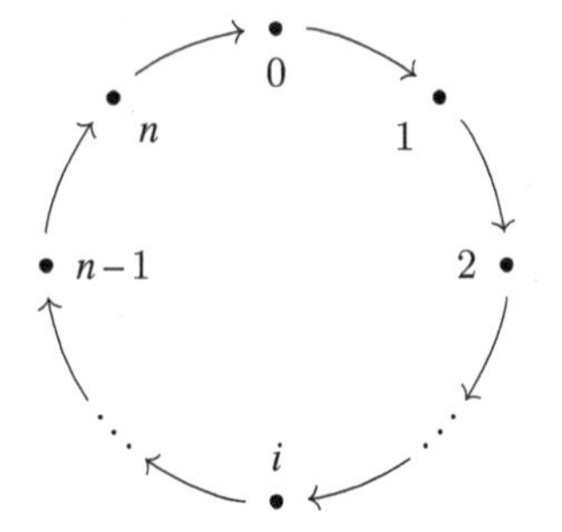

Such an order is designated Λ_n. If we take such Λ_n as our objects (one for each $n \in \mathbb{N}$), and we take as our morphisms (from an object Λ_m to Λ_n) the monotone functions, then we arrive at what is called the *cyclic category* (or *cycle category*), usually denoted by $\mathbf{\Lambda}$. Following Connes and Consani (2015),[14] it is more common to see this category $\mathbf{\Lambda}$ given a description that takes the standard orders as objects (one for each $n \in \mathbb{N}$) but where the morphisms (from Λ_m to Λ_n) of $\mathbf{\Lambda}$ are instead seen as functions f on the integers that satisfy both of the following properties:

- nondecreasing/order-preserving: if $x \leq y$, then $f(x) \leq f(y)$;
- periodicity/"spiral property": $f(x+m+1) = f(x) + n + 1$ for all $x \in \mathbb{Z}$.

Composition is then just the usual composition of functions. To complete the construction, one then takes equivalence classes of such functions under the relation $f \sim g$ whenever their difference is a constant multiple of $n+1$. This equivalence relation is compatible with the composition of functions, and there can be only finitely many such equivalence classes of functions for each pair of natural numbers m, n.

Orders, especially preorders and posets, are very important in category theory, and we will see a lot more of them throughout the book.

Example 5 A graph is typically represented by a bunch of dots or vertices together with edges between certain of the vertices, so that each edge is linking a pair of vertices, supplying what is called a relationship of *incidence* between the vertices and edges. More formally, a *(simple) graph* G consists of a set V of *vertices*, together with a collection of two-element subsets $\{x, y\}$ of V (where we generally assume $x \neq y$), called the *edges*. Sometimes the collection of two-element subsets of V is instead just represented by a set E that consists of the "names" of such pairings, where one then specifies an additional mapping that interprets edges as pairs of vertices. On this approach, a graph effectively consists of two sets—a "vertex set" V and an "edge set" E—together with a map that works to assign each edge to a pair of vertices (via a certain one-to-one function from E to 2-element subsets of V). The relevant notion of a map $G \to G'$ from a graph G to a graph G' is then given by a *graph homomorphism*, where this is a function $f : V \to V'$ on the vertices such that if $\{x, y\}$ is an edge of G, then $\{f(x), f(y)\}$ is an edge of G'.

As the pairs of vertices making up the edge set were implicitly defined to be *unordered*, since they were said to be *sets* of the form $\{x, y\}$ rather than ordered pairs, the graphs

14. Alain Connes was the first to describe this category; see Connes (1983), where it is given a somewhat different description.

we just defined are *undirected*. Moreover, assuming that the map interpreting edges as unordered pairs of vertices does so in a one-to-one way amounts to requiring that the graph be *simple* in the sense of having *at most one* edge between two vertices. Taking such unordered simple graphs together with the associated notion of a graph morphism, we have the materials of a category, namely the category of undirected (simple) graphs, **UGrph**, or more commonly **SmpGrph**. The objects of this category are what the graph theorist usually means, by default, when they speak of a "graph." One such graph is pictured below (the names of vertices and edges left out):

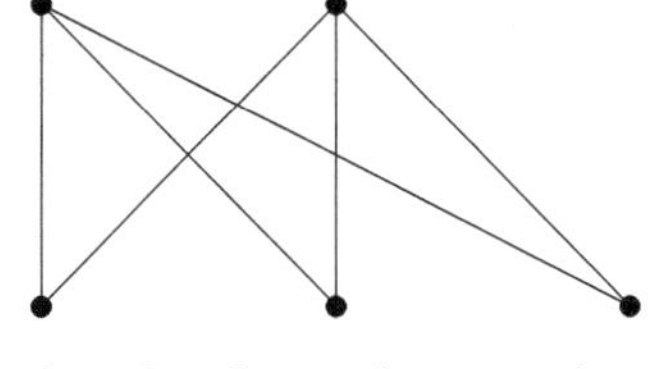

If we had instead allowed the function from edges to pairs of vertices to be many-to-one, so that for each unordered pair of distinct vertices there could be an entire set of edges between these, we would be left with undirected *multigraphs*, such as

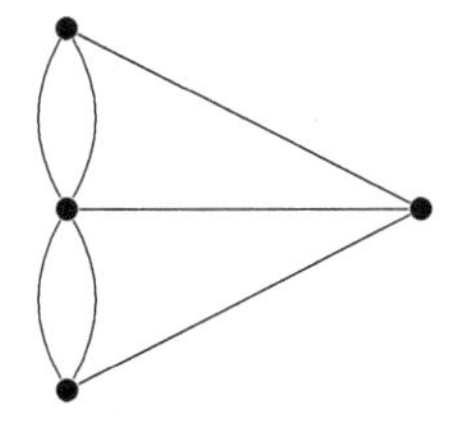

We can further define *directed graphs* (which are often thought of by the category theorist as *quivers*), where our edges now become arrows. These are the sorts of graphs that will be of most interest to us. Fundamentally, a (directed) *graph* $G=(V,A,s,t)$ can be seen as being comprised of two sets and two functions. Specifically, it consists of a set V of vertices, a set A of directed edges or arcs (arrows), and two functions

$$A \overset{s}{\underset{t}{\rightrightarrows}} V$$

that effectively act to pick out the *source* and *target* of an arc.

Then if $G=(V,A,s,t)$ and $G'=(V',A',s',t')$ are two graphs, the relevant notion of a morphism, namely a *graph homomorphism* $f:G \to G'$, is defined as a pair of morphisms $f_0: V \to V'$ and $f_1: A \to A'$ such that sources and targets are preserved, that is, both the diagrams

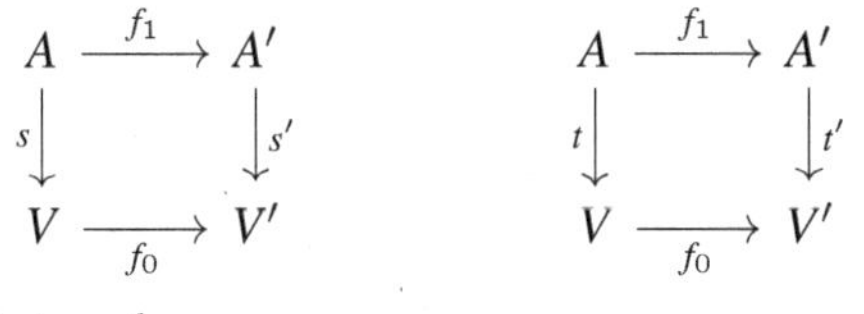

commute, in the sense that we have

$$s' \circ f_1 = f_0 \circ s \text{ and } t' \circ f_1 = f_0 \circ t.$$

One can verify that composing two such graph homomorphisms f, g will leave us with another morphism $g \circ f$ that is itself a graph homomorphism. Altogether, we have a category called **DirGrph** (or just **Grph**), which has directed graphs as objects and their directed graph homomorphisms as morphisms.

In general, as before, we may further allow there to exist several parallel arrows—that is, with the same source and same target—in which case we would be dealing with directed *multigraphs*. Furthermore, if we were to add closed arrows or loops—that is, arrows whose source and target are identical—then we would be dealing with *looped* (or *reflexive*) graphs. There is a lot more to say about distinctions between different graphs, the distinct categories each gives rise to, and their categorical features of interest; but we will postpone further discussion of such matters.[15]

Example 6 It is probably fair to assume that the reader is already very familiar with vector spaces and the associated ideas of linear algebra. But if not, or in case a refresher is needed, here is the definition.

Let $\mathbf{k}$ be a given field—these supply us with the *scalars* by which we *scale*, by multiplying with, our vectors; so think, for concreteness, of $\mathbf{k} = \mathbb{R}$, the real numbers under addition and multiplication. Let V be a (nonempty) set with addition and scalar multiplications that assigns, to any $u, v \in V$, a *sum* $u + v \in V$, and to any $u \in V$ and $k \in \mathbf{k}$, a *product* $ku \in V$. Then V is said to be a *vector space* (or *linear space*) *over* $\mathbf{k}$, the elements of V being called *vectors*, if the following axioms hold:

- A_1: For any vectors $u, v, w \in V$, we have $(u + v) + w = u + (v + w)$.
- A_2: There exists a vector in V, denoted 0 and called the *zero vector*, for which $u + 0 = u$ for any vector $u \in V$.
- A_3: For each vector $u \in V$, there exists a vector in V, denoted by $-u$, for which we have $u + (-u) = 0$.
- A_4: For any vectors $u, v \in V$, we have $u + v = v + u$.
- M_1: For any scalar $k \in \mathbf{k}$ and any vectors $u, v \in V$, we have $k(u + v) = ku + kv$.
- M_2: For any scalars $a, b \in \mathbf{k}$ and any vector $u \in V$, we have $(a + b)u = au + bu$.
- M_3: For any scalars $a, b \in \mathbf{k}$ and any vector $u \in V$, we have $(ab)u = a(bu)$.
- M_4: For the unit scalar $1 \in \mathbf{k}$, we have $1u = u$ for any $u \in V$.

Matrices furnish us with a basic example of such a thing. Letting $\mathbf{M}_{m,n}$ denote the set of all $m \times n$ matrices over an arbitrary field $\mathbf{k}$, then $\mathbf{M}_{m,n}$ will be a vector space over $\mathbf{k}$, with the usual operations of matrix addition and scalar multiplication. Other important examples are given by the space of all polynomials $P[x]$ with coefficients in some field $\mathbf{k}$, and function spaces, where the elements are all functions from some given nonempty set into some $\mathbf{k}$.

Mappings between two vector spaces V, W over the same field are given by *linear transformations* (or *vector space homomorphisms*), where these are defined as functions $F : V \to W$ that satisfy, for any vectors $v, u \in V$ and any scalar $k \in \mathbf{k}$, the following two conditions:

15. For now, just note that while *quiver* and *directed graph* are often used synonymously, technically the graph theorist expects of its directed graphs that there is at most one arc from one vertex to another, and the notion of a quiver allows for there to be multiple "parallel" arcs between vertices, that is, a quiver is a directed multigraph (where loops are also allowed).

- $F(v+u) = F(v) + F(u)$
- $F(kv) = kF(v)$.

In other words, F is linear if it "preserves" the two fundamental operations of a vector space, vector addition and scalar multiplication.

The category $\mathbf{Vect}_{\mathbf{k}}$ is the category of **k**-vector spaces (for a given field **k**, dropping the **k** when this is understood), which has vector spaces $V, W, \ldots$ over **k** for its objects and linear transformations for its morphisms. To see that this is indeed a category, suppose $F: V \to W$ and $G: W \to U$ are linear transformations between the specified vector spaces. Then, for any $v, v' \in V$ and any $a, b \in \mathbf{k}$, we must have

$$(G \circ F)(av + bv') = G(F(av + bv')) = G(aF(v) + bF(v'))$$
$$= aG(F(v)) + bG(F(v')) = a(G \circ F)(v) + b(G \circ F)(v'),$$

which just shows that the composite $G \circ F: V \to U$ must itself be a linear transformation. Moreover, for any vector space, the identity function will be a linear transformation. Finally, associativity of composition comes for free, since linear transformations are functions and function composition is always associative. If we restrict attention to just finite-dimensional vector spaces, this would yield the category **FinVect**, which is where most of linear algebra takes place.

In vector spaces, the scalars come from the given field and act on the vectors of the space by scalar multiplication (which then obeys the axioms). Every field is a ring, but there are rings that are not fields, as the notion of a ring generalizes that of a field (for a ring, multiplication need not be commutative and multiplicative inverses need not exist); for example, the ring of integers $\mathbb{Z}$ is not a field, since 2, for instance, has no multiplicative inverse in $\mathbb{Z}$. A *module* is just like a vector space, except the scalars need only come from a ring (with identity), and (left or right) multiplication is defined between elements of the ring and elements of the module. So while any ring R that is also a field recaptures the notion of vector spaces, there are plenty of modules that are not vector spaces. We call a module taking its coefficients in the ring R an R-module.

For a concrete illustration of a module, here is one that is often used by music theorists. First consider that, in general, for a set X and a vector space V, we can define the set of all functions from X to V. Then, for f, g in that set and for $c \in \mathbf{k}$ the field of scalars, addition is defined as $(f+g)(x) = f(x) + g(x)$ and scalar multiplication as $f(cx) = c(f(x))$ for all $x \in X$. It can be shown that this set of functions is itself a vector space (over the field **k**). And in fact, for X the set of n-tuples, it can be shown that this function vector space is isomorphic to the vector space of n-tuples. Applied to music, we might accordingly consider the space $\mathbb{R}^{\{O,P,L,D\}}$, by taking $X = \{O, P, L, D\}$, where O stands for the onset values, P for pitch, L for loudness, and D for duration, and where these give ordered quadruples taking values in O in the first component, P in the second component, and so on. In other words, the members f of this function space can be thought of as the "note event" vector

$$f = (f_O, f_P, f_L, f_D),$$

where, for simplicity, onset f_O may come in units of quarter notes ♩, pitch f_P in units of semitones, loudness f_L in units of cents, and duration f_D in units of quarter notes. This in fact forms an $\mathbb{R}$-module (obviously, since it is in fact a vector space), one that is isomorphic to $\mathbb{R}^4$, the vector space of all ordered quadruples of real numbers (x_1, x_2, x_3, x_4), and it happens to be of some use to music theorists.

Returning to the more general account: a map between two R-modules will be a map that satisfies the same conditions on a linear transformation (except that the scalar is an $r \in R$), that is, it is a function between modules that "preserves" the module structures. More precisely, if we let M and N be R-modules, an *R-homomorphism* (or *module homomorphism*) from M to N is a map $f : M \to N$ that satisfies for every $a, b \in M$ and $r \in R$

- $f(a+b) = f(a) + f(b)$, and
- $f(r \cdot a) = r \cdot f(a)$.

Modules over a fixed ring R, together with such R-module homomorphisms, assemble into a category $\mathbf{Mod}_R$.

Given a homomorphism f from an R-module M to an R-module N, we can define another map $g : M \to N$ by $g(x) = n + f(x)$, where $n \in N$. Such a map is called an *affine transformation*, and when the underlying modules are the same, that is, $M = N$, then such a g captures *symmetries*. Affine maps can be composed just as functions are composed, and thus by taking R-modules together with such maps between them we end up with another category, one that might be denoted by $\mathbf{ModAff}_R$. Incidentally, such a category—and, more broadly, the algebraic theory of rings and modules—supplies a natural setting for the treatment of many aspects of music theory. For instance, if you think of a score of music in terms of the module $\mathbb{R}^{\{O,P,L,D\}}$ representing notes with onset, pitch, loudness, and duration, then morphisms as affine maps or symmetries on this module capture all kinds of musically meaningful (and geometrically representable) transformations—such as pitch transposition, vertical and horizontal inversion and dilation, rhythmic shift, onset-pitch arpeggio, and so on—and a variety a fundamental results concerning things like harmonic analysis and modulation can be derived within this framework.[16]

Example 7 The category **Mon** (**Group**) of monoids (groups) has monoids (groups) for objects and monoid (group) homomorphisms for morphisms. (This example, together with the necessary definitions, will be discussed in more detail in a moment.)

The previous examples involve categories whose objects are sets equipped with some additional structure and whose morphisms are functions that preserve the underlying structure. Categories such as these effectively take given structures of the same sort, together with their structure preserving maps, and package them into a single structure. Conceptually, such categories do what they sound like they do: essentially *categorizing* existing mathematical structures. As such, these categories supply us with a few salient examples of

16. Modules play a large role in much of mathematical music theory, especially variants of $\mathbb{R}^{\{O,P,L,D\}}$ and the module of integers mod 12 (under addition), together with its affine transformations. At least as far back as Lewin (see Lewin (2010), originally published in 1987), transformations between musical elements (instead of musical objects like the C major chord) became the focus of many music theorists, where the transformations in question often form a group. Certain elements of algebra, especially group theory, have accordingly been used for some time to systematize the treatment of common operations on musical chords, and geometrical models of musical structure have also been considered by a few authors (such as Tymoczko (2011)). Approaches to questions of music theory that incorporate category theory (starting with Mazzola (1985)) are somewhat harder to come by, but the reader curious about more category theoretic takes on music may find interesting Popoff, Andreatta, and Ehresmann (2018), Noll (2005), and the work of Guerino Mazzola and followers. The module $\mathbb{R}^{\{O,P,L,D\}}$ presented above, together with further uses of it, is discussed in Mazzola and Andreatta (2006).

what you might think of as *categories of structures*. When Eilenberg and Mac Lane (1945) first defined categories and the related notions allowing categories to be compared (introduced in the next chapter), they stressed how it provided "opportunities for the comparison of constructions. . . in different branches of mathematics." But with Grothendieck's *Tôhoku* paper a decade later,[17] it started to become evident that category theory was not just a convenient tool for comparing different mathematical structures, but was itself a significant mathematical structure of its own intrinsic interest. While earlier uses of category theory treat categories as largely dispensable tools for helping to identify properties of given mathematical entities such as Abelian groups or certain modules, in Grothendieck's paper categories become objects of mathematics in their own right, whose common properties start to take on an intense mathematical interest. One way of starting to appreciate the sort of shift here is to realize that we do not just have categories consisting *of* mathematical structures of interest, but equally important are those categories that allow us to view categories themselves *as* mathematical structures of interest in their own right. The following two examples supply an initial way into this perspective of *categories as structures* (each of which example reveals crucial features of categories in general and is accordingly often said to supply us with a means of doing "category theory in the miniature").

Example 8 (*Each order is already a category*) Let $(X, \leq_X)$ be a given preorder (or, less generally, a poset). It is easy to verify that we can form the category $\mathcal{X}$ by

- taking for objects of $\mathcal{X}$ the elements of X; and
- declaring that there exists a morphism in $\mathcal{X}$ from a to b exactly when $a \leq b$ (and there is *at most* one such arrow, so this morphism will necessarily be unique).

Notice how transitivity of the relation $\leq$ will automatically give us the required composition morphisms, while reflexivity of $\leq$ just translates to the existence of identity morphisms. Thus, we can regard any given preorder (poset) $(X, \leq_X)$ as a category $\mathcal{X}$ in its own right.

In the other direction, if for every pair of objects A, B in a category $\mathbf{C}$, there is at most one morphism between A and B, then $\mathbf{C}$ in fact defines a presentation of a preorder (poset).

Example 9 (*Each monoid is already a category*) A *monoid* $\mathcal{M} = (M, \cdot, e)$ is a set M equipped with

- an associative binary multiplication operation $\cdot : M \times M \to M$, that is, $\cdot$ is a function from $M \times M$ to M (a *binary operation on* M) assigning to each pair $(x, y) \in M \times M$ an element $x \cdot y$ of M, where this operation is moreover associative in the sense that
$$x \cdot (y \cdot z) = (x \cdot y) \cdot z$$
for all $x, y, z \in M$; and
- a designated "unit" element $e \in M$, where this acts as a two-sided identity in that it satisfies
$$e \cdot x = x = x \cdot e$$
for all $x \in M$.

17. See Grothendieck (1957); it is now common to refer to this as the Tôhoku paper.

Comparing this definition to that of a category, it is straightforward to see how any monoid $\mathcal{M}$ can be regarded as a category of its own. Specifically, it is a category with just one object. It does not matter what this object is taken to be, and a priori it has nothing to do with sets and certainly does not come with any structure. To indicate as much, we can just represent it with an arbitrary symbol, such as $\bigstar$. For morphisms of this category we just use the elements $x \in M$, that is, for each $x \in M$ there will be a morphism

$$\bigstar \xrightarrow{x} \bigstar$$

The identity morphism $id_\bigstar$ is then taken from the monoid unit e and the composition formula for morphisms

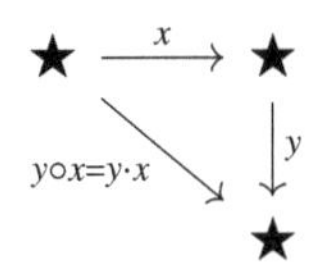

from the monoid multiplication. As this operation is associative and e acts as an identity, we do indeed have a category.

In the other direction, notice that if **C** is a category with only one object C—or just picking one object out from the category—and we let $\text{Hom}(C, C)$ denote its collection of morphisms (i.e., all the morphisms from C to itself, called "endomorphisms"), then $(\text{Hom}(C, C), \circ, id_C)$ will be a monoid.

Finally, an element $m \in M$ of a monoid is said to have an *inverse* provided there exists an $m' \in M$ such that $m \cdot m' = e$ and $m' \cdot m = e$. This lets us define a very important mathematical object: that of a group. A *group* is just a monoid for which every element $m \in M$ has an inverse. Moreover, if the group operation does not depend on the order in which two group elements are written, then the group is said to be *abelian*. Similar to what we saw with a monoid, any group itself can be shown to give rise to a category in which there is just one object, but where every morphism (given by the group elements) is now an isomorphism, in the following purely category theoretic sense:

Definition 10 In a category **C**, a morphism $f: A \to B$ for which there exists a morphism $g: B \to A$ in **C** such that $g \circ f = id_A$ and $f \circ g = id_B$ is called an *isomorphism.*

Inverses are unique, so we can write $g = f^{-1}$. Observe that the objects A and B are then said to be *isomorphic*, denoted $A \cong B$, whenever there exists an isomorphism between them. Such a notion is used to inform us about when we can regard two objects of some category as *the same*.

The previous two examples—orders and monoids—are not just examples of *any old* categories, but in an important sense, categories in general may be regarded as a sort of fusion of preorders on the one hand and monoids on the other. Over and above the fact that each monoid and each preorder is itself already a category in its own right, these two examples are special in that, through them, we can appreciate how categories more generally are exceptionally "monoid-like" and "preorder-like." We saw that every monoid can be exhibited as a single-object category. Seen from the other side, categories in general may be regarded as the *many-object* version of monoids. We saw that every preorder can be exhibited as a single-arrowed category, as between any two objects there is at most one arrow. Seen from the other side, categories in general may be regarded as the *many-arrowed* version of preorders. Sometimes, like Leibnizian monads, within a small part of the universe, set off on its own, we can see reflections of the whole. In this sense, individual monoids and preorders are like microcosms in which we can glimpse "in the miniature" the essential features of the general notion of a category. Monoids furnish us with not just a study of composition "in the miniature" (by collapsing down to a single object), but in a sense the associative binary operation and neutral or identity element that comprise the data of a monoid seem to provide a prototype for the general associativity and identity *axioms* of a category. Preorders, for their part, furnish us not just with a study of comparison of objects via morphisms "in the miniature" (by collapsing down to at most one morphism from any object to another), but in a sense the reflexivity and transitivity of the order seems to provide the model for the key *data* specifying a category, that is, the assignment of an identity arrow to each object (via reflexivity) and the composition formula (via transitivity).

There is another important (if more philosophical) way in which monoids in particular can shed light on categories. This has the added benefit of introducing the interesting notion of *oidification* and an alternative (philosophically appealing, if somewhat less useful) definition of categories. In order not to unduly distract the reader, we press on with some additional examples of categories and other more pressing matters, leaving a brief section on this topic to the very end of the chapter.

1.2 A Few More Examples

There are many more categories that we might mention, and that are important to mathematicians. However, we will instead just draw attention to a few more categories, and let the rest that will be of particular use to us emerge organically throughout the book.

Example 11 **Top** is the category that has topological spaces for objects and continuous functions for morphisms. This category, and topology in general, is discussed in detail in chapter 4.

Example 12 **Measure** is the category that has measure spaces for objects and (on one definition) appropriate equivalence classes of measurable functions for morphisms.

Example 13 **Man** is the category that has smooth (infinitely differentiable) manifolds for objects, smooth maps for morphisms.

Example 14 Suppose we are given V a vector space. Then we can define a category $\mathbf{V}$ as follows:

- for objects: $\mathbf{V}$ has only one object, called $*$;
- for morphisms (arrows from $*$ to $*$): the vectors v in V;
- for the identity arrow for $*$: the zero vector; and
- for composition of vectors v and v': their sum.

Let us now have a look at an example involving a rather important category, one that starts to make better sense of the idea that, in being visualized by arrows between dots, category theory might be regarded as some sort of graph theory, but with something extra (where this involves some extra structure regarding *composition* of arrows). In our earlier definition of a (directed) graph from example 5, observe that there were no conditions placed on arcs and vertices other than those involving the source and target functions, picking out the source vertex and the target vertex of a given arc a; in particular, there was no requirement regarding the composition of arcs. Thus, it is not the case that a category *is* a graph, for a directed graph in general has no notion of composition of arcs (and does not even have a notion of identity arrows). However, any category—well, any small category[18]—does have an underlying graph. While the converse does not hold, it is an important fact that every directed graph can be *made* into a certain category, via a special construction, discussed in the following example.

Example 15 Given a directed graph G, we first describe the notion of a *path* in G, as any sequence of successive arcs where the target of one arc is the source of the other. More explicitly, for each $n \in \mathbb{N}$, we define a *path* through G of length n as a list of n arcs,

$$i(0) \xrightarrow{e(1)} i(1) \xrightarrow{e(2)} i(2) \longrightarrow \cdots \qquad \xrightarrow{e(n)} i(n)$$

where the target of each arc is the source of the next one. A path of length 1 would then amount to a single arc, while a path of length 0 would be a vertex (node). We can create a category $\mathbf{Pth}(G)$, the *category of paths* through G, with objects the vertices of G and for morphisms from objects x to y all the paths through G from x to y. Given two paths,

$$i(0) \longrightarrow i(1) \longrightarrow i(2) \longrightarrow \cdots \longrightarrow i(n)$$

and

$$j(0) \longrightarrow j(1) \longrightarrow j(2) \longrightarrow \cdots \longrightarrow j(m),$$

with the end node of the first equal to the start node of the second, that is, $i(n) = j(0)$, we form the *composite path* by concatenating or sticking together the two paths along this identical node, that is,

$$i(0) \longrightarrow i(1) \longrightarrow \cdots \longrightarrow i(n) = j(0) \longrightarrow j(1) \longrightarrow \cdots \longrightarrow j(m)$$

resulting in a new path from $i(0)$ to $j(m)$. Then, concatenating paths end to end is associative, making composition in $\mathbf{Pth}(G)$ associative. As for ensuring that each object (vertex of G) has an identity arrow in $\mathbf{Pth}(G)$, we can observe that each vertex has an associated "length 0" path, and sticking such a path at the end of another path does nothing to change that other path. Thus, we can just take the paths of length 0 as our identity arrows, that is, the identity at an object x is given by the path of length 0 from x to x. Moreover, given a graph homomorphism $f : G \rightarrow G'$, every path in G will be sent under f to a path in G'.

We will have more to say about this category, and the construction that generates it, in the next chapter.

1.3 Returning to the Definition and Distinctions of Size

Before moving into consideration of how we can get some new categories from old, let us take the opportunity to make an important observation, leading to some important notions and distinctions having to do with size. We have stressed how many of the examples given thus far are categories whose objects are *structured sets* and where the morphisms are

18. The meaning of this distinction is discussed in a moment, in section 1.3.

(some appropriately "structure-preserving") *functions* between the underlying sets. Categories of this sort are sometimes called *concrete categories*. Beyond **Set** itself, of those categories introduced thus far, **Pre**, **Pos**, **Top**, **Mon(Group)**, $\mathbf{Mod}_R$, $\mathbf{Vect}_k$, **Grph**, and **Man** are all examples of concrete categories. Such examples might lead us to believe that all categories involve just considering some mathematical structures that are fundamentally set-like together with the appropriate notion of morphism between them (where this is some sort of function). However, in an arbitrary category, it cannot be assumed that objects are structured sets and morphisms structure-preserving functions, something that will become more evident especially as we consider more complicated categories. Moreover, the morphisms of a category are *not* always functions. For now, consider "nonconcrete" (or *abstract*) examples furnished by each poset (or preorder) regarded as a category, each group (or monoid) regarded as a category, and example 14. With a given monoid regarded as one object category, recall that the single object has nothing in principle to do with sets, and the morphisms don't carry any structure. Likewise, the morphisms of a given poset regarded as a category are just derived from the order relation, and a priori are not functions.

Considerations of this sort let us raise another important matter, taking us back to the original definition of a category we gave in definition 1. Recall how this definition mentioned a "collection" of objects and a "collection" of morphisms. While I originally suggested that the reader think of a "collection" roughly in terms of a "set," this is not in fact accurate, as the collection of objects may not be a set and arrows need not even be functions. In particular, some collections will not be sets for they are "too big" to be sets—an observation that leads to an important distinction of categories and a few useful definitions.

To better motivate discussion of these matters, first observe that for each of the concrete categories we have looked at, the "collection" of objects will not form a set. To appreciate this, it suffices to consider the case of the category **Set**. We said that this category was concrete, where this meant that the *objects* of the category are each themselves (perhaps structured) sets, and the arrows are functions (perhaps of a certain "structure-preserving" type). But notice what this does *not* say: it does not say that the *collection of all objects* of the category—which is what we are interested in, as far as the first data item of the definition is concerned—forms a set. Suppose that by "collection" in the definition of a category we *had* meant *set*. Then, as far as **Set** goes, the first data item of such a category would be a set of all sets. But for reasons having to do with Russell's paradox, which first emerged in the context of naive set theory, assuming such a set of all sets could even exist leads to a number of fatal foundational problems. So what is the status of the "collection" of objects?

The most direct way this sort of problem was addressed with set theoretical solutions involves prohibiting the unrestricted formation of sets so that such a *set* of all sets does not exist, and to introduce *classes* as a way of retaining "set-like" collections that can still be clearly defined by a property that all its members share, while being distinct from sets and so avoiding paradoxes of the sort first brought to light by Russell. In the Zermelo-Fraenkel (ZF) set theory the reader may be familiar with, one does not see a formal notion of classes, yet it is still common to speak informally of classes, referring to a class that is not a set as a *proper class* (or *large class*), while a class that is a set is called a *small class*.

The precise formal notion of "class"—and the consideration of whether a given entity constitutes a large class or not—ultimately depends on foundational context, that is, on the choice of a set theoretical framework that might allow for the formulation of such things. Strictly speaking, from the axioms of ZF, the class of all small classes (i.e., of all sets) cannot even be formally constructed and attempting to do so leads to contradictions of the sort involved in Russell's paradox. However, adopting certain extensions of ZF, such as the von Neumann-Bernays-Gödel axioms (NBG), allows us to formally distinguish classes and sets, and offers perhaps the most basic technical solution. Classes are taken to be the basic objects of the theory, and a set is formally defined in terms of it: a *set* is just a class that can be an element of other classes (and a *proper class*, then, is just a class that is not a set, in this sense). In such a context, the "collection" of all sets that supplies the objects of **Set** can be seen as a proper class, rather than a set—and the same would go for other similar categories, such as **Group**.

There are other ways of addressing the issues at stake here, including using "Grothendieck universes," but going deep enough into these matters to do them justice would take us too far afield.[19] Without stressing about which foundational framework we are using, let us agree to say that

Definition 16 A class is *small* if it is a set; it is *large* otherwise.

While any individual set—being a *set*!—is small, the point is that (within the foundational contexts that allow for the precise formulation of such notions) there are classes that are large, and the class of all sets is one example of such a thing.

Definition 17 A category **C** is said to be *small* if both the collection of objects of **C** and the collection of morphisms of **C** are sets; it is *large* otherwise.

Occasionally one will see a small category **C** defined as one for which the collection of all *morphisms* in **C** is small (large otherwise), that is, where there is no more than a set's worth of morphisms. Adopting the framework of NBG, then, we would be saying that a large category is one whose class of morphisms is a proper class; otherwise, the category is small. Such definitions that focus on the size of the collection of *morphisms* is ultimately the same as the one given in definition 17, using the fact that objects can be shown to correspond bijectively with identity morphisms, which morphisms of course form a subclass of the morphism class of a category, from which it follows that if **C** is small in the sense that its morphism class is small, then the class of objects of **C** will be small (i.e., a set) as well.

There are many categories that are not small, which can be seen by noting how the concrete categories introduced above each have too many objects to be small, and so are all examples of categories that are large. Yet, a category that is not small may locally—that is, between any pair of fixed objects—look like a small category. This notion is captured by the following definition:

Definition 18 A category is said to be *locally small* whenever between any pair of objects there is only a set's worth of morphisms.

19. Shulman (2008) is a nice reference for the reader interested in pursuing these size issues and the related foundational matters.

We could thus say that a category is small if and only if (iff) it is both locally small and its class of objects is small. Of course, any small category is thus locally small, but many large categories end up being locally small as well. For instance, the concrete categories we have seen thus far, while not small, are indeed locally small in this sense.[20] For a locally small category **C**, it is customary to write $\mathrm{Hom}_{\mathbf{C}}(A, B)$ or (dropping the subscript) $\mathbf{C}(A, B)$ for the collection of morphisms from A to B in the category **C**. As this collection is indeed a *set*, it is customary to call the set of morphisms between a pair of fixed objects a *hom-set* (regardless of whether these are literal *homomorphisms*). The reader should also be aware that it is now customary to use this hom-set notation even in the case of categories that are not necessarily locally small, a custom adopted in this book.

The distinctions just introduced will be useful to us going forward and issues related to the size of a category will occasionally resurface throughout the rest of the book. Before ending this section, though, let us address one potentially lingering issue. At this point, the reader who has followed this remark so far might be wondering:

> if removing any vagueness in the word "collection," thereby cashing in on the definition, is to make use of some rigorous foundation allowing for, say, the notion of *classes*, why not just say "class" and specify what is meant by this in the first place?

Adopting a rigorous foundation, such as NBG, capable of supporting the theory of classes, and then writing "class" everywhere we wrote "collection" in the definition is indeed a perfectly acceptable approach, and one will sometimes see authors do this. However, it is even more common to see the definition stated in terms of "collections," as we did. One might viably understand "collection" to encompass both sets and proper classes, as "class" does—but since the precise definition of "class" itself depends on the foundational context, one prefers not to commit to saying "class" (and having to specify the foundational context) and instead uses the more agnostic and deliberately vague "collection." Another reason for preferring the word "collection" is a little more philosophical: a category is really just *anything* that conforms to the definition's conditions. And if any rigorous foundation allowing us to make the relevant distinctions might be used in support of the definition, so that it is effectively independent of chosen foundation, then being overly specific about the particular set theoretical notion of class, for instance, seems to suggest that the concept of a category ultimately rests on some prior set theoretical framework and that any categorical treatment of set theory, for its part, may end up involving some circularity. But this is misleading and misrepresents the power of category theory. There are some thornier issues here, certainly, but let us instead end this section by noting that while **Set** is indeed a very important category, one is not to imagine that category theory somehow lives within set theory. In a way that we can make more precise in chapter 6, set theory itself can be thought of as doing "zero-dimensional" category theory. Moreover, since the 1970s with the work of Lawvere, efforts have been made to show that effectively all the mathematically significant portions of set theory and logic (in the narrow sense) can be seen as part of category theory.

20. In part because so many categories of interest are locally small, some authors accordingly even take local smallness as part of their definition of a category.

1.4 Some New Categories from Old

There are many important things one can do *to* categories, to generate new categories from old ones. Attention is confined, in this final section of the chapter, to those that will be most important for our purposes.

Definition 19 Let $\mathbf{C}$ be a category. The *dual* (or *opposite*) category $\mathbf{C}^{op}$ is then defined as follows:

- objects: same as the objects of $\mathbf{C}$;
- morphisms: given objects A, B, the morphisms from B to A in $\mathbf{C}^{op}$ are exactly the morphisms from A to B in $\mathbf{C}$ (i.e., just reverse the direction of all arrows in $\mathbf{C}$).

Identities for $\mathbf{C}^{op}$ are defined as before, composites are formed by interchanging the order of composition as one would expect, yielding a category. In more detail, for each $\mathbf{C}$-arrow $f : A \to B$, introduce an arrow $f^{op} : B \to A$ in $\mathbf{C}^{op}$, so that ultimately, these give all and only the arrows in $\mathbf{C}^{op}$. Then the composite $f^{op} \circ g^{op}$ will be defined precisely when $g \circ f$ is defined in $\mathbf{C}$, where for

$$A \underset{f^{op}}{\overset{f}{\rightleftarrows}} B \underset{g^{op}}{\overset{g}{\rightleftarrows}} C,$$

we have that $f^{op} \circ g^{op} = (g \circ f)^{op}$.

In slogan-form,

> *Given a category, just reverse all its morphisms and the order of composition, and you'll get another category (its* dual*)!*

With this seemingly innocuous construction, every result in category theory will have a corresponding dual, essentially got "for free" by simply formally reversing all arrows (and respecting the induced change in the order of composing arrows). In general, given a statement or construction framed in the language of category theory, when we refer to the *dual* of that statement or construction, we simply mean the statement or construction that is obtained by interchanging the source and target of each morphism as well as the order of composition of two morphisms. When a statement is true in a category $\mathbf{C}$, then its dual will be true in the dual category $\mathbf{C}^{op}$—"by duality" will refer to this invariance of truth under the operations involved in taking the opposite category. Such duality not only can clarify and simplify relationships that are often hidden in applications or particular contexts but it also *reduces by half* the proof of certain statements (since the other, dual statement will "follow by duality")—or, to see things another way, it *multiplies by two* the number of results, as each theorem will have its corresponding dual. Finally, for any $\mathbf{C}$, note that we will have that $(\mathbf{C}^{op})^{op} = \mathbf{C}$.

Next, we consider how, given a category $\mathbf{C}$, we can form a new category by taking as our objects all the *arrows* of $\mathbf{C}$.

Definition 20 For a category $\mathbf{C}$, we define the *arrow category* of $\mathbf{C}$, denoted $\mathbf{C}^{\to}$, as having

- objects: the morphisms $A \xrightarrow{f} B$ of $\mathbf{C}$;

- morphisms: from the $\mathbf{C}^{\rightarrow}$-object $A \xrightarrow{f} B$ to the $\mathbf{C}^{\rightarrow}$-object $A' \xrightarrow{f'} B'$, a morphism is a pair $\langle A \xrightarrow{h} A', B \xrightarrow{k} B' \rangle$ of morphisms from $\mathbf{C}$, making the diagram

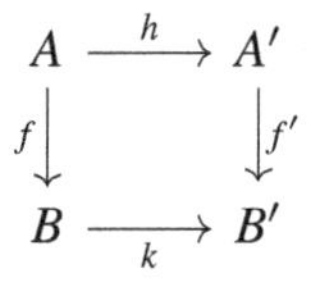

  commute (in $\mathbf{C}$).

Composition of arrows is then carried out by placing commutative squares side-by-side, that is, we put

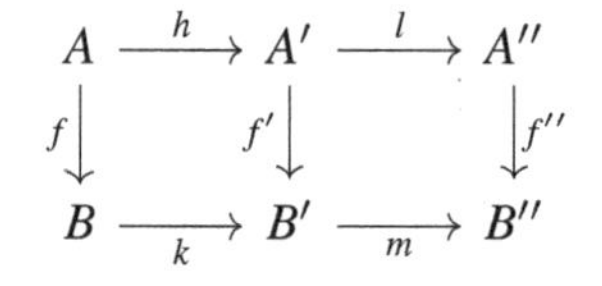

so that $\langle l, m \rangle \circ \langle h, k \rangle = \langle l \circ h, m \circ k \rangle$. The identity arrow for an object $A \xrightarrow{f} B$ is given by the pair $\langle \mathrm{id}_A, \mathrm{id}_B \rangle$.

With the arrow category, we are seeing *all* the arrows of the old category as our objects in the new category. The next construction instead looks at just *some* of the old arrows, where we restrict attention to arrows that have fixed domain (source) or codomain (target).

Definition 21 Given a category $\mathbf{C}$, and an object A of $\mathbf{C}$, we can form the two categories called the *slice* and *co-slice* categories, respectively denoted

$$(\mathbf{C} \downarrow A) \qquad (A \downarrow \mathbf{C}),$$

also called the category of

objects over A objects under A,

respectively.[21] The objects of the new category are given by

arrows to A arrows from A.

In other words, objects of the slice category are given by all pairs (B,f), where B is an object of $\mathbf{C}$ and $f : A \rightarrow B$ an arrow of $\mathbf{C}$, and of the co-slice category by all pairs (B,f) such that $f : B \rightarrow A$ is an arrow of $\mathbf{C}$.

Morphisms in the new category are given by $h : (B,f) \rightarrow (B',f')$, where this is an arrow $h : B \rightarrow B'$ of $\mathbf{C}$ for which the respective triangles

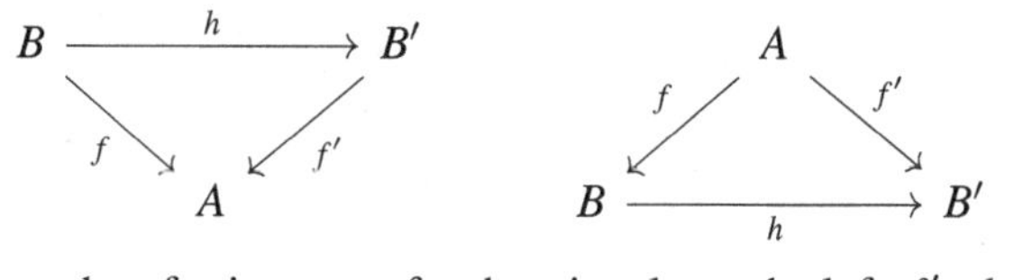

commute in the sense that, for instance, for the triangle on the left, $f' \circ h = f$.

21. These categories are also particular cases of a more general construction, known as *comma categories*. It is not uncommon to see the slice category of objects over $A \in \mathrm{Ob}(\mathbf{C})$ referred to as $\mathbf{C}/A$, and the co-slice category of objects under A referred to as $A/\mathbf{C}$.

Composition in $(\mathbf{C} \downarrow A)$ and $(A \downarrow \mathbf{C})$ is then given by composition in **C** of the base arrows h of such triangles.

Categories of this type play an important role in advancing some of the general theory, in addition to being of some intrinsic interest. For now, the slice category of *objects over A* might be thought of as giving something like a view of the category *seen within the context of A* (and the corresponding dual statement for the category of objects under A).

Finally, we define the following notion of a *subcategory*.

Definition 22 A *subcategory* **D** of a category **C** is got by restricting to a subcollection of the collection of objects of **C** (i.e., every **D**-object is a **C**-object), and a subcollection of the collection of morphisms of **C** (i.e., if A and B are any two **D**-objects, then all the **D**-arrows $A \to B$ are present in **C**), where we further require that

- if the morphism $f : A \to B$ is in **D**, then A and B are in **D** as well;
- if A is in **D**, then so too is the identity morphism id_A;
- if $f : A \to B$ and $g : B \to C$ are in **D**, then so too is the composite $g \circ f : A \to C$.

Moreover, we can also define the following:

Definition 23 Let **D** be a subcategory of **C**. Then we say that **D** is a *full subcategory* of **C** when **C** has no arrows $A \to B$ other than the ones already in **D**, that is, for any **D**-objects A and B,

$$\mathrm{Hom}_{\mathbf{D}}(A, B) = \mathrm{Hom}_{\mathbf{C}}(A, B).$$

Example 24 The category **FinSet** of finite sets—the category whose objects are all finite sets and whose morphisms are all the functions between them—is a subcategory of **Set**. In fact, it is a *full* subcategory.

The category of abelian groups **Ab** is a (full) subcategory of the category of groups **Group**. $\mathbf{Mod}_R$ is a subcategory of $\mathbf{ModAff}_R$.

If **C** is the category that has for objects those parts of $\mathbb{R}^n$ that are open—we will have more to say on this in chapter 4—and for morphisms those mappings between objects that are continuous, then a subcategory **D** of **C** is formed by restricting to mappings that have a derivative, where a rule of basic calculus shows that **D** has composition. A further subcategory of **D** could be got by further restricting to those mappings that have *all* derivatives (i.e., the *smooth* ones). There are many other important examples of subcategories that we will encounter throughout this book.

There are a number of other useful things one can do with categories, not to mention the important things one can do and find within categories. Discussion of such matters is taken up in the next chapters, and left to emerge organically throughout the book.

The real power of category theory, however, only really comes into its own once it is realized how, by putting everything on the same "plane," we can consider principled relations *between* categories. This is what we discuss in the next chapter.

1.5 Aside on "No Objects"

Box 1.1
No Objects

It is entirely common to name categories after their objects, and in most cases entirely natural to present categories in a "two-sorted" manner, with the two sorts *objects* and *morphisms*. In a given mathematical context, we are often already very familiar with the objects, and comfortable with seeing the relevant structure-preserving morphisms as entities that sit on top of, or are somehow secondary to, the objects. More generally, as human beings, we seem especially ready to divide up the world into objects, on the one hand, and processes or connections between those objects, on the other, where we take the latter to be somehow parasitic on the more primitive objects.

But as natural as this approach may seem, the objects of a category are in fact in bijective correspondence with (i.e., equivalent to) the identity morphisms—which, on account of one of the axioms, are uniquely determined by how they act as two-sided identities for composition. As such, it is really just the *algebra of morphisms* (without objects) that determines a category. Guided by this, one can give alternative definitions of a category that use only morphisms. The following presents such a single-sorted or "no objects" version of the definition of a category.

Definition 25 (*Category definition again ["no objects" version]*) A *category* (single-sorted) is a collection C, the elements or "individuals" of which are called *morphisms*, together with two endofunctions $s, t : C \to C$ (think "source" and "target") on C and a partial function $\circ : C \times C \to C$, where these satisfy the following axioms:

1. $x \circ y$ is defined iff $s(x) = t(y)$;
2. $s(s(x)) = s(x) = t(s(x))$ and $t(t(x)) = t(x) = s(t(x))$ (so s and t are *idempotent* endofunctions on C with the same image);
3. if $x \circ y$ is defined, then $s(x \circ y) = s(y)$ and $t(x \circ y) = t(x)$;
4. $(x \circ y) \circ z = x \circ (y \circ z)$ (whenever either is defined);
5. $x \circ s(x) = x$ and $t(x) \circ x = x$.

Notice how the elements of the shared image of s and t—that is, the x such that $s(x) = x$ (equivalently, $t(x) = x$)—are the *identities* (or what we would normally construe as the *objects*).

It is likely that the "punch" of this definition is lost on a reader seeing it for the first time. The important thing to realize is that behind this presentation is the idea that each object in the usual definition of a category can in fact be *identified* with its identity morphism, allowing us to realize an arrows-only (or "object-free") definition of a category.

Moreover, referring back to example 9, it is in the context of such an arrows-only version that we can even more easily see how monoids are just one-object categories—which really matters because it ultimately lets us better appreciate how categories in general are just many-object monoids. From a given monoid, we would obtain a category by defining $s(x) = t(x) = e$, where e is the monoid constant (identity) element. Going the other way, given a (nonempty) category satisfying any of

- $s(x) = s(y)$,
- $s(x) = t(y)$, or
- $t(x) = t(y)$,

we can define e as the (unique) identity morphism, and thus obtain a monoid. Note, by the way, how this says that s is a constant function (and thus, so is t, and they are in fact equal).

In this way, single-sorted categories can be seen to emerge via what is sometimes called "*oid*ification" (in this case, of monoids), where this describes a general twofold process whereby

1. some construction is realized as equivalent to a certain category with a *single* object; and then
2. the construction is generalized ("oidified") by moving to a further instantiation or version of that same category type that now has more than one object.

As the nLab highlights, in terms of nomenclature, categories give a pretty notable exception to this general rule of appending "oid" to a concept as we move to its many-object version, and perhaps we should all be greatly relieved that enough people did not succumb to the temptation to replace the term *category* with what this process suggests we should call such many-object monoids: a *monoidoid*!

We will see more examples of this process later on. For now, let us remark briefly on the broader significance of this object-free perspective. Consider how, in the context of graphs and graph theory, the novice will likely see arcs (arrows) as secondary to vertices (objects), for the arcs are frequently construed as just pairs of vertices. It also seems plausible that "psychologically" it is somehow more natural or easier for many of us to begin with objects (as the irreducible "simples") and then move on to *relations* between those objects. But in more general treatments of graphs, dealing with directed multigraphs or quivers for instance, one begins to appreciate that this proclivity really gets things backward: in fact, in more general settings, arcs are more naturally seen as primary and vertices can be seen as degenerate sorts of arcs, or as equivalence classes of arcs under the relations "has the same source (target) as."

In a similar fashion, one might argue that our default object-oriented mindsets can get things backward, in terms of what is really fundamental conceptually. It is often said in category theory that "what matters are the arrows/relations, not objects"—which is the substance of the observation that it is the algebra of morphisms that really determines a category, but it goes far beyond this as well. This is a very powerful idea, one that seems to permeate many aspects of category theory, and it even resurfaces in a particularly poignant way with one of the key results in category theory (the Yoneda results, covered in chapter 6). The object-free definition of a category given above is not typically the one seen in an introduction to categories, perhaps because it seems to complicate the presentation of many classical examples of categories, whose presentation is comparatively more straightforward using the standard two-sorted definition of a category. However, the object-free approach is arguably even more fundamental conceptually, and well attuned to the core philosophy of much of the categorical approach—which insists, in a number of ways, that what really matters is how objects and structures interact or relate—so it is worthwhile to at least be familiar with the existence of such a definition.

2 Prelude to Sheaves: Presheaves

In which we climb up the ladder of abstraction—scaling up the "morphisms are what's important" paradigm—by introducing morphisms between categories (functors) and morphisms between those (natural transformations), and then develop a store of examples of, and perspectives on, certain sorts of functors called presheaves, *a construction on which the central object of this book depends.*

It is often said that category theory privileges relations over objects, and in the last chapter you were exposed to what we might call the "morphisms are what's really important" perspective. As you meet more and more examples of categories, and especially as you begin to appreciate how a category is itself a mathematical construction with characteristic structure of its own intrinsic interest, a natural set of questions arise: if we treat categories themselves as objects, do we have a notion of morphisms between categories? And if so, what do the morphisms between categories look like? Is the model of preservation of structure still a good one? And if so, what is the relevant structure of a category that will have to be preserved by morphisms between categories?

Morphisms between categories are supplied by *functors*. Here is where the remarkable power of these categorical notions really starts to kick in, and where we lean even further into the "morphisms are what's really important" philosophy. To begin to appreciate where this might lead us, recall the arrow category, defined last chapter in definition 20, where one just takes the morphisms of a given category as the objects of a new category. What if, extending this idea, we now regarded functors as objects? Is there a notion of morphisms between functors? Yes, these are *natural transformations*. These constructions interact in an astoundingly rich manner, and we can continue further with this same controlled scaling procedure.

This chapter introduces functors and natural transformations, considers a wide variety of examples of functors and exhibits the many things we can do with them, introduces the concept of a *presheaf* (another name for a certain type of functor) and a category built out of presheaves, and uses a variety of examples of presheaves to further reflect on the different classes of things the presheaf does. The central object of this book, the sheaf, depends on this more general concept of the presheaf. In the introduction, we mentioned that sheaves effectively attach information locally to regions of some "space," doing so in a way that permits passage from local to global. Slightly more formally, the space can be made into a category, and so can the data. Then a presheaf will be a certain principled association of the first category to the second one, where a sheaf will amount to a presheaf that satisfies some further conditions. This will all be substantially improved upon. For the present chapter, we focus on developing a good understanding of functors and presheaves.

2.1 Functors

If a category is a context for studying a specific type of mathematical object and the network of relations entertained between those objects, a *functor* is a principled way of comparing categories, translating the objects and actions of one category into objects and actions in another category in such a way that certain structural relations are preserved through this translation. As a way of moving in a controlled way *between* categories, one can initially think of a functor as doing any of the following things: specifying data locally; producing a "picture" of the source category inside the target category, modeling the one category or some aspect of that category within another; taking advantage of the methods available in the target category to analyze the source category; converting a problem in one category into another where the solution might be more readily apparent; realizing an abstract theory of some structured notion (such as a group) in a certain background or on a specific "stage"; forgetting or deliberately losing some information, perhaps in order to examine or identify those features more robust to variations or to ease computation. But underneath these different interpretations or uses is a very simple requirement: a functor just transforms objects and maps in the source category into objects and maps in the target category, in such a way that two equations (amounting to the preservation of the structure supplied by identities and composites) are satisfied. Functors also come in two flavors, depending on their direction or variance. Formally,

Definition 26 A *(covariant) functor* $F:\mathbf{C}\to\mathbf{D}$ between categories $\mathbf{C}$ and $\mathbf{D}$ is an assignment of

1. an object $F(c)\in\mathbf{D}$ for every object $c\in\mathrm{Ob}(\mathbf{C})$; and
2. a morphism $F(f):F(c)\to F(c')$ in $\mathbf{D}$ for every morphism $c\to c'$ in $\mathbf{C}$,

which assignments satisfy the following two axioms:

1. For any object c in $\mathbf{C}$, $F(\mathrm{id}_c)=\mathrm{id}_{F(c)}$ ("F of the identity on c is the identity on $F(c)$");
2. For any composable pair f, g in $\mathbf{C}$, $F(g)\circ F(f)=F(g\circ f)$.

Observe that this last condition just states that F of a composite of two morphisms in $\mathbf{C}$ is the composite (in $\mathbf{D}$) of their images under F, that is, whenever we have

$$\begin{array}{ccc} c & \xrightarrow{f} & c' \\ & \searrow^{g\circ f} & \downarrow g \\ & & c'' \end{array}$$

commutative in $\mathbf{C}$, then the induced diagram

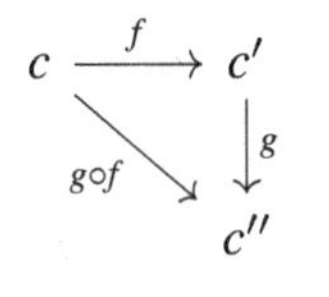

commutes in $\mathbf{D}$.

While such functors preserve the "direction" of morphisms—since the source of a morphism in $\mathbf{C}$ is assigned to the source of the image morphism in $\mathbf{D}$, and the same for targets—there are plenty of constructions throughout mathematics that would supply us with examples of functors, in that they seem to do everything a functor does, except that they reverse direction by taking sources to targets and targets to sources. The notion of a *contravariant functor* lets us accommodate such things.

A *(contravariant) functor* F from category $\mathbf{C}$ to category $\mathbf{D}$ is defined in the same way on objects, but differently on morphisms (where the source and target are swapped). Explicitly, to each morphism $f: c \to c' \in \mathbf{C}$ a contravariant functor F assigns a morphism $F(f)$: $F(c') \to F(c) \in \mathbf{D}$. This assignment must satisfy the same identity axiom $F(\text{id}_c) = \text{id}_{F(c)}$ as above, but for any composable pair f, g in $\mathbf{C}$, we must now have $F(f) \circ F(g) = F(g \circ f)$ (note the change in order of composition).

All the information of this definition is displayed below (the covariant case on the left and contravariant case on the right, and with identity maps omitted except for on one of the objects):

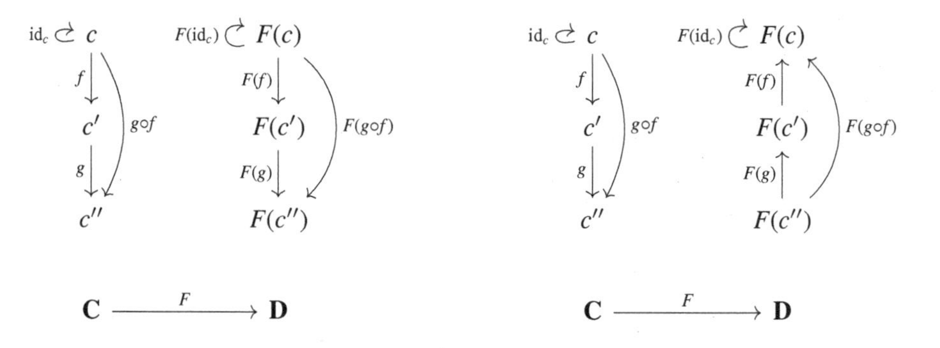

We will write $F: \mathbf{C} \to \mathbf{D}$, or $\mathbf{C} \xrightarrow{F} \mathbf{D}$, to indicate that F is a functor from $\mathbf{C}$ to $\mathbf{D}$. Functors will usually be denoted with upper-case letters (F, G, etc.), though we may occasionally use a more evocative name to indicate what the functor does.

Remark 27 Observe that by simply reversing the direction of all the morphisms in the category $\mathbf{C}$—that is, using the opposite category $\mathbf{C}^{op}$, as defined in definition 19—and then just using a *covariant* functor, we recover the notion of a contravariant functor on $\mathbf{C}$. On account of this, in principle a contravariant functor $F: \mathbf{C} \to \mathbf{D}$ can always be replaced by a *covariant* one $F: \mathbf{C}^{op} \to \mathbf{D}$, using the opposite category for the source category. Accordingly, whenever we speak of "functor," we will by default mean *covariant functor*, and to handle contravariant functors we will just speak of (implicitly covariant) functors and use the opposite category for our source category.

The truth of the frequently cited claim of Eilenberg and Mac Lane that "the whole concept of a category is essentially an auxiliary one; our basic concepts are essentially those of a functor and of a natural transformation"[22] proves itself in time to anyone who works with

22. Eilenberg and Mac Lane (1945, 247). Natural transformations, the morphisms between functors, are introduced in the next section of this chapter.

categories. Before helping the reader to better appreciate this for themselves by exploring a variety of examples of functors, let us use the notion of a functor to introduce a few other items of interest.

In addition to their intrinsic interest, functors are of special interest to us because of their essential role in the definition of *presheaves*—a concept that, as the name suggests, will be rather important to the development of sheaves.

Definition 28 A (set-valued) *presheaf* on $\mathbf{C}$—where $\mathbf{C}$ is assumed to be a small category—is a functor $\mathbf{C}^{op} \to \mathbf{Set}$.[23]

As we will see, a presheaf can often be thought of as consisting of some specification or assignment of local data, according to the "shape" of the domain category; a sheaf will emerge as a special sort of presheaf in that its local data can be glued or patched together locally. Before addressing in more detail the nature of presheaves, we give another important definition and then provide some examples of functors in general.

By now the reader should be comfortable with the core idea of taking mathematical objects of a certain type together with their (possibly structure-preserving) morphisms and assembling this into a category. In this same spirit, functors define morphisms between categories, there is a natural notion of composition of functors as well as a functor that acts as the identity, and the composition of functors can be shown to respect the axioms on associativity and identities. So categories and functors can be assembled into a category of their own!

Definition 29 The *category of (small) categories*, denoted **Cat**, is the category that has

- objects: small categories;
- morphisms: functors between them.

To verify that this is indeed a category, we need to establish identity morphisms (functors), that the composition of two functors (when defined) leaves us with a functor, and that all this data satisfies the axioms on associativity and identity. Explicitly, we will need the following very boring, but ultimately quite useful, functor.

Definition 30 Given a category $\mathbf{C}$, the *identity functor* is the functor $\text{id}_{\mathbf{C}} : \mathbf{C} \to \mathbf{C}$ that does what you would expect it to do. Explicitly, it takes an object to itself and a morphism to itself, that is,

- $\text{id}_{\mathbf{C}}(c) = c$,
- $\text{id}_{\mathbf{C}}(f) = f$.

As for composites: letting $\mathbf{C}, \mathbf{D}$, and $\mathbf{E}$ be (small) categories, and $F : \mathbf{C} \to \mathbf{D}$ and $G : \mathbf{D} \to \mathbf{E}$ be (covariant) functors, we can then define the composition of G with F, or composite functor $G \circ F : \mathbf{C} \to \mathbf{E}$, on objects c of $\mathbf{C}$ by $(G \circ F)(c) := G(F(c))$, and on morphisms $f : c \to c'$ of $\mathbf{C}$ by $(G \circ F)(f) := G(F(f))$. We can then show that the composition of two functors

23. Incidentally, as a presheaf is just a contravariant *functor* from a category $\mathbf{C}$ to **Set**, the reader may wonder why we give it two names. Such a reader might find useful the fun notion, used by the nLab authors, of a *concept with an attitude*—meant to capture those situations in math when one and the same concept is given two different names, one of the names indicating a specific perspective or attitude suggesting what to do with the objects, or the sorts of things one might expect to be able to do with them. In renaming a (set-valued) contravariant functor as a presheaf, then, we have a concept with an attitude, specifically looking forward to *sheaves*.

is a functor. Letting f, g be morphisms of $\mathbf{C}$ such that their composite $g \circ f$ is defined, we have

$$\begin{aligned}(G \circ F)(g \circ f) &= G(F(g \circ f)) \\ &= G(F(g) \circ F(f)) \\ &= G(F(g)) \circ G(F(f)) \\ &= (G \circ F)(g) \circ (G \circ F)(f),\end{aligned}$$

where the first and last lines are by definition of the composition of functors and the middle two use the fact that F and G are assumed to be functors themselves. Moreover, observe that for any object c of $\mathbf{C}$, we have

$$\begin{aligned}(G \circ F)(\mathrm{id}_c) &= G(F(\mathrm{id}_c)) \\ &= G(\mathrm{id}_{F(c)}) \\ &= \mathrm{id}_{G(F(c))} \\ &= \mathrm{id}_{(G \circ F)(c)}.\end{aligned}$$

This shows that the composite $G \circ F$ of two functors is itself a (covariant) functor.[24] It is straightforward to show that, as morphisms, functors obey the associativity and identity axioms, thus ensuring that **Cat** indeed forms a category.

Remark 31 In section 1.3, we raised some size issues and important distinctions related to size. In considering possible definitions of the category of categories, the same sorts of size issues take on an even greater significance. In definition 29, we implicitly tried to get ahead of some of these lurking size issues by only counting small categories among our objects. Observe that the resulting category **Cat** itself will be locally small, yet not small. This is good, since it means that, being large itself, **Cat** won't be an object of itself, and so we skirt any issues analogous to Russell's paradox. However, in taking only small categories for objects, categories like **Set**, **Pos**, **Mon** will not be found among the *objects* of **Cat**. (Though they are subcategories.)

Naturally, this raises a question: if such nonsmall (large) categories cannot be found among the objects of **Cat**, is there a category that *does* include them among its objects? We will denote such a category, which admits large categories as objects and the functors between them as morphisms, by **CAT**. Again hoping to avoid problems analogous to the paradox discussed in section 1.3, we are inspired to stipulate that the objects in **CAT** still be locally small; defined thus, **CAT** for its part would not be locally small, and so it is not in danger of being an object of itself.

In this book, we won't worry too much further about differences between these two categories, and will mostly just find ourselves dealing with **Cat**.

2.1.1 Examples of Functors

Functors appear all over mathematics. But perhaps the lowest-hanging examples of functors can be found by looking at those established mathematical structures of a certain type

24. Relying on this fact, and in order to avoid certain notational infelicities like $(G \circ F)(g \circ f)$, we will occasionally use the juxtaposition notation GF for composite functors, where this is understood to be the same as $G \circ F$.

that individually assemble into a category and thereby supply us with a particularly simple way in to categories as objects of study in their own right. In these cases, we would expect that a functor between such objects, now each regarded as an individual category in its own right, would recover the usual important structure-preserving relations that are expected to obtain among such objects as they appear in their native setting. This is what is illustrated by the following two examples.

Example 32 Recall from example 3 (chapter 1) that a preorder $\mathcal{X} := (X, \leq)$ is traditionally defined as a set X together with a reflexive and transitive binary relation $\leq$. Recall also that we can transform a given preorder into a category $\mathcal{X}$ by defining, for every pair of objects $x, x' \in X$, the hom-set $\mathrm{Hom}_{\mathcal{X}}(x, x')$ as either empty (in case the pair (x, x') is not related by $\leq$) or as consisting of the unique morphism $x \to x'$ (just in case $x \leq x'$), making the composition formula completely determined. In other words, it is a category that has at most one morphism between any two objects.

If $\mathcal{X} := (X, \leq_X)$ and $\mathcal{Y} := (Y, \leq_Y)$ are two preorders, regarded as categories, then what is a functor $F : \mathcal{X} \to \mathcal{Y}$? First of all, it assigns an object x in the set $\mathrm{Ob}(\mathcal{X}) = X$ to the object $F(x) \in Ob(\mathcal{Y}) = Y$. In other words, it acts as a function on objects. Given a morphism $f : x \to x'$ in $\mathcal{X}$ (from which, from the traditional perspective of the preorder, we are just saying that $x \leq x'$), by the definition of a functor this will get sent to $F(x) \to F(x')$ in $\mathcal{Y}$, which will moreover be unique (and corresponds, at the level of the preorder, to saying that $F(x) \leq F(x')$). But this just says, at the preorder level, that $x \leq x'$ implies $F(x) \leq F(x')$. In other words, with the notion of a functor between preorders treated as categories we have recovered the usual notion of maps between preorders, namely a monotone map, as defined in definition 4. Moreover, it is easy to see that a contravariant functor between preorders regarded as categories will just recover the notion of an *antitone (order-reversing) map*, that is, whenever $x \leq x'$, then $f(x') \leq f(x)$.

Example 33 Recall from example 9 (chapter 1) that each monoid (and each group) can be regarded as its own category. Explicitly, we saw that a monoid $(M, e, \cdot)$ can be considered as a category $\mathcal{M}$ with one object and with hom-set equal to M, where the identity morphism comes from the monoid identity e and the composition formula from the monoid multiplication $\cdot : M \times M \to M$. Given two monoids $(M, e, \cdot)$ and $(N, e', \star)$, where these are regarded as one-object categories $\mathcal{M}$ and $\mathcal{N}$, we might hope that a (covariant) functor from one to the other would just recover the usual notion of a morphism between monoids, a *monoid homomorphism*, where this is defined as a map $\phi : M \to N$ that respects the structure in the sense that

$$\phi(m \cdot m') = \phi(m) \star \phi(m') \text{ and } \phi(e) = e'.$$

One can immediately see that the above equations are the same as defining a covariant functor between $\mathcal{M}$ and $\mathcal{N}$, when these are each regarded as a category. A contravariant functor from $\mathcal{M}$ to $\mathcal{N}$, for its part, is exactly a monoid morphism that flips the elements, that is, $\phi(m \cdot m') = \phi(m') \star \phi(m)$.

Since a group is just a monoid in which every element is invertible, a similar story can of course be told using groups. A group can be regarded as a category with one object such that every morphism is an isomorphism, and then a functor between such categories recovers the notion of a group homomorphism.

Before leaving this example, we might also mention a few other prominent functors relating the mathematical structures under consideration. There exists a functor *Core*: **Mon** $\to$ **Group** that ingests a monoid $(M, e, \cdot)$ and spits out the subset of invertible elements of that monoid—which of course leaves us with a group, typically called the *core* of the monoid M. There is a related functor **Cat** $\to$ **Grpd** sending a category **C** to the largest groupoid inside **C**, also called its core.[25] It's also worth mentioning that when we said that each monoid could be regarded as its own category we were really just appealing to the fact that there is a functor **Mon** $\to$ **Cat** that takes a monoid to its corresponding category!

Moving beyond examples where the functors pass between categories that are fundamentally the same type of structure, we need to begin to appreciate some of the other important things functors do.

Example 34 In many settings, one might want to transfer one system of objects that present themselves in a certain way in one context to another context where irrelevant or undesirable (e.g., noisy) features are suppressed, while simultaneously preserving certain basic qualitative features. In the definition of a category, and indeed in the definition of many mathematical objects, typically one specifies (1) underlying data, together with (2) some extra structure, which in turn may satisfy (3) some properties. One obvious thing to do when considering some category **C** is to deliberately lose or ignore some or all of the structure or the properties carried by the source category. This process informally describes *forgetful functors*, which provide us with a large source of examples.

There are many examples where **Set** is the target category, since many important categories are sets with some structure; however, forgetful functors need not have **Set** for the target category. For instance, since a group is just a monoid $(M, e, \cdot)$ with the extra property that every element $m \in M$ has an inverse, this means that to every group we can assign its underlying monoid and every group homomorphism will get assigned to a monoid homomorphism between its underlying monoids—this is carried out simply by forgetting the extra conditions on a group. Thus, there is a forgetful functor U : **Group** $\to$ **Mon**.

While the "forgetting" terminology might suggest some sort of (possibly pejorative) loss of information, another way of looking at the same process is that it extracts and emphasizes only the important features of the objects under study. An illustration of this comes from *detectors*, which in practice often act to forget or lose information carried by a signal, while preserving fundamental features of the underlying signal. This is exactly what is useful about such tools, since what is removed is clutter, leaving us with a compressed representation of the original information (with the effect that the result of applying the functor might be more robust to variations, more relevant to a particular application, simpler for computation, etc.).

A *signal* is essentially a collection of (local) measurements related to one another, and the topology associated with these measurements tells us how a measurement is affected by noise; for instance, a signal over a discrete set is typically either not changed by noise at all or it changes drastically, while a signal over a smoother space may depend less drastically

25. A groupoid is just like a group except that it can have more than one object, as the discussion of oidification in section 1.5 would suggest. More formally, a *groupoid* is a category such that every morphism is an isomorphism; a morphism between groupoids is also just a functor. As such, we could then define a group as a groupoid with only one object.

on perturbations.[26] As already anticipated, as a forgetful functor, a detector acts to *remove* something—specifically, it acts to remove topological structure from the signal (which may have the effect of quantizing signals)—but, as a functor, it should also preserve certain features of the signal.

As a specific and simple instance of this, consider a *threshold detector*.[27] A detector can just be regarded as a functor from some category of signal data into some subcategory of **Set**. We can describe a threshold detector as a detector that ingests a continuous (real-valued) function $f \in \mathbf{Cont}(\mathbb{R})$ and spits out the open set on which $f(x) > T$ for a given threshold $T \in \mathbb{R}$. The domain of this functor will be the category $\mathbf{Cont}(\mathbb{R})$ which has for objects the continuous real-valued functions and for morphisms $f \to g$ whenever $f(x) > g(x)$ for all $x \in \mathbb{R}$. The threshold detector is thus a functor D that assigns to each $f \in \mathbf{Cont}(\mathbb{R})$ the open set $D(f) = \{x \in \mathbb{R} \mid f(x) > T\}$, that is, it lands in the category $\mathbf{Open}(\mathbb{R})$ of open sets of $\mathbb{R}$, whose morphisms are given by subset inclusion. Moreover, one can see that if $f \to g$, then we will have $D(g) \subseteq D(f)$, making D a *contravariant* functor from $\mathbf{Cont}(\mathbb{R})$ to $\mathbf{Open}(\mathbb{R})$, that is, our threshold detector D is a functor $D : \mathbf{Cont}(\mathbb{R})^{op} \to \mathbf{Open}(\mathbb{R})$.

In general, forgetful functors frequently can tell us interesting things about the source category. For instance, we have a functor $U : \mathbf{Cat} \to \mathbf{Grph}$, which informs us that categories have underlying graphs. Recall from the definition of a directed graph (see example 5, chapter 1) that a graph $G = (V, A, s, t)$ just consists of a set V of vertices, a set A of edges (which will be directed, so we also call them arrows), and a pair of functions $s, t : A \to V$ codifying the direction of the edges (arrows) by assigning to each $a \in A$ its source vertex $s(a)$ and target vertex $t(a)$. In general, we have been displaying the objects and morphisms of a category as the vertices and edges of a directed graph, and we draw directed graphs the way we draw categories—which may superficially suggest graphs and categories are fundamentally "the same" sort of thing. But observe that, as defined, directed graphs just consist of vertices and edges—where the directedness of the edges is codified by information regarding designated source and target vertices—and a priori there is no notion of composition of edges and no identities to speak of.

To appreciate how the functor in question works, consider that, when defining a category, we could have equally defined a (small) category by saying that the data of the category involves a set of objects (let's denote this by C_0), a set of morphisms (denoted C_1), and a diagram $C_1 \overset{s}{\underset{t}{\rightrightarrows}} C_0$, together with some structure (composition and identities) and properties (identity and associativity axioms). Really, the additional structure and properties concerning composition and identities can be codified by supplementing the previous diagram with another map $i : C_0 \to C_1$ assigning identity arrows to each object, and another set $C_2 := \{(f, g) \in C_1 \times C_1 \mid t(f) = s(g)\}$ together with a partial operation capturing composition $C_2 \overset{\circ}{\to} C_1$. In this way, we are effectively describing a small category $\mathbf{C}$ with the diagram

$$C_2 \xrightarrow{\ \circ\ } C_1 \overset{s}{\underset{t}{\overset{\longleftarrow i \longrightarrow}{\rightrightarrows}}} C_0$$

26. All matters of topology—including the open sets and related topological matters invoked in the next paragraphs—are discussed extensively in chapter 4.

27. This idea for this threshold example comes from Robinson (2014).

where the expected behaviors of compositions and identities are codified by further equations. Comparing this formulation to the definition of a directed graph, it should come as no surprise that there is a functor sending categories to their underlying directed graph. What the functor U does is take the objects of a category to vertices of its underlying graph and the morphisms to the graph's directed edges (arrows), where the associated source and target functions of the graph agree with the source (domain) and target (codomain) assignments of the category. As there is no further structure or conditions having to do with composition of arrows or with identities in a graph, U basically "forgets" anything having to do with specifications of identities and composition. More explicitly, U can be seen as taking the category **C**—described by the diagram above—to the underlying graph

$$C_1 \underset{t}{\overset{s}{\rightrightarrows}} C_0$$

and then acting on morphisms (functors) by ignoring parts of the category diagram that contain information about i and $\circ$ (since graph homomorphisms, as simply a mapping of edges to edges and vertices to vertices that preserves sources and targets, effectively have nothing to say about matters of identities and compositions).

Forgetful functors often come paired with corresponding *free functors*. For instance, corresponding to the "underlying graph" functor U, there exists the *free category functor* $F: \mathbf{Grph} \to \mathbf{Cat}$, which we have in fact already met in example 15 (chapter 1). Recall that, given a directed graph G, we can create a category $\mathbf{Pth}(G)$, the category of paths through G, with objects the vertices of G and morphisms the paths through G. The resulting category of paths of a graph G, $\mathbf{Pth}(G)$, gives us the *free category generated by* G, which can be thought of as the result of freely adding to a given directed graph all paths (all possible composite arrows) as well as all the identity arrows. The resulting category has the same set of objects (i.e., vertices) as the original graph, but it will in general have a larger set of morphisms, for the hom-set $\mathrm{Hom}(v, v')$ in the resulting category will consist of all the paths in the graph G from v to v', which may include arrows that were not in the original graph. Graph homomorphisms then extend into unique (covariant) functors.

In short, this path construction gives rise to a functor $F: \mathbf{Grph} \to \mathbf{Cat}$, called the *free category functor*. $\mathbf{Pth}(G)$ is always the *largest* category generated by G. On the other hand, G also generates a *smallest* category by taking the quotient of $\mathbf{Pth}(G)$ by the relation that identifies two paths that share the same source and the same target. In this connection, any category **C** can be obtained as a quotient of the corresponding category of paths of its underlying graph, under the equivalence relation identifying two paths if and only if they have the same composite in **C**.[28]

Altogether, constructions and results established in the context of graphs can be applied to categories, once we forget about composition; and conversely, results concerning categories can be applied to graphs by simply replacing a graph by its category of paths. And these things are codified by the existence of the functors just described.

28. Getting ahead of ourselves somewhat, it is worth noting that the pair of functors $\mathbf{Cat} \underset{F}{\overset{U}{\rightleftarrows}} \mathbf{Grph}$ are related in a very special way, one that will be explained in the treatment of *adjunctions* in chapter 7. This relation is especially significant, for it forms an important adjunction that gives rise to a particular construction called a *monad* that is a starting point for the generalization to n-categories.

Example 35 We just saw that categories have underlying graphs. There is the important related notion of a *diagram* in a category $\mathbf{C}$, a notion that in some sense captures a generalized idea of a subgraph of a given category's underlying graph.[29] A diagram is defined as a functor $F : \mathbf{J} \to \mathbf{C}$ where the domain, called the *indexing category* or *template*, is a small category. Typically, one thinks of the indexing category as a directed graph, that is, some collection of nodes and edges that serves as a template defining the shape of any realization of that template in $\mathbf{C}$ and that may also specify some commutativity conditions on the edges which are to be respected by $\mathbf{C}$. Then a diagram can be regarded as something like an instantiation or realization in $\mathbf{C}$ of a particular template $\mathbf{J}$. Each node in the underlying graph of the indexing category is instantiated with objects of $\mathbf{C}$, while each edge is instantiated with a morphism of $\mathbf{C}$. If we write the objects in the index category $\mathbf{J}$ as $i, j, \ldots$, and the values of the functor $F : \mathbf{J} \to \mathbf{C}$ in the form $F(i), F(j), \ldots$, then a diagram amounts to a family of objects $F(i)$ of $\mathbf{C}$ indexed by the nodes of $\mathbf{J}$ and a family of arrows $F(e)$ of $\mathbf{C}$ indexed by the edges of $\mathbf{J}$. Accordingly, one sometimes speaks of a diagram F as a $\mathbf{J}$-indexed set, or $\mathbf{J}^{op}$-parametrized set (depending on the variance of the functor). Functoriality demands that any of the composition relations (in particular, commutative diagrams) that obtain in $\mathbf{J}$ carry over (under the action of F) to the image in $\mathbf{C}$.

We will have a lot more to say about this perspective later in this chapter. For now, let us look at a few concrete illustrations of this. First consider the category

$$\begin{array}{ccc} id_0 & & id_1 \\ \circlearrowright & & \circlearrowright \\ \bullet 0 & \longrightarrow & \bullet 1 \end{array}$$

of two objects and a single non-trivial morphism, often called $\mathbf{2}$.[30] With such a category for our indexing category, a (set-valued) diagram yields a category that has as objects all the functions from one set to another set, and as morphisms the commutative squares between those arrow-objects. In more detail: a morphism from object $f : A \to B$ to object $g : C \to D$ will be a pair of functions $\langle h, k \rangle$ such that

$$\begin{array}{ccc} A & \xrightarrow{h} & C \\ {\scriptstyle f}\downarrow & & \downarrow{\scriptstyle g} \\ B & \xrightarrow[k]{} & D \end{array}$$

commutes. Composition is component-wise, that is, $\langle j, l \rangle \circ \langle h, k \rangle = \langle j \circ h, l \circ k \rangle$, and the identity arrow for $f : A \to B$ will be the function pair $\langle \mathrm{id}_A, \mathrm{id}_B \rangle$. Does it look familiar? It should. This is just the *arrow category* introduced earlier, in definition 20!

Suppose instead we take for our indexing category $\mathbf{3}$, or $\mathbf{[2]}$, the linear order category with length 2

29. A *subgraph* is just what it sounds like: a subset of a (directed) graph's edges (and associated vertices) that constitutes a (directed) graph. For undirected graphs, the notion is even more straightforward: graph G' is a subgraph of graph G when the vertex set V' of G' is a subset of the vertex set V of G and the edge set E' of G' is a subset of the edge set E of G. In other words, a subgraph is essentially a graph within a larger graph.

30. $\mathbf{2}$ is isomorphic to the linear order $\mathbf{[1]}$, so one will occasionally see it go by that name.

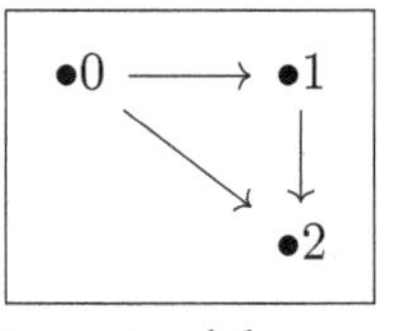

Then a diagram on this category just acts to pick out as objects commutative triangles. As a final example, taking the category $\mathbf{2} \times \mathbf{2} \times \mathbf{2}$ as our indexing category just serves to pick out as objects commutative cubes

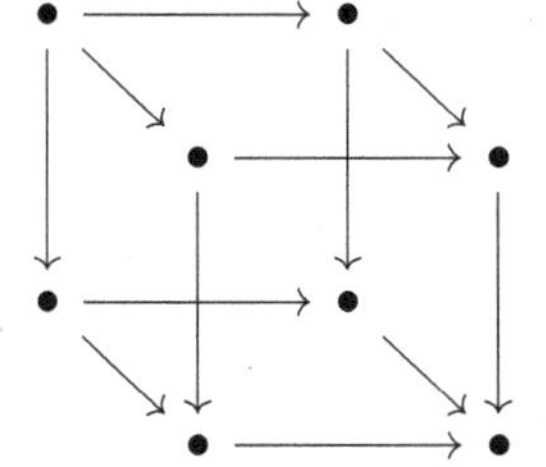

in the target category.

As we will see, this diagram approach can be significantly generalized and can even be used to provide definitions of n-categories, specifying the data for an n-category as a diagram (presheaf) $A : \Sigma^{op} \to \mathbf{Set}$, where Σ is some category of shapes and the functor yields, for each shape, a set of "cells" of that shape.[31]

Example 36 Expanding on the previous perspective, assume we are given as indexing category $\mathbf{J} :=$

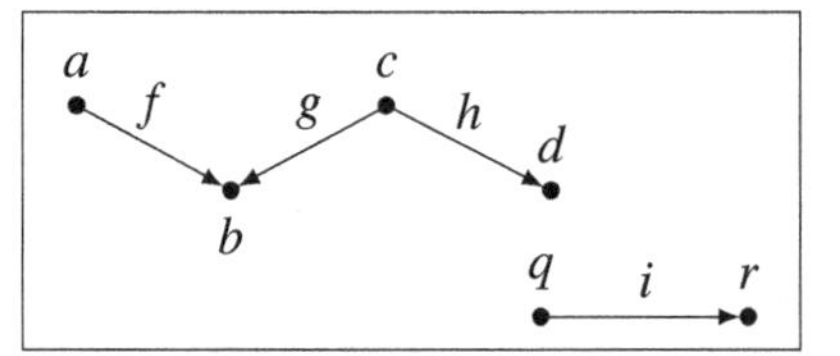

Now let the diagram $F : \mathbf{J} \to \mathbf{Set}$ be given on objects by

$$F(a) = \{1, 2\}, \quad F(b) = \{1, 2\}, \quad F(c) = \{1, 2, 3\},$$
$$F(d) = \{1, 2, 3, 4\}, \quad F(q) = \{1, 2, 3\}, \quad F(r) = \{1, 2\}$$

31. Treatment of n-categories is beyond the scope of this book. Instead, let us just observe how this diagram approach already suggests a more general definition of presheaves: for categories $\mathbf{C}$ and $\mathbf{J}$, a $\mathbf{C}$-presheaf on $\mathbf{J}$ can be defined as a contravariant functor from $\mathbf{J}$ to $\mathbf{C}$. Instead of taking presheaves to be functors taking values in **Set**, we can thus use other target categories, like the category of groups, rings, vector spaces, modules, and so on. While this more general definition is perfectly coherent (and is useful for achieving greater generality), presheaves are classically regarded as valued in **Set**. While this is not entirely necessary, as we just saw, there is also good reason for it. In brief, it has to do with the fact that the category of sets occupies a somewhat special place: as we will explore in chapter 6, the usual categories are *enriched* over sets, by which we effectively mean that given a pair of objects $X, Y \in \mathbf{C}$, we can form $\text{Hom}_{\mathbf{C}}(X, Y)$, an object of **Set**. Moreover, another factor here has to do with the fact that only set-valued functors are *representable*. Both of these matters—that of enrichment, and representable functors—are covered in chapter 6. For now, however, it is worth noting that in most categories $\mathbf{C}$, the hom-sets $\text{Hom}_{\mathbf{C}}(X, Y)$ are richer than just sets.

and on morphisms by

$$F(f) = 1 \mapsto 1, 2 \mapsto 2;$$
$$F(g) = 1 \mapsto 1, 2 \mapsto 2, 3 \mapsto 1;$$
$$F(h) = 1 \mapsto 1, 2 \mapsto 2, 3 \mapsto 4;$$
$$F(i) = 1 \mapsto 2, 2 \mapsto 1, 3 \mapsto 1.$$

This can be pictured as follows:

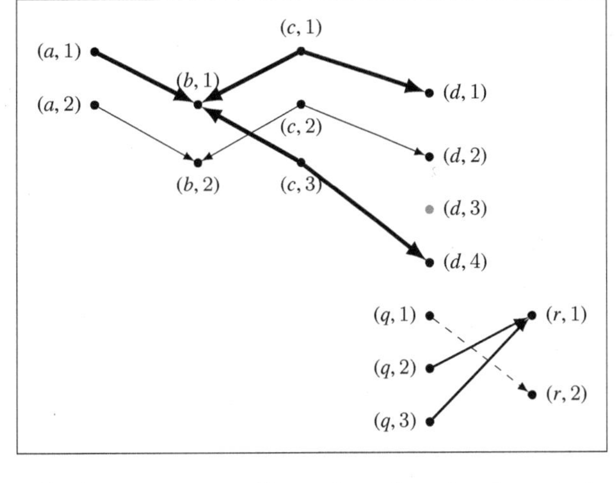

This realization affords us a concrete illustration of another important construction, the *category of elements*, which will be defined in chapter 3 and used to explain why there are different thicknesses of arrows in this picture.

There are also many functors that recover important established constructions that appear within the context of more specialized study of certain mathematical structures. The following is an example of that.

Example 37 *Graph coloring problems* are a commonly discussed class of problems in graph theory having to do with assigning colors to certain components of a graph subject to certain constraints. Such problems can ultimately be formulated as a problem of *vertex coloring*, where this is an assignment of "colors" (or any label) to a graph's vertices such that no two adjacent vertices share the same color, that is, so that whenever an edge connects two vertices, those vertices are assigned distinct colors. Moreover, one speaks of a coloring of a graph G by n colors as an *n-coloring* of the graph G. For an application of such a problem, the vertices of a graph might represent radio stations, where two vertices are adjacent (i.e., connected by an edge) whenever the stations are near enough to cause interference, so that a coloring would then amount to an assignment of noninterfering frequencies to the stations.

Recall the category of undirected (simple) graphs, **SmpGrph**, introduced in example 5 (chapter 1). The objects of such a category are the graphs that a graph theorist usually means by the word, and the morphisms are the usual (undirected) graph homomorphisms, that is, a graph homomorphism from a graph $G = (V, E)$ to a graph $G' = (V', E')$ is a function $f : V \to V'$ on the vertices such that if $\{x, y\}$ is an edge of G, then $\{f(x), f(y)\}$ is an edge of G'. There is a contravariant functor from this category of undirected graphs to the category of sets—that is, a functor $nColor : \mathbf{SmpGrph}^{op} \to \mathbf{Set}$—that takes a graph to the set of n-colorings of its vertices, so that for any graph G, $nColor(G)$ will consist of the set of all n-colorings of G. Observe how an n-coloring of a graph G' together with a graph homomorphism $G \to G'$ will give rise to an n-coloring of G, illustrating the contravariance of this functor. We will return to this functor in later chapters.[32]

Here is an example of a different flavor, one that also ties together a number of constructions introduced thus far.

Example 38 There are natural language expressions that we use all the time to express that someone or something has a certain property *qua* (or *as*) one thing but not *qua* some other thing. For instance, one might say

John is fair *qua* businessman, but not *qua* politician,

or

Maria is inspirational *qua* teacher, but not *qua* basketball player.

We often make use of judgments involving the logic of *qua*. Suppose someone asks you whether your friend Abe is honest. You might answer, "Well, it depends: in some respects/aspects, Abe is honest; in other respects/aspects, not so much." Perhaps you know him to be an honest friend, and have no reason to suspect his dishonesty as a businessman, but you have serious doubts about his honesty as a card player. In attempting to form a judgment about your friend's honesty, you can argue about which of the aspects are relevant, or most relevant, and also about how Abe's behavior, under a particular aspect, should be interpreted (as honest or dishonest). However, in general, this sort of answer—"In some aspects, yes; in others, not so much"—and the ensuing discussion or debates make sense. Once agreement about these matters (which aspects are relevant, etc.) has been achieved, we may use these assessments to arrive at a global judgment about Abe's honesty.

We might conceptualize this situation category theoretically, using a particular *category of aspects* or *qua* category.[33] In this setting, we will be able to model things like the honesty of Abe under aspect A, and moreover model the assembly of global judgments of the type "Abe is honest," "Abe is not honest," "Abe is dishonest," and so on, from this data of judgments about Abe's honesty *qua* the various relevant aspects.

Let us first define something we could call the *nominal category* **CN**.

Definition 39 The *nominal category*, **CN**, has for

32. Looking ahead, a sheaf will be defined as a particular presheaf satisfying certain properties with respect to "covers" of the objects of the domain category (which we will meet more formally in chapter 4, dedicated to topology). Anticipating this, the *nColor* presheaf functor defined on a related subgraph category will turn out to give us a sheaf, since if some subgraphs G_i cover G, and if $\{c_i \in nColor(G_i) \mid i \in I\}$ is a family of colorings such that the colorings agree on intersections among the G_i, then the c_i's induce a unique coloring of the entire graph G, which essentially is what it means to have a sheaf in this instance.

33. The idea for this, and the key definitions provided below (as well as the example, with mostly trivial modifications), comes from Reyes et al. (1999).

- objects: CNs (count nouns) relevant to discussion, for example, "a student," "a coworker," "a husband," "a parent," "a family man," "a student, a coworker, and a family man," written as

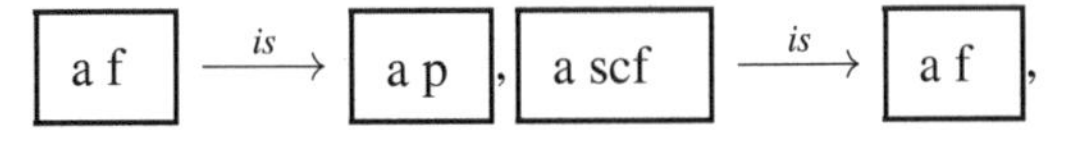

- morphisms: "identification" postulates of the form (copula connecting nouns)

a f $\xrightarrow{is}$ a p , a scf $\xrightarrow{is}$ a f ,

where these are meant to capture the identifications frequently used in natural languages, such as "a family man is a parent," "a student, a coworker, and a family man is a family man," "a dog is an animal," and so on.

Identity morphisms are those particular axiomatic identifications of the form

a f $\xrightarrow{is}$ a f .

Composition is given by stringing together identifications in the obvious way, that is, whenever we have two arrows, the codomain of one as the domain of the other, we complete the graph by adding an arrow that is the composition of the two arrows, and where this corresponds to the common rule of inference in natural languages from things like "a human is a primate" and "a primate is a mammal" to "a human is a mammal."

Arrows of this category can be thought of as supplying a *system of identifications*, where this replaces a notion of *equality* between kinds (since equality is a relation that might be argued to obtain only between the members of a given kind). This category is assumed to be posetal, where this means there is at most one arrow between two objects.

We know from definition 20 (chapter 1) that for a category **C**, we can form the *arrow category* of **C**, denoted $\mathbf{C}^{\rightarrow}$. Moreover, it turns out that we can *identify* any object A of **C** with the object $A \xrightarrow{id_A} A$ in $\mathbf{C}^{\rightarrow}$. Using this arrow category construction, we can define our main category of interest.

Definition 40 The *qua category*[34] of **CN**, **Qua(CN)** (or just **Qua**), is defined as $(\mathbf{CN}^{\rightarrow})^{op}$.

Because of how **CN** was defined, we will have at most one morphism from an object A to an object B. When such a morphism exists, we can see $A \rightarrow B$ as A *qua* B, for example,

will be to look at "a family man *qua* a parent." Identifying $A \xrightarrow{id_A} A$ with A, we are thus identifying the count noun A with its global aspect A *qua* A, that is, *as itself*.

In modeling consideration of the various aspects of people, we will be interested in a particular subcategory of $(\mathbf{CN}^{\rightarrow})^{op}$, namely the co-slice category of objects under the global object. For instance, the category

34. This is called the *aspectual category* in Reyes et al. (1999).

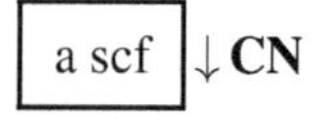

of objects under "a student, a coworker, a family man"—where this is identified with the global aspect 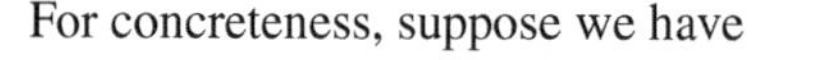

—forms a subcategory **A** of the *qua* category $(\mathbf{CN}^{\rightarrow})^{op}$.

For concreteness, suppose we have

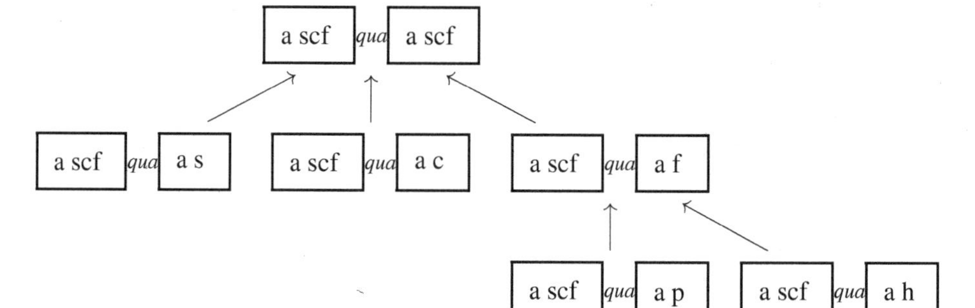

where the identity maps and those induced by composition are left implicit. In this way, such an **A** will thus serve as a way of representing those aspects of a person, including for instance Abe, relevant to whether or not a certain predicable holds of them, like "honesty." This is something that will be evaluated "*qua* scf" (in terms of all their "hats," via the global aspect), "*qua* student," "*qua* family man," and so on, where this latter has two subaspects: "*qua* parent" and "*qua* husband." Abbreviating these aspects, then, we could have displayed **A** as

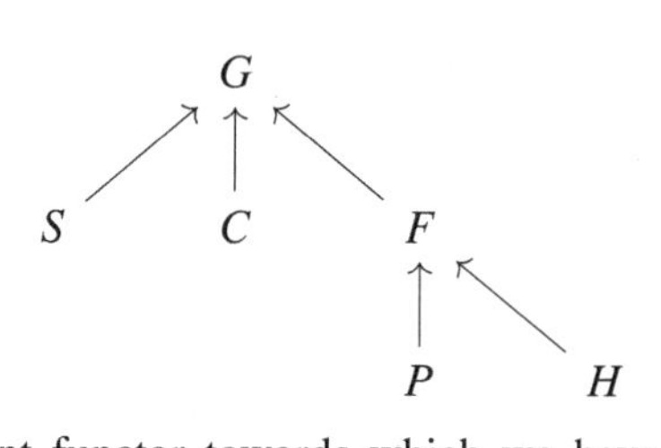

Before defining the relevant functor towards which we have been building, let us also record the following definition of a concept we will use here and throughout this book.

Definition 41 A functor $F : \mathbf{C}^{op} \to \mathbf{Set}$ is a *subfunctor* (*subpresheaf*) of a functor $G : \mathbf{C}^{op} \to \mathbf{Set}$ if

- for all $c \in \mathbf{C}$, $F(c) \subseteq G(c)$; and
- for all morphisms $f : c \to c'$ of $\mathbf{C}$, $F(f) : F(c') \to F(c)$ is the restriction of $G(f)$ to $F(c)$.

In other words, put more abstractly, for any $f : c \to c'$ in $\mathbf{C}$ there exists a commutative diagram

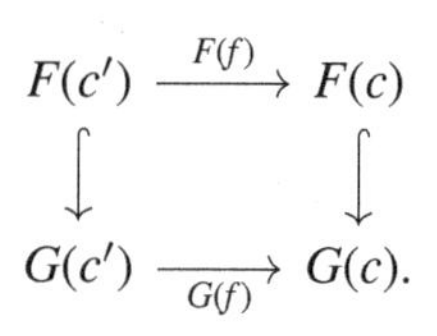

For reasons that will be better appreciated after we have introduced a few other notions, we sometimes write $F \hookrightarrow G$ to indicate that F is a subfunctor of G.

Now, given a *qua* category and $\mathscr{P}$ a set of predicables that are applicable to the count nouns of **CN**—where predicables may be thought of for now as just involving grammatical expressions consisting of adjectives, verbs, or adjectival and verb phrases, including expressions such as "mortal" or "honest," and "to be a person," where this is derived from or sorted by a count noun—we define an *interpretation* of (**Qua**, $\mathscr{P}$) as a functor

$$X : \mathbf{Qua}^{op} \to \mathbf{Set}$$

together with a set

$$\{X_\phi \hookrightarrow X \mid \phi \in \mathscr{P}\}$$

of subfunctors of X that satisfy the following conditions:

1. $X\left(\boxed{\text{A}}\,{}^{qua}\boxed{\text{B}}\right) = X\left(\boxed{\text{A}}\,{}^{qua}\boxed{\text{A}}\right)$; and
2. $X_\phi\left(\boxed{\text{A}}\,{}^{qua}\boxed{\text{B}}\right) = X_\phi\left(\boxed{\text{B}}\,{}^{qua}\boxed{\text{B}}\right) \circ X\left(\boxed{\text{B}}\,{}^{qua}\boxed{\text{B}} \to \boxed{\text{A}}\,{}^{qua}\boxed{\text{A}}\right)$

Notice that since $\mathbf{Qua} = (\mathbf{CN}^{\to})^{op}$, then its dual, $\mathbf{Qua}^{op}$, is just $\mathbf{CN}^{\to}$, making X equivalently expressible as a functor

$$X : \mathbf{CN}^{\to} \to \mathbf{Set}.$$

That we can compare the interpretations of count nouns in fact forms the basis of the possibility of comparing the corresponding interpretations of predicables that are functorial. Using this notion of interpretation, and given a subcategory of **Qua** such as **A** above, we can restrict the interpretation to the subcategory. Fundamentally, the functoriality here can be understood as saying, for instance, if Abe is honest *qua* family man, then he is honest *qua* parent. Moreover, $\boxed{\text{a p}}$ will be interpreted as a set of parents, $\boxed{\text{a scf}}$ as a set of students who are also coworkers and family men. We will return to this example later in the chapter, and again in later chapters.

The next example is very important for the general theory that will be developed in later chapters, especially chapter 6.

Example 42 Let **C** be a category, and fix an object a of **C**. Then we can form the (covariant) *hom-functor* $\mathrm{Hom}_{\mathbf{C}}(a, -) : \mathbf{C} \to \mathbf{Set}$, which takes each object b of **C** to the set $\mathrm{Hom}_{\mathbf{C}}(a, b)$ of **C**-morphisms from a to b, and takes each **C**-morphism $f : b \to c$ to the following map between hom-sets:

$$\mathrm{Hom}_{\mathbf{C}}(a, f) : \mathrm{Hom}_{\mathbf{C}}(a, b) \to \mathrm{Hom}_{\mathbf{C}}(a, c), \tag{2.1}$$

which outputs $f \circ g : a \to c$ for input $g : a \to b$. In other words, the action on morphisms is given by *postcomposition*. This hom-functor will be defined for any object whenever the hom-sets of **C** are *small*, that is, whenever **C** is *locally small*. Intuitively, the set $\mathrm{Hom}_{\mathbf{C}}(a, b)$ can be thought of as the set of ways to pass from a to b within **C**, or the set of ways a "sees" b within the context or framework of **C**. Then, refraining from filling in the object b, it should be obvious how $\mathrm{Hom}(a, -)$ can be thought of as representing in a rather general fashion "where and how a goes elsewhere" or "how a sees its world." Given an object $a \in \mathbf{C}$, we say that the covariant functor $\mathrm{Hom}(a, -)$ is *represented by a*; for reasons we will explore in chapter 6, this functor is also denoted Y^a (or sometimes h^a). It will turn out to be an important observation that instead of restricting ourselves to the hom-functor on a given a, we can assign to *each* object $c \in \mathbf{C}$ its hom-functor $\mathrm{Hom}(c, -)$, and then collect all these together.

We can also form the *contravariant hom-functor* $\mathrm{Hom}_{\mathbf{C}}(-, a) : \mathbf{C}^{op} \to \mathbf{Set}$, for a fixed object a of $\mathbf{C}$, which takes each object b of $\mathbf{C}$ to the set $\mathrm{Hom}_{\mathbf{C}}(b, a)$ of $\mathbf{C}$-arrows from b to a, and takes each $\mathbf{C}$-arrow $f : b \to c$ to $\mathrm{Hom}_{\mathbf{C}}(f, a) : \mathrm{Hom}_{\mathbf{C}}(c, a) \to \mathrm{Hom}_{\mathbf{C}}(b, a)$, that is, outputting $g \circ f : b \to a$ for input $g : c \to a$, acting by *precomposition*. This functor can be thought of as representing "how a is seen by its world." Given an object $a \in \mathbf{C}$, we say that the contravariant functor $Y_a := \mathrm{Hom}(-, a)$ is *represented by a*. As above, instead of restricting ourselves to the hom-functor on a, we can ultimately let this functor vary over all the objects of $\mathbf{C}$.

Throughout this book, we will see many more examples of functors. For now, though, we can continue to develop the main concepts, ascending once more in generality, now regarding functors as objects in a category, with morphisms given by certain *transformations* between the functors.

2.2 Natural Transformations

Functors are important for many reasons. In particular, as we will explore in chapter 3 and beyond, universal properties are given in terms of functors. Moreover, it is possible to use two functors to do a variety of important things, such as produce a new category from old categories. However, perhaps most important for our present purposes is the fact that functors can be composed, and there is a nice notion of comparing functors.

There may exist a variety of ways of embedding or modeling or instantiating one category within another, that is, there may exist many functors from one category to another. Sometimes these will be equivalent, but sometimes not. Moreover, the same blueprint may be realized in different ways, that is, there can be different functors that act the same way on objects. *Natural transformations* enable us to compare these realizations. If functors allow us to systematically import or transform objects from one category into another and thus translate between different categories, natural transformations allow us to compare the different translations in a controlled manner.

Definition 43 Given categories $\mathbf{C}$ and $\mathbf{D}$ and functors $F, G : \mathbf{C} \to \mathbf{D}$, a *natural transformation* $\alpha : F \Rightarrow G$, depicted in terms of its boundary data by the diagram

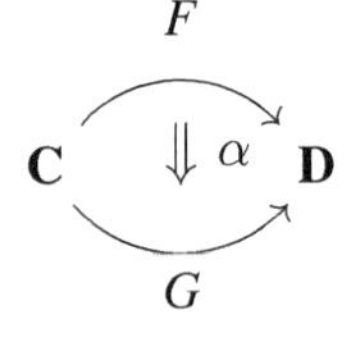

consists of the following:

- for each object $c \in \mathbf{C}$, an arrow $\alpha_c : F(c) \to G(c)$ in $\mathbf{D}$, called the c-component of α, the collection of which (for all objects in $\mathbf{C}$) defines the *components* of the natural transformation; and

- for each morphism $f: c \to c'$ in $\mathbf{C}$, the following square of morphisms (depicted on the right), called the *naturality square* for f, must commute in $\mathbf{D}$:

$$\begin{array}{ccc} c & \quad & F(c) \xrightarrow{\alpha_c} G(c) \\ {\scriptstyle f}\downarrow & & {\scriptstyle F(f)}\downarrow \qquad \downarrow{\scriptstyle G(f)} \\ c' & & F(c') \xrightarrow[\alpha_{c'}]{} G(c') \end{array}$$

The collection of natural transformations from F to G is sometimes denoted by $Nat(F, G)$.

Let's step back and unpack this a little. In the general context of a category, we think of each arrow of a category as a way of comparing two objects. As we climb the ladder of abstraction and introduce functors, we think of functors as distinct ways of comparing two categories. As we suggested earlier—though we will refine this in the coming sections—one way of viewing a functor from **C** to **D** is as supplying some sort of "picture" of **C** within the world of **D**. Taking the next step on the ladder of abstraction, a natural transformation effectively compares two functors. Following Goldblatt (2006), one way of building intuition of the idea of a transformation from F to G is to imagine that, within **D**, you have to superimpose or slide the picture of **C** given by F onto the picture of **C** given by G, where you make use of the structure of **D** in carrying out this translation. Minimally, to compare a picture given by F to one given by G, we ought to assign to each object c of **C** an arrow in **D** going from the image $F(c)$ of c under F to the image $G(c)$ of c under G. For a given c, we could name this "component" of the transformation by $\alpha_c : F(c) \to G(c)$, to track which object we are attending to. Collecting these together, for all the objects of **C**, gives us the *components* of our transformation α. But there surely needs to be something else to this process, beyond just an indexed family of arrows. As functors, we also have information about how F and G act on the morphisms of **C**, and we will need to use this to complete the pictures supplied by F and G. Incorporating this information, for each morphism $f: c \to c'$ of **C**, we automatically end up with diagrams like the naturality square depicted above. As a final step, asking that such diagrams *commute* (in **D**) is exactly the "natural" thing to do, if we want to continue with the same sort of notion of structure-preservation that we have been developing on lower rungs of the ladder of abstraction.

We will see a variety of examples of natural transformations in the coming sections. For now, let us make a few other useful observations and definitions. First of all, observe that the same notion works for functors of the other common variance as well. If F and G are both contravariant functors, then the same definition applies, except the vertical arrows are reversed in each of the naturality squares.

Now, if natural transformations are ways of comparing functors, at the extreme we ought to be able to use this notion to develop a refined idea of when two functors are fundamentally the *same* functor. The following notion helps us do just that—and, as we will see, a lot more.

Definition 44 A *natural isomorphism* is a natural transformation $\alpha: F \Rightarrow G$ for which every one of the components $\alpha_c : F(c) \to G(c)$ is an isomorphism (in the target category). In other words, each α_c has an inverse $\alpha_c^{-1} : G(c) \to F(c)$, where these inverses form the components of a natural transformation α^{-1} from G to F.

When α is such a natural isomorphism, we denote this by writing $\alpha : F \cong G$.

Let's push this further. We already know that we can compose functors, and we have also met the identity functor on a category. Combining these ingredients together with the notion of a natural isomorphism provides us with a more refined notion of when two categories are *equivalent.*

Definition 45 An *equivalence of categories* consists of a pair of functors $F:\mathbf{C}\to\mathbf{D}$, $G:\mathbf{D}\to\mathbf{C}$ together with the natural isomorphisms $\eta:\mathrm{id}_{\mathbf{C}}\cong G\circ F$ and $\epsilon:F\circ G\cong\mathrm{id}_{\mathbf{D}}$. Another way of saying this is that the functors are inverse to each other "up to natural isomorphism of functors." The categories $\mathbf{C}$ and $\mathbf{D}$ are then said to be *equivalent* if there exists an equivalence of categories between them, and in such cases we write $\mathbf{C}\simeq\mathbf{D}$.

Remark 46 Recall the category-theoretic notion of an isomorphism from definition 10 (chapter 1). By running that definition on the category of categories (say, **Cat**), we get the concept of an *isomorphism of categories*, where this is a pair of functors $F:\mathbf{C}\to\mathbf{D}$, $G:\mathbf{D}\to\mathbf{C}$ such that their composites *equal* the respective identity functors, that is, $G\circ F=\mathrm{id}_{\mathbf{C}}$ and $F\circ G=\mathrm{id}_{\mathbf{D}}$. This will moreover induce a bijection between the objects of each category as well as between their morphisms.

The overly restrictive notion of an isomorphism of categories should be compared to the more relaxed (and ultimately superior) criterion of identity or "sameness" of two categories we find in the notion of an equivalence of categories. Even rather simple categories can be found to illustrate the point that, as far as categories are concerned, the notion of isomorphism of categories is not the right thing to consider—many categories may indeed appear "the same" in all important respects, yet they won't be isomorphic (often for reasons that appear to be trivial or irrelevant). Grothendieck, who first introduced the notion of equivalence of categories, accordingly stressed the differences between the two notions, and how the notion of an isomorphism of categories is, in many cases of practical interest, not at all a useful notion. He was in part motivated to this because he was dealing mainly with *functor categories* (introduced below), and among such categories it is common to find categories that appear to be the same in every important categorical respect, yet are not isomorphic.

If we know from constructions like the arrow category that it is sensible to take things like functors as objects, and if natural transformations can be regarded as morphisms between functors, this data should let us form a new category! Indeed, this gives us the following important category.

Definition 47 For any fixed pair of categories $\mathbf{C}$ and $\mathbf{D}$, we can form the *functor category*, denoted by $\mathbf{D}^{\mathbf{C}}$ (or, less commonly, by $Fun(\mathbf{C},\mathbf{D})$), where this has for

- objects: all the functors from $\mathbf{C}$ to $\mathbf{D}$;
- morphisms: all the natural transformations between such functors.

Of course, to exhibit this as a category, we need identities and composites, and to show that the relevant axioms are satisfied. For a functor $F:\mathbf{C}\to\mathbf{D}$, we can assign its identity natural transformation $\mathrm{id}_F:F\Rightarrow F$ to be the natural transformation whose components are each identities. For composition of morphisms in $\mathbf{D}^{\mathbf{C}}$, we use the following notion:

Definition 48 Let $\alpha:F\Rightarrow G$ and $\beta:G\Rightarrow H$ be natural transformations between the parallel functors F, G, and H, from $\mathbf{C}$ to $\mathbf{D}$, as depicted by the diagram:

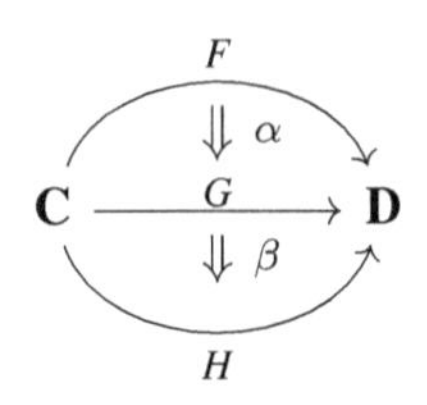

There is a natural transformation $\beta \circ \alpha : F \Rightarrow H$, where this is defined on components: components $(\beta \circ \alpha)_c := \beta_c \circ \alpha_c$ are taken to be the composites of the components of α and β.

You might think of this, at the level of the components and their naturality squares, as a matter of pasting the naturality squares together into rectangles.

For reasons discussed briefly in the remark to follow, this sort of composition is usually referred to as *vertical composition*.

One can verify that this composition operation satisfies the two axioms of a category, by checking it on components and using the fact that composition automatically satisfies these properties in **D** (since it is a category). We thus have a category, and a rather important one at that, as we will see.

Remark 49 After you have sat with categories and natural transformations, it may occur to you that the notion of natural transformations seems to support another kind of composition, distinct from that given in definition 48. Indeed, imagine we move "horizontally" via functors from one category **C** to **D** and then again via functors from **D** to **E**, as depicted in the following diagram:

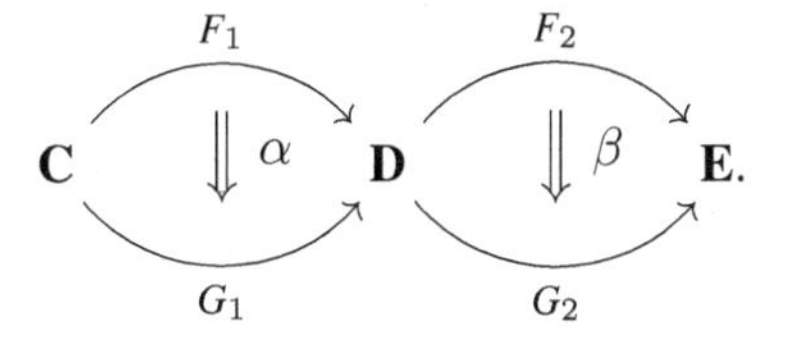

Horizontal composition uses the symbol $\diamond$ and gives $\beta \diamond \alpha : F_2 \circ F_1 \Rightarrow G_2 \circ G_1$, whose component at c of **C** is defined as the composite of the following commutative square:

$$\begin{array}{ccc} F_2F_1(c) & \xrightarrow{\beta_{F_1c}} & G_2F_1(c) \\ {\scriptstyle F_2(\alpha_c)}\downarrow & \searrow^{(\beta\diamond\alpha)_c} & \downarrow{\scriptstyle G_2(\alpha_c)} \\ F_2G_1(c) & \xrightarrow[\beta_{G_1c}]{} & G_2G_1(c). \end{array}$$

While we will not be needing this notion of horizontal composition in this text, it is used to further generalize the definition of a category, defining a *2-category*, a starting point for higher and higher climbs up the ladder of abstraction.

For our purposes, perhaps the most important thing to note here is that since presheaves are just (contravariant) functors, so that we are given a notion of a morphism of presheaves from F and G as just a natural transformation $\alpha : F \Rightarrow G$, we can form the presheaf functor category.

Definition 50 The *presheaf category*, denoted $\mathbf{Set}^{\mathbf{C}^{op}}$ or **PreSh**(**C**), is the (contravariant) functor category having for objects all functors $F : \mathbf{C}^{op} \to \mathbf{Set}$, and for morphisms $F \to G$

all natural transformations $\theta : F \Rightarrow G$ between such functors. As a natural transformation, such a θ will assign to each object c of $\mathbf{C}$ a function $\theta_c : F(c) \to G(c)$, and do so in such as way as to make all diagrams

$$\begin{array}{ccc} d & \qquad & F(d) & \xrightarrow{\theta_d} & G(d) \\ {\scriptstyle f}\uparrow & & {\scriptstyle F(f)}\downarrow & & \downarrow{\scriptstyle G(f)} \\ c & & F(c) & \xrightarrow[\theta_c]{} & G(c) \end{array}$$

commute for each $f : c \to d$ in $\mathbf{C}$.

Example 51 For $\mathbf{J}$, an arbitrary category viewed as a template or indexing category for $\mathbf{C}$, we can define another sort of functor category by looking at the category $\mathbf{C}^{\mathbf{J}}$ of $\mathbf{J}$-diagrams in $\mathbf{C}$, where each object is a functor $F : \mathbf{J} \to \mathbf{C}$, and for two such objects F, G, a morphism of $\mathbf{C}^{\mathbf{J}}$ from F to G is a natural transformation between the functors.

Especially on account of the important place that presheaf categories will occupy in our story, we will see many more examples of natural transformations in action, in a variety of contexts. For now, let us delve a little deeper into what it is to be a presheaf.

2.3 Seeing Structures as Presheaves

When we work with a mathematical structure, it is common to try to approach it in terms of its elements. In general, it is very natural to want to break things down by decomposing more complicated structures into their components—and elements, like points, are one sort of component we seem especially ready to recognize as such. But in certain settings, one needs to consider *figures of a more general shape* than points. Points, after all, might be regarded as just a particularly simple kind of "shape." For instance, suppose you are presented with the structure X:

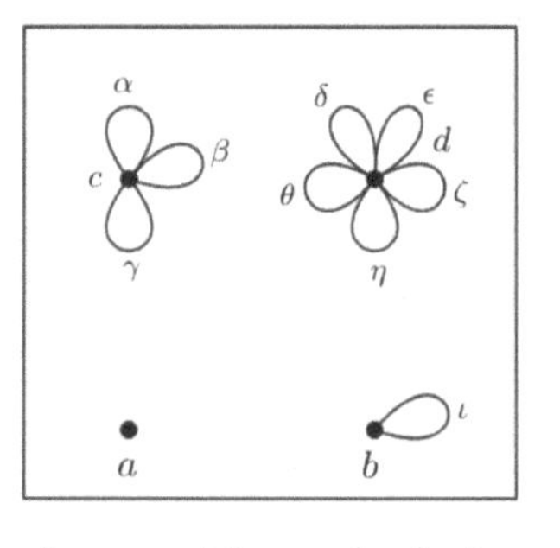

This X depicts what is called a *bouquet*. Figures in the bouquet X with the shape "point" can be regarded as maps $\bullet \to X$, each of which map names a point in X:

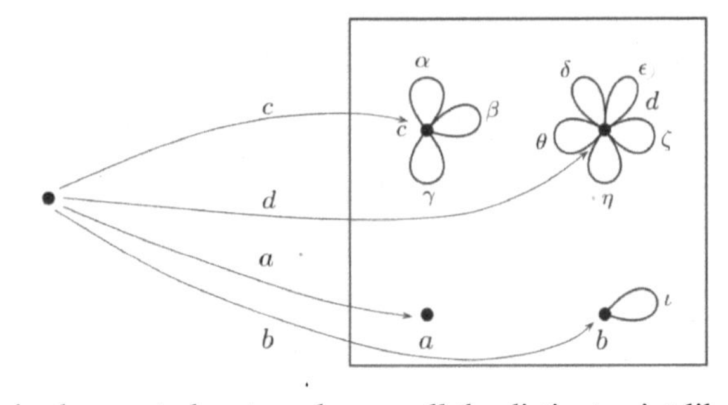

We use the generic shape • to locate and name all the distinct point-like figures of X, that is, via a particular map $\bullet \xrightarrow{a} X$ we name with a one of the various •-like components of X. Altogether, the data of our point-like figures in X really just amounts to a set

$$X(\bullet) = \{a, b, c, d\},$$

which you might read as saying "X realizes a, b, c, and d as its figures of shape •."

But you could not hope to understand all that this structure X is just by considering the point-like figures! After all, points are not the only sort of figural component in X. Thus, there are in fact many distinct bouquets that may even have the same set of points (yet will look rather different!). So we also need a way of picking out those figures in X whose shape is that of a "loop." Similar to our point-like figures in X, we can pick out and name loops via maps from the generic shape $\circlearrowleft$ into X:

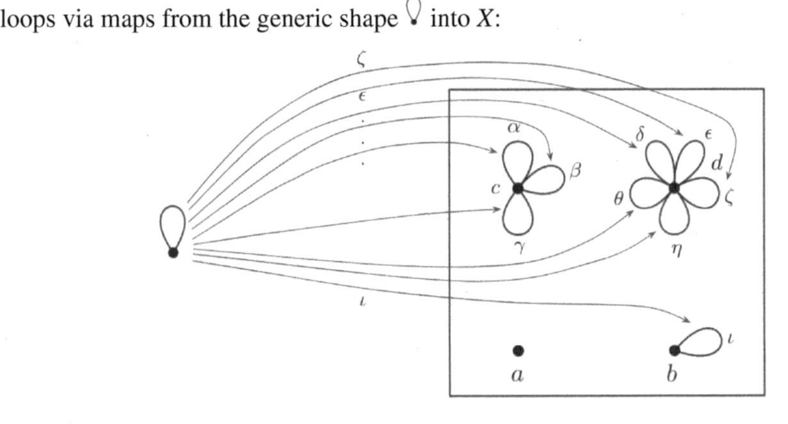

In other words, the data here is captured by the set

$$X(\circlearrowleft) = \{\alpha, \beta, \gamma, \delta, \epsilon, \zeta, \eta, \theta, \iota\},$$

which you might read as saying "X realizes $\alpha, \beta, \gamma, \ldots$ as its figures of shape $\circlearrowleft$."

But do we then have enough information to reconstruct X? Well, many bouquets may have the same set of loops, but the home point at which they live may be different. You would not regard these as the same thing. To fully capture X, then, we also need a way of extracting the data of where each loop-shaped figure lives, that is, *how the loop-shaped figures relate to the point-shaped figures*. Corresponding to the inclusion of the generic shape "point" in the generic shape "loop"

$$\bullet \xrightarrow{i} \circlearrowleft,$$

we should then have a map

$$X(\circlearrowleft) \xrightarrow{X(i)} X(\bullet)$$

$$l \mapsto p,$$

taking a loop l to the point p at which it is stationed, and so informing us about which loops get stationed at which points. For instance, this will tell us that

$$X(i)(\alpha) = c,$$

or "the loop-shaped component named α lives at the point-shaped component named c." The equations telling us which point each of the loops are assigned supply us with what are called the *incidence relations*.

With all this information—the set $X(\bullet)$ and $X(\circlearrowleft)$, together with a map describing how the loop elements in the latter set are sent to the points in the former set—it would seem that we will be able to recover the whole of the information of the bouquet X itself.

But described in this way—and this is the point!—what else have we been saying but that X itself can just be regarded as a presheaf

$$X : \mathbb{B}^{op} \to \mathbf{Set},$$

where the indexing category $\mathbb{B}$ is

$\mathbb{B} :=$ 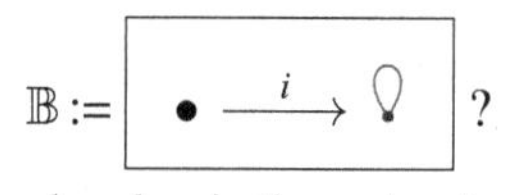 ?

In general, what we are starting to develop is the extension of an idea whereby we consider maps with codomain X as *figures* in X, where the domain of a figure is regarded as its *shape* or *type*, and where the incidence relations giving such an X its structure, and describing how the figures of distinct shapes relate, are then specified in terms of a map between the figures. In this same manner, if our domain or source category is regarded as consisting of some generic "shapes," related in some particular way (such as points included in pointed loops), then the result of realizing or figuring those shapes, via the image of a specified presheaf X, can be imagined as a "container" holding onto the various concrete realizations or instantiations of the different generic shapes supplied by the source category, where the natural relation that obtains between the underlying shapes is respected by the associated figures realized in the target category.[35]

In a similar way, suppose we instead take for our indexing category of shapes the single object $*$ and all the morphisms generated by iterations of σ, that is, the free monoid on one generator (σ),

$\mathbb{E} :=$ 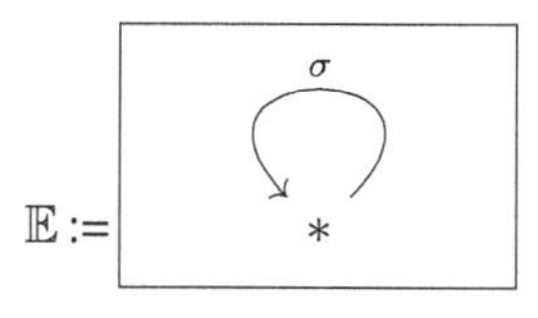

35. This *figure* perspective, applied to presheaves, is advocated by Lawvere, especially in Lawvere and Schanuel (2009). This approach is taken even further in the delightful Reyes, Reyes, and Zolfaghari (2008), which much of the remainder of this chapter is inspired by and which is highly recommended to the reader who finds the topics and discussions of this chapter of particular interest.

Then presheaves X on this arise as dynamical systems (evolutive sets) or automata, where X supplies the set of possible states, and the given endomap σ gives rise to the evolution of states (think the change in internal state that results after the passage of one unit of time, or as a result of pressing the "button" σ on the outside of a machine). In other words, if X is a presheaf on $\mathbb{E}$, we think of its image as a container containing a set of figures—shaped in the form of dots, corresponding to various instantiations of the object $*$ of $\mathbb{C}$, and in the form of arrows between certain of those dots, corresponding to the endomap σ of $\mathbb{E}$—with a process taking each element to a next stage or next element. In this way, we might end up with an X such as

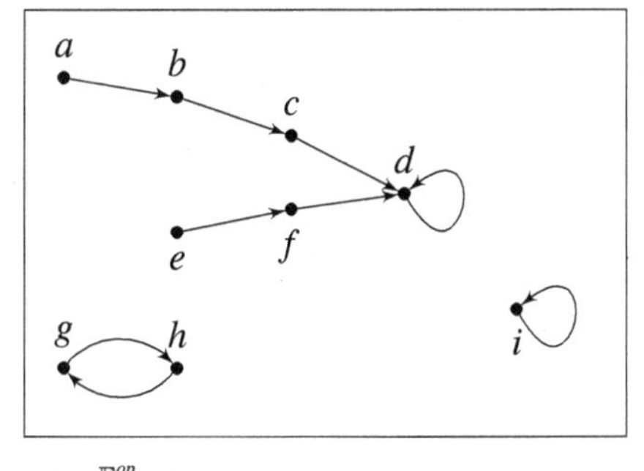

Altogether, as we will see, $\mathbf{Set}^{\mathbb{E}^{op}}$, the category of presheaves on $\mathbb{E}$, with objects like X, will itself be none other than the category of evolutive sets or dynamical systems.

Similarly, if we instead took as our indexing shape category the category of n-evolving sets, that is, $\mathbb{E}_n$, freely generated by n nonidentity morphisms:

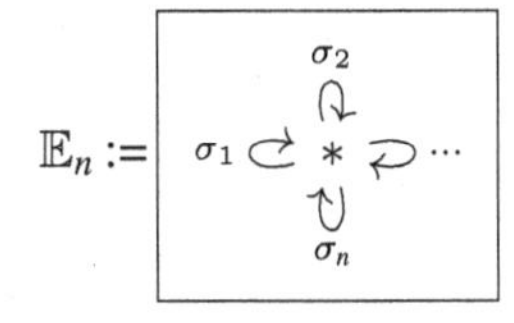

then the container of $\mathbb{E}_n$-shaped figures would have figures similar to the picture of a presheaf on $\mathbb{E}$, except with (up to) n different processes carrying one $*$-figure to the next, for example,

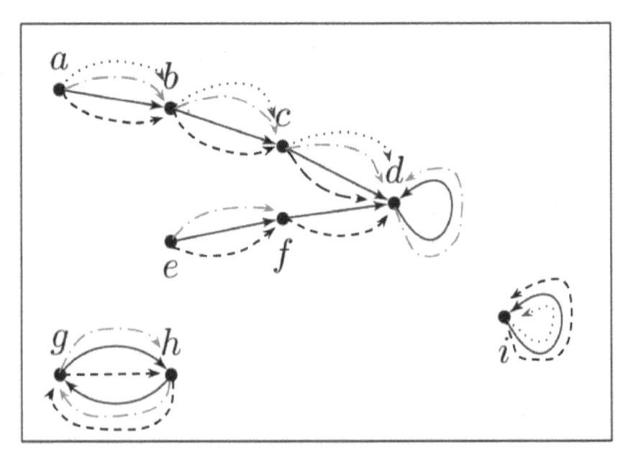

A similar approach can also be taken when using more distinct indexing categories. For instance, recall the particular subcategory **A** of **Qua** from example 38. Applying an interpretation functor $X : \mathbf{A}^{op} \to \mathbf{Set}$ then results in a container of **A**-shaped figures, for example,

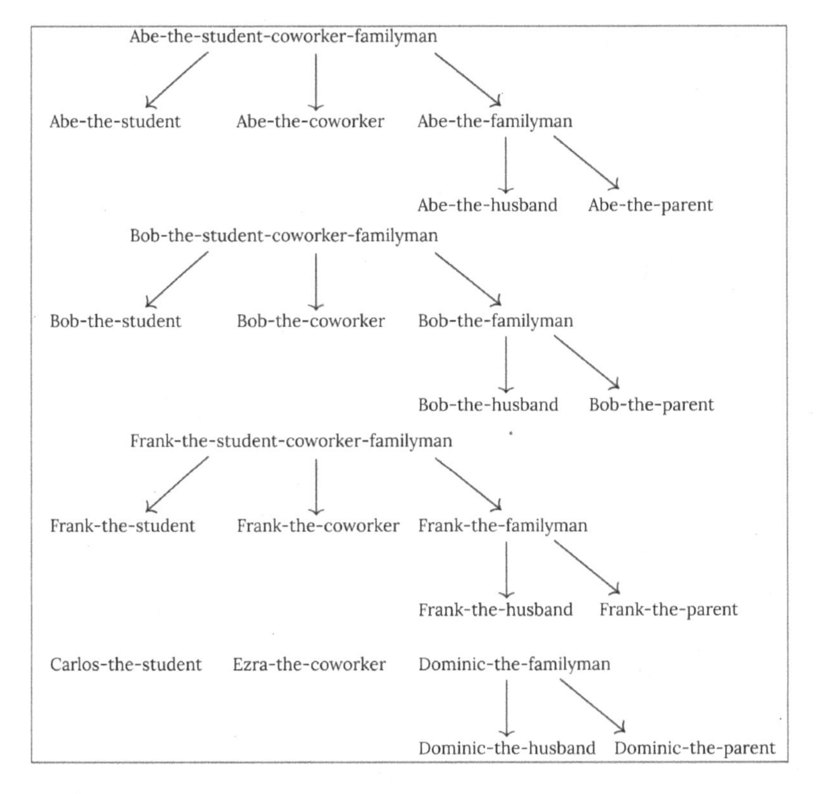

As before, we could regard $X(S)$, for instance, as picking out or naming those who are "shaped like," or of type, *student*. If we organize things a bit, grouping together those people that are picked out as conforming to the same shape (in this case, their role), then what is fundamentally going on here can be displayed more sensibly as

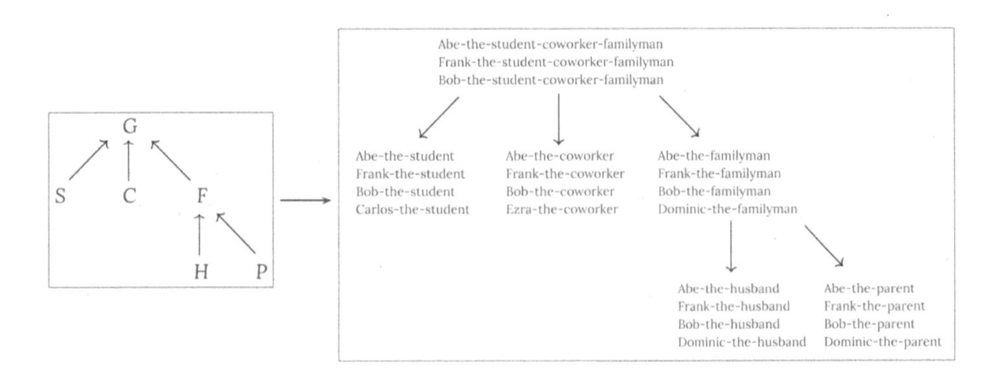

An interpretation X of such an **A** is really just an object of the presheaf category $\mathbf{Set}^{\mathbf{A}^{op}}$ together with a set of subfunctors corresponding to the predicables of $\mathscr{P}$. Morphisms of this category are just natural transformations from an interpretation X to an interpretation Y such that the restrictions to $X_\phi \hookrightarrow X$ can be factored through $Y_\phi \hookrightarrow Y$.

As in the previous examples, the idea here is that the domain category supplies the generic shape according to which figures or instantiations of such shape are organized, and where the overall realization of such figures as figures of such a shape, constrained to relate in a way that respects the underlying relations of the shapes of the domain, is accomplished via the functor.

Generalizing from such examples, the story we are starting to tell is that for a presheaf $P: \mathbf{C}^{op} \to \mathbf{Set}$, for certain **C** it is often natural to think of the result of applying P as leaving us with some sort of container of **C**-shaped figures, where the various objects c of **C** are thought of as supplying the *generic figures*, or templates of a figure, that are then *instantiated* or *figured* or *realized* in **Set**; for example, $P(c)$ is some particular *set* of instantiations or figures of c-shape. A functor is fundamentally a way of turning objects and structured connections in one world **C** into objects and maps in another world **D**, and doing it in such a way that certain equations are satisfied (where these equations code for the preservation of structure, or compatibility of the transformation with the composition of maps in **D**). Another way of thinking of a presheaf is thus as a *realization* of **C** in **Set**.

So far, we have mostly dwelt on how the presheaf operates *on objects*. Obviously, as a functor, we must also consider the action specified by the (contravariant) functor, that is, how it acts *on morphisms*. The basic idea will be that, following the figures of generic shape c interpretation, a morphism in **C** from one object to another will give rise to a *change of figures*, where this means, more precisely, that if we have a figure x of shape c (i.e., $x \in P(c)$) and a figure y of shape c' (i.e., $y \in P(c')$), then asking about the effect of changes of figures amounts, at the level of the presheaf, to asking to what extent the figures are incident or overlap (and what this overlap structure looks like) or otherwise relate. But the same idea would apply to less geometrical settings, for instance using an indexing category that is more time-like, and adopting the interpretation of $P(c)$ as a set existing *at* *stage c*. Then the morphisms of **C**, when acted on by the presheaf, would model *varying the stage*, or transitions between stages—so that, overall, the functor can be interpreted as supplying a picture of a set *varying through time*. In the next section, we start to look more closely at interpretations such as the incidence relation, as well as some others such as the variable-set interpretation—as always, via examples. Such examples will enable us to start to think more systematically about presheaves and their action. Moreover, in building on some of the examples thus far, looking closer at the resulting presheaf category, we will see some further examples of natural transformations.

2.4 The Presheaf Action

It is not uncommon to see in the literature on presheaves references to *right* **C**-sets (which are the same as *left* $\mathbf{C}^{op}$-sets). Similarly, one will sometimes hear talk of a presheaf's *right action*. We will think of there being *four characteristic kinds of cohesivity or variability* presented by presheaf categories in accordance with four main ways the right action of the presheaves in question can be found to operate. But before discussing these interpretations of a presheaf (illustrating them each through select examples), it may be useful to further explain the reference to the presheaf action as a *right action*, in case it is not already clear why one can see this being referred to as an *action* (and, moreover, why the action is *right*).

2.4.1 Right Action Terminology

A presheaf is ultimately just a *functor* (one with a particular variance). At least with the usual set-valued presheaves, in applying a given functor P to each of the objects of the domain category **C**, we just get a bunch of *sets*, $P(c), P(c')$, and so on, indexed by the objects of **C**. The functoriality of the given presheaf P, then, just means that for every map $f : c' \to c$ in **C**, we will have a *function*—since we are landing in **Set**, after all!—$P(f) : P(c) \to P(c')$ going the other way. So we just have a function that takes the elements x of the *set* $P(c)$, that is, the set P *seen at stage* c (or "seen in the shape of" c), to elements of the *set* $P(c')$, that is, P *seen at stage* c' (or "seen in the shape of" c'). In other words, for each element x of $P(c)$ and each map $c' \xrightarrow{f} c$ of **C**, there is an associated element xf of $P(c')$.[36]

Now, the *contravariance* of the functor of course means that the functor applied to a composite $f \circ g$, where $c'' \xrightarrow{g} c' \xrightarrow{f} c$, should be the same as the functor first acting on f then acting on g. In other words, in terms of the elements $x \in P(c), f, g$, and xf as above, whenever $c'' \xrightarrow{g} c' \xrightarrow{f} c$, we must have $x(f \circ g) = (xf)g$ in $P(c'')$. Moreover, functors must respect identities. But all this data essentially means that we are dealing with what, in other settings, one would call a *right action* of **C** on the underlying set (formed by the presheaf P), and where this right action expresses the incidence relations or transitions among the various figures x, x', and so on.

In those other contexts, if we have some mapping $X \times A \to X$, it is common to refer to such a map as a *right action* of A on X. Usually A is some monoid or group (and X is a set). The basic idea is that A is thought of as furnishing a set of "buttons" that control the states

36. The reason for writing the element xf this way is explained below.

of X, while the given action $X \times A \xrightarrow{\alpha} X$ is regarded as supplying us with the data of a state-machine or automaton. Considering a particular "button" a then gives rise to an endomap of X, specifically $\alpha(-, a)$, where this means that for each element x of X, its image $\alpha(x, a)$ under the action map α is just a new element of X. "Pressing" a once takes a particular state x into the state $\alpha(x, a)$; pressing it twice takes x to $\alpha(\alpha(x, a), a)$; and so on. Of course, we can also press a different button (i.e., take a different object a' of A). Combining things, we can press one button and then another. This will mean: suppose we are in state x and button a is pressed and then button a', the resulting state will be $\alpha(\alpha(x, a), a')$. As is common, we can choose, notationally, to represent the result of the action $\alpha(x, a) = x \cdot a$, which you might read as "having pressed a on state x."

In a similar fashion, with presheaves we speak of a *right action* of $\mathbf{C}$ on a set P that is partitioned into sorts coming from the objects of $\mathbf{C}$ (i.e., parameterized by the objects of $\mathbf{C}$). Being a "right action" here means that whenever we have an arrow $f : c' \to c$ in $\mathbf{C}$ and an element $x \in P(c)$—that is, an element of the set P of sort c—then xf yields an element of P of sort c' subject to the conditions

$$x\, \mathrm{id}_c = x;$$

$$x(f \circ g) = (xf)g \text{ whenever } c'' \xrightarrow{g} c' \xrightarrow{f} c \in \mathbf{C}.$$

We may write the action in the form of concatenation; that is, xf is short for $x \cdot f$ where the action $\alpha : \mathbf{Set} \times \mathbf{C} \to \mathbf{Set}$ is defined as $\alpha(x, f) = x \cdot f$ and the set in question is actually just the disjoint union $\amalg_{c \in Ob(\mathbf{C})} P(c)$.

The idea, then, is that given an element $x \in P(c)$ for some $c \in \mathbf{C}$, such an x will be acted on by all the morphisms $c' \xrightarrow{f} c$ in $\mathbf{C}$, and in such a way that composite morphisms act as above. In asking what the value of a function $f : c' \to c$ in $\mathbf{C}$ at such an element x will look like, we are asking about $P(f)(x)$. Regarding this in terms of a right action $\alpha(x, f) = x \cdot f$, we have for composite maps (which we write here with juxtaposition notation), $\alpha(x, fg) := x \cdot (fg) = \alpha(\alpha(x, f), g) = (x \cdot f) \cdot g$. If we agree, notationally, then, to let $x \cdot f = P(f)(x)$,[37] it is evident that the contravariance of the functor P is equivalent to specifying that $\mathbf{C}$ acts (and does so *on the right*) on P (regarded as a set). This is evident since

$$\alpha(x, fg) = \alpha(\alpha(x, f), g)$$

$$x \cdot (fg) = \alpha(x \cdot f, g)$$

$$x \cdot (fg) = (x \cdot f) \cdot g$$

$$P(fg)(x) = P(g)(P(f)(x))$$

$$P(f \circ g)(x) = P(g) \circ P(f)(x).$$

Having established the reasoning behind that terminological and notational choice, let us now consider more closely the various interpretations this presheaf action takes on in practice and explore what the natural transformations will look like for various presheaf categories we have already introduced.

37. That f gets written on the right of x here not only is meant to reveal the underlying right action, but it is a good notational choice since it accords with the induced notation for a composite arrow $f \circ g$ as $x \cdot (f \circ g) = (x \cdot f) \cdot g$.

2.4.2 Four Ways of Acting as a Presheaf

We will think of there being *four characteristic kinds of cohesivity or variability* presented by presheaf categories in accordance with four main ways the right action can be found to operate:

1. As *processual*, for example, as passing from sets indexed by one stage to sets indexed by another. Here, objects of $\mathbf{C}$ play the role of stages; for every c in $\mathbf{C}$, the set $P(c)$ is the set of elements of P *at stage* c, while the morphisms model transitions between stages.
2. As *extracting boundaries* (or picking out components), for example, using the source and target map to pick out the vertices of a graph's edge, picking out lower-dimensional boundaries of simplices (generalizations of triangles or tetrahedrons to arbitrary dimensions). For something like a space that consists of points, edges, triangles, and so on, in *changing figures* we pass from higher-dimensional figures to lower-, so that, for instance, the action works by extracting the endpoints of an edge or extracting the edges of a triangle.
3. As *conditions on how different "probes" of a space relate to each other*, for example given a category $\mathbf{C}$ of geometrical figures or spaces of some sort, a presheaf X is regarded as a rule assigning to each object U (each "test space") of $\mathbf{C}$ the set $X(U)$ of admissible maps from U into a generalized space or geometrical object X—giving "probes of X by U"—and the presheaf action then concerns how maps from one test space U to another test space V induce maps of sets $X(V) \to X(U)$, ultimately codifying how probes of X by V transform into probes of X by U.
4. As *restriction*, for example, whenever some sort of topology is involved, where the data specified over or about a larger region can be restricted to the data specified over a region included in the former region.[38]

We illustrate these four action perspectives, in order, via specific examples.

Example 52 We discussed earlier how presheaves on $\mathbf{C}$, as functors, can be thought of as providing a set of figures with the shape of the indexing category for each object in $\mathbf{C}$ and a *process* operator for each morphism in $\mathbf{C}$. For each a in $\mathbf{C}^{op}$, the resulting set $F(a)$ is a set of elements of F at stage a, while each arrow between objects in $\mathbf{C}^{op}$ induces a transition map between the varying set F at stage a and the varying set F at stage b (for an arrow from b to a), so that, altogether, we are regarding the objects of $\mathbf{C}$ as playing the role of stages of $F : \mathbf{C}^{op} \to \mathbf{Set}$ and F itself as a *set that varies through the stages*.

This perspective of the action as exemplifying a kind of *process* is nicely illustrated by considering presheaves on a variety of finite indexing categories. For instance, consider the case of finitely free monoids. We said in section 2.3 that if $\mathbb{E}$ is free monoid on one generator σ, or the additive monoid of natural numbers, then $\mathbf{Set}^{\mathbb{E}^{op}}$, the category of presheaves on $\mathbb{E}$, is none other than the category of evolutive sets or dynamical systems. Objects of

38. It is not uncommon to see presheaves and sheaves introduced exclusively via this fourth approach—and, indeed, our first look at sheaves in chapter 5 falls under this umbrella—but the first three perspectives are also important to consider, especially since the first two often involve examples with finitely generated categories (and, as such, provide a good stock of simple and computationally tractable examples) and the third achieves a level of generality that, were it pursued to its end, would ultimately let us speak of sheaves in "higher dimensions."

$\mathbf{Set}^{\mathbb{E}^{op}}$ consist of a set X equipped with a "process" endomap. For objects X of $\mathbf{Set}^{\mathbb{E}^{op}}$, in other words, X supplies the set of possible states, and the given endomap σ gives rise to the evolution of states. Referring back to our earlier such X,

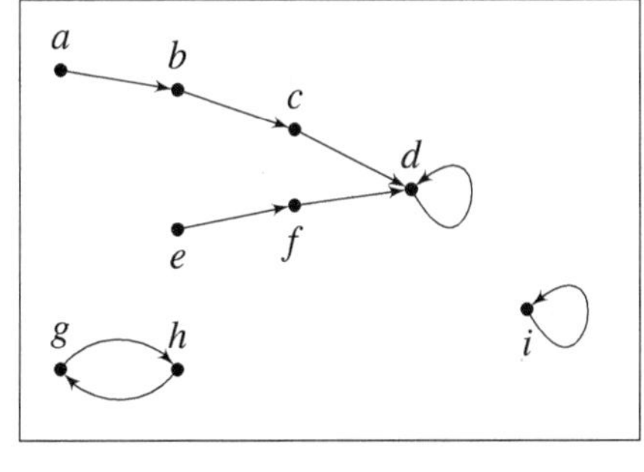

the idea is that, with this picture, we are displaying the presheaf consisting of $X(*) = \{a, b, c, d, e, f, g, h\}$ and where σ (i.e., $* \to *$) acts for instance on the figure a (i.e., on $* \xrightarrow{a} X$) to produce the figure $b : * \to X$; that is, $X(\sigma) : X(*) \to X(*)$ takes the particular $*$-figure given the name "a" to the particular $*$-figure given the name "b."[39]

Then a morphism in this entire presheaf category from a presheaf X (with endomap named α) to another object (presheaf) Y (with endomap β) will be the expected "equivariant map" (the usual map of relevance for such mathematical structures) in $\mathbf{Set}^{\mathbb{E}^{op}}$, that is, just a natural transformation $(X, \alpha) \xrightarrow{f} (Y, \beta)$, where this preserves the structure in the sense that $f \circ \alpha = \beta \circ f$.

We also saw how the same story is easily generalized to the category of n-evolving sets, that is, $\mathbb{E}_n$, freely generated by n nonidentity morphisms, so that the container of $\mathbb{E}_n$-sets would have figures similar to the above picture, except with (up to) n different processes carrying one $*$-figure to the next. We could also further consider finitely generated monoids, such as $\mathbb{E}_{1,R}$, where certain relations are imposed on the indexing category. For instance, taking $\mathbb{E}_{1,R}$ with one object and nonidentity morphism σ obeying some relation R—say the relation $\sigma^2 = \mathrm{id}_*$—the resulting category of presheaves on $\mathbb{E}_{1,R}$ gives rise to what is usually called, in other contexts, the category of *involution sets*. We could generalize this to any presentation of a monoid $\mathcal{M}$, $\mathbb{E}_{n,R}$, for n generators (i.e., sigmas), and R relations, ultimately leading us to show that the usual *Cayley graph* for a group is nothing other than a presheaf on the category $\mathbb{E}_{n,R}$.

In this context, we can take the opportunity to highlight that presheaves on a monoid are just equivalent to the usual *right actions* on a set by a monoid (which is in part responsible for the "right action" terminology). Recall that a monoid, viewed as a category $\mathcal{M}$ with just one object $*$, will imply that a set-valued functor on $\mathcal{M}$ yields just *one set*, $F(*) \in \mathrm{Ob}(\mathbf{Set})$. We must also supply, though, a function from $\mathrm{Hom}_{\mathcal{M}}(*, *)$ to $\mathrm{Hom}_{\mathbf{Set}}(F(*), F(*))$, that is, from M to $\mathrm{Hom}_{\mathbf{Set}}(F(*), F(*))$. In general, given a set A, for any sets X, Y, there is a

39. Incidentally, the notation $* \xrightarrow{a} X$ will be fully justified by the Yoneda results, covered in chapter 6. Looking ahead to that, for any object of $\mathbf{Set}^{\mathbf{C}^{op}}$, that is, some presheaf F, and any object c of $\mathbf{C}$, the set of elements of F of sort or type c can be naturally identified with the set of $\mathbf{Set}^{\mathbf{C}^{op}}$-morphisms from $\mathrm{Hom}_{\mathbf{C}}(-, c)$ to F, which is precisely what justifies the abuse of notation that alternately treats the elements of F of sort c as a morphism $c \to F$ in $\mathbf{Set}^{\mathbf{C}^{op}}$, letting any $c \to F$ be interpreted as a particular figure in F of sort c.

bijection

$$\mathrm{Hom}_{\mathbf{Set}}(X \times A, Y) \xrightarrow{\cong} \mathrm{Hom}_{\mathbf{Set}}(X, Y^A)$$

where $Y^A := \mathrm{Hom}_{\mathbf{Set}}(A, Y)$ the set of functions from A to Y. Moving between these two equivalent formulations via the bijection is sometimes called "currying." Our function from M to $\mathrm{Hom}_{\mathbf{Set}}(F(*), F(*))$ just belongs to $\mathrm{Hom}_{\mathbf{Set}}(M, F(*)^{F(*)})$. Currying, this is the same as a function $M \times F(*) \to F(*)$. Functors preserve identities by definition, so the monoid action law concerning the unit element e is satisfied, while the composition law for functions provides the other monoid action law. This shows that each monoid action is nothing other than a set-valued functor. Depending on the variance of the functor from a monoid M to **Set**, we get the left (covariant) or right (contravariant) M-sets.[40]

The variability provided by the right action in each of the above examples is fundamentally *processual*. This perspective is even clearer in an important related example, where we consider sets varying over some time-like linearly ordered category, such as over **N** (the linearly ordered set of natural numbers $\mathbb{N}$, regarded as a category), in terms of a functor. With such a category as our indexing category, objects in $\mathbf{Set}^{\mathbf{N}}$ are just sets varying through n successive stages, that is, $(n-1)$-tuples of maps:

$$X : X_0 \xrightarrow{f_0} X_1 \xrightarrow{f_1} X_2 \xrightarrow{f_2} \cdots X_{n-2} \xrightarrow{f_{n-2}} X_{n-1}.$$

The functor $\mathbf{N} \to \mathbf{Set}$ picks a sequence $X_0 \to X_1 \to \cdots$ of sets X_n and functions $X_n \to X_{n+1}$. A morphism between two such objects (sequences) is a sequence of functions, for example,

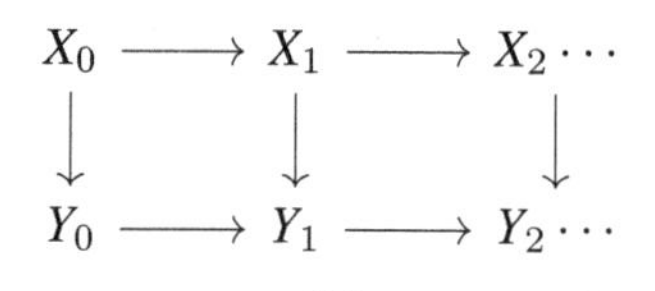

such that each individual square commutes. More generally, we have an **N**-indexed family of functions $(f_i : X_i \to Y_i)_{i \in N}$ compatible with the maps, that is, whenever $i \leq j$, this square commutes:

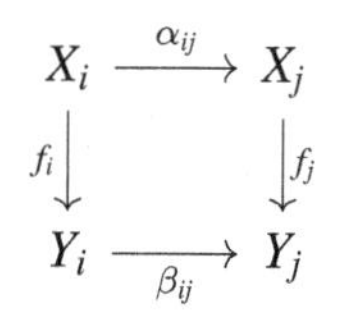

This is equivalently just to describe a natural transformation between X and Y viewed as functors. As such, these transition functions $\alpha_{ij} : X_i \to X_j$ for each $i \leq j$ should satisfy

- $\alpha_{ik} = \alpha_{jk} \circ \alpha_{ij}$ whenever $i \leq j \leq k$
- $\alpha_{ii} = \mathrm{id}_{X_i}$ for all i.

Composing would look just as you would expect:

40. Just as for monoids, a *group action* on a set $S \in \mathrm{Ob}(\mathbf{Set})$ is just a functor $G \to \mathbf{Set}$ that sends the single object of G to the set S. Right G-sets are the same as the presheaf category $\mathbf{Set}^{G^{op}}$.

$$\begin{array}{ccccccccc}
X & & X_0 & \xrightarrow{\alpha_{01}} & X_1 & \xrightarrow{\alpha_{12}} & X_2 & \xrightarrow{\alpha_{23}} & X_3 \cdots \\
{\scriptstyle f}\downarrow & & \downarrow & & \downarrow & & \downarrow & & \downarrow \\
Y & & Y_0 & \xrightarrow{\beta_{01}} & Y_1 & \xrightarrow{\beta_{12}} & Y_2 & \xrightarrow{\beta_{23}} & Y_3 \cdots \\
{\scriptstyle g}\downarrow & & \downarrow & & \downarrow & & \downarrow & & \downarrow \\
Z & & Z_0 & \xrightarrow{\gamma_{01}} & Z_1 & \xrightarrow{\gamma_{12}} & Z_2 & \xrightarrow{\gamma_{23}} & Z_3 \cdots
\end{array}$$

where we require that each individual square commutes. The basic idea here is that once an element is in a set, for example, $x \in X_t$, it *remains there*, that is, $\alpha_{tt'}(x) \in X_{t'}$. However, certain elements $a, b \in X_t$ could become identified in the long run (so the αs do not have to be injective); additionally, new elements can appear over time (something that is expressed by the fact that the maps do not have to be surjective).

The resulting category $\mathbf{Set}^{\mathbf{N}}$ that we have been describing is just a presheaf category $\mathbf{Set}^{\mathbf{C}^{op}}$, taking $\mathbf{N}^{op}$ as our $\mathbf{C}$, where $\mathbf{N}^{op}$ of course has natural numbers for objects and for morphisms $n \to m$ the pairs $\langle n, m \rangle$ such that $n \geq m$. Part of the power of this way of seeing this construction is that there is not really any need to restrict attention to linear orders. Thus, we could similarly consider the functor category $\mathbf{Set}^{\mathscr{P}}$ of sets varying over a preorder or poset $\mathscr{P}$, something we take up in later chapters. Here, too, it is entirely sensible to regard the resulting functor objects as $\mathscr{P}$-variable sets, since we have sets varying according to the shape of the order supplied by $\mathscr{P}$. For instance, the category $\mathbf{Set}^{\mathbf{R}}$, with $\mathbf{R}$ the ordered set of real numbers regarded as a category, has for objects sets varying through real time. More generally, the idea of a set varying over an ordered (poset or preordered) set is really all just a specialization of the general idea of a set "varying over" some arbitrary small category.

Example 53 Moving beyond examples using single-object categories such as the monoids introduced in the previous example, we might also consider presheaves on categories with more than one object. For instance, consider the presheaf from the very beginning of section 2.3, where the indexing category was

$$\mathbb{B} := \boxed{V \xrightarrow{i} L},$$

the object V standing for vertex and L for loop, and the single (nonidentity) morphism i including the vertex in the vertex of the loop. The presheaves on this $\mathbb{B}$ yielded bouquets, or those structures with any number of loops stationed at vertices. A particular presheaf X on this indexing category, then, gives all the data of a particular bouquet X, entirely described by a set $X(V)$ of vertices and a set $X(L)$ of loops, together with a function $X(L) \xrightarrow{X(i)} X(V)$ that acts to pick out the vertex of each of the loops. The action $\gamma \cdot i = c$ or $X(i)(\gamma) = c$, where $\gamma \in X(L)$, just extracts the appropriate vertex (boundary) of the loop in question. Finally, a natural transformation from one bouquet (presheaf on $\mathbb{B}$) X to another bouquet (presheaf on $\mathbb{B}$) Y will just amount to a rule that sends loop-figures in X to loop-figures in Y, point-figures of X to point-figures of Y, and does so in such a way that it preserves the incidence relations. In other words, $\tau : X \Rightarrow Y$ is a natural transformation making the diagram

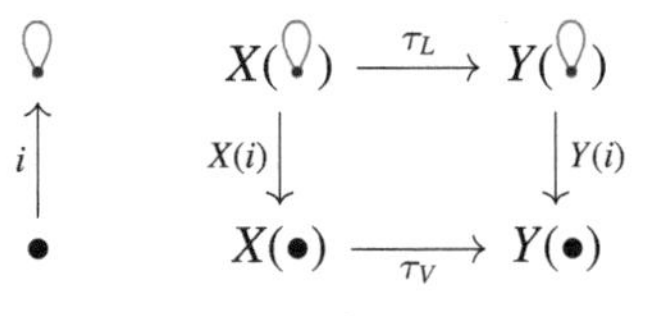

commute, recovering the appropriate notion of a mapping between bouquets.

For our purposes, the thing to note in the above example is how the presheaf action is one that amounts to an operation of *boundary extraction*. The presheaf action operates by extracting from a loop-figure the vertex-figure to which it is attached, an operation it is very natural to think of as *taking the boundary*, or extracting the simpler elements that form the components of given (higher-dimensional) figures.

For another example of this type, consider a (directed, multi)graph X

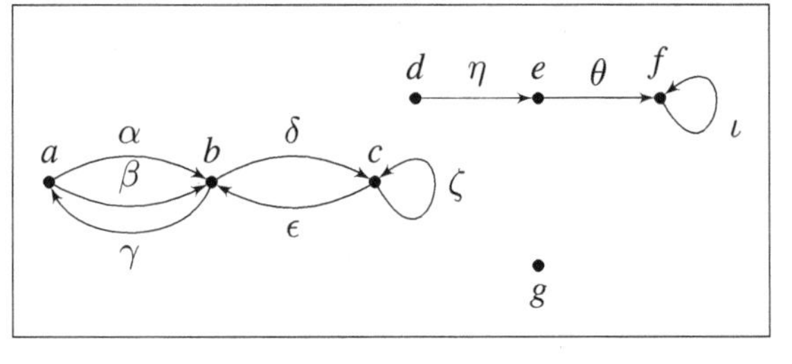

As we have been doing with the other examples, we can regard this as a functor, that is, as being generated by a presheaf on a particular indexing category. Moreover, there is then the obvious action representing the "boundary extraction" of the source and target vertices (boundaries) from a given arrow. More explicitly, take for indexing category the category consisting of two nonidentity arrows (the identities again left implicit),

$$\mathscr{G} := \boxed{V \overset{s}{\underset{t}{\rightrightarrows}} A}$$

where the arrows s, t go from an object V (think vertex) to another object called A (think arrow). Regarding X as a presheaf on $\mathscr{G}$, then, is straightforward: the presheaf $X : \mathscr{G}^{op} \to \mathbf{Set}$ just assigns a set of vertex-shaped objects, a set of arrow-shaped objects, and functions $X(A) \xrightarrow{X(s)} X(V)$ and $X(A) \xrightarrow{X(t)} X(V)$, where the function $X(s)$ just assigns to each arc its source vertex and the function $X(t)$ picks out each arc's target vertex, thus giving us a presheaf action that can naturally be thought of as performing a sort of boundary extraction.

More explicitly, for our given graph X displayed above, the data (including some of the action data) is

$$X(V) = \{a, b, c, d, e, f, g\}$$

$$X(A) = \{\alpha, \beta, \gamma, \delta, \epsilon, \zeta, \eta, \theta, \iota\}$$

$$X(s)(\alpha) = X(s)(\beta) = a, X(s)(\gamma) = X(s)(\delta) = b, X(s)(\epsilon) = X(s)(\zeta) = c, \ldots.$$

By taking presheaves like X for our objects, and natural transformations between such functors for our morphisms (which preserve the incidence relations), we recover the usual graph morphisms, that is, graph homomorphisms—showing that the presheaf category $\mathbf{Set}^{\mathscr{G}^{op}}$ is none other than **Grph** (actually, provided we have multigraphs, it is **Quiv**), consisting of irreflexive directed graphs. We can perform a similar analysis for other sorts of graphs, for instance reflexive graphs. A graph is reflexive provided each vertex v is assigned a designated edge $v \to v$. Equivalently, in terms of quivers, a reflexive quiver has a designated identity edge $\mathrm{id}_X : X \to X$ on each object X. For reflexive graphs, we would take for our indexing category

$$\mathcal{G}' := \boxed{V \underset{t}{\overset{s}{\leftleftarrows}} \!\!\xrightarrow{l}\! A}$$

which consists of two nonidentity arrows, just as in $\mathcal{G}$, but now with an extra, third arrow (l for "looped edges") going in the other direction from the two already given. This indexing category is subject to the following equations:

$$l \circ t = \mathrm{id}_V = l \circ s.$$

Using this as our indexing category, $\mathbf{Set}^{\mathcal{G}'^{op}}$ recovers the category of reflexive (directed, multi)graphs, **rGrph** (or **rQuiv**).[41] Maps of reflexive graphs, that is, natural transformations between the presheaf objects of $\mathbf{Set}^{\mathcal{G}'^{op}}$, must not only respect the source and target maps, but also the extra piece of structure given by l.

As a final observation on this sort of example, let us briefly note that we could generalize all this to (n-uniform) "hypergraphs" taking values in "multisets"—where a *hypergraph* is a generalization of a graph, where edges can join any number of vertices, and where a *multiset* is a generalization of the standard "set," where there can be multiple occurrences of a given element—or still other graph structures. Moreover, as we discussed in example 34, each category can be regarded as a directed graph with some structure. In order to begin to appreciate the third perspective, we could generalize this and consider the n-dimensional analogue of a directed graph, that is, via so-called globular shapes.[42]

Definition 54 For $n \in \mathbb{N}$, an *n-globular set* X is a diagram

$$X(n) \underset{t}{\overset{s}{\rightrightarrows}} X(n-1) \underset{t}{\overset{s}{\rightrightarrows}} \cdots \underset{t}{\overset{s}{\rightrightarrows}} X(1) \underset{t}{\overset{s}{\rightrightarrows}} X(0)$$

of sets and functions such that $s(s(x)) = s(t(x))$ and $t(s(x)) = t(t(x))$ for all $m \in \{2, \ldots, n\}$ and $x \in X(m)$.

But an n-globular set can also be defined as a presheaf on the category $\mathbb{G}_n$ generated by the objects and arrows

$$n \underset{\tau_n}{\overset{\sigma_n}{\leftleftarrows}} n-1 \underset{\tau_{n-1}}{\overset{\sigma_{n-1}}{\leftleftarrows}} \cdots \underset{\tau_2}{\overset{\sigma_2}{\leftleftarrows}} 1 \underset{\tau_1}{\overset{\sigma_1}{\leftleftarrows}} 0$$

which moreover satisfy the equations $\sigma_m \circ \sigma_{m-1} = \tau_m \circ \sigma_{m-1}$ and $\sigma_m \circ \tau_{m-1} = \tau_m \circ \tau_{m-1}$ for all $m \in \{2, \ldots, n\}$.

In short, the category of n-globular sets can also be defined as the presheaf category $\mathbf{Set}^{\mathbb{G}_n^{op}}$. For X an n-globular set, we call elements of $X(m)$ the m-cells of X: for instance, $a \in X(0)$ is a dot or vertex labeled by a; $f \in X(1)$ is an arrow with a source and target boundary;

41. Observe that there is an obvious forgetful functor $U : \mathbf{rQuiv} \to \mathbf{Quiv}$ from reflexive graphs to irreflexive graphs, where this acts by neglecting the structural component l.

42. The following definition and page or so of discussion can be skimmed or skipped on a first reading and resumed with the next example.

$\alpha \in X(2)$ looks just like a natural transformation arrow satisfying certain relations; a three-cell $x \in X(3)$ an arrow between arrows of the natural transformation type, and so on. In this way, various prominent mathematical constructions including the likes of simplicial sets, cubical sets, and globular sets can be construed as examples of presheaf categories. The basic idea in all this is that one selects a category **C** of cell shapes with morphisms "face inclusions" and "degeneracies"; then, as above, one produces a presheaf category $\mathbf{Set}^{\mathbf{C}^{op}}$, and the boundary extraction action perspective will generally fit such situations. But such constructions also encourage the (third) view that for a category **C**, whose objects can be regarded as certain geometrical figures or spaces of a certain sort and its morphisms as structure-preserving morphisms between those spaces, presheaves on such a category give rise to spaces modeled on **C** in the sense that they are probed by the objects of **C**.

The idea with this third perspective can be roughly sketched as follows.[43] If **D** is taken to be, for example, the category of sets or certain topological spaces,[44] and if we regard the indexing category **C** as some category supplying the shapes or generic (geometrical) figures, then the presheaf category $\mathbf{D}^{\mathbf{C}^{op}}$ will be a (generally large) category that will include more general geometrical or spatial objects that are probable or testable with the help of **C**. Altogether, this provides a perspective according to which a presheaf on **C** can be seen as a very general space *modeled on* **C**, where this otherwise unknown space emerges from "probing" it with the known objects of **C**.

In referring to "probes" of a hypothetical space X (for now, just think of some generic space, not necessarily a topological space in the strict sense), we are really thinking of all the ways of mapping into X using the objects of **C**. In other words, if you start with a test space U (an object in **C**) and are returned a set $X(U)$, we are thinking of this set $X(U)$ as designating the set of ways U can be mapped into X, supplying the probes or ways of testing X with U. However, these probes alone will not usually suffice to give you a very discriminating or complete understanding of the space X. To attain a more complete picture, you also need to know about how the different tests or probes of the space relate to one another. This is what the presheaf action takes care of. If you have a map $f : U \rightarrow V$ in **C**, then given some (probe) element $V \xrightarrow{p} X$ of $X(V)$, precomposing with f (acting on the right) will induce a map going in the other direction $X(V) \xrightarrow{X(f)} X(U)$ that will tell you how probes of the space X by V change into probes by U; and with *that* information, you can get an accurate picture of what X itself is. In this way, in describing a generalized space modeled on the objects of **C**, we are in fact describing nothing but a presheaf X on **C**, where each presheaf is a rule assigning to each $U \in \mathbf{C}$ the set $X(U)$ of admissible maps from U into the space X, and where this comes with the information of how certain probes of X change into other probes of X. One of the purposes of doing this is that by probing a big space with a number of smaller or simpler test spaces, not only can we model parts of the space into which we are mapping but we can ultimately look to piece together the small or partial tests into information about tests with bigger test spaces, arriving at a picture of the overall space of interest. This is of particular importance since the information such probes

43. This general perspective largely follows Lawvere; see, for instance, Lawvere (2005).

44. One usually starts with thinking about topological spaces, that is, the category **Top**, but really we just need it to be a category and for this category to support some notion of how certain objects can be *covered* by other objects. There is much more on this in chapter 4 and in the final chapters.

gather turns out to be most useful precisely when the presheaves are in fact sheaves, that is, satisfy some further consistency conditions.[45]

Example 55 To illustrate the last (but arguably most significant) perspective on the presheaf action—namely, action by *restriction*—we can begin by considering the construction of a presheaf on the partial order of open sets $\mathcal{O}(X)$ (or **Open**), for X a topological space.[46] A presheaf on X is just a functor $F : \mathcal{O}(X)^{op} \to \mathbf{Set}$. For each open $U \subseteq X$, we then think of the set $F(U)$ as the set that results from assigning set-values or data throughout or over all of U. An open subset $V \subseteq U$ can be seen in terms of an inclusion arrow $V \hookrightarrow U$ when regarded in the poset $\mathcal{O}(X)$ seen as a category, so applying the (contravariant) functor F will give us a function that passes from the data assigned throughout or specified over U (the generally "larger" region) to the data assigned throughout the subregion V, in a process aptly called the *restriction*, and typically denoted by $\rho_V^U : F(U) \to F(V)$ (or just $F(V \hookrightarrow U)$). Especially when the particular application involves looking at all the *functions* of a certain type (e.g., continuous functions) defined throughout that region U, given an element $f \in F(U)$, one sometimes denotes $\rho_V^U(f)$ by $f|_V$ and speaks of the *restriction* of f from U to V, treated like the usual restriction of a function along a part of its domain.

As a first illustration of such a functor, we can consider the set of all continuous real-valued functions, that is, functions from $U \subseteq X$ to $\mathbb{R}$. Importantly, when there is an inclusion of opens $V \subseteq U$, we will have a restriction $\rho_V^U : \mathbf{Top}(U, \mathbb{R}) \to \mathbf{Top}(V, \mathbb{R})$, which just sends $f : U \to \mathbb{R}$ to $f|_V : V \to \mathbb{R}$. The presheaf here thus acts to *restrict* the collection of functions given over some region (say, $(0, 6)$) down to the open subsets of that region (say, $(2, 4)$ in particular), as suggested by the following picture:

45. The reader intrigued by this admittedly more subtle perspective may find the extended discussion in nLab Authors (2019) particularly illuminating; the paragraph above leans on this discussion.

46. Topological spaces and the poset of open sets of a space are discussed in ample detail in chapter 4.

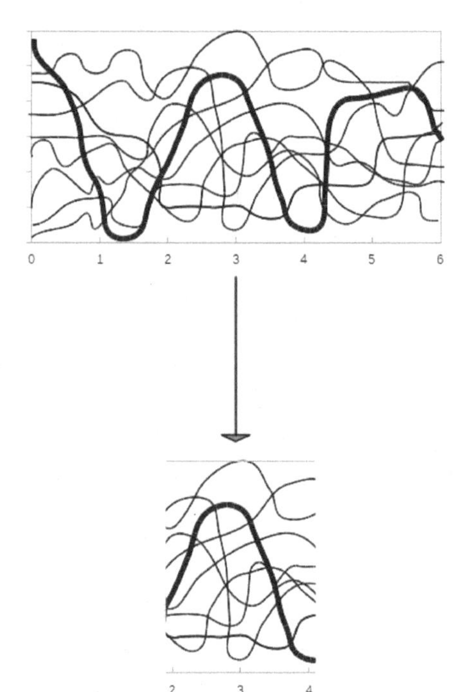

The action of this presheaf is thus given by *restriction*, an action that is clearly functorial. Observe, also, how each of the collections $\mathbf{Top}(U, \mathbb{R})$ of continuous functions in fact assembles into a ring structure, meaning that the restriction functions will in fact be ring homomorphisms—as such, this presheaf $\mathbf{Top}(-, \mathbb{R})$ is actually a presheaf of rings (as opposed to being valued in **Set**).[47]

For another restriction-type example, but of a rather different flavor, start by considering for regions the set J of jurisdictions with their subjurisdictions, that is, $(J, \subseteq)$ a preorder.[48] We can consider that within the set of *possible laws*—where laws are just treated as propositions, that is, objects of the preorder **Prop** regarded as a category whose objects are logical formulas or propositions and whose morphisms are (equivalence classes of) *proofs* or derivations that one proposition implies another—some of these laws are being followed by all people in the region. To each jurisdiction V, then, we can assign a set $R(V)$ consisting of whatever laws are being respected by all the people throughout V. In other words,

47. Another standard way of producing presheaves on a space arises by taking the *local sections* of a continuous function $p: E \to X$, via the "local section functor." A local section of p is a continuous function $s: U \to E$ from an open subset U of X to E, such that $p \circ s(x) = x$ for all $x \in U$. If we let $\Gamma(p)(U) = \{s: U \to E \mid s \text{ is a local section of } p\}$, then by considering that whenever $V \subseteq U$ we can restrict local sections over U to local sections over V, we see that this defines a presheaf on X. Don't worry if this doesn't make sense yet. We have a lot more to say about this local section approach in chapters 5 and 8.

48. This example comes from Spivak (2014, 263-264). See also Awodey (2010) for more on the category of propositions that is used in the example.

laws are being assigned *locally* to each jurisdiction; after all, a law is dictated to be valid only within a specific region. If V is a subjurisdiction of U, that is, $V \subseteq U$, then any law respected throughout U is obviously respected throughout V, so we can *restrict* from the laws respected throughout U to those respected throughout V. Clearly any law respected throughout the state of Illinois will be respected throughout any county in Illinois, so we can regard such a law given over Illinois from the restricted perspective of a county in that state. But observe that the converse is not true! A law respected throughout a part of Illinois need not be respected throughout all of Illinois.

Here we have a local assignment of data to the space of jurisdictions that moreover obeys the property that whenever we have a region V included in U, then the action of the presheaf works in the opposite direction: it takes data assignments given throughout U and *restricts* them to (the same) data assignments now given throughout V, the smaller region. The idea to keep in mind here is this: if you have some data (like a list of those laws being respected by everyone) assigned to some region (like Illinois), and you have another list of those laws being respected by everyone in some subregion included in Illinois (like Cook County), then you will expect that the list of laws respected by everyone throughout Cook County will be (equal if not) larger than the list of laws respected by everyone throughout Illinois. In a larger region, there are more chances for the data not to fit—for example, for someone to fail to respect that law—than there is in a smaller region.

The main takeaway is that in the previous construction, we have made use of two key ingredients: (1) a local assignment of data to a space (each of the laws in the "respected laws data" is expected to hold throughout all of the jurisdiction region to which it is assigned); and (2) a natural operation of restriction (induced by the natural inclusion relation governing the overall space of jurisdictions) allowing us to move from the data assigned throughout a region to data assigned throughout subregions. Formally, these two ingredients just specify what we need to have a functor R that is *contravariant*, that is, we have been describing a presheaf $R : J^{op} \to \mathbf{Prop}$.

With an eye toward sheaves, the restriction-style action is in some sense the most decisive of the four perspectives considered, or at least the most immediately relevant to the subsequent initial presentation of sheaves in terms of sheaves on topological spaces. Thus, it pays to understand it well. We could dwell at length on the notion of restriction and its relation to some of the other key basic categorical concepts that we will meet—for instance, that restrictions are not "right cancellable" in general, and that it is easy to construct many examples of *distinct mappings* that have equal restrictions to the same part, that is, mappings f, g such that $f|_i = g|_i$ but $f \neq g$. For now, a few general observations concerning restriction may be worth stressing.

Given an inclusion $V \hookrightarrow U$, restriction tells us that we can take some $x \in F(U)$ and restrict that data assignment to the part of U that makes up V, and this will leave us with a viable data assignment on V. It should be easy to see, both intuitively and precisely (on the model of function restriction), how this amounts to a restriction. However, it is important to recognize what is *not* going on: it would be a mistake to take the language of restriction as implying that, at the level of the presheaf itself, we are passing from a (generally) bigger set of data to a smaller one—"restricting our attention" as it were. Strictly speaking, at the level of the maps between the presheaf data $F(U)$ and $F(V)$, this is *not* what is going on.

This should already be evident from close consideration of the laws example, where the set of respected laws $R(U)$ specified over a larger region U will actually typically be *smaller* than the set of respected laws $R(V)$ specified over the subregion $V \subseteq U$. The same is true of the continuous functions: it is easier to be continuous on a smaller region; that is, over a bigger region there will be more opportunities to fail to be continuous. It is perhaps an unnecessary warning, but the point is that, at the level of the presheaf maps themselves, for example, moving from $R(U)$ to $R(V)$, we are not generally dealing with a restriction in the sense of moving from a bigger *dataset* (set of value assignments) to a smaller dataset. Confining our attention to U and V as regions or components of a space, it makes sense to think of these objects (regions) as *constraints* of sorts, according to which V, a subregion, amounts to a weaker constraint on any data specified locally over the regions. It should be evident that given any inclusion of a smaller region into a larger region, more data will generally be able to satisfy the weaker constraint (corresponding to the smaller region) than will satisfy the stronger one (corresponding to the larger region). Yet, at the level of a particular data assignment, we can regard the presheaf maps as amounting to a restriction of that data along inclusions of subregions. The next and final example of this chapter should help to further clarify this.

Example 56 Consider time intervals as objects and morphisms given by inclusions, yielding a category we will denote $\mathcal{T}$. For concreteness, suppose we consider the period spanning from January 1 (at midnight) of 2018 until June 1 (at midnight) of 2018. Then, suppose this period is decomposed into various two-month intervals

$$[Jan1, March1], [Feb1, April1], [March1, May1], [April1, June1],$$

that together "cover" the entire half-year period from $Jan1$ through to the end of May.[49] The natural overlapping subintervals produce the following overall structure on the system of intervals ordered by inclusion (as indicated by the inclusion arrows):

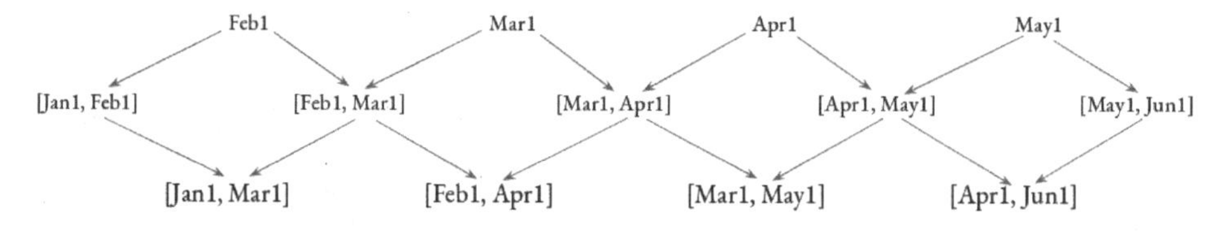

where the single dates like Feb1 are short for the degenerate interval $[Feb1, Feb1]$ representing the instant of midnight on February 1.

Now, for a particular company with some (generally fluctuating) stockpile of products, we can define a contravariant functor $S : \mathcal{T}^{op} \to \mathbf{Set}$ that assigns to each time period $[t, t']$ the products that are in stock throughout the *entire interval* $[t, t']$ (where, for the moment and for simplicity, we just imagine that a product is simply present or absent, say, as if the only important question was whether the company had at least one item of the product or had none, ignoring the question of quantity). $S([t, t'])$ is thus just the set of products the company has in stock throughout the entirety of the time period $[t, t']$.

49. For now, you can think of this notion of "covering" in a naive way. We will be precise about this sort of thing below, starting in chapter 4.

Then, for any inclusion of time periods $i:[t,t'] \hookrightarrow [u,u']$, the functor S acts (contravariantly) *by restriction*, mapping each stocked item onto itself. Clearly, any product present in the company's store throughout the bigger time interval $[u,u']$ must be present as well throughout any subinterval $[t,t']$. But this tells us that the "list" of products recorded as present throughout the bigger time interval $[u,u']$ is in general likely to be shorter or smaller than the list of products assigned to the smaller time interval $[t,t']$.

A particular presheaf on such a $\mathcal{T}$ might then be given by something like the following (where each of the A, B, C, and so on, sitting over each interval-object, represents one of the products held by the company throughout the entire interval):

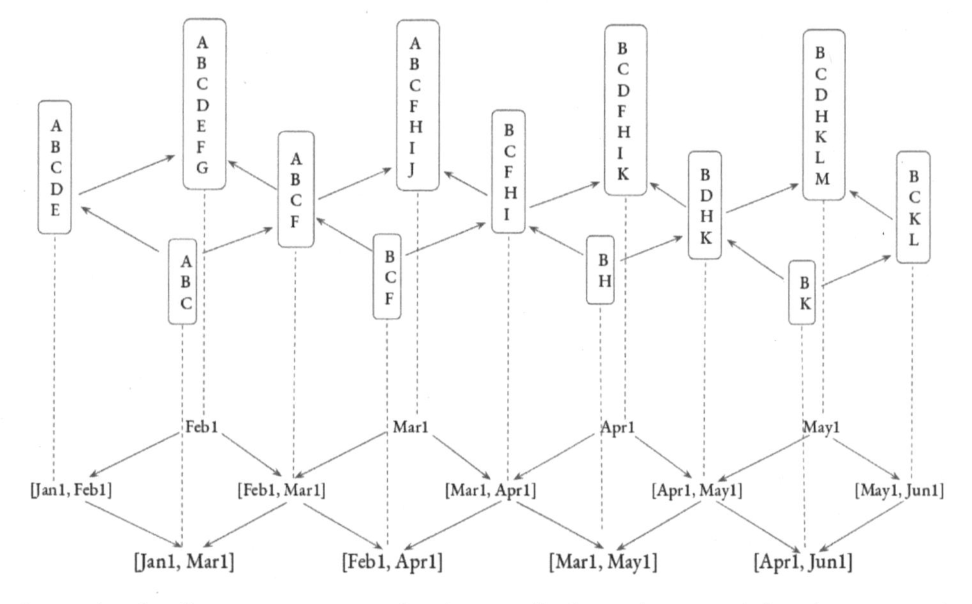

Inspecting the diagram, one can see that the sets of value assignments (of products present) throughout each interval are generally smaller over larger regions (time intervals); thus, strictly attending just to the part of the diagram sitting on top of the network of time intervals, the presheaf arrows in fact generally go from smaller sets to larger ones.[50]

A final thing to realize is that, in general, restriction along an inclusion is *not necessarily either surjective or injective* (despite what a naive understanding of the language of restriction might seem to imply, for instance, suggesting at least surjectivity). An easy counterexample is provided by the following.[51] As was seen in an earlier example, restriction of continuous functions is continuous. However, take the (poset) category $\mathbf{A}$ that consists of just two objects, U and C with the single nonidentity (inclusion) map $U \to C$, where U is the open interval $(0,1)$ and C is the closed interval $[-1,1]$. Now let the presheaf $F:\mathbf{A}^{op} \to \mathbf{Set}$ act on objects as follows: $F(U)$ is the set of all continuous real-valued functions given over the open interval $(0,1)$ and $F(C)$ is the set of all continuous real-valued functions over the closed interval $[-1,1]$. Then the induced presheaf action $F(C) \to F(U)$ is clearly by restriction; yet, it should be evident that this particular restriction process can

50. This relates to the warning discussed just before this example.

51. This counterexample is derived from Lawvere and Rosebrugh (2003).

be neither surjective nor injective. It is not surjective since there are functions that remain continuous over $(0,1)$ while having discontinuities at either or both "end points"—in particular, at 0—so that such functions cannot come from any continuous functions specified over all of $[-1,1]$. It is not injective since there exist distinct continuous functions given over all of $[-1,1]$, each of whose restrictions to $(0,1)$ are identical.

2.5 Philosophical Pass: The Four Action Perspectives

Box 2.1
On the Presheaf

The general understanding of presheaves developed in section 2.3 might be thought of in terms of Plato's notion of the *form* or *shape* (*eidos*) of something, that structural schema according to which the concrete realizations thereof are organized. This form also supports a great variety of realizations or "manifestations" (*phantasmata*), and the plurality of particular manifestations of it populates a world that acts as some sort of "receptacle" for such instantiations of the forms. The process by which the manifestations are unfolded according to the structural schema of the form is what Plato would call the *participation* (*methexis*) of the form. The form is held to be invariant, its components sufficiently generic, and altogether it is fundamentally simpler (and so, in the end, more *intelligible*) than its many realizations or manifestations.

Applied to presheaves, the gist here is that the "generic figures" supplied by the domain category $\mathbf{C}$ act as something like the form, while the value assignments $P(c)$ for each object of $\mathbf{C}$ supply something like the concrete appearances or manifestations of the static components of the form, and the presheaf action enforces dynamic relationships between the various manifestations modeled on the invariant generic relationships between the components of the form. The presheaf P itself, on this way of seeing things, would then be nothing other than the process of manifestation or participation of the form in concrete particulars, and to understand how the concrete manifestations and their components respect among themselves the relations that obtain between the components of the form itself is just to understand the general functoriality of the functor P.

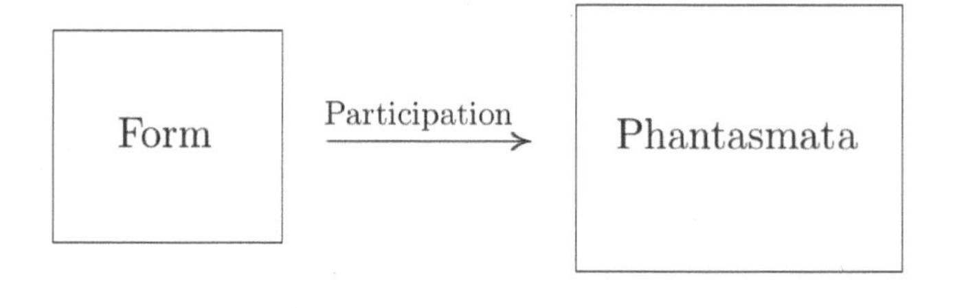

Presheaves accordingly supply a uniform framework for capturing, in an at once compressed and illuminating way, many structures that appear throughout math and that can otherwise appear, in their traditional presentation, rather complicated or haphazard (frequently leading to a complicated or haphazard description).

Many important mathematical structures and categories—including some of those already discussed, such as dynamical systems, bouquets, graphs, hypergraphs, and more—arise as a presheaf category consisting of contravariant functors on some simple indexing category, where the result of applying the functor to the objects of the indexing category yields what we naturally think of as containers (in **Set**) holding on to various manifestations or figures each of which conforms to the shape or form determined by the generic figures populating the indexing category (one for each of its objects), and where the changes of figure indicated by the indexing category (given by its morphisms) are respected by the figures of the container. While this perspective is perhaps most appropriate, or easy to countenance, when the objects and morphisms of the indexing category **C** have some sort of geometrical interpretation, it is a surprisingly useful perspective even in more general cases.

As for the four perspectives on the presheaf action, the fundamental idea that they share is that the domain category **C** plays the role of specifying the general internal structure or schema—in the form of the figure-types or shapes, the glue, the nature of the internal dynamic, or the locality of data assignments—in which all the sets in $\mathbf{Set}^{\mathbf{C}^{op}}$ must participate. The resulting presheaf category in each case has for its objects different instantiations or realizations of the general form supplied by **C**. The other way of looking at this is that the domain category plays the role of a parameter specifying in an invariant form how (temporal, dynamic, geometric) variation or cohesion is to take place, while the target category (**Set**) serves as the container or arena holding on to all the particular values or results of trying to "participate in" or "realize" this form of variation. One might accordingly think of a presheaf itself as mediating between the invariance or fixedness of a structure "outside of time" belonging to the domain category, on the one hand, and its multifarious concrete presentations or manifestations "in time," on the other.

Most of the presheaf functors above are valued in **Set**, which can be useful in taming many problems or otherwise complicated structures, and there are good reasons for the central role **Set** plays in classical category theory, largely accounting for why presheaves classically take values in **Set**. (Again, these reasons have to do mainly with the Yoneda results, covered in chapter 6.) However, it is worth emphasizing that, philosophically speaking, presheaves are anything but the static and qualitatively barren objects the usual set-theoretical perspective on sets as "bags of fixed objects" might encourage us to believe.

Considering presheaves as coming equipped with an action that is *processual* recaptures a dynamic perspective in which objects are not regarded as static collections; instead, the sets are seen as evolving through stages, either merely temporally or in accordance with an internal dynamic (as in cases of evolutions subject to certain equations). Against the generally discrete and static context of sets, this restores a more continuous and dynamic perspective. The *boundary-extraction* perspective, for its part, reveals the incidence relations, relations that together describe something like how the overall structure "holds together" or coheres. Against the usual set-theoretical perspective of sets of objects grouped together more or less arbitrarily into a set that cannot internally differentiate objects or discern important qualitative (or dimensional) features of those objects or their modes of relation, this perspective restores a kind of continuity in telling us how the various components of a structure can be regarded as glued or stitched together from other (lower-dimensional) components. The third perspective lets us regard a space in terms of all the ways of probing it from the outside and thinking about the entire space in terms of how these various probes behave with respect to one another. This perspective, similar to that of the *relationism* of Yoneda (discussed in chapter 6), is more "continuous" in the sense that it insists that we understand something in terms of all the relations or perspectives on it, instead of as something set off on its own or intelligible by itself. Finally, acting via *restriction*, presheaves open onto a range of relationships between the parts of a whole. In general, such a relationship emerges as "regular" in the sense that in passing from data specified over some containing region to data over a subregion, there remains a kind of conformity or identity of the data given over the parts in relation to the same rule or function describing the containing region. This perspective thus opens onto the notion of a conformity of parts of a whole to a single rule or idea (as opposed to the usual set-theoretical consideration of a whole independently of the specific way, beyond whether or not a part *belongs*, the whole enforces relationships among the parts).

In short, while we are able to benefit from certain tame properties of **Set**, the presheaf perspective lets us recapture a generally more dynamic, nuanced, and continuous (in a general sense) perspective on many structures of interest.

3 Universal Constructions

In which we learn about the decisive category-theoretic notion of universality (universal properties), *familiarize ourselves with a variety of notable examples of this phenomenon, present definitions of other important notions making use of these properties, and consider the broader significance of the notion of universality developed here.*

3.1 Limits and Colimits

In category theory there is a very important notion of limits and colimits. These terms really codify and powerfully generalize certain decisive constructions that had been noticed in many concrete cases all over mathematics well before category theory was born. Throughout mathematics, we are constantly building new mathematical objects from old ones. Categorical limits and colimits are, in an important sense, the *most* efficient way of doing so. Examples include many familiar constructions—like taking disjoint unions or the intersections of sets, the least upper bound (supremum) or greatest lower bound (infimum) of a set of numbers, forming Cartesian products, direct sums, kernels and cokernels, forming the coarsest topology making a map continuous, and more. Limits capture a wide array of constructions where a certain subcollection of given objects is isolated, while colimits capture something like the amalgamation or gluing together of given objects. What specific instances of each construction have in common—usually indicated by the use of a superlative adjective, like the *largest*, *least*, *coarsest*, and so on—is that the construction or object satisfies a certain *universal property* in relation to other components of the category.

To carry out these constructions, we need to know *what we are taking the (co)limit of.* Rather than confining our attention to special objects—like numbers whose maximum or whose least common multiple we are interested in taking—the right way to develop this is to give the most general account, allowing for potentially complex input data. The input data for both notions will be a *diagram*, that is, some collection of objects in a category and some morphisms between them. In chapter 2, we met and began to explore the notion of a *diagram* (as a functor), where this was thought of as involving an instantiation of a particular template supplied by the shape or generic figures of an indexing category. There, you were invited to regard the indexing category as a template consisting of some nodes together with directed edges between certain nodes, that is, as a directed graph, on which the instantiations were patterned. A diagram instantiates (within the target category) each node of the template with an object of the target category and each edge with an arrow of the target category, thus yielding a diagram built out of shapes whose generic form is provided by the template category.

A diagram in a given category can be thought of as posing two problems, the left and right problems, the solutions to which are supplied by certain objects, together with a collection of morphisms, that "complete" the diagrams at either end. Such gadgets formed by the object and arrows of the left solution—that is, a special object C together with a collection of arrows, one for each object in the diagram, such that for any arrow between objects of the diagram there are arrows from C that make the resulting triangles commute—are called *cones*. Similarly, for a right solution in which all arrows terminate, such a thing is often called a *cocone*.

In general, on any particular side, a solution need not exist at all (or it may exist on one side but not on the other); on the other hand, each problem may have many solutions. A *universal* solution is one through which each (left or right) solution must pass by means of a (fundamentally) unique mediating arrow. In other words, if there are solutions (of the relevant handedness), then the universal solution is one that is abstractly "nearest" to the diagram and, as such, is the best or most optimal solution to the problem ("better" than any other object that could be used to complete the diagram). A limit is just a universal left solution, a colimit a universal right solution. If a diagram has a (co)limit, this (co)limit will be essentially unique, so that whenever such a solution does exist, we can in fact speak of *the* (co)limit. In short, and in the general case, a limit and a colimit are given by nominated objects among the (co)cones, that "universally complete" the diagram on the left and on the right (respectively).

Let us be more explicit about all this. Recall the notion of **J**-shaped diagrams $F : \mathbf{J} \to \mathbf{C}$, first introduced in example 35. Using this construction and natural transformations, we can introduce the concepts of cones and cocones of a diagram, thereby characterizing the limit and colimit of a diagram as the universal such (co)cone.

3.1.1 Cones and Cocones: Limits and Colimits Defined

Suppose

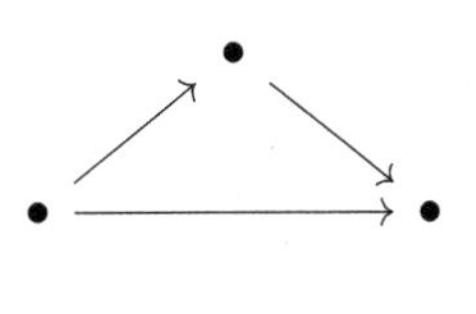

is the template for the diagram

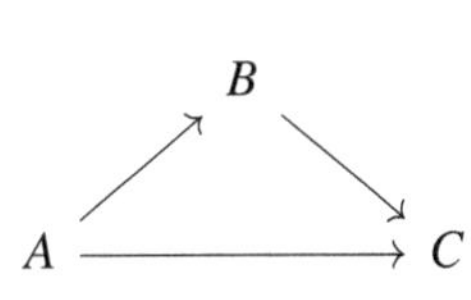

in a category **C**. Observe that, in general, for any given object X in a category **C**, there will be a *constant functor*—one that we will just call by the name X—from **J** to **C**. This functor acts to send every object to X and every morphism to the identity map on X. This lets us view the object X itself as a diagram, $X : \mathbf{J} \to \mathbf{C}$.

Thus, given any functor (diagram) $F : \mathbf{J} \to \mathbf{C}$, we can consider a natural transformation between X (now viewed as a functor) and the diagram F. This natural transformation will just consist of a collection of morphisms between X and the objects found in the diagram F, such that these morphisms moreover commute with the morphisms found in the diagram. When the arrows go from X to the diagram F, this construction is called a *cone over* D; when the arrows go the other way, now from the diagram F to X, then the construction is called a *cone under* D (or a *cocone*). In this way, using the diagram introduced above, a natural transformation $X \Rightarrow F$ will look like the image on the left, while a natural transformation $F \Rightarrow X$ will look like the image on the right:

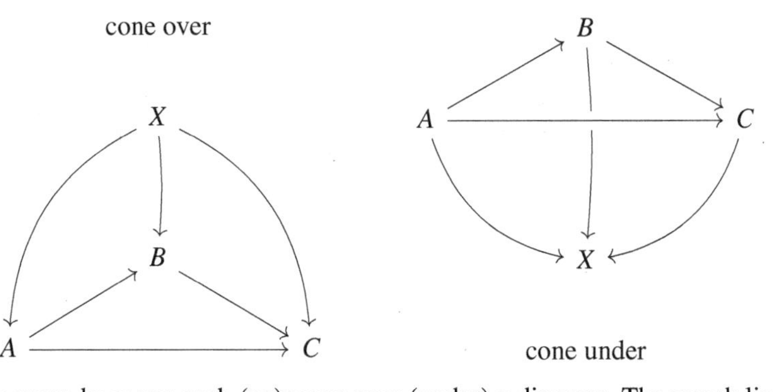

Now, there many be many such (co)cones over (under) a diagram. The punch line toward which we will build is that

> the *limit* of a diagram F is just a special (optimal) cone over F, in that it is the cone that "gets as close as possible" to the diagram F (where this means that any other cone will have to factor or pass through it);

and

> the *colimit* of a diagram F is just a special (optimal) cone under (cocone for) F, in that it is the cocone that "gets as close as possible" to the diagram F (where this means that any other cocone will have to factor or pass through it).

Let us now do this more formally, starting with the limit. There are a few steps. First, let us be more precise about how to regard any given object of a category in terms of a cone. In a category $\mathbf{C}$, a *terminal object* is a special object, usually denoted 1 (owing to the fact that in **Set**, it is just a one-element set), satisfying a certain universal property:

> for every object X of $\mathbf{C}$, there exists a unique morphism $! : X \to 1$.

If such a terminal object exists, it will be unique (up to unique isomorphism). Dually, an *initial object* in a category $\mathbf{C}$ is an object $\emptyset$ such that for any object X of $\mathbf{C}$, there is a unique morphism $! : \emptyset \to X$. Similarly, an initial object, if it exists, will be unique up to unique isomorphism, letting us speak of *the* initial object. Note that an initial object in $\mathbf{C}$ is the same as a terminal object in $\mathbf{C}^{op}$.

But **Cat** is a category, and we thus speak of the terminal object in **Cat** as the *terminal category*. This is just the category with a single object and a single morphism (necessarily the identity morphism on that object). We denote this $\underline{1}$ (or sometimes $\mathbf{1}$).

Let $t:\mathbf{J}\rightarrow\underline{1}$ denote the unique functor to the terminal category. Suppose, given a category $\mathbf{C}$, we are given an object $X\in \mathrm{Ob}(\mathbf{C})$. Such an object can be represented by the functor $X:\underline{1}\rightarrow\mathbf{C}$. Then, precomposing this functor with t to get $X\circ t:\mathbf{J}\rightarrow\mathbf{C}$ will just give us the *constant functor* at X, where this sends each object in $\mathbf{J}$ to the same $\mathbf{C}$-object X and every morphism in $\mathbf{J}$ to the identity id_X on that object. In other words, composing with t induces a functor $\mathbf{C}\cong Fun(\underline{1},\mathbf{C})\rightarrow Fun(\mathbf{J},\mathbf{C})$, which is commonly denoted $\Delta_t:\mathbf{C}\rightarrow Fun(\mathbf{J},\mathbf{C})=\mathbf{C}^{\mathbf{J}}$. Altogether, this actually gives $\Delta:\mathbf{C}\rightarrow\mathbf{C}^{\mathbf{J}}$ that takes an object X to the constant functor at X and a morphism $f:X\rightarrow Y$ to the *constant natural transformation*, where each component is defined to be the morphism f. One can observe that each arrow $f:X\rightarrow Y$ in $\mathbf{C}$ will induce a natural transformation $\Delta(X)\xrightarrow{\Delta(f)}\Delta(Y)$ such that

$$\begin{array}{ccc} (\Delta X)(i) & \xrightarrow{\Delta(f)_i} & (\Delta Y)(i) \\ {\scriptstyle(\Delta X)(e)}\downarrow & & \downarrow{\scriptstyle(\Delta Y)(e)} \\ (\Delta X)(j) & \xrightarrow[\Delta(f)_j]{} & (\Delta Y)(j) \end{array} \qquad \begin{array}{c} i \\ \downarrow{\scriptstyle e} \\ j \end{array}$$

commutes for each edge e of the indexing category $\mathbf{J}$. But recall that the constant functor just sends every object to itself and assigns the identity map to each edge, so the previous diagram in fact just reduces to

$$\begin{array}{ccc} X & \xrightarrow{\Delta(f)_i} & Y \\ {\scriptstyle \mathrm{id}_X}\downarrow & & \downarrow{\scriptstyle \mathrm{id}_Y} \\ X & \xrightarrow[\Delta(f)_j]{} & Y \end{array}$$

which obviously commutes.

If we now consider, for an arbitrary $\mathbf{J}$-diagram $F:\mathbf{J}\rightarrow\mathbf{C}$ and for $X\in\mathbf{C}$, the arrows (which are in fact natural transformations)

$$\Delta X\longrightarrow F \qquad\qquad F\longrightarrow\Delta X,$$

we get that a typical arrow in $\mathbf{C}^{\mathbf{J}}$ corresponding to these arrows is just a natural transformation, that is, a family of arrows of $\mathbf{C}$,

$$(\Delta X)(i)\xrightarrow{\xi(i)}F(i) \qquad\qquad F(i)\xrightarrow{\xi(i)}(\Delta X)(i)$$

indexed by the various objects or nodes of $\mathbf{J}$ and such that

$$\begin{array}{ccc} (\Delta X)(i) & \xrightarrow{\xi(i)} & F(i) \\ {\scriptstyle(\Delta X)(e)}\downarrow & & \downarrow{\scriptstyle F(e)} \\ (\Delta X)(j) & \xrightarrow[\xi(j)]{} & F(j) \end{array} \qquad \begin{array}{c} i \\ \downarrow{\scriptstyle e} \\ j \end{array} \qquad \begin{array}{ccc} F(i) & \xrightarrow{\xi(i)} & (\Delta X)(i) \\ {\scriptstyle F(e)}\downarrow & & \downarrow{\scriptstyle(\Delta X)(e)} \\ F(j) & \xrightarrow[\xi(j)]{} & (\Delta X)(j) \end{array}$$

commute for each such edge $e:i\rightarrow j$ in $\mathbf{J}$. But when we apply the functor Δ, these commutative squares collapse to the commutative triangles

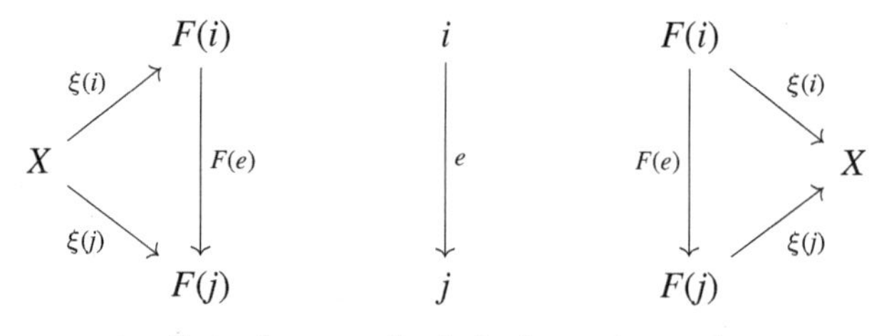

The definitions guarantee that whenever the indexing category has composable edges, the corresponding composite triangles commute. The natural transformations represented by the triangles on the left give a *left solution* for the diagram in **C**, sometimes also called a *cone over* the diagram F with *summit* vertex X. The natural transformations represented by the triangles on the right give a *right solution* for the diagram, also called a *cocone for* (or *cone under*) the diagram F with *nadir* X.[52]

We can then take such gadgets and use them to form the category of cones, where an object in the category of cones over F will be a cone over F, with some summit, while a morphism from a cone $\xi : X \Rightarrow F$ to a cone $\mu : Z \Rightarrow F$ is a morphism $f : X \to Z$ in **C** such that for each index $j \in \mathbf{J}$, $\mu_j \circ f = \xi_j$, that is, a map between the summits such that each leg of the domain cone factors through the corresponding leg of the codomain cone.

Using these notions, we can define the *limit* of F in terms of a universal cone, where a cone $\alpha : L \to F$ with vertex L is universal with respect to F provided for every cone $\Delta X \to F$, there is a unique map $g : \Delta X \to L$ making

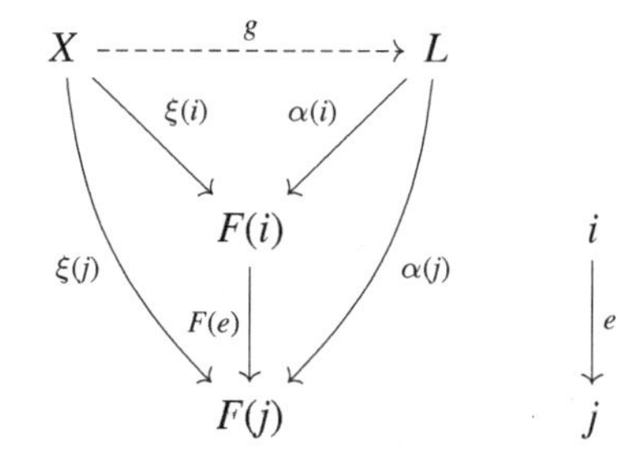

commute. In such a case, one usually refers (somewhat improperly) to the universal cone by just the vertex L, and calls this the limit of F.

For reasons that will become clearer after chapter 6 on the Yoneda results and representability, we can also see a limit for a diagram $F : \mathbf{J} \to \mathbf{C}$ as a representation for the corresponding functor $Cone(-, F) : \mathbf{C}^{op} \to \mathbf{Set}$, sending $X \in \mathbf{C}$ to the set of cones over F with summit X.[53] The limiting cone will fundamentally be *universal* in the sense that for any other cone over F, there will exist a unique arrow from the summit of that cone to the summit of the limiting cone, that is, any other cone must pass uniquely through the limiting cone if it wants to pass down to F. In short,

52. I hope this is already clear, but in case it is not: the terminology of "over" and "under" has to do with the fact that the above triangles can be presented as rotated clockwise 90 degrees, as we did earlier.

53. While we could consider limits and colimits in any category, something called the Yoneda lemma (discussed in chapter 6) assures us that the constructions of (co)limits of diagrams valued in the category **Set** suffice to provide formulae for (co)limits in any category. To ensure that we have a *set* of cones, we need only assume that the diagram is indexed by a small category **J** and that **C** is locally small, thereby guaranteeing that the functor category $\mathbf{C}^{\mathbf{J}}$ is locally small. Under certain conditions, we can weaken these restrictions.

Definition 57 The *limit* of a diagram $F : \mathbf{J} \to \mathbf{C}$ is an object $\lim F$ in $\mathbf{C}$ together with a natural transformation $\eta : \lim F \Rightarrow F$ satisfying the following universal property:

> for any object X and for any natural transformation $\alpha : X \Rightarrow F$, there is a unique morphism $g : X \to \lim F$ such that $\alpha = \eta \circ g$.

The dual construction leaves us with a category of cocones $\mathbf{CoCones}(F)$, wherein the universal cocone emerges as the *colimit* of the diagram F, denoted colim F, where this can be seen as a representation for $Cone(F, -)$, forcing all cocones to receive maps from the colimiting cone if they want to receive maps from F. Explicitly,

Definition 58 The *colimit* of a diagram $F : \mathbf{J} \to \mathbf{C}$ is an object colim F in $\mathbf{C}$ together with a natural transformation $\epsilon : F \Rightarrow \text{colim } F$ that satisfies that for any object X and any natural transformation $\beta : F \Rightarrow X$, there is a unique morphism $h : \text{colim } F \to X$ such that $\beta = h \circ \epsilon$.

Altogether, in terms of where we began this discussion, for such a diagram, we can picture the limit and colimit as follows:

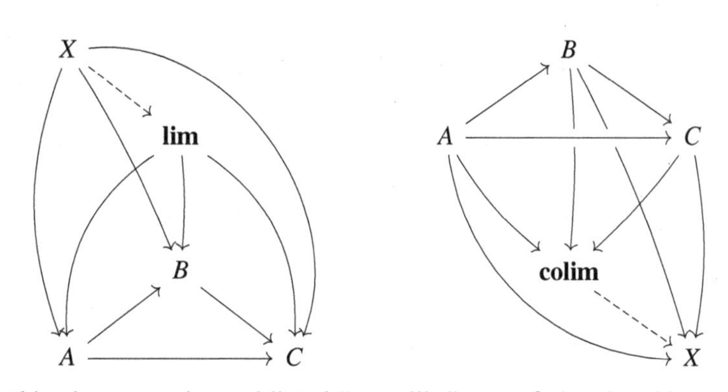

But while this gives us a nice and literal "cone-like" way of picturing things, there is no reason why we should have to restrict ourselves to diagrams of precisely such a shape—by considering diagrams of different shapes, taking limits and colimits of such diagrams will again recover a number of important constructions found across mathematics. Let us start with some examples of limits.

3.1.2 Examples of Limits

Example 59 An extreme case, where we take the limit of the *empty diagram*—the diagram that has no objects and no morphisms—yields a construction we have in fact already met: the *terminal object*. One can check that the limit of the empty diagram will be just one object, if one exists, that has the property that there is a unique morphism to it from every object in the category. While not every category has a terminal object, if it does, then it is unique.

Concrete instances of this include the following:

- In **Set**, the terminal object is given by the *singleton set* $*$, for if X is any set, then there is only one possible function $X \to *$, the one that just takes everything to $*$.

- In a poset $\mathcal{P}$, the terminal object will be its *greatest element*, when such a thing exists. In other words, it will be an element l such that $p \leq l$ for all $p \in \mathcal{P}$. In particular, then, in the poset $[0, \infty]$, the terminal object is ∞, as $x \leq \infty$ for all $x \in [0, \infty]$.
- For bouquets, the terminal object will be given by a single loop stationed at a vertex.
- In **Top**, the terminal object is the *one-point space*, that is, the one-point set $*$ equipped with the indiscrete topology (described explicitly in chapter 4); note that if X is any space, then there will be only one possible continuous function from X to this one-point space.
- In **Group**, the terminal object is given by the *group with one element* $\{e\}$; note that if G is any group, then there will be only one possible group homomorphism $G \to \{e\}$.

Example 60 Suppose we instead use a *discrete diagram*, where this is one that consists of just objects (dots), with only identity morphisms. Taking the limit of such a diagram yields some very familiar "product-like" constructions, when specialized to familiar categories. Again, products need not exist—for instance, in particular, not every poset has all products. For some concrete examples of this construction, we have the following:

- In **Set**, the limit of a discrete diagram consisting of sets $X_1, X_2, \ldots$, is the *Cartesian product* $\prod_{i \in I} X_i$, where this construction comes with projection maps $\prod_{i \in I} X_i \to X_i$ onto each of the factors.
- In a poset $\mathcal{P}$, the limit of a discrete diagram (set of elements $p_1, p_2, \ldots$) is their *infimum* (or *greatest lower bound*) $\bigwedge_{i \in I} p_i$.[54]
- In **Top**, the limit is given by the Cartesian product $\prod_{i \in I} X_i$ equipped with the product topology.

Example 61 Suppose we have a diagram of shape

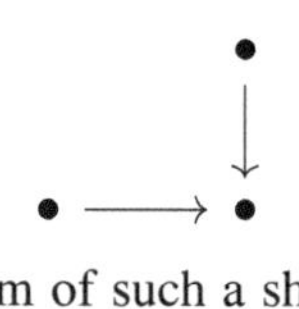

In a category **C**, the limit of a diagram of such a shape is called the *pullback* (or *fibered product*)—this will consist of an object together with morphisms satisfying the stipulated universal property. Explicitly, we can define this as follows:

Definition 62 Given any two maps with common codomain, as in

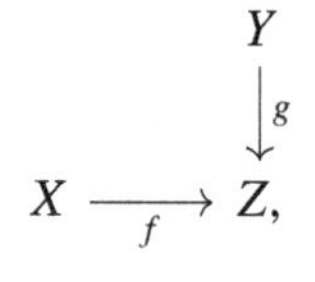

by their *pullback* (or *fibered product*) we mean a pair of maps π_0, π_1 with common domain P that

54. In general, in the context of posets, given a subset $S \subseteq P$, we say that $p \in P$ is a *lower bound* for S if for all $s \in S$, $p \leq s$. The *infimum* inf(S) of S, provided it exists, is then an element p such that (1) p is a lower bound for S and (2) if p' if another lower bound for S, then $p' \leq p$. It is common to write $\bigwedge_{a \in A} a$ for inf(A); we also use $\wedge$, especially when considering the infimum of a pair of elements x and y, written $x \wedge y$, where this is referred to as the *meet* of x and y.

1. forms a commutative square $f \circ \pi_0 = g \circ \pi_1$,

$$\begin{array}{ccc} P & \xrightarrow{\pi_1} & Y \\ {\scriptstyle \pi_0}\downarrow & & \downarrow{\scriptstyle g} \\ X & \xrightarrow[f]{} & Z \end{array}$$

and also

2. is universal among all such commutative squares, that is, for any T, x, y

$$\begin{array}{ccc} T & \xrightarrow{y} & Y \\ {\scriptstyle x}\downarrow & & \downarrow{\scriptstyle g} \\ X & \xrightarrow[f]{} & Z \end{array}$$

if $f \circ x = g \circ y$, then there exists a unique map $k : T \to P$ such that $x = \pi_0 \circ k$ and $y = \pi_1 \circ k$, as in

$$\begin{array}{ccc} P & \xrightarrow{\pi_1} & Y \\ {\scriptstyle \pi_0}\downarrow & & \downarrow{\scriptstyle g} \\ X & \xrightarrow[f]{} & Z. \end{array}$$

(with $T \xrightarrow{y} Y$, $T \xrightarrow{x} X$, and $T \overset{k}{\dashrightarrow} P$)

In a poset $\mathscr{P}$, the pullback of a diagram

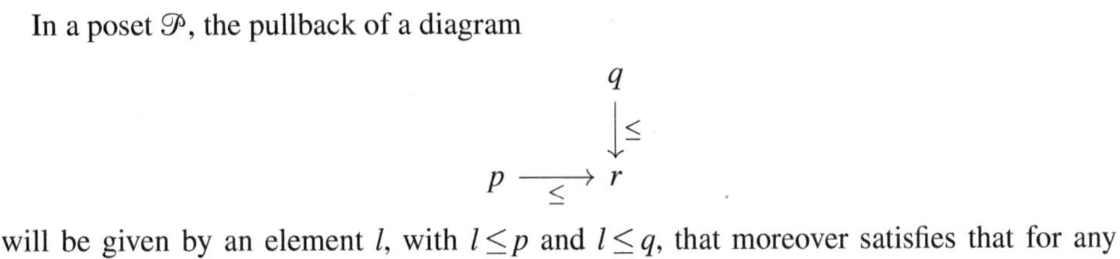

will be given by an element l, with $l \leq p$ and $l \leq q$, that moreover satisfies that for any element s for which it is also the case that $s \leq p$ and $s \leq q$, we have $s \leq l$.

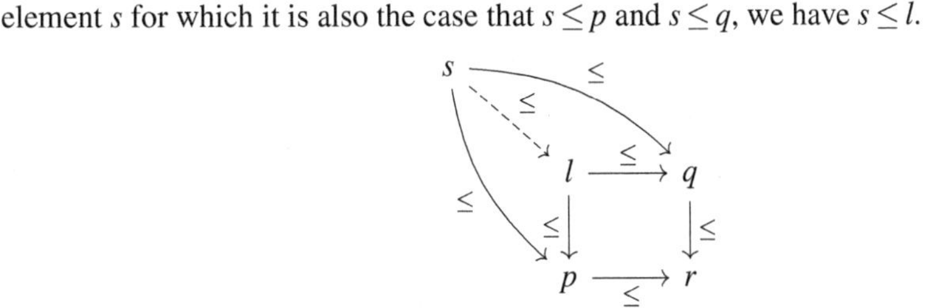

In other words, l will be the *greatest lower bound* of p and q.

In a poset, it turns out that the pullback reduces to the product. However, in more general categories, the two need not coincide.

Working in **Set**, **Top**, **Group**, in particular, the pullback will consist of (1) the subset (or subspace, or subgroup) of the product $X \times Y$ that comprises pairs (x, y) such that $f(x) = g(y)$, together with (2) two projection morphisms $X \times_Z Y \to X$ and $X \times_Z Y \to Y$, the first mapping (x, y) onto the first factor x, and the second onto the second factor y. What, then, will this pullback be like? The short answer is that it depends not just on the category we are working within but also on how the data of this category is defined: in particular, on what the given objects are and how the morphisms are defined. By adopting certain objects, or choosing certain given morphisms, this construction recovers a number of other familiar constructions. For instance, even just confining our attention to **Set**, the following can all be constructed using the pullback construction:

- Suppose $Z = *$, the singleton set. Then, as $*$ is the terminal object in **Set**, both $X \xrightarrow{f} *$ and $Y \xrightarrow{g} *$ are the unique functions taking everything to $*$. The pullback of such a diagram would then be all pairs (x, y) such that both x and y are sent to $*$ under the (unique) functions f and g—but there is no pair that does not satisfy this requirement. Thus, we recover *all* pairs (x, y), making the pullback of such a diagram the entire set $X \times Y$, the usual binary Cartesian product.
- Now suppose $Y = *$, while Z is any set. A function $* \xrightarrow{g} Z$ just picks out an element $z \in Z$. Then, for any function $X \xrightarrow{f} Z$, the pullback will be the subset of elements in X that are sent to z by f, recovering the usual *preimage* (or *fiber*) of f over g construction.
- Now suppose that X and Y are subsets of Z, making f and g inclusions,

$$\begin{array}{ccc} & & Y \\ & & \big\downarrow g \\ X & \xhookrightarrow[f]{} & Z \end{array}$$

  In such a case, the pullback will consist of pairs (x, y) such that x and y are *equal* on being included into Z. In other words, the pullback will consist of the elements $x = y$ of X that are also in Y (and vice versa)—and this is just the *intersection* $X \cap Y$.

Example 63 Suppose we have a diagram of shape

$$\bullet \longleftarrow \bullet \longleftarrow \bullet \longleftarrow \bullet \cdots$$

In a category **C**, the limit of such a diagram is called an *inverse limit*. It will comprise an object together with morphisms from that object to each $\bullet$ such that each of the resulting triangles commutes, and the universal property of the limit is satisfied.

Explicitly, supposing $X_1, X_2, \ldots$ are objects in **C**, then the inverse limit, denoted $\varprojlim X_i$, of the diagram

$$X_1 \xleftarrow[f_1]{} X_2 \xleftarrow[f_2]{} X_3 \xleftarrow[f_3]{} X_4 \cdots$$

would amount to an object with maps into each of the X_i, making the resulting triangles commute. In **Set**, for instance, this would be a subset $\varprojlim X_i$ of the product $\prod_{i \in I} X_i$ containing all sequences $(x_1, x_2, x_3, \ldots)$ where the i-th factor is such that $f_i(x_{i+1}) = x_i$.

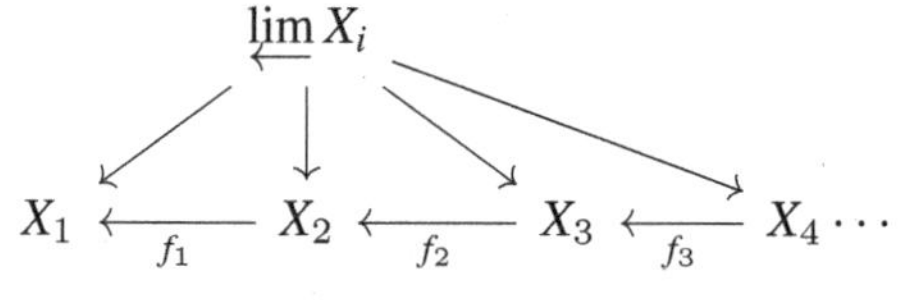

In **Top** and **Group**, among other categories, the inverse limit would also look just like this. Note that in **Set**, if we have sets related by inclusion, so that $X_1 \supseteq X_2 \supseteq X_3 \supseteq X_4 \supseteq \cdots$, then the inverse limit is just the intersection $\bigcap_{i \in I} X_i$.

Example 64 Suppose we have a diagram indexed by the category consisting of two objects and two parallel nonidentity morphisms,

$$\bullet \rightrightarrows \bullet$$

Of course, a diagram in **C** of this shape will just be a parallel pair of morphisms

$$X \overset{f}{\underset{g}{\rightrightarrows}} Y$$

living in the target category **C**. What is a cone over this diagram? Well, a cone with summit C will consist of a pair of morphisms $h: C \to X$ and $i: C \to Y$ such that $f \circ h = i$ and $g \circ h = i$, which together just assert that $f \circ h = g \circ h$. In short, a cone over such a parallel pair of arrows f, g can be represented by a single morphism $h: C \to X$ such that $f \circ h = g \circ h$. We can then define E together with $e: E \to X$, called the *equalizer of* f and g, as the universal arrow with this same property, that is, $f \circ e = g \circ e$. To be more explicit, the universal property in question asserts that given any $h: C \to X$ such that $f \circ h = g \circ h$, there exists a unique $k: C \to E$ that factors the morphism h through e in the sense that $e \circ k = h$, as summed up in the diagram

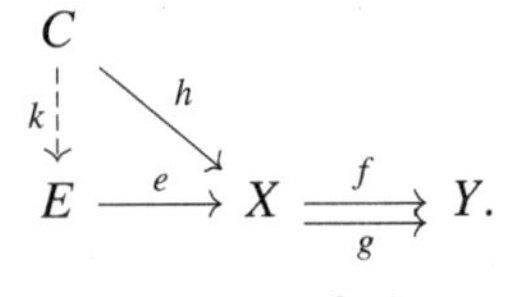

In **Set**, the equalizer of f, g is then a subset of elements of X for which the two given functions coincide,

$$E = Eq(f, g) := \{x \in X \mid f(x) = g(x)\},$$

which set is accompanied by the natural *inclusion* $e: Eq(f, g) \to X$ into X.

In terms of graphs, since a graph can be defined using a pair of functions $A \overset{s}{\underset{t}{\rightrightarrows}} V$, where A stands for arrows and V for vertices, and where s just picks out the source vertex of an arrow and t the target vertex, consider that for each graph G we can find its set of length one loops via the equalizer construction $Eq(G)$:

$$Eq(s, t) \xrightarrow{e} A \overset{s}{\underset{t}{\rightrightarrows}} V.$$

Observe that the equalizer assignment is in fact functorial, thus incidentally furnishing us with another example of a functor, since given a graph homomorphism $G \to G'$, there will be an induced function $Eq(G) \to Eq(G')$.

Let us also take this opportunity to define the following notion, capturing when arrows can be canceled on one side:

Definition 65 A morphism $i: B \to C$ in a category is called a *monomorphism* (or *monic* morphism) provided for any A with parallel morphisms f, g as in

$$A \overset{f}{\underset{g}{\rightrightarrows}} B \xrightarrow{i} C,$$

$i \circ f = i \circ g$ implies $f = g$.[55]

55. This is a categorical generalization of the set theoretic notion of an injective function.

Observe that an equalizer of two morphisms is automatically a monomorphism. To emphasize that a particular morphism is a monomorphism, it is common to use the decorated arrow $\rightarrowtail$ (or $\hookrightarrow$ when it is an inclusion).

Certain conditions can ensure that a given mapping is a monomorphism. One such condition is that the mapping has something called a *retraction*, another useful notion we record here (together with its dual notion of a *section*):

Definition 66 For mappings r, s in any category, r is said to be a *retraction* for s provided $r \circ s$ is an identity mapping. In such a situation, s is said to be a *section* for r.

So if i is a morphism from X to Y, r is a retraction provided we have

$$X \underset{i}{\overset{r}{\rightleftarrows}} Y \qquad \text{with } r \circ i = \mathrm{id}_X.$$

A given i may or may not have a retraction, and if there is such a map, there may in fact be many retractions. Likewise, a given morphism need not have sections. However, if there is at least one retraction r for a given i, then i will be a monomorphism.

Finally, observe also that any section will be a monomorphism. A section for a mapping f might be thought of as a procedure that picks out an element from each of the fibers of f. We will have more to say on this in subsequent chapters.

3.1.3 Examples of Colimits

Limits and colimits are dual notions, meaning that colimits in a category are just limits in the opposite category, so by dualizing the above constructions (same definitions, but with all arrows reversed), we can thus expect to get specific examples of colimits. These will include: initial objects, coproducts (disjoint unions), pushouts, direct limits (denoted $\varinjlim$), and coequalizers. We will focus on explicitly constructing just a few of these.

Example 67 First consider the coproduct. This is the colimit of the discrete diagram of shape

$$\bullet \qquad \bullet$$

and is usually written $\coprod$, so that with a diagram with objects $X(i)$ and $X(j)$, the colimit of such a diagram is written $X(i) \coprod X(j)$. This binary case can be generalized to more objects, $\coprod_i X(i)$. For some concrete examples of this construction, we have the following:

- In **Set**, the colimit of a discrete diagram consisting of sets $X_1, X_2, \ldots,$ is the *disjoint union* $\bigsqcup_{i \in I} X_i$, where this construction comes with injective functions of each X_i into the coproduct set $X = \bigsqcup_{i \in I} X_i$ such that each element of X belongs to exactly one of the images of the injections. If the sets being summed are pairwise disjoint, the disjoint union just becomes the standard union $\bigcup$.
- In a poset $\mathcal{P}$, the colimit of a discrete diagram (set of elements $p_1, p_2, \ldots$) is their *supremum* (or *least upper bound*) $\bigvee_{i \in I} p_i$, where this construction also consists of the

inequalities $p_i \leq \bigvee_{i \in I} p_i$.[56] Guided by this, and considering the importance of order theory as a microcosm for the more general categorical notions, one can see colimits as a generalization of suprema or joins (just as limits generalize infima or meets).

Example 68 Dual to inverse limits are *direct limits*, where this is the colimit of a diagram indexed by the ordinal category ω. In other words, for a diagram

$$X_1 \longrightarrow X_2 \longrightarrow X_3 \longrightarrow X_4 \longrightarrow \cdots$$

its colimit is the direct limit $\varinjlim X_n$, defining a diagram of shape $\omega + 1$:

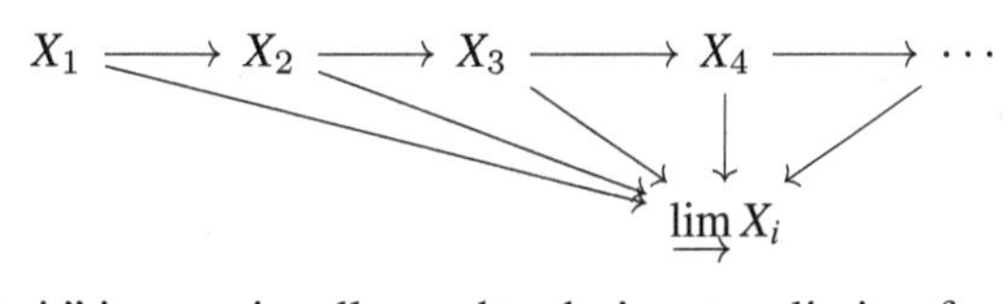

The term "direct limit" is occasionally used to designate colimits of any shape.

Observe then that the colimit of a sequence of sets with the inclusions

$$X_0 \hookrightarrow X_1 \hookrightarrow X_2 \hookrightarrow \cdots$$

recovers their union $\bigcup_{n \geq 0} X_n$.

Example 69 We have already met the construction called an *equalizer*, where this is the limit of a diagram of shape

$$\bullet \rightrightarrows \bullet.$$

In a similar fashion, we can form the dual notion of a *coequalizer* by taking the colimit of the diagram consisting of two objects X, Y and two parallel morphisms $f, g : X \to Y$, which gives us something like the categorical generalization of taking a quotient by an equivalence relation. More explicitly, a coequalizer is defined as an object Q (sometimes denoted $Coeq(f, g)$, wanting to stress the arrows it coequalizes) together with a morphism $q : Y \to Q$ such that $q \circ f = q \circ g$, where the pair (Q, q) must moreover be universal in the sense that given any other pair (Q', q') there exists a unique morphism $u : Q \to Q'$ so that $u \circ q = q'$, as in the diagram

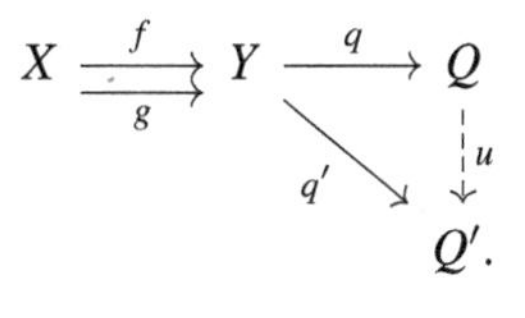

In **Set**, the coequalizer of two functions $f, g : X \to Y$ is just the *quotient* of Y by the smallest equivalence relation $\sim$ such that for all $x \in X, f(x) \sim g(x)$.

56. In general, just as with the notion of greatest lower bounds, given a subset $S \subseteq P$ of a poset, we say that $p \in P$ is an *upper bound* for S if for all $s \in S$, $s \leq p$. The *supremum* sup(S) of S, provided it exists, is then an element p such that (1) p is an upper bound for S and (2) if p' if another upper bound for S, then $p \leq p'$. It is common to write $\bigvee_{a \in A} a$ for sup(A); we also use $\vee$, especially when considering the supremum of a pair of elements x and y, written $x \vee y$, where this is referred to as the *join* of x and y. Incidentally, we can use this notion (together with that of meets) to define an important entity we will meet throughout the book: a *lattice* is a poset for which every pair of elements has a join and a meet. We will wait until chapter 7 to more formally introduce lattices and explore them in greater depth.

In the context of bouquets, the coequalizer of two inclusion morphisms of a vertex into an object consisting of individual loops stationed at different vertices would be given by an object that glued those loops together onto a single (the same) vertex—an example that makes it evident how colimits can be seen as involving some sort of gluing.

Various categories—including graphs, reflexive graphs, discrete dynamical systems, the category of elements (see below, definition 71)—support a construction that allows us to count the *connected components* in that category. For concreteness, we stick for now with the case of the category of graphs, **Grph**, and consider the "connected components" functor $\Pi_0 : \mathbf{Grph} \to \mathbf{Set}$. This is actually obtained via the coequalizer construction:

$$A \underset{s}{\overset{t}{\rightrightarrows}} V \overset{q}{\dashrightarrow} Coeq(s,t).$$

$Coeq(s,t)$ is defined as $V/\sim$, where $\sim$ is an equivalence relation on V, that is, $s(x) \sim t(x)$ for all $x \in A$, and where q is the quotient function $q : V \to V/\sim$. This construction accordingly acts to identify all arrows where the source of one arrow is equal to the target of the other. In other words, all we are doing is picking out the connected components of the graph. This assignment of the set of connected components of a graph can be shown to be functorial as well. In a moment, we will see more of this construction in action.

Finally, let us also take the opportunity to define the dual of the notion of a monomorphism.

Definition 70 $f : X \to Y$ is said to be an *epimorphism* (or *epi* for short) if for all B, $h, h' : Y \to B$,

$$X \xrightarrow{f} Y \underset{h'}{\overset{h}{\rightrightarrows}} B$$

$h \circ f = h' \circ f$ implies $h = h'$.

Observe that every coequalizer is automatically an epimorphism, and that moreover every retraction is automatically an epimorphism. For graphical emphasis, epimorphisms are sometimes displayed using $\twoheadrightarrow$.

Before moving on, let us now consider an alternative way to view things.

Definition 71 Let **C** be a category, and let $F : \mathbf{C} \to \mathbf{Set}$ be a (covariant) functor. Then the *category of elements of* F, denoted $\int_{\mathbf{C}} F$ (or just $\int F$ if the context is clear), is defined:

$$\mathrm{Ob}\left(\int F\right) = \{(c,x) \mid c \in \mathbf{C}, x \in F(c)\}$$

$$\mathrm{Hom}_{\int F}\big((c,x),(c',x')\big) = \{f : c \to c' \mid F(f)(x) = x'\}.$$

Similarly for the contravariant case: for $F : \mathbf{C}^{op} \to \mathbf{Set}$ the *category of elements of* F, denoted $\int_{\mathbf{C}^{op}} F$ (or just $\int F$), is defined:

$$\mathrm{Ob}\left(\int F\right) = \{(c,x) \mid c \in \mathbf{C}, x \in F(c)\}$$

$$\mathrm{Hom}_{\int F}\big((c,x),(c',x')\big) = \{f : c \to c' \mid F(f)(x') = x\}.$$

Associated with these constructions are the functors $\pi_F : \int F \to \mathbf{C}$, called the projection functors, sending each object $(c,x) \in \mathrm{Ob}(\int F)$ to the object $c \in \mathrm{Ob}(\mathbf{C})$ or $\mathrm{Ob}(\mathbf{C}^{op})$, and each morphism $f : (c,x) \to (c',x')$ to the morphism $f : c \to c'$, that is, $\pi(f,(c,x),(c',x')) = f$.

As a concrete instance of this, recall the vertex coloring functor *nColor* from example 37 (chapter 2). An object in the category of elements $\int nColor$ of this functor *nColor* will be a graph together with a chosen n-coloring, that is, objects are n-colored graphs. A morphism $\phi : G \to G'$ between a pair of n-colored graphs will be a graph homomorphism $\phi : G \to G'$ so that the induced function $nColor(\phi) : nColor(G') \to nColor(G)$ takes the chosen coloring of G' to the chosen coloring of G, that is, the graph homomorphism ϕ will preserve the chosen colorings in the sense that each red vertex of G will be carried to a red vertex of G'. In short, then, $\int nColor$ is the category of n-colored graphs and the color-preserving graph homomorphisms between them.

For another example, recall the hom-functors, first introduced in example 42 (chapter 2). Objects in the category of elements of $\mathrm{Hom}_{\mathbf{C}}(c, -)$ are the morphisms $f : c \to d$ in $\mathbf{C}$. A morphism from $f : c \to d$ to $g : c \to e$ is then a morphism $h : d \to e$ such that $g = h \circ f$. h is said to be a morphism *under* c because of the diagram attached to this condition:

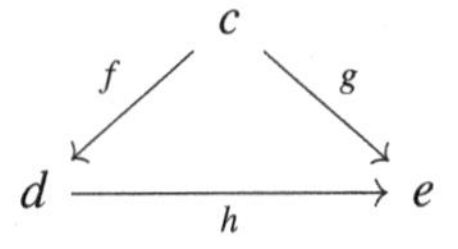

This category is none other than the *co-slice category* of objects *under* the $c \in \mathbf{C}$. Note that the forgetful functor $U : c/\mathbf{C} \to \mathbf{C}$ sends an object $f : c \to d$ to the codomain, and takes a morphism (a commutative triangle) to the arrow opposite the object c, that is, to h in the above instance. We could also construct the dual category of elements $\int \mathrm{Hom}_{\mathbf{C}}(-, c)$ in terms of the *slice category* $\mathbf{C}/c$ *over* the object $c \in \mathbf{C}$.[57]

The category of elements is rather significant because any universal property can be seen as defining an initial or terminal object in this category. In particular, it turns out that for any small functor[58] $F : \mathbf{C} \to \mathbf{Set}$, we have

$$\mathrm{colim}\, F \cong \Pi_0\left(\int F\right),$$

where Π_0 operates by picking out the connected components and $\int F$ is the category of elements of F. So, in other words, the set of connected components of the category of elements of a functor F, $\Pi_0(\int F)$, is isomorphic to the colimit of F. Alternatively, in general, we could just have said that a colimit is an initial object in the category $\int Cone(F, -)$, and we note that the forgetful functor $\int Cone(F, -) \to \mathbf{C}$ will take a cone to its nadir.

To see this in action, recall the functor (diagram)

57. In this connection, we can mention the important result that for $\mathbf{C}$ small and P a presheaf on $\mathbf{C}$, one can show an equivalence of categories

$$\mathbf{Set}^{\mathbf{C}^{op}}/P \simeq \mathbf{Set}^{(\int_{\mathbf{C}} P)^{op}}.$$

58. A functor or diagram is small if its indexing category is small.

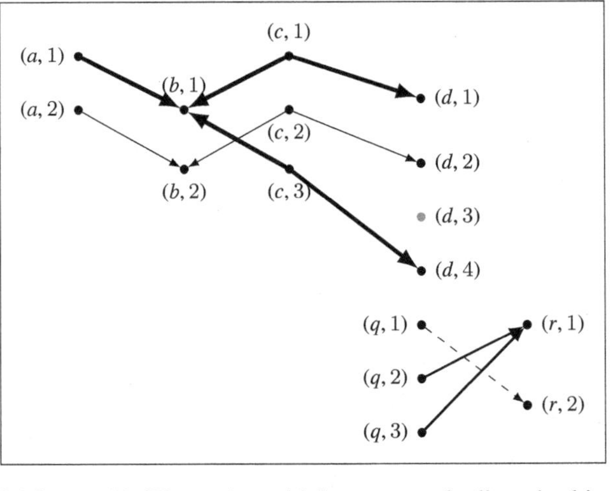

from example 36 (chapter 2). The various thicknesses or shadings in this picture can now be explained. The picture above is in fact a representation of the category of elements of F. The various thicknesses or shadings depict the result of taking its connected components. By inspection, one can verify that this is just the set

$$\{[(a,1)],[(a,2)],[(d,3)],[(q,1)],[(q,2)]\},$$

where each element is a representative of one of the components, which is in turn isomorphic to a set of cardinality 5, entailing that $\mathrm{colim}_{\mathbf{J}}F \cong$ a set with 5 elements.

As for limits, we could also show that the *limit* of any small functor $F:\mathbf{C}\to\mathbf{Set}$ is isomorphic to the set of functors $\mathbf{C}\to\int F$ that define a *section* to the canonical projection $\pi:\int F\to\mathbf{C}$. Alternatively, we can define the limit as a terminal object in the category of elements of cones over F, that is, in $\int Cone(-,F)$. Note also that the forgetful functor $\int Cone(-,F)\to\mathbf{C}$ will send a given cone to its summit.

3.1.4 Further Notions: (Co)Complete, (Co)Continuous

We have seen how both the limiting and the colimiting cones are universal in the sense of acting as a kind of doorkeeper or special mediator for all other cones. Such universal objects need not exist. However, we define a category $\mathbf{C}$ as *complete* if it admits limits of all small diagrams valued in $\mathbf{C}$, and as *cocomplete* if it admits all colimits of all small diagrams valued in $\mathbf{C}$.

We should also take the opportunity to supply the important definition:

Definition 72 A functor is *(co)continuous* if it preserves all small (co)limits,

where *preservation* is defined as follows: for any class of diagrams $K:\mathbf{J}\to\mathbf{C}$ valued in $\mathbf{C}$, a functor $F:\mathbf{C}\to\mathbf{D}$ is said to *preserve* limits if for any diagram K and limit cone over K, the image of this cone under the action of the functor defines a limit cone over the (composite) diagram $F\circ K:\mathbf{J}\to\mathbf{D}$.[59]

59. Importantly, in chapter 6 on the Yoneda results, we will learn about *representable functors*, and how covariant representable functors preserve all limits, taking limits in $\mathbf{C}$ to limits in $\mathbf{Set}$; while contravariant representable functors preserve all limits in $\mathbf{C}^{op}$, taking colimits in $\mathbf{C}$ to limits in $\mathbf{Set}$. We will learn about something called the Yoneda embedding $\mathbf{y}:\mathbf{C}\hookrightarrow\mathbf{Set}^{\mathbf{C}^{op}}$, which preserves all limits that exist in $\mathbf{C}$; and the dual embedding $\mathbf{y}:\mathbf{C}^{op}\hookrightarrow$

Intuitively, this concept of a (co)continuous functor as a special sort of functor that takes universal objects in the source category to universal objects in the target category can be thought of as follows: whatever else the functor does to objects as it sends them from one category to another, a (co)continuous functor will send the entity that acted as a privileged mediator or doorkeeper ((co)limit) in relation to the rest of the objects of the source category to an entity that plays the similar role of privileged mediator or doorkeeper for its fellows in the target category. We could label this characterization of continuity via the preservation of the roles of privileged mediators or doorkeepers/gatekeepers as *katholic continuity* (after the Greek word *katholou* meaning "universal," and perhaps also evoking connections with *katechon*, sometimes identified in theological contexts with the figure of the gatekeeper who indirectly enforces lawfulness by restraining chaos or lawlessness).

3.2 Philosophical Pass: Universality and Mediation

The katholic understanding of (co)continuity as a preservation of (co)limits is very important for certain definitions of sheaves to follow. The pivotal notion of a *(co)continuous functor* as a special sort of functor that takes universal objects in the source category to universal objects in the target category, moreover enables us to regard *continuity* as a special kind of passage or translation from one world of objects and relations to another world of objects and relations—*special* in that, in passing between worlds, it takes care to preserve the role of those that act as privileged or optimal mediators or gatekeepers for the rest of the entities of their world. Finally, the universal constructions introduced in this chapter seem to be of some philosophical interest in their own right in light of the particular conception of universality that is ushered in.

Box 3.1
On Universality

One could argue that basic category theory is the study of universal properties. Within a category, for some structure of a given kind, there may indeed be a family of such structures, that is, a number of objects of the category related in such a way that they exhibit the structure in question. A very natural question to ask, in the presence of a family of structures of the same kind, is: *Which is the optimal (most efficient) one?* (And what does "optimal" here mean?) It is here that universal properties come into play. Broadly, universality has been understood here to mean that a certain gadget occupies a privileged position in relation to other gadgets of its world that are of the same type, in that it serves as a special gatekeeper or go-between: "You have to pass (factor) through me if you want to relate in this way to anything else of this type." If such universality is thus thought of in terms of a designated object's privileged role as mediator or gatekeeper for all other objects of the same type trying to relate or interact, katholic continuity can moreover be understood as explaining how, in passing from one network of objects and relations to another, those objects that occupy the position of mediators *preserve* their roles within their respective networks; that is, in passing from one world to another, the special mediator or gatekeeper in the one world is sent, or given a direct line, to the special mediator or gatekeeper in the other world.

$\mathbf{Set}^{\mathbf{C}}$ that preserves limits in $\mathbf{C}^{op}$. For our purposes, we can highlight that it will turn out that a sheaf is a special sort of presheaf that preserves limits in this way.

It is curious that, long before category theory, a number of philosophers—such as Aristotle, Hegel, and Charles Peirce—suggested, and attempted to think through, close connections between universality, mediation, and continuity. Charles Peirce was perhaps the most insistent on the connections between the general (universality) and mediation, brought together in his notion of "thirdness," one of his three "categories" (in his own sense of the word, having nothing to do with category theory). Peirce was a systematic philosopher who developed a theory of three main categories, first designed to help develop his theory of signs (semiotics) and classify the sciences and human knowledge, but gradually extended to be of grander and grander scope. Peirce came to have rather ambitious aspirations for these categories, such that, as he thought, they furnished a classification applicable to systems of all sorts, so that ultimately all conceptions and phenomena—even elements of cosmology and physiology—at the most fundamental level could be broken down in terms of these three general categories and their interplay. As such, Peirce would come to admit that these three categories "are excessively general ideas, so very uncommonly general that it is far from easy to get any but a vague apprehension of their meaning" (Peirce 1997, 4.3). Roughly, we could describe these three categories as follows:

- *Firstness*: being considered simply in itself, or independent of anything else, as a *unit*—this involves immediacy and a kind of naive realism (insofar as it relates to perceivers and knowers like us); the paradigm is brute quality "free of relations";
- *Secondness*: being correlative to, dependent on, an effect or result of, in reaction with, or limited by something else—this involves dyadic relations corresponding to some "brute action" or finite process of reaction or resistance;
- *Thirdness*: involving mediation bringing something into relation to another. (See, for instance, Peirce (1997, 6.32 [1891]).)

There is an element of givenness or sheer fact to phenomena that fall within the scope of either Firstness and Secondness. Secondness is a dyadic relation that is irreducible to a single part. Thirdness, for its part, is generally some ternary relation that (provided it is a "genuine Third") cannot be reduced to two terms, to two-part relations. (Peirce isolated two sorts of "degenerate Thirds," which could indeed be reduced to other categories; see, for instance, "A Guess at the Riddle" [Peirce 1997, 1.3.3].) Peirce held that all other more complex relations could be reduced ultimately to combinations of triads. While, again, the true scope of applicability of the categories was meant to be extremely broad, in terms of Thirdness in particular, Peirce would speak of a Third as "every kind of sign, representative, or deputy, everything which for any purpose stands instead of something else, whatever is helpful, or mediates." For Peirce, in short, "Thirdness is nothing but the character of an object which embodies Betweenness or Mediation in its simplest and most rudimentary form; and I use it as the name of that element of the phenomenon which is predominant wherever Mediation is predominant" (Peirce 1997, 5.77 [1903]).

Here is an initial example, of a somewhat phenomenological flavor, that might help to start to give some shape to these notions:

- Firstness: the visible light of the sun;
- Secondness: a child turning its eyes to the sun and the surprise and pain that is the event of being hurt;
- Thirdness: a child watching someone else look at the sun and forming the judgment, "Oh, that must have hurt!"

As such an example is meant to suggest, the "theory of mind" and empathy involved in what is described as Thirdness above is not reducible to Secondness—the child itself being painfully affected by the light and the other person themself being painfully affected by the light. Taking either of those individually, or even simply adding them together, cannot hope to account for anything involved in the capacity for forming the empathic judgment.

An example often used by Peirce to better illustrate Thirdness and its distinction from Secondness in particular is found in the act of *giving*:

> Analyze for instance the relation involved in "*A gives B to C*." Now what is giving? It does not consist in A's putting B away from him and C's subsequently taking B up. It is not necessary that any material transfer should take place. It consists in A's making C the possessor according to Law. There must be some kind of law before there can be any kind of giving—be it but the law of the strongest. But now suppose that giving did consist merely in A's laying down the B which C subsequently picks up. That would be a degenerate form of Thirdness in which the thirdness is externally appended. In A's putting away B, there is no thirdness. In C's taking B, there is no thirdness. But if you say that these two acts constitute a single operation by virtue of the identity of the B, you transcend the mere brute fact. (Peirce 1997, 8.331 [Letter to Lady Welby])

As he would stress elsewhere,

> [W]e cannot build up the fact that A presents C to B by any aggregate of dual relations between A and B, B and C, and C and A. A may enrich B, B may receive C, and A may part with C, and yet A need not necessarily *give* C to B. For that, it would be necessary that these three dual relations should not only coexist, but be welded into one fact. (Peirce 1997, 1.3 ["A Guess at the Riddle"])

Still another related example might be found in that of a legal contract. Such a thing cannot be accounted for just by the combination of two dyadic relations: the first being A's signature on document C and the second being B's signature on document C. As Peirce would stress, the nature of such a contract in fact lies in the intent and existence of the contract, which amounts to certain conditional rules governing the future behavior of A and B (see Peirce 1997, 1.475 [ca. 1896]), where this is not reducible to the component dyads (signatures), but amounts to a bringing together of these two dyads into a relationship binding for the future.

To stress another aspect of the difference between Second and Third, Peirce sometimes used the example of a jurisdiction's law enforcement. Here, an uninterpreted feeling of fear might be First, and the law court's injunctions and judgments would involve Thirdness. But once "I feel the sheriff's hand on my shoulder, I shall begin to have a sense of actuality" (Peirce 1997, 1.24 [1931–1958])—here we have Secondness. Whereas action confronting one with another is Secondness, Thirdness is to be understood as involving rule-governed conduct, predictions on future behavior, and habit-formation legislating over potential actions and interactions. As such, the lawfulness of any phenomenon exhibiting Thirdness has a decidedly *general* flavor.

Peirce would come to think of Thirdness as present in any case of generality (universality). Strictly speaking, both Firstness and Thirdness involve elements of generality, yet of a different sort: Firstness involves a latent potentiality, a sort of vague generality, the generality of qualitative immediacy and the indeterminacy involved in this. Peirce calls the generality of Firstness *negative generality*. The generality of Thirdness, on the other hand, is one of *necessity*, which is why Peirce made Thirdness the domain of law of nature or rule—this generality of Thirdness is called *positive generality*.

While there is a great deal more that could be said of all this, I will just also note that it is curious that, for Peirce, the sort of generality of Thirdness was, fundamentally, "nothing but a rudimentary form of true continuity," forming a basis for the most demanding conception of continuity, whose articulation he struggled with for most of his life. (See also Peirce's remark that continuity "represents Thirdness almost to perfection" [Peirce 1997, 1.337]. Zalamea [2012] has a number of fascinating discussions of Peirce's evolving understanding of continuity, which the reader is encouraged to consult.)

While, of course, there is no expectation or desire here of trying to map the categories of Peirce's classification onto the mathematics we have been developing, I think the connections between *generality* (of which universality is an extreme form) and *mediation*, as intimated by Peirce, provide an interesting perspective from which to view the category-theoretic formulation of universality in terms of this special mediator or gatekeeper property. It is interesting, as well, that in making this connection more exact, continuity (of the katholic sort) also reemerges in a decisive way. At another level, I think it is also useful to consider how, for a given category, it does often appear that the objects of a category are something like Firsts, while the morphisms are plausibly something like Seconds—where there is even an element of *givenness* to these ("these are just the objects, the units; and these are just how they happen to relate or act on one another"). Universal constructions—limits and colimits—for their part, do seem to fall into something like Thirdness: they cannot be reduced to a single object or isolated morphism relating two objects, but would seem to essentially involve an act of mediation relating or factoring any other morphisms or patterns to others within the category. And it is with such constructions, perhaps more than anywhere else, that the "law-governedness" of a category often emerges.

Altogether, with the category-theoretic notions of (co)limits, we are given a more precise way of thinking about universality as a form of *mediation*—an *optimal* mediation. With the category-theoretic notion of (co)continuous functors, built out of the material of such universal constructions, we are given a more precise way of thinking about continuity in terms of a process or transformation that preserves the special role played by those optimal mediators or gatekeepers.

4 Topology: A First Pass at Space

In which we cover all the elements of general (point-set) topology needed for sheaves on a topological space, introduce the relevant categories of interest, and raise some questions (properly addressed in the appendix) regarding some of the fundamental ingredients of topology.

The classical presentation of sheaves begins by looking at sheaves on topological spaces. As we will explore in the final chapters of the book, there are powerful elaborations of sheaves to more general settings than topological spaces. However, the right place to start to introduce the sheaf notion is with sheaves on topological spaces. While a few important topics of basic category theory remain to be covered, we already have the main category-theoretic ingredients for defining sheaves in the context of topological spaces. In order to take our first look at sheaves, then, all that remains is to cover the necessary elements of general topology. That is the task of the present chapter. In the next chapter, we take a first look at sheaves. Chapters 6 and 7 then return to two last important topics and ingredients of basic category theory, matters worth exploring in their own right but also ones that will allow us to significantly enrich the treatment of sheaves that will emerge in later chapters.

4.1 Motivation

The standard story told, when introducing point-set topology, presents the concepts of topology as abstractions of features of metric spaces, specifically those nearest and dearest to the heart of analysts: the Euclidean space $\mathbb{R}^k$, especially $\mathbb{R}$, the real line, and $\mathbb{R}^2$, the plane. It is true that point-set topology has its roots in analysis, and while today we develop these notions on their own terms, historically the staples of topology developed in order to meet the needs of analysis, not as part of a separate subject. A metric space is basically just a set of points and a relation on those points that acts as a certain quantitative measure of the degree of closeness or nearness of pairs of points. Equipped with a notion of how far any two points are, the decisive concerns of the analyst—such as continuity and approximation, among others—can be easily characterized. But as far as continuity and approximation (together with a number of other key notions of analysis) are concerned, one is never really concerned with points per se, but rather with regions that include everything nearby a point of interest. For now, think of these as *regions of approximation*.

In the general algebra of sets, we work with the set-theoretic operations of union, intersection, and complementation—the fundamental operations for relating and combining existing sets, allowing us to generate new ones from old. By attending to the behavior of the regions of approximation in relation to one another, as we take their union and intersection, it can be observed that these objects—whatever their shape or whatever notion of distance we use—always behave in a characteristic way: they are stable under intersection and union, but in an asymmetrical fashion as concerns the finiteness of these operations.

The passage from metric spaces to topological spaces in general was largely aided by the realization that if we get rid of the distance function that helps define a metric space and gives us our very sense of nearness but retain the properties concerning how the pertinent objects are stable under intersection and union (in an asymmetrical way with respect to finiteness), not only can we recover the same metric notions, but we are left with a new and more general notion that can capture a wide variety of further structures of interest.

Given a set, we can observe certain things about it, as far as its constituent parts are concerned—for instance, we can look at its cardinality: How *many* elements are in it? But we might care less about something like the number of parts and more about the relations between the parts. In asking questions about the interrelations between the parts of a set, we would appear to need an answer to the questions:

- What kind of parts do we allow?
- What sorts of relations between those parts are allowed?

As it turns out, in the more general setting, questions about what kind of objects we have (e.g., how the regions of approximation are characterized) can be reduced to specifying how they relate to one another. A topology fundamentally consists of a collection of subsets of a set X, a certain structure endowing the constituent objects with some coherence, meaning that it makes sense to determine when things are nearby or close together; this can be articulated entirely in terms of how these subsets satisfy certain conditions, specifically conditions on how the subsets relate to one another.

The standard story told to motivate general topological notions and arrive at the key axioms of a topology proceeds by

1. observing certain features of approximating objects native to Euclidean space; then
2. abstracting from Euclidean space to metric spaces in general; then
3. abstracting again from metric spaces to certain properties of general *open sets*, using these as the axioms determining a space in general.

The standard objects or parts we work with in topology are *open* and *closed* sets. The canonical example of an open set, from standard metric space settings, is an open interval or open disk—where this subset characteristically contains none of its boundary. The canonical example of a closed set is a closed interval or closed disk, where this by contrast contains all of its boundary. In general, when we take the intersection of two sets, say A and B, denoted $A \cap B$, we end up with the set that contains all elements of A that also belong to B (or, equivalently, all elements of B that also belong to A). Suppose we are working with some closed intervals. Clearly, the intersection of any closed intervals will itself be a closed set, and this can be extended to arbitrarily many such closed sets. By contrast, if we start with open intervals, notice how the intersection of any two open intervals will itself be an open interval—yet this *cannot* be extended to an arbitrary number of open intervals. After all, the intersection of an infinite number of shrinking open intervals centered on a point will just be the point itself—yet points are boundary-like objects and cannot be considered open without things becoming rather degenerate. In a similar way, we can consider what happens when we take unions of sets. Suppose we have open intervals. The union of any number of open intervals will itself be an open interval. But if we start instead with closed intervals, it is not the case that the union of an arbitrary number of closed intervals will be closed—so, to ensure that we are left with another closed set after we deploy the operation of union on closed sets, we must restrict the union to the finite case.

At bottom, these two dual conditions are what define a topology:

- for open sets: stability under finite intersection and arbitrary union;
- for closed sets: stability under arbitrary intersection and finite union.

Open and closed sets are related to one another in a particularly nice way, so in principle it does not really matter which sorts of object we use. Rather, the essence of the topology notion is the characteristic asymmetrical finiteness conditions on how subsets behave when unioned and intersected. These conditions on the interrelations of subsets are what ultimately constitute openness versus closedness, rather than any intrinsic property of some "open" or "closed" entity in itself.

There are aspects of the standard story—where topological spaces are presented as a natural generalization of conspicuous properties of especially familiar metric spaces—that can appear compelling to newcomer and seasoned mathematician alike. However, the confidence with which this story is typically told tends to leave unanswered—or, worse, even obfuscate—a number of legitimate and lingering questions. These important (if somewhat more philosophical) questions are raised at the end of this chapter, but treatment of them is relegated to the appendix, in order not to distract from the march onward toward sheaves. The appendix is perhaps best read after chapter 7 on adjunctions—or, if the reader is already familiar with such matters and is willing to postpone arriving at sheaves, after the present chapter.

In addition to being vital to the first presentation of the sheaf concept, general topology is one of the fundamental branches of modern mathematics, next only to set theory. The concepts one meets in point-set topology provide us with a framework for expressing ideas that extend to nearly all branches of mathematics. For these reasons, this chapter devotes considerable time to developing the core notions of general topology (henceforth, in this chapter, we'll just say "topology"), and refers to the appendix for a firmer understanding of the "essence" of topology by exploring potential answers to the three questions raised at the end of this chapter.

The rest of this chapter is structured as follows. Section 4.2 first motivates and describes, in a more informal fashion, via a dialogue, a number of the decisive concepts explored and dealt with in general topology. Section 4.3 then explores these in a more formal and detailed fashion, providing a self-contained account rooted in examples and worked-out exercises, of all that is needed from topology for the purposes of sheaf theory. The Philosophical Pass at the end of the chapter (section 4.4) raises three lingering questions about topological matters, questions whose treatment is relegated to the appendix.

4.2 A Dialogue Introducing the Key Notions of Topology

Suppose you have in front of you a sheet of paper on which a blot of ink has been dropped:

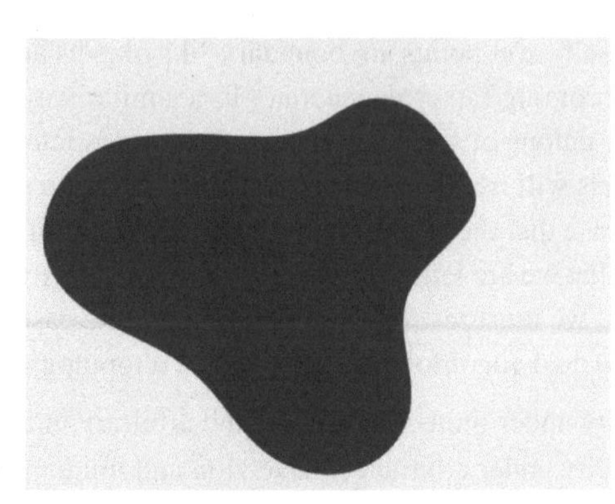

A bright but mischievous young student points to somewhere well within the ink blob, announcing that the paper is black there. You agree. The student asks you,

> "But how do you know that the point I pointed to is black?"

You take them to be asking,

> "How do you know that the point (call it *b*) is within the region of the ink blob (call it *R*)?"

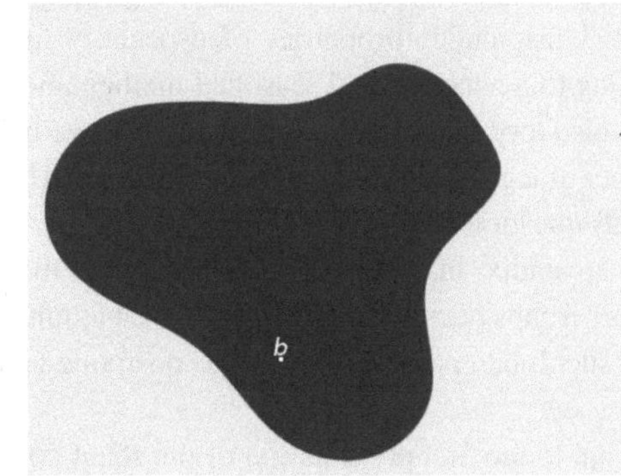

How will you answer them?

Perhaps you are feeling rather lazy at first, and you are not "on duty" at the moment, and so you attempt to meet their queries with some reference to how one can "just see" that the point in question has to be black, adding some tautological remark about the color black while gesturing to the point. But the student is not deterred—sensing that some evasion is happening, they grow even more eager to address the issue head on:

STUDENT: You said that I can observe that *b* is black (in *R*). But *what* exactly am I being asked to observe? I have no problem agreeing that, whatever the black region *R* is, *R* is itself black—this is trivial. But I wanted to know about *b*.

MATHEMATICIAN: Yes, well *b* is in *R*, as I said.

STUDENT: Yes, and as I said: *How* do you know that? You cannot say that I can observe *b* itself—I cannot. It seems to me that whatever is given to observation will be extended. Plus, in my geometry classes they say that points have no size, no extension. So even though we use dots to represent points, isn't this just an idealization, just like the mythical "instant"—there is no such extended point, so there can be no features *at the point*; in particular, I can observe no color there.

MATHEMATICIAN: Interesting. And you seem to be suggesting that this would not just be a limitation on your part: If it is assumed that points are not extended, but observables are extended (in space or time or both, say), then you would be correct in asserting that points would (at the very least) not be the primary bearers of spatial properties or relations—no property, such as color, could be observed of, or primarily ascribed to, a point itself. No matter how fine-grained an observation could get—no matter how much you could "zoom in"—it could not zoom in *exactly onto the point* itself and directly observe a property of it.

STUDENT: Yes, that is basically what I was thinking.

MATHEMATICIAN: In that case, it seems like we have two options: (1) give up on saying anything about points at all or (2) declare that whatever holds at a point really pertains to what holds of some extended region or some non-point-like part of a space *surrounding* the point, so of the *nearby* points (where perhaps we can then use what is nearby to say something about our point).

With the second option, we might even initially relinquish the idea that there are zero-sized points or instants—instead, we might take ourselves to be working with nonzero or nondimensionless regions, while allowing that such regions could be made *arbitrarily small*. The wager here is that, even though we are apparently relinquishing the idea of points, we may be able to recover everything we might want to say about points via such regions.

STUDENT: The first option seems silly. After all, we say things about points all the time, and it seems unnecessarily pessimistic to prohibit ourselves from making claims about points. And I was being sneaky before. I *do* see that some points of the paper are black and others gray (the color of the paper). And I see that where I pointed earlier, at b in particular, is clearly within R—I just want to know how to articulate in a less suspect or vague way this word "clearly."

MATHEMATICIAN: Indeed. And the second option seems sensible in its own right: after all, we have been asking about how b relates to its environs.

This also gets at an important feature of the overall problem, one that relates to your observations about points. In settings where mathematicians first clearly formulated the notions needed to address our problem, along the lines of route 2, a primary concern was with *continuity* and related matters. Strictly speaking, the sorts of things that can be continuous—motions, velocities, and so on—cannot be seen as an *intrinsic* property of some object like a point or instant. Motion, for instance, is something that involves a relationship between the state of an object at one instant and the state of that object at another instant or collection of instants. In concerning ourselves with such things, and whether or not they are *continuous*, we are concerning ourselves with something that ultimately involves a relationship between *multiple* moments, instants, or points.

Before we get too carried away with any of this, though, I bet we can start to get a better handle on your earlier questions by returning to this idea of "zooming in," since I suspect that we will find that, even if we cannot directly observe a property exactly at a point, there is something we can use in this notion of zooming.

STUDENT: So we are thinking about having something like lenses or magnifying glasses, each one of which gives us a distinct magnification?

MATHEMATICIAN: Yes, that works. Each lens's distinct magnification could then just be uniquely described by the radius of the resulting window of observation that we would get, upon looking through the lens. In that way, we can give each lens with a distinct magnification its own name, depending on the radius of its window of observation.

STUDENT: Okay, so for a given point, we can center a lens on that point and take a look at it with that lens. For instance, for our point b, a medium-sized lens might show us

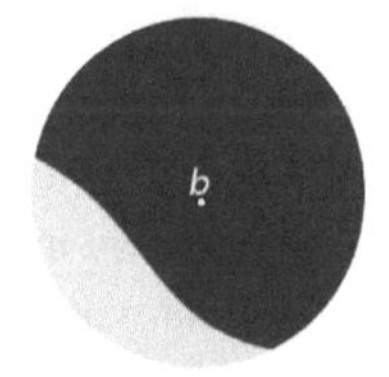

while a lens that can zoom in further might instead show us

MATHEMATICIAN: That is the right idea. And what do these lenses tell you?

STUDENT: Well, so far, I'm not really sure they tell me anything. Not anything definitive, at least. I would like to say that the second one is somehow "better" though.

MATHEMATICIAN: If it is better, as you say, you could try zooming in even further.

STUDENT: That is exactly what I was thinking. Suppose we end up with a small enough region of observation such that all around us we see only black,

MATHEMATICIAN: It seems that, with this lens, we can now declare without hesitation that b itself is black, that is, inside the ink blob, no?

STUDENT: I agree. And I don't have to zoom in any further, right?

MATHEMATICIAN: Let's give this a name.

Paradigm/Principle of Truth Continuity: an observable property will be true of a point whenever it continues to be true in all better approximations.

STUDENT: And this would have worked just as well on, say, the point a, where this is decidedly in the gray paper region. It seems that whenever, around a point a, there is a small enough region of observation such that all around us we see only gray, we can likewise declare without hesitation that a itself must be gray, that is, in the paper region.
It seems like I can use these lenses to tell when any point is black or not black.

MATHEMATICIAN: Hold on. Consider

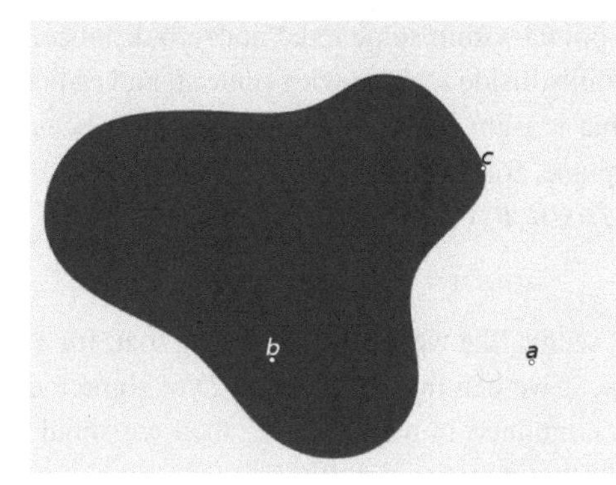

For a and b the concepts we are building would indeed suffice to tell us unequivocally. Imagine zooming in further and further to b. Clearly we can zoom in close enough such that "we only see black." For a, similarly, we can zoom in to a small enough region around a such that "we only see gray." But what about c?

STUDENT: Well, as I zoom in further and further to c, *if we are assuming that the region R includes all its outer edges*, so that c is exactly *on* the edge of R, there is evidently no way of zooming in *close enough* that we are surrounded by only black. Likewise, there is no way of zooming in close enough that we are only surrounded by only gray. However small we make our zoomed-in region around c, we will see both black and not-black!

MATHEMATICIAN: Yes, exactly, and we can use such trouble cases to define a new notion: that of the *boundary*.

STUDENT: Doesn't that mean, by the way, that what we have been calling "lenses" are being thought of as *not* including their boundary? After all, if they had included their boundary, then there would be points within its observation window that we could zoom in on with more and more refined lenses and yet for which we could never say unequivocally whether or not they were black or not-black. It seems like this could further force us to give up on your Paradigm of Truth Continuity, which seemed sensible enough to me.

MATHEMATICIAN: That is exactly right. Incidentally, what we have been attempting to name with our "lenses"—or, really, the various windows of observation that result from using such lenses—could instead be thought of in terms of what mathematicians would call an *open disk* in a plane (or, if we had been restricting our attention to the real line, an *open interval*).

STUDENT: Yes, I know about these. In $\mathbb{R}$, we're thinking about intervals of the form $(a, b) = \{x \in \mathbb{R} \mid a < x < b\}$:

And in terms of the plane $\mathbb{R}^2$, we are thinking of objects like the following:

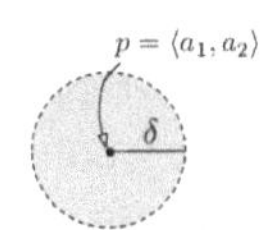

that is, the set of all the points within some fixed nonzero distance. For instance, with such disks, we have all the points inside a circle with center p and radius $\delta > 0$.

MATHEMATICIAN: That is right. Here, of course, we have been implicitly using a particular distance. In the plane, for instance, $d(p, q)$ denotes the usual distance between two points $p = \langle a_1, a_2 \rangle$ and $q = \langle b_1, b_2 \rangle$ in $\mathbb{R}^2$, that is,

$$d(p, q) = \sqrt{(a_1 - b_1)^2 + (a_2 - b_2)^2}$$

STUDENT: Okay, so it seems like we have been saying that, for a given region (say R), in examining some point x, if we can find such a disk D of some radius $\delta > 0$ around x such that the disk itself D is contained in the region R, then we should be able to confidently assert of the point x itself that it is "inside" R.

MATHEMATICIAN: Yes, you have effectively described here what mathematicians call an *interior point*: for a subset A of $\mathbb{R}^2$, a point $p \in A$ is an interior point of A precisely when p belongs to some open disk D_p that is contained in A, that is, $p \in D_p \subseteq A$.

STUDENT: And is there a name for the inverse of the relation "x in an interior point of A"?

MATHEMATICIAN: Yes, you were simultaneously describing the notion of a *neighborhood*: for a point $x \in \mathbb{R}^2$, a subset N of $\mathbb{R}^2$ is a neighborhood of x precisely when it contains an open disk that contains x, that is, $p \in D \subseteq N$.

STUDENT: Wait, that's all great, but getting back to R and what we were talking about a moment ago, how do these notions apply?

MATHEMATICIAN: Well, let's look closer at c. If c is assumed to be on the boundary of the region R, will wc able able to form an open disk around it such that that open disk is entirely contained in R?

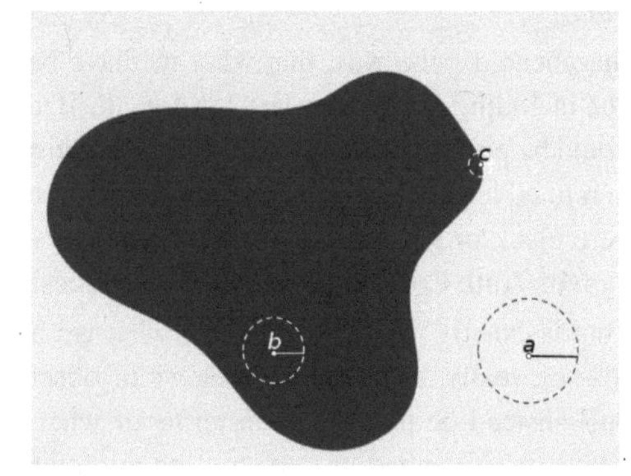

STUDENT: No. Any open disk around c would have to include something that was *not* R. So c cannot be an interior point of R.

MATHEMATICIAN: Exactly.

STUDENT: But b is surely an interior point—and a, for its part, may be an interior point of the gray paper region (whatever we want to call that).

MATHEMATICIAN: That is right. Suppose, then, that *each* of the points of R is an interior point. In this case, mathematicians would call R itself an *open set*.

STUDENT: It seems like this is all very wrapped up with the notion of boundary. For instance, to say that R is open—so that every point of R is an interior point—seems to just say that R does not include any of its boundary.

MATHEMATICIAN: Right! You have hit on another formulation of this notion of openness.

STUDENT: Okay, but when we move from the real line $\mathbb{R}$ to the plane $\mathbb{R}^2$, the same notions seem to carry over without a problem. I was wondering: What happens as we move to other spaces? And I have heard that there are still other notions of distance. What happens then? Also, is there something special about disks?

MATHEMATICIAN: Great observations! The construction we have been using works for any notion of nearness or distance, as long as we have specified what this is. The relevant notion of nearness is specified by the underlying "metric." So far, we have really just been working with a particularly familiar metric, and the familiar space over which this reigns, so we have not bothered to be explicit about this. But let's correct that now.

Definition 73 A *metric space* (X,d) consists of X a nonempty set, the elements of which are called "points," and a function $d: X \times X \to \mathbb{R}_{\geq 0}$ called a *metric* or *distance* that associates to any two points $x, y \in X$ a point $d(x,y)$ in such a way as to satisfy the following properties for all $x, y, z \in X$:

1. $0 \leq d(x,y)$ (or just $d(x,x) = 0$);
2. if $d(x,y) = 0$, then $x = y$;
3. $d(x,y) = d(y,x)$;
4. $d(x,y) + d(y,z) \geq d(x,z)$.

We can then use this notion of a metric to define a more general version of our open disks (and open intervals) from before.

Definition 74 For (X,d) a metric space, with $a \in X$ and $r > 0$, an *open ball around a of radius r* is the set

$$B_r(a) = \{x \in X \mid d(x,a) < r\}.$$

As you might have expected, these open balls can then be used to define the following:

Definition 75 For (X,d) a metric space, and $x \in X$, a subset $A \subseteq X$ is called a *neighborhood* of x if there exists an $\epsilon > 0$ such that

$$B_\epsilon(x) \subseteq A,$$

that is, provided an open ball around x can be contained in A.

Finally,

Definition 76 For (X,d) a metric space, subset $O \subseteq X$ is said to be *open* in (X,d) whenever for every $a \in O$, there exists an $\epsilon > 0$ such that

$$B_\epsilon(a) \subseteq O.$$

STUDENT: So what we have been calling lenses, and thinking of as "wiggle rooms," in the context of $\mathbb{R}^2$ with the usual sense of distance, are just particular instances of this general notion of *open balls*, which is like the distance-agnostic definition of the same thing?

MATHEMATICIAN: Precisely. Now observe how as we have been thinking about open sets thus far, in terms of in our real line $\mathbb{R}$ or plane, with the usual sense of distance between points, we have been thinking that

A set is open if whenever it contains a number a, it also contains all numbers "sufficiently close" to a.

In other words, a subset A of $\mathbb{R}$ is open if for every $a \in A$, there exists a number $\epsilon > 0$ such that the open interval

$$(a-\epsilon, a+\epsilon)$$

is a subset of A, where of course the interval $(a-\epsilon, a+\epsilon)$ just consists of all numbers within (the usual) distance ϵ of a, that is,

$$\{x \in \mathbb{R} \mid |x-a| < \epsilon\}.$$

But the point is that we need not confine ourselves to the usual notion of what counts as close—as long as there is a notion of distance, we can use that and the same objects will be available to us. A set A is open if, for each point in the set, we can find a little "wiggle room"—defined in terms of the prescribed distance—around the point, without having to leave A.

STUDENT: And, on this more general account, we need not confine attention to discs or open sets of a certain shape?

MATHEMATICIAN: That's correct. They need not be (literal) balls or circles at all. Even in the plane, especially as we introduce different metrics, open sets will include a variety of things, like open discs $D = \{\langle x, y\rangle \mid (x-a_1)^2 + (y-a_2)^2 < \delta^2\} = \{q \in \mathbb{R}^2 \mid d(p, q) < \delta\}$, open half-planes $\{\langle x, y\rangle \mid y < a\}$, open rectangles, and so on.

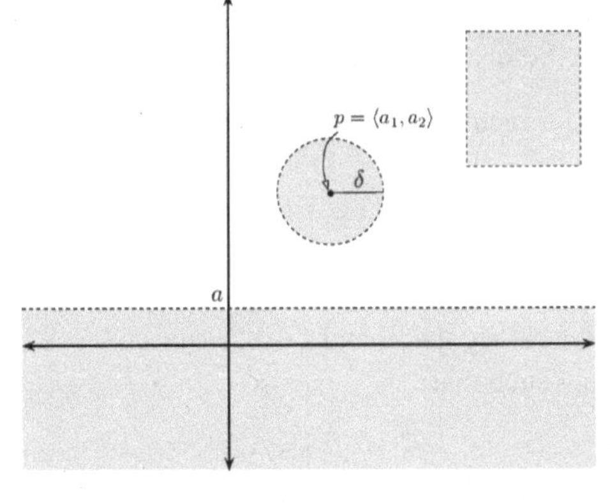

The only thing we require, for a set to be open, is that it be a neighborhood of *each of* its points.

Observe, though, that an open ball is itself an open set. And since each open ball is itself an open set, it is routine to show that a set U is open precisely when it is a union of open balls.

STUDENT: It seems like so much of this depends fundamentally on how we treat the boundary. None of these so-called open sets includes their boundaries.

MATHEMATICIAN: Yes, as I mentioned before, this gets at an important alternative characterization of open sets.

STUDENT: What if we had decided to treat boundaries differently?

MATHEMATICIAN: Great question. If we instead work with objects that include their boundaries—think of the circle including its boundary, or an interval including its end points—we are exhibiting instances of the notion of a *closed set*. And we could just as well have told this story using them!

STUDENT: Okay. Now we have a lot of words to describe when parts of a space equipped with a notion of distance are "near," which we can also use to home in on points. What's the big deal?

MATHEMATICIAN: Actually, we do not even need a notion of distance to develop such ideas.

STUDENT: What?! Explain!

MATHEMATICIAN: So far, we have basically been assuming that we already understand the underlying "space," the points of which we were comparing for nearness, interiority, and so on. But let us step back a bit.

Let us take these open set objects as our primitives and see what happens when we relate them. The most obvious thing we can try, for any two open sets, is to take their union or their intersection.

As for union, if we take two open disks and union them together, there is clearly no way of introducing any boundary, so the result will itself be open. But we can go further than this: we can union together any two open disks, and as we take the union of an arbitrary number of open disks, still no boundary can be introduced, so the union of any open disks must itself remain open. Moreover, trivial as it seems, notice the degenerate case as well: if we take the empty union—the union of no open disks, which is still technically a union!—this can only be the empty set $\emptyset$ itself. So this must also be open! Moreover, we can accordingly describe a set as open precisely when it is a union of open balls.

When taking intersections, the result will include the parts of the space that all of them have in common. First of all, notice the degenerate case here as well: if we take an empty intersection—the intersection of *no sets*—the result will be the entire space! In our case of the plane, this is all of $\mathbb{R}^2$, making the entire plane itself open.

Suppose we now take the intersection of two open disks:

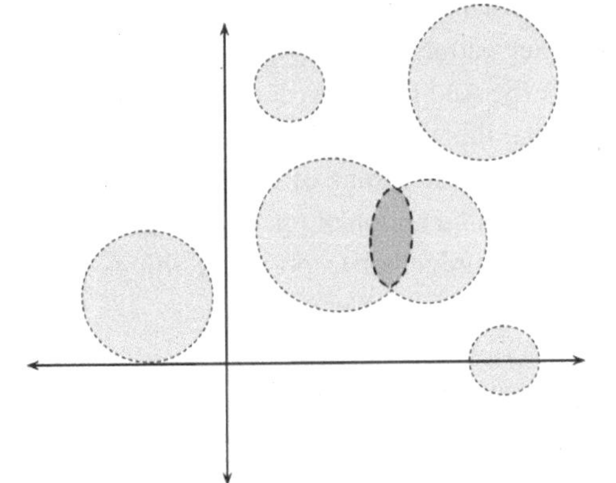

Observe how, for a point $p_0 \in D_1 \cap D_2$ in the intersection of the two open disks D_1 and D_2—the ones that are pictured as overlapping—we will have that $D(p_1, p_0) < \delta_1$ and $d(p_2, p_0) < \delta_2$. So if we just set $r > 0$ as

$$r = \min\{\delta_1 - d(p_1, p_0), \delta_2 - d(p_2, p_0)\},$$

and let

$$D = \{q \in \mathbb{R}^2 \mid d(p_0, q) < \frac{1}{2}r\},$$

then clearly $p_0 \in D \subseteq D_1 \cap D_2$, making p_0 an interior point of the intersection $D_1 \cap D_2$. This shows that the intersection of any two open disks will also be open.

STUDENT: And, as we could do with unions of opens, if we intersect an arbitrary number of open sets, then the result will be open?

MATHEMATICIAN: I was just about to speak to this. We cannot! This can easily be seen by looking at open intervals of the real line (exactly the same sort of argument applies to the plane).

Suppose, for some point $x \in \mathbb{R}$, I tell you that I can supply you with a "magical lens" (call it L_∞)

$$\bigcap_{n=1}^{\infty}\left(x - \frac{1}{n}, x + \frac{1}{n}\right).$$

For concreteness, let's just consider the point $x = 0$, so that this becomes

$$\bigcap_{n=1}^{\infty}\left(\frac{-1}{n}, \frac{1}{n}\right).$$

As n gets larger, this is of course like zooming in closer and closer, giving smaller and smaller windows of observation, around the point 0. As the windows of observation around the point 0 are getting arbitrarily small, the only thing that can be found in *every one* of the abstract lenses of the form $(\frac{-1}{n}, \frac{1}{n})$ is the *point* $\{0\}$ itself. If *this*—the point $\{0\}$ itself—was allowed to be declared open as well, as you were suggesting, then we would have some problems.

There is no such problem in the case of restricting ourselves to *finite* intersections, for any finite intersection of open balls will be itself open.

STUDENT: I see. It does seem like it would be strange to countenance the idea that points themselves would be open, as there is no "room" in a point at all, let alone wiggle room.

MATHEMATICIAN: Right.

STUDENT: So, I see that, focusing on the plane for concreteness: any union of open sets will be open, and finite intersections of open sets will be open.

MATHEMATICIAN: Exactly. And recall the degenerate cases of union and intersection as well, which informed us that the whole plane $\mathbb{R}^2$ and the empty set must also be open.

STUDENT: Sure. But what is the point of all this again? How does this answer my question about how to develop the notion of open sets even without any notion of distance?

MATHEMATICIAN: Well, in order not to prejudice things, forget everything you know about "open discs" and "open intervals" and all that.

STUDENT: I'll try.

MATHEMATICIAN: Suppose only that we have some featureless objects—they are sets, but for now don't attribute any properties to them. To insist on this, let us just call such things *blahs*. Here is the punch line. Suppose

> arbitrary unions of blahs results in a blah; and
> finite intersections of blahs results in a blah.

In precisely this case—needing nothing more, knowing nothing of distances—we agree to call such blahs *open sets*.

STUDENT: Fine. Of course our familiar open intervals and balls from before will be accommodated by this definition. But so what?

MATHEMATICIAN: Well, two things are of importance to note. First, observe that we have now given a definition that does not rely on any intrinsic properties of certain special objects, but the objects in question are themselves entirely determined by the sorts of relations they entertain (specifically with respect to the operations of union and intersection).

STUDENT: Like recovering the space of the plane itself just by looking at how the approximating lenses must relate to one another?

MATHEMATICIAN: That is a good way of thinking about it.

And this lends itself to the second point. In our earlier special cases of the real line and plane, the open sets were characterized in terms of the distance function we were assuming. And such metrics in fact induced the structure of the opens, according to which the two conditions were met. One can appreciate this by considering, for instance, how our sketch of a proof of the finite intersection property made constant use of the distance function.

As it turns out, despite the interest we have in Euclidean spaces, like the plane, there are a number of interesting and conspicuous examples of structures of subsets of a given set that meet the two defining conditions supplied above. And some of these have no notion of distance at all. For such things, the structure of the open sets is not induced by any metric.

Mathematicians call any such structure of subsets of a set X that act as open sets, in the sense that the two conditions above are met—even where there is no attendant notion of distance—a *topology*, and the set X together with this topology forms what is called a *topological space*.

STUDENT: So even though we began with a notion of distance as we formulated our approximating lenses—which happened to be useful for telling us when things are near, and so useful ultimately for formulating other things, presumably like a viable concept of continuity—with this new and more general formulation, we see that the notion of distance must be auxiliary even to the notion of continuity (even though we first learn to think about it in terms of "sending nearby points to nearby points").

MATHEMATICIAN: Yes. But there is a great deal more. Many topological spaces are massively important in their own right, as any student of any branch of mathematics must come to appreciate.

STUDENT: Okay, I can believe that. But what if we had used closed sets instead of open sets, that is, if we wanted to explicitly reason with boundaries?

MATHEMATICIAN: Good question. Earlier, I hinted that closed sets and closed set topologies can just be described in a dual fashion, and this will require that their general definition be in terms of any subsets that respect

arbitrary intersections of closed sets results in a closed set; and
finite unions of closed sets results in a closed set.

STUDENT: I see. So, thinking in terms of boundaries, you could say

open set version: *without introducing boundaries, an arbitrary number of boundaryless objects can be joined together (without restriction), while only finite intersections of boundaryless objects are permitted (otherwise boundaries can be introduced where there were none before).*

closed set version: *without introducing boundarylessness, an arbitrary number of boundary-containing objects can be intersected together (without restriction), while only finite unions of boundary-containing objects are permitted (otherwise boundarylessness can be introduced where there were none before).*

MATHEMATICIAN: That is an interesting way to put things.

STUDENT: In dealing with a topological space, does it matter, then, whether we use open sets or closed sets to describe it?

MATHEMATICIAN: Yes and no. We can convert—in a purely formal way—back and forth between open and closed sets, appropriately dualizing the pairs of axioms as we do so, and arriving at the alternative definitions. So while closed and open sets will in general be different—and one must of course respect that difference—any topology that we can describe in terms of open sets can be given an alternative closed set formulation.

On the other hand, this is hardly the end of the story. And there are some structures that seem much more amenable to description in terms of one or the other. Moreover, some mathematicians seem to act as if they believe that, for the most part, open sets might be better or more natural to work with than closed—in part, perhaps because it can appear more useful, in certain conspicuous contexts (such as when proving important theorems of analysis and of familiar topological spaces), to have arbitrary unions of open sets being open instead of having arbitrary intersections of closed sets being closed. It might in the end have something to do with the nature of boundaries—but this is a more complicated matter that we should postpone for the time being.

STUDENT: It does seem like all this is really about the structure of approximations and the implications of certain decisions about how to treat boundaries. But it is okay with me if we leave that aside for now. There is something else I have been wondering about anyway, and the "dual" description in terms of closed sets only makes me wonder even more.

MATHEMATICIAN: Shoot.

STUDENT: If all we need to formulate the notion of topologies and topological spaces—in all the full glory of their generality—is a pair of axioms regarding behavior of the subsets with respect to their union and intersection; and if topologies really are so integral to so much of modern mathematics; can you help me better see *why* these governing axioms, as opposed to some other axioms describing some other governing features of collections of subsets, are the decisive ones?

MATHEMATICIAN: Well, recall how the open intervals and open discs of our familiar and beloved spaces behaved. It was there that we first observed these features, and there that so much of advanced inquiry into continuous phenomena using such approximating tools first flourished.

STUDENT: Hmm. But I feel that this doesn't really address my question.

MATHEMATICIAN: [*Awkward silence*]

STUDENT: For instance, in my physics class, we learned about light, how it behaves in certain characteristic ways—for instance, as it moves from one medium to another, it changes direction, and so on. Later, we learned that we had been regarding light as a wave, and that this was part of a bigger story about *waves in general* and *wavelike behavior*. While light exhibited wavelike behavior, we learned about other waves that were not light waves and began to look at this in a more general way.

Later, as our teacher was describing waves and considering some of the laws governing the behavior of waves in general, one student asked the teacher to justify, or at least to better clarify, the representation of such phenomena in terms of waves, the fully general formulation of waves and why the defining wavelike behavior was the way it was.

If our teacher had simply appealed back to the example of light and how light's behavior can be observed as conforming to the general pattern of wavelike behavior, I would have felt that this answer had failed to address what the student had been wondering.

MATHEMATICIAN: Hmm.

STUDENT: Open sets or closed sets, I still don't see why these two properties—the ones involving the asymmetry in the finiteness of the conditions on stability with respect to unions and intersections—are the key defining features of such an important concept. It does not seem that the *fact* that such features govern the usual objects of our familiar Euclidean metric spaces offers much in the way of justification or clarification of these axioms, especially as they are apparently governing for such a wide spectrum of structures, well beyond the confines of our familiar plane.

MATHEMATICIAN: [*Somewhat flustered*] I see. Well, we may just have to come back to this question. Before we take it up, let us take a closer and more structured look at all these, and some further, notions of topology. Then, we may return to address your very legitimate questions.[60]

4.3 Topology and Topological Spaces More Formally

As we have started to see, a topology on a set X is just a collection of subsets of X satisfying certain properties. There are several equivalent definitions of the notion of a topology, but the following is the one most commonly used:

Definition 77 Given a set X, a *topology* on X is a collection τ of subsets of X—that is, a subset $\tau \subseteq \mathbb{P}(X)$ of the power set—that satisfies the following properties:

1. The empty set $\emptyset$ and X are in τ.
2. Any union of elements in τ is also in τ, that is, whenever $\{U_i\}_{i\in I}$ is a family (finite or no) of subsets of X such that $U_i \in \tau$ for all $i \in I$, then $\bigcup_{i\in I} U_i \in \tau$.
3. Any finite intersection of elements in τ is also in τ, that is, whenever $U_1, U_2 \in \tau$, then $U_1 \cap U_2 \in \tau$ (and this can be extended to the finite case).

Elements of the topology τ are called *open sets*, while a set is called *closed* if and only if its complement is open.

A *topological space* is then just a pair (X, τ), where X is a set equipped with τ a topology on X.[61]

Observe that, strictly speaking, we do not actually need to write the first condition, in the above definition. Regarding the redundancy of requiring $\emptyset \in \tau$: consider the second axiom (stability under arbitrary union), and take the degenerate case of the empty family of subsets, that is, $I = \emptyset$, so that $\bigcup_{i\in\emptyset} U_i$, where this is just the set of all points x such

60. The interested reader will find these (and a few other) questions raised in the final section of this chapter, and a dedicated treatment of them in the appendix.

61. When the context is clear, it is not uncommon to refer to a topological space by just its carrier set, X, a practice we occasionally adopt.

that $x \in U_i$ for some $i \in \emptyset$. But *there are no such points x*—thus, we already have that $\bigcup_{i \in \emptyset} U_i = \emptyset$, which amounts to ensuring that $\emptyset$ is in our collection.

As for $X \in \tau$: consider intersections $\bigcap_{i \in I} U_i$ of subsets $U_i \subseteq X$, where these are again indexed by the empty set $I = \emptyset$. The intersection of the empty family of subsets of X is just the set of all $x \in X$ such that $x \in U_i$ for $i \in \emptyset$—but *every* x satisfies this property, so $\bigcap_{i \in \emptyset} U_i = X$.

In short, then, a topology on a set X can be defined as a collection of subsets of X that is stable under arbitrary unions and finite intersections.[62]

For open sets A, B, we might intuitively read the relationship $A \subseteq B$ as "*A approximates B*"—in other words, A is only a "partial specification" of what B specifies.

Example 78 The so-called *usual (or standard) topology* on $\mathbb{R}^n$ is the topology we get by taking for open sets all possible unions of open balls. So, specializing for instance to the real line $\mathbb{R}$, this is given by the collection of intervals of the form (a, b) along with arbitrary unions of such intervals.

Recall that an open ball is defined using d, the usual Euclidean metric, the notion of distance we experience every day. Accordingly, this topology is also sometimes called the *Euclidean topology*.

Example 79 The two examples described here represent extreme cases, and might seem uninteresting at first but they are rather important nonetheless.

First, take the collection $\mathbb{P}(X)$ of all subsets of X. This forms a topology—called the *discrete topology*—on X. In other words, with the discrete topology, *every* subset of X is open.

Applied to $X = \mathbb{R}$, the discrete topology can be associated with a particular metric (very different from the usual Euclidean metric!), namely the *discrete metric*, defined thus:

$$d(x, y) := \begin{cases} 1 & \text{if } x \neq y \\ 0 & \text{if } x = y \, . \end{cases}$$

For the second example, at the other extreme, consider the following. By the first property in the definition of a topology, we know that a topology on X must contain both X and $\emptyset$. But notice that the set $\{\emptyset, X\}$, on its own, already forms a topology on X, called the *indiscrete* (or *trivial*) *topology*. In other words, with the indiscrete topology, the *only* sets counted as open sets are X and $\emptyset$. In particular, observe that this entails that the entire set X must be the only open set containing any point $p \in X$. In terms of the intuition, this latter consequence means that all points of the space are smushed together in such a way that, from the perspective of the topology, they cannot be distinguished.

Observe that the trivial topology on a set X will always be the topology with the *least* possible number of open sets—after all, to be a topology at all, the empty set and the entire set must be open, and the trivial topology just takes these as its *only* opens. By contrast, the discrete topology on a set X is the topology that has the *greatest* possible number of open sets—this is the topology that makes *every* subset open!

62. Instead of saying "stable," it is common to say that it is *closed* under arbitrary union and finite intersection. This terminology is entirely sensible, but it has nothing to do with "closed" in the topological sense, so in order to avoid confusing the reader this terminology is generally avoided in this chapter.

As the previous example already suggested, occasionally two topologies on the same set will be comparable. Topologies need not be comparable, but when they are, the following language is useful:

Definition 80 For two topologies $\mathcal{T}$ and $\mathcal{T}'$ on the same set, if $\mathcal{T} \subseteq \mathcal{T}'$, we say that the topology $\mathcal{T}$ is *coarser* (smaller) than $\mathcal{T}'$—or, what is the same, that the topology $\mathcal{T}'$ is *finer* (larger) than $\mathcal{T}$.

Here, to remember the terminology—where "coarse" means a smaller number of subsets in the collection, and "finer" means a larger number of subsets in the collection—topologists sometimes propose the following sort of analogy: think of some sort of grinder breaking up the pieces of something, like coffee beans. If the grinder is set to "fine" grind, then it will break things up into a great number of pieces; on the other hand, if one uses a "coarse" setting, one will be left with a smaller number of pieces.

Exercise 1 Take $X = \{a, b, c, d, e\}$. Establish whether or not each of the following collections of subsets of X forms a topology on X:

1. $\mathcal{T}_1 = \{X, \emptyset, \{a\}, \{a, b\}, \{a, c, d\}, \{a, b, c, d\}\}$.
2. $\mathcal{T}_2 = \{X, \emptyset, \{a, b, c\}, \{a, b, d\}, \{a, b, c, d\}\}$.

Solution

1. Yes, this is a topology, since it satisfies the three axioms of the definition, as you can check manually.
2. No, this is not a topology, since $\{a, b, c\}$ and $\{a, b, d\}$ each belong to $\mathcal{T}_2$, yet their intersection does not, that is, $\{a, b, c\} \cap \{a, b, d\} = \{a, b\} \notin \mathcal{T}_2$, violating the third axiom.

Exercise 2 Take $X = \mathbb{Z}$, the set of integers. Then the collection $\mathscr{C}$ of all finite subsets of the integers plus $\mathbb{Z}$ itself is *not* a topology. Why not?

Solution Well, in particular, the union of all finite subsets of $\mathbb{Z}$ not containing zero or any negative numbers

$$\{1\} \cup \{2\} \cup \cdots \cup \{n\} \cup \cdots = \{1, 2, \ldots, n, \ldots\}$$

will not be finite (and so cannot be in the given collection), and it is also not all of $\mathbb{Z}$. Thus, while each of the members of the above union are in $\mathscr{C}$, the union itself does not belong to $\mathscr{C}$. Thus, $\mathscr{C}$ does not satisfy axiom 2—the property of stability under arbitrary union—disqualifying it from being a topology.

The dialogue in the previous section hinted that the notion of a topological space was truly more general than that of a metric space. We speak of a space as *metrizable* if there is a metric that induces the topology, that is, the topology can be regarded as coming from a particular metric. If not, then it is said to be *nonmetrizable*. To show the topology notion to be a proper generalization, we need only exhibit a space that is nonmetrizable. There are in fact a great many nonmetrizable spaces!

Exercise 3 Take any set X that consists of more than one element. Take the indiscrete topology on X. Is the resulting space metrizable?

Solution It is not! For instance, take the set $X = \{1, 2\}$ and let τ be the indiscrete topology on X. We can show that τ is not induced by any metric one could put on X (and the same would be true for any arbitrary set with two or more elements).

Suppose we had a metric d on X. We can set $r = d(1, 2) > 0$, from which we will have that the open ball $B(1, r) = \{1\}$. Thus, in the topology induced by the metric d, the set $\{1\}$ will have to be open in X. Yet $\{1\}$ is not open in the indiscrete topology!

On the other hand, if X had only one element, the space would be trivially metrizable.

Exercise 4 Find the smallest topological space that is neither trivial nor discrete.

Solution On a set with only one element $X = \{x\}$, there is only one topology it can admit, namely $\{\emptyset, \{x\}\}$—and here, the trivial and discrete topologies coincide.

So take a two-point set $\{0, 1\}$. Take for the collection of open sets

$$\{\emptyset, \{1\}, \{0, 1\}\}.$$

This forms a topology, and one that is neither trivial nor discrete.

This particular space is rather special, and usually goes under the name of the *Sierpiński space* $\mathbb{S}$.

Example 81 Let X be the set $\mathbb{R}$, but for open sets all possible unions of (what, in the usual context of $\mathbb{R}$, we typically think of as) "half-open" intervals of the form

$$[a, b)$$

for $a, b \in X$.

This forms a topology called the *lower limit topology*, yielding a space sometimes called the *Sorgenfrey line*. Observe that this topology is finer, that is, has more open sets, than the usual topology on $\mathbb{R}$ (generated by the open intervals as basis).

Notice that, with respect to the usual topology on $\mathbb{R}$, with $a < b, a, b \in \mathbb{R}$, these intervals $[a, b)$ are neither open nor closed! However, with respect to the Sorgenfrey space, they are open (and closed!).

This example helps to reinforce an essential idea: *openness* (*closedness*) depends entirely on the topology in question. We should not treat these notions as involving some "inherent properties" of a particular set. A single set can carry many different topologies, and strictly speaking it is important to remember that there is no such thing as an "open set" in itself—only an open subset, that is, a set open in relation to some set. A given subset U of a space X is (or is not) a subset *open in* (or *open with respect to*) X. In practice though, which topological space we are working with will generally be obvious or clearly stated, so we will often simply speak (somewhat misleadingly) of certain sets as being "open."

Let us now investigate a little more closely a few other notions that arose in the dialogue of the previous section.

Definition 82 For p a point in a topological space X, we say that a subset $N \subseteq X$ is a *neighborhood* of p if and only if N is a superset of an open set G that contains p, that is,

$$p \in G \subseteq N \text{ where } G \text{ is an open set.}$$

Exercise 5 Suppose, on $X = \{a, b, c, d, e\}$, we have the following topology:

$$\mathcal{T} = \{X, \emptyset, \{a\}, \{a, b\}, \{a, c, d\}, \{a, b, c, d\}, \{a, b, e\}\}.$$

Supply all the neighborhoods of

1. the point e;
2. the point c.

Solution

1. What are the open sets that contain e? These are $\{a, b, e\}$ and the entire set X. A neighborhood of e is just any superset of an open set that contains e. Thus, we must look for the supersets of $\{a, b, e\}$ and X.

   The supersets of $\{a, b, e\}$ are $\{a, b, e\}$, $\{a, b, c, e\}$, $\{a, b, d, e\}$, and X, while the only superset of X is of course X. Thus, the neighborhoods of e are given by

   $$N_e = \{\{a, b, e\}, \{a, b, c, e\}, \{a, b, d, e\}, X\}.$$

2. What are the open sets that contain c? These are $\{a, c, d\}$, $\{a, b, c, d\}$, and X. Again, supersets of these sets will give us the neighborhoods, that is,

   $$N_c = \{\{a, c, d\}, \{a, b, c, d\}, \{a, c, d, e\}, X\}.$$

Instead of considering the relation

N is a neighborhood of point p,

we might look at things from the point's perspective. Doing so gives us an inverse relation

p is an *interior point* of N.

We can accordingly define a point x of a subset U of a topological space X to be an *interior point* for U if and only if U is a neighborhood of x. By taking the set of all interior points of U—that is, taking the union of all open subsets of U—we get the *interior of* U, which is denoted **int**(U).

Exercise 6 Take the topology from the previous exercise. Find the interior points of the subset $A = \{a, b, c\} \subseteq X$.

Solution Since each of

$$a, b \in \{a, b\} \subseteq A,$$

with $\{a, b\}$ an open set that is contained in A, a and b are each interior points of A. What about c? Well, c does not belong to any open set contained in A, and so it is not an interior point of A. Thus, **int**(A) $= \{a, b\}$.

Observe that the interior of a set A is the union of all open subsets of A. Moreover, one can show that

- **int**(A) is itself open;
- **int**(A) is the *largest* open subset of A; and
- A is open if and only if $A =$ **int**(A).

The third fact is especially decisive. After all, a topology is just determined by those sets declared open. Thus, the fact that a set will be open precisely when it is equal to its interior supplies us with an alternative way of construing a topology. It is important enough that we set it apart:

Theorem 83 A subset U of a topological space X is open precisely when $\mathbf{int}(U) = U$.

In this way, open set topologies can be determined by *taking the interior*, where there will be certain conditions on how this must behave. Specifically, construing things in these terms, we get an alternative (but ultimately equivalent) definition of a topological space.

Definition 84 A topological space is a pair $(X, \mathbf{int})$, for X a nonempty set and $\mathbf{int} : \mathbb{P}(X) \to \mathbb{P}(X)$ an operation satisfying the four so-called Kuratowski axioms:

1. **(i1)** $\mathbf{int}(X) = X$ (it preserves the total space);
2. **(i2)** $\mathbf{int}(A) \subseteq A$ (it is intensive);
3. **(i3)** $\mathbf{int}(\mathbf{int}(A)) = \mathbf{int}(A)$ (it is idempotent);
4. **(i4)** $\mathbf{int}(A \cap B) = \mathbf{int}(A) \cap \mathbf{int}(B)$ (it preserves binary intersections).

Observe that the family of open sets can then just be defined by setting $\tau = \{A \subseteq X : \mathbf{int}(A) = A\}$. We will have much more to say about this operator throughout this chapter and the appendix.

For now, though, let us continue to introduce further essential notions of basic topology. In the elementary examples presented thus far, we could generally specify a topology by simply explicitly describing the entire collection of open sets. But such a specification will typically be unfeasible for many topologies you might meet out in the wild. In most cases in practice, one instead specifies a smaller collection of subsets of X that then "generate" the topology in question and so can be used to define that topology. And even when we could explicitly specify the open sets of a topology, it is sometimes easier to understand, and work with, a description in terms of the smaller collection. The following notion serves such purposes.

Definition 85 For X a set, a *basis* for a topology on X is a collection $\mathcal{B} \subseteq \mathbb{P}(X)$ of subsets of X, called the *basis elements*, satisfying

1. ($\mathcal{B}$ "covers" X) for each $x \in X$, there is at least one basis element $B \in \mathcal{B}$ that contains x—in other words, $X = \bigcup\{B : B \in \mathcal{B}\}$; and
2. if x belongs to the intersection of two basis elements, that is, $x \in A \cap B$ where $A, B \in \mathcal{B}$, then there is at least one basis element $C \in \mathcal{B}$ such that $x \in C \subseteq A \cap B$.

Then, the topology τ *generated* by the basis $\mathcal{B}$ is defined as the coarsest topology containing $\mathcal{B}$.

Another way of writing the first of the conditions placed on a basis of a topology τ on a set X would be to say that every open set $U \in \tau$ is the union of members of $\mathcal{B}$—so that, in particular, $X = \bigcup\{B : B \in \mathcal{B}\}$. In short, a basis specifies a topology by taking unions, that is, starting with a basis on X, by adding to it all possible unions of basis elements, the collection we end up with will be a topology on X. A basis for a topological space X is thus just a collection $\mathcal{B}$ of open subsets of X, such that every open subset of X is a union of sets in $\mathcal{B}$.

Note, though, that the expression for an open U as a union of basis elements is not unique—a topological space can have several different bases. Thus, the terminology of "basis" here is not to be conflated with the use of the same term in linear algebra, for instance, where the expression of a vector as a linear combination of basis vectors is indeed unique.

Example 86 The basis that consists of the open intervals in $\mathbb{R}$ (or open discs in $\mathbb{R}^2$) generates the usual topology on $\mathbb{R}$ (on $\mathbb{R}^2$).

For any metric space X, the open balls also form a basis for the induced topology on X.

Example 87 The collection of all the singletons (one-element subsets) of a set X is a basis for the discrete topology on X: after all, they are all open in this topology, and an arbitrary open set is the union of its singleton subsets.

Example 88 Let $\mathscr{B} = \{[a,b) \subseteq \mathbb{R} : a < b\}$. This forms a basis on $\mathbb{R}$. The topology that it generates is the lower limit topology on $\mathbb{R}$, presented in example 81—resulting in the topological space called the Sorgenfrey line.

Exercise 7 Let $X = \{a,b,c\}$ and consider the collection $\mathscr{B} = \{\{a,b\},\{b,c\}\}$. Is $\mathscr{B}$ a basis for a topology on X?

Solution No. While of course $\{a,b\} \cup \{b,c\} = \{a,b,c\}$, if $\mathscr{B}$ were a basis, then $\{a,b\}$ and $\{b,c\}$ would each have to be open, and thus their intersection

$$\{a,b\} \cap \{b,c\} = \{b\}$$

would also be open. Yet $\{b\}$ is not the union of any members of $\mathscr{B}$.

Finally, given a topological space $(X,\mathcal{T})$ and a basis $\mathscr{B}$ on X, it is often convenient to use the following to check whether $\mathscr{B}$ indeed generates $\mathcal{T}$:

Corollary 89 For $(X,\mathcal{T})$ a topological space, and $\mathscr{B}$ a basis on X, $\mathscr{B}$ generates $\mathcal{T}$ if and only if

1. $\mathscr{B} \subseteq \mathcal{T}$; and
2. for every open set $U \in \mathcal{T}$ and every $x \in U$, there exists a $B \in \mathscr{B}$ such that $x \in B \subseteq U$.

Finally, it is worth mentioning that we do need not restrict the above notions only to open sets—they apply just as well to closed sets, which we take up more explicitly in the section to follow.

Exercise 8 What topology $\mathcal{T}$ on the real line $\mathbb{R}$ is generated by the collection $\mathscr{A}$ of all closed intervals $[a, a+1]$ of length 1?

Solution Consider any point $p \in \mathbb{R}$. Then the closed intervals $[p-1, p]$ and $[p, p+1]$ both belong to $\mathscr{A}$. Thus, their intersection $[p-1,p] \cap [p,p+1] = \{p\}$ must also belong to the topology $\mathcal{T}$. Thus, all singleton sets $\{p\}$ are open in this topology, giving us again the discrete topology on $\mathbb{R}$.

4.3.1 Closed Sets

In discussing topologies, we have confined ourselves thus far, as is common, to specifying the *open* sets. But if we appropriately dualize things, we can arrive at an alternative description of things in terms of closed sets and closed set topologies. Fundamentally, open and closed set topologies are formally distinct only in how nonfinite collections of elements of the topology are treated. Just building on what we already know of open sets,

Definition 90 A subset A of a topological space $(X, \mathcal{T})$ is *closed* if and only if its complement

$$X \setminus A$$

(also denoted A^c when the overall space is understood) is an open subset of $(X, \mathcal{T})$, that is, when $A^c \in \mathcal{T}$.

Example 91 Since $X \setminus \emptyset = X$ and $X \setminus X = \emptyset$, both $\emptyset$ and X are always closed, in any topology.

Example 92 For the real number line $\mathbb{R}$ with the usual topology, the canonical closed intervals $[a, b]$ of real numbers

$$[a, b] = \{x \in \mathbb{R} \mid a \leq x \leq b\}$$

give us closed sets. To see this, using the definition just presented, one need only recognize that the complement

$$\mathbb{R} \setminus [a, b] = (-\infty, a) \cup (b, \infty) = \{x \in \mathbb{R} \mid x < a \text{ or } x > b\}$$

is open. Yet this can be expressed as a union of open intervals, for example, $(-\infty, a) = \bigcup_{n \in \mathbb{N}} (a - n, a)$, and similarly for (b, ∞).

Moreover, with such closed intervals, we can also take $a = b$, resulting in the closed interval $[a, a]$ or the single point $\{a\}$, making $\{a\}$ itself a closed subset of $\mathbb{R}$. Indeed, for most topological spaces, it is typical to find that single points are closed subsets. In particular, in $\mathbb{R}$ with the usual topology, each singleton set $\{a\}$ is closed, for the complement of $\{a\}$ is the union of the two open sets $(-\infty, a)$ and (a, ∞), which is open.

Similarly, the set $\mathbb{Z}$ of all integers is a closed subset of $\mathbb{R}$—which fact can be seen by acknowledging that the complement of $\mathbb{Z}$ is the union $\bigcup_{n=-\infty}^{\infty} (n, n+1)$ of the open subsets of the form $(n, n+1)$, which is itself open in $\mathbb{R}$.

Example 93 Given $X = \{a, b, c, d, e\}$ with the topology

$$\mathcal{T} = \{X, \emptyset, \{a\}, \{c, d\}, \{a, c, d\}, \{b, c, d, e\}\},$$

the closed subsets of this topological space are given by the complements of the open subsets of the space. Thus, for closed sets we have

$$\emptyset, X, \{b, c, d, e\}, \{a, b, e\}, \{b, e\}, \{a\}.$$

As you can observe, there are subsets of X, like $\{b, c, d, e\}$, that are both open and closed; while there are subsets, like $\{a, b\}$, that are neither open nor closed.

Example 94 Take X to be a discrete topological space, that is, one where every subset of X is open. But then, every subset of X will also be closed, since its complement will automatically be open. Thus, in a discrete space, all subsets of X are both open and closed.

Exercise 9 Given the set $X = \{a, b, c\}$, come up with a topological space that is not discrete, where the closed sets and open sets are the same sets.

Solution Here is one:

$$\mathcal{T} = \{X, \emptyset, \{a, b\}, \{c\}\},$$

which you can manually check has the stipulated properties.

The example that follows is an unusual one, meant only to suggest that things don't always behave as one might expect, especially in relation to distances.

Example 95 One might expect that closed sets that were disjoint would have to be some distance apart. But we can describe two disjoint closed sets in the plane, that are at zero distance apart.[63] Let $C_1 = \{\langle x, r\rangle \mid xy = 1\}$ and $C_2 = \{\langle x, y\rangle \mid y = 0\}$ = the x axis. Then $C_1 \cap C_2 = \emptyset$ (they are disjoint), and both are closed. Now, for any $\epsilon > 0$, the points $\left(\frac{2}{\epsilon}, \frac{\epsilon}{2}\right) \in C_1$ and $\left(\frac{2}{\epsilon}, 0\right) \in C_2$ are at distance $\frac{1}{2}\epsilon < \epsilon$.

Because we have merely used the operation of complement to arrive at the corresponding notions of closed sets, we should not be surprised that, by some elementary facts of set theory, we can arrive at a "dual" definition of a topology in terms of closed sets. In particular, first recall the following:

Lemma 96 (deMorgan's Laws: *taking complements turns intersections into unions and unions into intersections*) Let X be a set and $(U_i)_{i \in I}$ a collection of subsets of X. Then,

$$X \setminus \left(\bigcup_{i \in I} U_i\right) = \bigcap_{i \in I} (X \setminus U_i)$$

and

$$X \setminus \left(\bigcap_{i \in I} U_i\right) = \bigcup_{i \in I} (X \setminus U_i).$$

Given a topology built from open sets, then, by taking complements of open sets and using deMorgan's laws, we can define a topology in terms of closed sets, a *closed set topology*.

Theorem 97 Take X a topological space. Then the collection of closed subsets of X will have the following properties:

1. The empty set $\emptyset$ and X are closed sets.
2. The intersection of any number of closed sets is closed.
3. The union of any finite number of closed sets is closed.

Observe how unions became intersections and intersections unions, yielding alternate finiteness conditions than what we had for open sets.

Exercise 10 It is fairly routine to check the statements made above, for instance that the intersection of any number of closed sets is closed. For the condition on unions: you should at least try to give an example of an infinite union of closed sets that is not closed, helping you see why we must restrict to finite unions when dealing with closed sets.

Solution One such example is given by taking thc union of all closed intervals of the form $\left[\frac{1}{n}, 1 - \frac{1}{n}\right]$. Observe that the infinite union equals $(0, 1)$, which is surely not a closed set.

Another example could be given by taking an infinite union of singleton sets $\left\{\frac{1}{n}\right\}$, which you can check will not be closed.

63. This example was taken from Gelbaum and Olmsted 1962.

With such a result, we can give an alternative definition of a topology, now in terms of closed sets: namely as a set X together with a collection τ of closed subsets of X. Moreover, because $X \setminus (X \setminus U) = U$, so that $(\tau^c)^c = \tau$, if you know what the closed sets are, you also know what the open sets are (and conversely).

In certain cases, however, it may be more natural to check the dual topology axioms using closed sets instead of open sets; and in some settings, it may simply be more natural to work with, or define things in terms of, closed sets.

Example 98 In general, a subset A of a set X that is such that its complement in X is a finite set is called a *cofinite* subset. For any set X, we can define a topology that takes for its open sets the empty set and all cofinite subsets of X, that is,

$$\tau = \{A \subseteq X \mid A = \emptyset \text{ or } X \setminus A \text{ is finite}\}.$$

In this topology—called the *cofinite topology*—all finite sets are closed (and they, together with all of X, are the only subsets that are closed). In other words, a subset $Z \subseteq X$ is closed in the cofinite topology if and only if it is finite or equal to X. And it is straightforward to verify that

1. the empty set is finite, and X is equal to X;
2. an arbitrary intersection of finite sets is finite;
3. a finite union of finite sets is finite.

Accordingly, such a topology appears to have a more natural or direct description, and the topology axioms are more easily verified, when we work with closed sets instead of open sets.

This topology is rather interesting and does indeed show up a fair amount in the wild. Moreover, it is the coarsest topology that satisfies the natural condition that singleton sets are regarded as closed. Before moving on, let us explore a few other features of this topology.

First of all, observe that the definition of a cofinite topology does *not* stipulate that every topology on X that has X and the finite subsets of X as closed is automatically the cofinite topology, as can be appreciated by considering that in the discrete topology on a set X, the set X and all finite subsets of X will be closed (yet many—all!—other subsets of X are closed as well). Rather, the definition says that X and the finite subsets of X have to be the *only* closed sets.

Second, it is not the case that infinite subsets will necessarily be open sets, as the following illustrates. Taking $\mathbb{N}$ the set of all positive integers, then of course sets such as

$$\{1\}, \{4, 6, 8\}, \{5, 6, 7, 8\}$$

are finite and thus closed in the cofinite topology. Therefore, their complements

$$\{2, 3, 4, 5, \dots\}, \{1, 3, 5, 7, 9, \dots\}, \{1, 2, 3, 4, 9, 10, \dots\}$$

will be open sets in the cofinite topology. However, the set of even positive integers, for instance, is not a closed set in this topology, since it is not finite. Thus, its complement, the set of odd positive integers, is *not* an open set in this topology. In other words, while it is true that all finite sets are closed in this topology, not all infinite sets will be open.

Again reinforcing a point we made earlier, observe how a set such as $\{n \mid n \geq 12\}$ is open in the cofinite topology on $\mathbb{N}$. However, this same set is *not* open in the indiscrete topology, for instance. Similarly, the set of even natural numbers is open in the discrete topology on the natural numbers, but it is *not* open in the cofinite topology. Again, it is easy to forget that the notions of "open" and "closed" entirely depend on the topology—we should not treat these notions as involving some "inherent properties" of a particular set.

Moreover, it is important to realize that, for a given topology, sets can be both open and closed, neither open nor closed, open but not closed, or closed but not open. So one should be sure to appreciate that we cannot prove that a set is open by proving that it is not closed.

As we did for open sets, we can define further notions corresponding to those we had for open sets. For instance, while open sets had interior points, closed sets have *limit* (or *accumulation*) *points*—which notion gives us another way of conceptualizing being closed.

Definition 99 We call a point x of a subset A of X a *limit* (or *accumulation*) *point* of A iff every neighborhood of x contains at least one point of A different from x. In other words, a point $p \in X$ is a limit point of a subset A of X iff

$$G \text{ open}, p \in G \text{ implies } (G - \{p\}) \cap A \neq \emptyset.$$

Thus, to show that a point y is *not* a limit point of A, it is enough to find even just one open set that contains y but no other point of A.

Example 100 Take the topological space (X, τ), where $X = \{a, b, c, d, e\}$ with the topology

$$\tau = \{X, \emptyset, \{a\}, \{c, d\}, \{a, c, d\}, \{b, c, d, e\}\}.$$

Consider the set $A = \{a, b, c\}$. The points a and c are *not* limit points of A—for, the set $\{a\}$ is open in τ and contains no other point of A, and the set $\{c, d\}$ is open in τ and contains c but no other point of A. On the other hand, b, d, and e are each limit points of A. To show that a point is a limit point of A, we just have to show that every open set containing that point contains a point of A other than it. So, for example, b is a limit point of A, as the only open sets containing b are X and $\{b, c, d, e\}$, each of which contains another element of A. Notice, finally, that d and e are indeed limit points of A, even though they are not in A.

Example 101 If (X, τ) is a discrete space, and A any subset of X, then A will have no limit points. Observe that for each $x \in X$, the singleton $\{x\}$ is an open set that contains no point of A different from x.

Example 102 In $\mathbb{R}$ with the usual topology, every element in the interval $[a, b]$ is a limit point of $[a, b]$.

Similar to what happened with open sets and their interior points, this notion of "limit point" gives us a useful way of characterizing which sets are closed.

Corollary 103 A subset A of a topological space (X, τ) is closed if and only if A contains all of its limit points.

The set of limit points of a set A is sometimes called the *derived set* of A, denoted A'. Using this notation, the above just says that a set A is closed if and only if $A' \subseteq A$.

Example 104 The set $[a, b]$ is closed in $\mathbb{R}$ with the usual topology, since all the limit points of $[a, b]$ are in $[a, b]$. By contrast, the set $[a, b)$ is not closed in $\mathbb{R}$, since b is a limit point, yet $b \notin [a, b)$.

Here is a strange and somewhat counterintuitive example.

Example 105 Suppose X is any set with more than one point. As we observed in example 79, if we let (X, d) be the metric space equipped with the discrete metric

$$d(x, y) := \begin{cases} 1 & \text{if } x \neq y \\ 0 & \text{if } x = y, \end{cases}$$

this will induce the discrete topology on X. Now suppose x is any point of X, and let O be the open ball (in this space), with center x and radius 1, while C is the closed ball with center x and radius 1. Observe that we must then have that $O = \{x\}$, while $C = X$. But, as the topology is necessarily discrete, $\mathbf{cl}(O) = O \neq B$.

Altogether, this shows that we can produce a space for which we have open and closed balls, O and C, each with the same center and same radius, such that C is *not* equal to the closure of O!

In many other familiar spaces, such an example would be impossible.

Observe how on the real number line, there is a natural notion of "closeness." For instance, each point in the sequence

$$0.1, 0.01, 0.001, 0.0001, \ldots$$

is closer to the point 0 than the previous point in the sequence. 0 is clearly a limit point of this sequence, so an interval such as $(0, 1]$ cannot be closed in $\mathbb{R}$, since it does not contain the limit point 0. However, in general topological spaces, we have seen that they need not be accompanied by a distance function. Thus, we make use of the notion of "limit point" that does not rely on distances. Yet even with this more general definition, we can still ensure that, in cases such as the above, the point 0 will remain a limit point of $(0, 1]$, as we would have expected.

Just as, earlier, we were able to describe an interior operator **int** and use this to present an alternative definition of a topology, there is a dual "closure" operator, **cl**, where for any $A \subseteq X$, $\mathbf{cl}(A)$ is defined as the union of A and its limit points, that is,

$$\mathbf{cl}(A) = A \cup A'.$$

It should be apparent that $\mathbf{cl}(A)$ is itself a closed set. An important observation is that for S, T nonempty subsets of a topological space (X, τ), with $S \subseteq T$, if p is a limit point of S, it is straightforward to show that p must also be a limit point of the set T. But this means that every closed set containing A must also contain A'. Thus, $A \cup A' = \mathbf{cl}(A)$ must in fact be the *smallest* closed set containing A—in particular, $\mathbf{cl}(A)$ will be the intersection of all closed sets containing A. Altogether, a set A will then be closed (relative to a topology) precisely when $\mathbf{cl}(A) = A$. In this manner, we can begin to appreciate that closed set topologies can be determined by closure operators, just as we were able to give an alternative definition of an open set topology in terms of the interior operator.

Exercise 11 Recall the definition of an interior operator. I have suggested that there is an operator dual to the interior operator—and this is the closure operator. Referring back to the four axioms governing the interior operator, introduced in definition 84, how might you define the dual notion of a closure operator?

Solution Recalling that one can convert between open and closed sets by taking complements, we arrive at the following definition:

Definition 106 A *closure operator* on a set X is a mapping $\mathbf{cl}: \mathbb{P}(X) \to \mathbb{P}(X)$ that satisfies the four so-called Kuratowski closure axioms, for all $A, B \subseteq X$:

1. **(k1)** $\mathbf{cl}(\emptyset) = \emptyset$ (it preserves the empty set);
2. **(k2)** $A \subseteq \mathbf{cl}(A)$ (it is extensive);
3. **(k3)** $\mathbf{cl}(A) = \mathbf{cl}(\mathbf{cl}((A)))$ (it is idempotent);
4. **(k4)** $\mathbf{cl}(A \cup B) = \mathbf{cl}(A) \cup \mathbf{cl}(B)$ (it preserves binary unions).

Moreover, a fact we alluded to earlier, that

- **(k4′)** $A \subseteq B \Rightarrow \mathbf{cl}(A) \subseteq \mathbf{cl}(B)$,

is a consequence of the fourth axiom.

Exercise 12 Returning to example 93, where the closed sets were

$$\emptyset, X, \{b, c, d, e\}, \{a, b, e\}, \{b, e\}, \{a\},$$

what is the closure of the sets $\{b\}$, the set $\{a, c\}$, and the set $\{b, d\}$?

Solution The closure $\mathbf{cl}(A)$ of any set A is the intersection of all closed supersets of A. To find the closure of a particular set, we need only find all the closed sets containing that set and then pick the smallest. Thus, in our present case, $\mathbf{cl}(\{b\}) = \{b, e\}$, while $\mathbf{cl}(\{a, c\}) = X$, and $\mathbf{cl}(\{b, d\}) = \{b, c, d, e\}$.

Example 107 In any discrete space, as every set is closed (and open!), every set is equal to its closure.

Exercise 13 Let $S = \left\{1, \frac{1}{2}, \frac{1}{3}, \frac{1}{4}, \ldots, \frac{1}{n}, \ldots\right\}$. Show that S is not closed in the usual topology on $\mathbb{R}$.

Solution 0 is a limit point of S, but $0 \notin S$, so S cannot be closed. If, instead, we had considered the same set, but now with 0 added, then it would of course be closed in $\mathbb{R}$.

Note also that S is not open either. Check that you can see why.

As should be evident from some of the examples just considered, the closure of a set U is in general bigger than U itself. As such, it is natural to want to consider what is in the closure of a set without being in the set itself. This gives rise to an operator ∂, called the *boundary*:

$$\partial(U) := \mathbf{cl}(U) \setminus U.$$

Note that for a set in a topology, we could also say that

$$\partial(U) := \mathbf{cl}(U) \setminus \mathbf{int}(U)$$

or

$$\partial(U) := \mathbf{cl}(U) \cap \mathbf{cl}(U^c),$$

where this last formulation describes the boundary of a set U as consisting of those points interior neither to U nor to its complement U^c (i.e., $X \setminus U$). As such, the intuition here is that the boundary of a subset $S \subseteq X$ consists of those points in X that are neither "fully in" S nor "fully not in" S. We typically think of the boundary of a set as the exterior surface or skin of a body. Yet the boundary of a set may be larger than the set itself—for example, since it can be shown that the rationals $\mathbb{Q}$ are dense in $\mathbb{R}$, $\partial\mathbb{Q} = \mathbb{R}$.

Another valuable way of defining the boundary of a subset S of a topological space X—now seeing things from the perspective of points—is as consisting of the set of points $p \in X$ such that every neighborhood of p contains at least one point of S *and* at least one point not of S. Here, such an element of the boundary of S is called a *boundary point* of S.

Example 108 Taking $\mathbb{R}$ with the usual topology,

$$\partial((0,3)) = \partial([0,3)) = \partial((0,3]) = \partial([0,3]) = \{0,3\},$$

just as one might have imagined.

While one might typically think of a boundary in the way the previous example supports, or imagine the boundary in intuitive terms of the edges of a circle or figure in the plane, boundaries can be strange, as the following example illustrates.

Example 109 One can construct three disjoint subsets of the plane, where these share a common boundary. (In fact, this works for any finite number of disjoint regions.)[64]

A few other things are worth observing at this point. In general, a set U may have limit points x where $x \notin U$, so $\partial(U)$ may in general be nonempty. Moreover, the boundary of a set is always closed, and a closed set will contain its own boundary. Altogether, the notion of boundary gives us yet another characterization of a closed set.

Definition 110 A set is closed if and only if it contains all of its boundary points.

Note also that the closure of a set can thus be expressed as the union of the set with its boundary,

$$\mathbf{cl}(S) = S \cup \partial(S),$$

the smallest closed set that contains S.

The above characterization of closed sets in terms of boundary points moreover informs us that a set is open if and only if it is disjoint from its boundary. Altogether, this characterization leaves us with a very useful way of thinking about closed versus open sets, one that was already anticipated in the earlier dialogue: a set is closed iff it contains its boundary, and thus open iff it is disjoint from its boundary.

At this point, the reader might be wondering: we have seen that we call a set closed if it contains all its limit points, and we have just seen that we can also express a set as closed if it contains all its boundary points—does this mean that these are two words for the same thing? No! Limit points and boundary points are not reducible to one another. In particular, a limit point can be a boundary point, but it need not be a boundary point, as a limit point can also be an interior point (recall that a boundary point is not an interior point). One can

64. See Gelbaum and Olmsted (1962, 138) for a description of this, and a nice story to help with the (counter)intuition.

also appreciate the difference by considering the interval $S = [0, 1]$ with the usual topology. Here, each element of $[0, 1]$ is in fact a limit point of S, while there are only two boundary points—namely, 0 and 1.

Likewise, in general, a certain point may be a boundary point, yet not a limit point. For instance, in $\mathbb{R}$ with the usual topology, the point 0 is a boundary point of the set $\{0\}$, but it is not a limit point of that set. Note that there may also be points that are both a limit point and a boundary, and points that are neither. However, again in general, it is often useful to realize that a limit point of a set that is not an element of the set itself will always be a boundary point.

In short, while a point $p \in X$ is a limit point of a subset A of X iff every neighborhood of p also contains a point of A other than p itself (which is something that can hold for an interior point as well), a point $p \in X$ is a boundary point of a subset A of X iff every neighborhood of p contains at least one point of A and at least one point not of A. While the notions are not the same, we can characterize closed sets in terms of either notion.

4.3.2 Covers

There is a final basic notion of topology, one that will be especially important for us in the development of sheaves: that of a *cover*. A common thing to do in mathematics is to approximate complicated objects or structures by means of simpler, more basic ones. The topological notion of a cover is a powerful way of doing so, allowing us to shift from a topological perspective to a more combinatoric perspective. Moreover, the notion of a cover will play a key role in the definition of sheaf, so we close out the treatment of the basic notions of general topology by devoting some attention to this notion.

A *cover* of a given subset S of a topological space X is any collection of subsets of X whose union contains S. In other words, a cover for a set S is just a particular bunch of (possibly overlapping) sets such that S is completely contained in that bunch of sets. Purely in terms of set theory, a cover of a set X is thus just a family $\mathscr{C} \subseteq \mathbb{P}(X)$ of subsets of X such that $X = \bigcup \mathscr{C}$. The set being covered can of course be the entire space itself X, in which one speaks of a cover of the space. But more generally, we can just define a cover for any subset of a space.

Definition 111 Let X be a topological space and $S \subseteq X$. Then an *(open) cover* of S is a collection $\{U_i\}_{i \in I}$ of (open) subsets U_i whose union contains S, that is, $S \subseteq \bigcup_{i \in I} U_i$. Note that if $S = X$, then this is just to require that

$$\bigcup_{i \in I} U_i = X.$$

If we are working with open sets, given an open set topology, we will often just speak of an *open cover*, where this is exactly the same thing as a cover, except we specify that all the members of our collection of sets doing the covering are themselves open sets. But observe that this definition would work just as well for a *closed cover*, where this would just be a collection $\{C_i\}_{i \in I}$ of closed subsets of X such that $\bigcup_{i \in I} C_i = X$. Closed covers can be got from open covers by taking the closure of each of the open subsets, which incidentally makes it so that every point $x \in X$ is in the interior of one of the closed subsets C_i (a further stipulation sometimes placed on closed covers).

Example 112 Consider $\mathbb{R}$ with the usual topology. Let

$$U_n = (n, n+2), \text{ where } n \in \mathbb{Z}.$$

Then the collection $\{U_n\}_{n\in\mathbb{Z}}$ forms an open cover of $\mathbb{R}$—one where, incidentally, we will have many overlaps.

Example 113 Again consider $\mathbb{R}$. Let

$$U_n = (-n, n), \text{ where } n \in \mathbb{Z}.$$

Then the collection $\{U_n\}_{n\in\mathbb{Z}}$ forms an open cover of $\mathbb{R}$—a nested cover where we will have many more overlaps than in the previous example, since each set contains all the preceding ones.

Example 114 We can have a cover that has considerable redundancy. For instance, take the set $(0, 1)$ in $\mathbb{R}$. This set has an open cover with the collection

$$\left\{\left(0, \frac{3}{4}\right), \left(\frac{1}{4}, \frac{1}{2}\right), \left(\frac{1}{4}, 1\right)\right\}.$$

Observe, though, that we do not need all these subsets. Already with the first and last members of the cover we will have a cover of $(0, 1)$, that is, with

$$\left\{\left(0, \frac{3}{4}\right), \left(\frac{1}{4}, 1\right)\right\}.$$

In such case, we speak of the latter cover as a *subcover*.

As such examples can help one see, be sure to observe how, in general, a subcover consists of *fewer* open sets, not *smaller* subsets.

Example 115 Take $S = \mathbb{R}$. The family $\{U_n\}$ of open intervals

$$U_n = (n-1, n+1), \text{ where } n \in \mathbb{Z}$$

forms an open cover of $\mathbb{R}$ that contains no nontrivial subcovers.

Example 116 Suppose our space is $\mathbb{R}$ and $S = (0, 1)$. Then, the collection

$$\left\{\left(-1, \frac{1}{2}\right), \left(0, \frac{3}{2}\right)\right\}$$

is an open cover of S. This is an example of a *finite cover*, as it consists of only a finite number of sets (in this case, just two).

Example 117 (Nonexample) A *Cantor set* is constructed by iteratively deleting the open middle third from a set of line segments. For instance, one can first delete the open middle third $\left(\frac{1}{3}, \frac{2}{3}\right)$ from the interval $[0, 1]$, giving us the two line segments $\left[0, \frac{1}{3}\right] \cup \left[\frac{2}{3}, 1\right]$. One then deletes the open middle third of each of those remaining segments, leaving us with

$$\left[0, \frac{1}{9}\right] \cup \left[\frac{2}{9}, \frac{1}{3}\right] \cup \left[\frac{2}{3}, \frac{7}{9}\right] \cup \left[\frac{8}{9}, 1\right].$$

One can continue indefinitely with this process. The Cantor ternary set then contains all points of the interval $[0, 1]$ that are *not* deleted at any step in this infinite process.

Such a set cannot cover $[0, 1]$, and Cantor sets in general will fail to cover $\mathbb{R}$.

Exercise 14 Given X a topological space, is $\{X\}$ an open cover of X?

Solution Yes, a rather trivial one, but it is indeed a cover—specifically, it is the coarsest cover of the space. Obviously, this fact would apply to *any* open subset as well: namely, for an open subset U of a space X, the coarsest cover of U is just the cover given by one element $\{U\}$.

There are generally going to be many different ways to cover a given set, that is, there may exist many open covers for subsets of a topological space. For instance, still working with $S=(0,1)$, consider the following.

Example 118 Let $U_n=\left(-\frac{n}{3},\frac{n}{3}\right)$. Then

$$\bigcup_{n\in\mathbb{N}} U_n$$

clearly contains $(0,1)$, making it an open cover of S. But observe that the subcover that contains only the first three sets U_1, U_2, U_3 already covers $(0,1)$—and so too would any larger subcover of the original cover! This means that the original open cover has a finite subcover that consists of only three open sets. As such, this makes the original cover a rather inefficient one.

As it turns out, not every cover has a finite subcover, as one can see in a number of ways. For instance, consider the set

$$S=(-\infty,-1]\cup[1,\infty).$$

Then S is a closed subset of $\mathbb{R}$, as the complement of S is $(-1,1)$, which is an open set of $\mathbb{R}$. The collection

$$\{(n-1,n+2)\mid n\in\mathbb{Z}\}$$

is an open cover of S. Yet it has no finite subcover.

Exercise 15 Take $U_n=\left(\frac{1}{n},1\right)$, an open cover of $(0,1)$. Does this cover have a finite subcover?

Solution It is straightforward to show that

$$\bigcup_{n\in\mathbb{N}}\left(\frac{1}{n},1\right)=(0,1).$$

However, we can also show that no finite collection of the $\{U_n\}_{n\in\mathbb{N}}$ can act as a cover of $(0,1)$.

Those sets for which every cover has a finite subcover are rather special, and so are given a special name, namely *compact*. Compactness is a very important topological property, one that plays a central role in much of topology. Open covers are not terribly interesting on their own, without further properties or without engaging more advanced constructions or properties. It is generally fairly simple to find an open cover of a set. Properties like compactness, by contrast, are less trivial because they involve saying something about *every* open cover of a set. Metric spaces, for their part, are rather significant spaces, in part because of the fact that they are *paracompact*—where this is a property that can be defined in terms of open covers, and appears to account for certain of the "nicer" properties of such spaces.

One further notion concerning covers that will be especially useful to us in the subsequent story we tell of sheaves is that of *refinement of covers*. Subcovers are an example of this, as every subcover of a cover will be a refinement of that cover. As we advance in the development of ideas, and especially as we arrive at the sheaf notion, it will often be natural to ask the question: If a certain property holds for some cover, for what other covers can we expect it to hold?

As we already began to appreciate, there are potentially many different covers of a set. When we formally introduce sheaves on topological spaces in the next chapter, we will see that this sheaf notion effectively just combines the data of a functor (presheaf) with the notion of a cover. In the definition of a sheaf, the sheaf axioms will be required to hold for all covers. But we will also see that if F is a sheaf for some cover $\mathcal{V}$, then it is a sheaf for every cover that $\mathcal{V}$ refines. Since we will have to verify the sheaf axioms on spaces with covers, whenever such spaces have a finest cover, verifying that the sheaf axioms hold on such a space will amount to just checking it on the finest cover, since this will guarantee it for all covers by the result we just mentioned. The following notion will thus be rather useful in the formulation of such things.

Definition 119 Suppose we have two covers $\mathcal{U}$ and $\mathcal{V}$ of a subset S. We say that $\mathcal{V}$ *refines* (or is a *refinement* of) the cover $\mathcal{U}$ if for each $V \in \mathcal{V}$ there is a $U \in \mathcal{U}$ such that $V \subseteq U$.

Exercise 16 Which covers, if any, are refinements of the trivial cover $\{U\}$ of an open subset U?

Solution This one should be easy. Without knowing anything else, you should have been able to say that *every cover* of U will refine the trivial cover $\{U\}$.

We can note that the collection of all covers of a set U in fact forms a category $\mathbf{Cov}(U)$: its objects are covers and its morphisms are supplied by the refinement relation, that is, there is a unique morphism $\mathcal{V} \to \mathcal{U}$ whenever $\mathcal{V}$ refines $\mathcal{U}$.

There is a final notion related to covers and their refinements that we will occasionally make use of: that of the *nerve* of a cover. The nerve of a cover gives us something like a combinatorially friendly representation of the data of a cover, one where we can effectively forget about points in the space and instead get a simplified "approximation" of the space by working instead with a structure that represents the abstract relations between elements of a cover. While the notion of a nerve can be useful for formulating the sheaf definition, we postpone its formal definition, and further discussion, until chapter 9.

4.3.3 Category of Topological Spaces

So much of topology is driven by the need to study continuous functions. From an early age, we become accustomed to the idea of a function and functional dependence, where change in one variable quantity relates to change in another variable quantity, thus establishing a functional dependence between one variable (input) and another (output). Such dependence can unfold in a number of general ways, of course, but we typically learn to think of the property of continuity of functions in terms of

> a big change in the output implies that there must have been a big change in the input.

The standard ϵ-δ definition of continuity one learns in calculus is one way to get a better handle on the implied question of "how big?" There, of course, we say that

> a change in the output greater than ϵ implies a change in the input greater than δ,

or,

> if the change in input is bounded by δ, then the change in output is bounded by ϵ.

More explicitly, a function $f:\mathbb{R}\to\mathbb{R}$ is defined as continuous provided for every $x\in\mathbb{R}$ and every real number $\epsilon>0$, another $\delta>0$ can be found such that

$$|f(x)-f(y)|<\epsilon \text{ for every } x\in\mathbb{R} \text{ with } |x-y|<\delta.$$

Regardless of which maximal error $\epsilon>0$ is required, the idea is that there is always an interval around x—which is $(x-\delta, x+\delta)$, with size $\delta>0$—where all approximated function values $f(y)$ deviate by less than ϵ from the function value $f(x)$ being approximated.

In the context of the real line or plane, where we first learn these notions, we think about it in terms of an implied metric, so that continuity involves something like a control problem: we can control the error $f(x)$ to be lower than ϵ by keeping the error in the argument sufficiently small, that is, smaller than $\delta>0$. If you are measuring some x, using it to compute $f(x)$ where f is a continuous function, this ϵ-δ criterion allows you to find the maximal error δ in x (i.e., $|y-x|<\delta$), which guarantees that the final error $|f(y)-f(x)|$ will be smaller than ϵ. Such a δ may be found only if small changes around the argument x also determine small changes around the function value $f(x)$. Thus, for functions continuous at x, we must have

$$y\approx x \implies f(y)\approx f(x),$$

which just says that whenever y is sufficiently close to our point of interest x, then $f(y)$ will be approximately $f(x)$. Such an idea can of course be described using the notion of an ϵ-neighborhood: for every ϵ-neighborhood $(f(x)-\epsilon, f(x)+\epsilon)$ around $f(x)$, there is always a δ-neighborhood $(x-\delta, x+\delta)$ around x, whose function values are all mapped into the ϵ-neighborhood.

As we ascend beyond the narrower context of Euclidean spaces, and even ultimately beyond metric spaces, we get a more general treatment of the notion of continuity as well.

Definition 120 A function $f:X\to Y$ between two topological spaces X, Y, is *continuous* iff $f^{-1}(U)$ is open in X for every subset U open in Y.

This shows how we can determine whether or not a function is continuous without using any information about a metric. We need only know which subsets of X and Y are declared open. Notice that we could just as well have defined continuity using closed sets, as $f^{-1}(Y\setminus A)=\{x\in X \mid f(x)\notin A\}=X\setminus f^{-1}(A)$.

Now, one can easily verify that for any topological space X, the identity map $\mathrm{id}_X : X\to X$ is continuous; that for any topological spaces X, Y, Z, and any continuous functions $f:X\to Y$ and $g:Y\to Z$, the composition

$$g\circ f : X\to Z$$

is itself continuous; and this composition will be associative.

Altogether, this informs us that topological spaces, together with continuous functions, form a category, one we call **Top**. More explicitly, the category **Top** has topological spaces for objects, and continuous functions for morphisms. In more detail: we have seen that a *topological space* is a pair $(X, \mathcal{O}(X))$—usually abbreviated by the carrier set X—where X is a set and $\mathcal{O}(X)$ are the open sets of the topology on X. Described in a more categorical way, the morphisms here are just functions $f: X \to Y$ such that for every $V \in \mathcal{O}(Y)$, the preimage $f^{-1}(V)$ in the order of all subsets of X is in $\mathcal{O}(X)$, that is, so that there exists an arrow along the top of the following square making the diagram commute (where the vertical arrows are just inclusions):

$$\begin{array}{ccc} \mathcal{O}(Y) & \longrightarrow & \mathcal{O}(X) \\ \downarrow & & \downarrow \\ \mathbb{P}(Y) & \xrightarrow[f^{-1}]{} & \mathbb{P}(X). \end{array}$$

In other words, a continuous map $f: X \to Y$ gives rise to a function $f^{-1}: \mathcal{O}(Y) \to \mathcal{O}(X)$ that carries an open subset $U \subseteq Y$ to its preimage $f^{-1}(Y)$, open in X—recapitulating the usual notion of continuity, as defined in general topology.

Notice how in the setup above, morphisms $(X, \mathcal{O}(X)) \to (Y, \mathcal{O}(Y))$ in **Top** already include a morphism between orders, just one that goes in the opposite direction $\mathcal{O}(Y) \to \mathcal{O}(X)$. In other words, there is a functor $O: \mathbf{Top}^{op} \to \mathbf{Pos}$ that takes a space X to its underlying poset $\mathcal{O}(X)$ of open subsets.

Given a topological space X, the open sets $\mathcal{O}(X)$, ordered among themselves by inclusions, forms a poset. As such, we can describe the following:

Definition 121 For a topological space X, the *category of open subsets* $\mathcal{O}(X)$ (or, if you prefer, **Open**(X)) of X is the category that has

- for objects: open subsets $U \hookrightarrow X$ of X; and
- for morphisms: inclusions $V \hookrightarrow U$ of open subsets $V, U \subseteq X$.

Given a topological space, this category of open subsets of that space will prove to be a category of great interest to us, as we can examine data assigned to a space in terms of what it does to the open subsets.

When we think about covers in terms of $\mathcal{O}(X)$, the poset of open subsets of X, ordered by inclusion, we are saying that an I-indexed family of open subsets $V_i \hookrightarrow U$ *covers* U provided the full diagram consisting of the sets V_i together with the inclusions of all their pairwise intersections

$$V_i \hookleftarrow V_i \cap V_j \hookrightarrow V_j$$

has U for its colimit. Roughly, one can think of a covering of a given object U as some sort of *decomposition* of that object into simpler ones, the resulting simpler pieces of which, when taken altogether, can be used to recompose all of U. At the outset, it is reasonable to just think of this in terms of specifying a collection of subregions that can be laid over a given region in such a way that the entire region is thereby covered, where an entirely obvious but still decisive observation is that such subregions making up the cover can *overlap* one another. The naive image to keep in mind is that we have a region U that we want to cover with some collection of pieces into which it may be regarded as being decomposed. Suppose we have some $V_1 \subseteq U$ and $V_2 \subseteq U$:

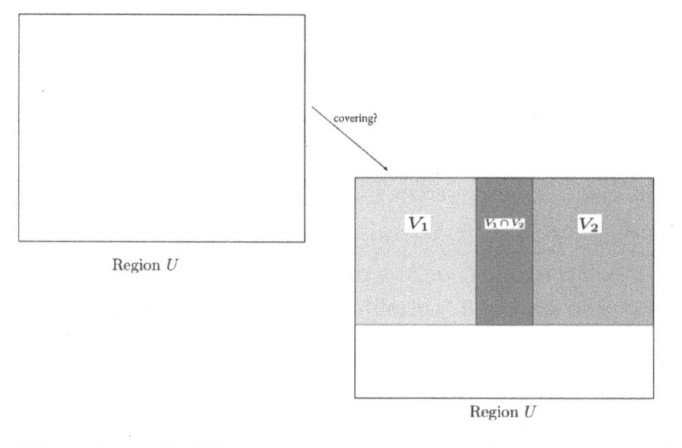

Clearly, V_1 and V_2 collectively fail to cover U, yet we can observe that there is a subregion where V_1 and V_2 overlap, which we call $V_1 \cap V_2$ and regard as specifying another "piece." Since V_1 and V_2 collectively cover more of U than either does individually, we should also consider the larger region (the entire northern half of U) that results from joining V_1 and V_2. We might continue in this manner, working our way up to a collection of subregions of U that actually cover all of U. For instance, we might have another V_3, laid on top of the entire southern half of the region (and partly overlapping with each of V_1 and V_2), such that the entire region U is now covered by the collection $\{V_1, V_2, V_3\}$. Altogether, the data of such a system of open sets, ordered by subset inclusion, will have the structure of a poset (this means, in particular, that we can regard $\mathcal{O}(X)$ as a category). In our particular case, this could be displayed by the diagram:

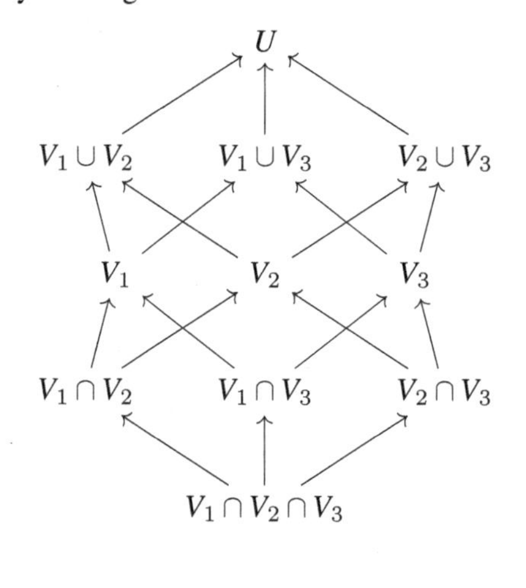

$\mathcal{O}(X)$

revealing the components of the space, together with their relevant inclusion relationships as members of a cover of the entire space.

Sheaves on a topological space can be described as particular presheaves on the open subsets $\mathcal{O}(X)$, presheaves that satisfy a further property. What ultimately will distinguish presheaves and sheaves is that sheaves are a special kind of presheaf—one that is "sensitive to" the information or structure of a cover.

In some sense, the sheaf notion is one that unfolds in an analogous fashion to something we see in analysis and in the context of metric spaces. In such settings, one first learns how to think about the continuity of a function f in terms of one that commutes with limits—where the limits come from sequences of points $\{a_n\}_{n=1}^{\infty}$ converging to a point a—in the sense that $\lim_{n\to\infty} f(a_n) = f\left(\lim_{n\to\infty} a_n\right) = f(a)$, and where $f(a)$ is ultimately independent of the sequence chosen to approximate a. At a much higher level of abstraction, the notion of a sheaf almost seems to amount to a retelling of this story. We saw in the last chapter how functors that commute with (the categorical version of) limits are said to be continuous. Letting the relevant limits now come from open covers, on this general categorical version a sheaf is fundamentally just a functor that commutes with limits—and, as such, we appear to have described something like a purified version of the continuity we first came to know in studying continuous functions on the real line and plane. The ingredients we will need to develop this notion of a sheaf will just be that of a functor (presheaf) and the notion of a cover of a space, so we have everything we need to jump right in.

4.4 Philosophical Pass: Open Questions

Box 4.1
Questions Concerning Elements of Topology

In the usual presentations of general topology, one notable oversight is that the characteristic conditions constituting a topology are almost never justified or properly clarified. One is left wondering why these conditions are so important, why they are what they are. Moreover, one is told that open and closed sets are basically formally dual to one another—yet nearly all treatments of topology go on to work almost exclusively with open sets, treating them as somehow primitive or indirectly alluding to their specialness. Finally, in the standard accounts, the true scope or activity of topology seems to be artificially limited, or misunderstood. In order to help rectify these things, the appendix takes up and addresses the following questions:

1. **Why are the axioms of a general topology what they are?** The standard account informs us that notions of general topology and the axioms of a topology itself arise from an abstraction from Euclidean space to metric spaces, followed by a further abstraction of certain properties of the open sets as definitive for a topology. While the Euclidean space motivation is easy enough to understand on its own, this often masks *why* the properties abstracted in the second step are the ones that survive the abstraction and come to constitute a general topology, arguably one of the most powerful notions in all of mathematics. One would like a clarification of these axioms and some account of why they are what they are. If the notion of a topological space really is some sort of generalization of features found in metric spaces, simply referring back to the example or instance of particular metric spaces, and the appearance of such features therein, cannot do anything to clarify why the axioms of a topology are what they are.

   Suppose some concept X, defined by a collection of properties T, is a generalization of concept Y, for which the same properties can be, and were first, observed to hold as a matter of fact. As a generalization, this means in particular that Y is an instance of X, yet concept X captures or describes phenomena or instantiations that are not Y. As such, an *explanation or justification* of why X is not only governed, but even characterized, by properties T instead of others, cannot be given by appealing to the simple *fact* that such properties can also be found governing instance Y.

2. **Why opens?** Nearly every modern text on general topology appears to give some sort of precedence to open sets, treating them as the primitives, even while assuring us that we could equally well have used closed sets, simply by appropriately dualizing the open set account. The status quo perspective on such matters seems to have little problem equivocating between taking open sets as primitive (presumably for some sensible, though never forthcoming, reason), on the one hand, and assuring us that "it's all the same, purely a matter of convention whether we use opens or closed sets" (which seems at times to involve a fundamental misunderstanding of what is and what is not entailed by such formal duality).

   This equivocation has historical roots. Historically, the first of the topological ideas to arise, deeply rooted as they were in problems in analysis, was that of the limit point of a set, used in formulating the notion of a closed set, which arose shortly thereafter. The very first formulation of what is now universally called a topological space, given by Kuratowski, is given in terms of the primitive operation of the closure of a set. Surprisingly, at least given its current status, the idea of an open set emerged last, and mathematicians appear to have been quite fine without it for longer than you might assume. (Moore (2008) has some useful history on these matters—especially on how many prominent mathematicians of the nineteenth century seem to have had little use for "open sets," and instead were initially, and for some time, apparently far more interested in defining and working with boundary points, limit points, and closed sets. The Sierpińksi quote below is taken from this paper.)

   It was not until the 1920s that definitions of a topology in terms of open sets even appeared. In his 1928 book *Introduction to General Topology*, Sierpiński paved the way for taking the notion of "open set" to be primitive, offering an axiomatic definition in terms of open sets that was close to the standard definition now given for a topological space. In the preface he wrote:

   > The axiomatic development based on the concept of an open set (as a basic concept) seemed to us simpler and more intuitive than other axiomatic treatments which will be mentioned. (Sierpiński 1934, iii)

   Later textbooks followed suit, usually with little more of substance than the fairly unconvincing justification given above.

   In short, we would like to know if there really are any mathematical—or even just more persuasive philosophical—reasons for treating open sets as primitive, as somehow more desirable to work with than their closed set counterparts. Or perhaps this is all confused and we need to do things differently.

3. **What is topology really about?** The standard story lends itself to a not unreasonable story about what we are doing when we are doing topology. While plausible on its own terms, it does seem that topology is about something slightly different, and more general, than the usual account would have us believe.

5 First Look at Sheaves

In which we meet sheaves, present a few intuition-building examples, and consider the idea of a sheaf.

The notion of a presheaf, together with the notion of a cover of a topological space, supplies us with all the ingredients needed to offer a first pass at the sheaf concept. A number of the examples of presheaves already introduced—including, in particular, those falling under the heading of *restriction*, such as the continuous functions, laws respected throughout jurisdictions, and the company stockpile examples—are, in fact, *sheaves*. This chapter turns, at long last, to a first presentation of sheaves, in the course of which will be seen a number of intuition-building examples. The chapter ends with a more philosophical discussion of the *idea* of a sheaf. The full definition supplied in this chapter—of a sheaf in the case of topological spaces—will be generalized in later chapters. As the book unfolds, we will build up gradually toward more and more involved examples.

5.1 Sheaves: The Topological Definition

The definition of a sheaf that we will presently be concerned with—sheaves on topological spaces—is a definition very much motivated by the action-as-*restriction* presheaf perspective. It may be thought of as involving a choice of data from each of the sets (as assigned by the presheaf to each piece of a cover of the space), that moreover forms a locally compatible family (meaning that it respects the restriction mappings and the chosen data items agree whenever two pieces of the covering overlap) and together induce or extend to a *unique* choice over the entire space being covered.

Since our sheaf candidate will already be a presheaf, to determine whether or not a given presheaf is a sheaf will just amount to testing for certain properties of a set-valued functor $F : \mathcal{O}(X)^{op} \to \mathbf{Set}$. That we have a presheaf on $\mathcal{O}(X)$ means, first of all, that to every open set U in $\mathcal{O}(X)$, we will have a set $F(U)$, where this is the value of the functor F on U. For reasons we will explore in more detail in chapter 8, we also sometimes call $F(U)$ the *set of sections over* U, and an element s in this $F(U)$ a *section over* U.[65] Moreover, corresponding

65. Strictly speaking, elements of $F(U)$, that is, value assignments specified over $U \subseteq X$, are called *local sections* of the sheaf F over U, to distinguish them from elements of $F(X)$, i.e., value assignments given over the *entire* space, which are called *global sections* of F. More broadly, whenever local information, such as elements like functions f and g given over certain domains, restricts to the same element in the intersection of their domains, then such f and g are called *sections*. We will develop this approach in terms of sections in more detail in chapter

to every inclusion of open sets $V \hookrightarrow U$ of the space, that F is a presheaf means that we will have a *restriction* $F(U) \xrightarrow{\rho_V^U} F(V)$. Recall also how, given $s \in F(U)$, it is common to denote its restriction to V by $s|_V$, that is, $\rho_V^U : F(U) \to F(V)$ takes $s \mapsto s|_V$ for each $s \in F(U)$, where we treat this like the usual restriction of a function. Observe that whenever we have three nested open sets $W \subseteq V \subseteq U$, restriction will be transitive, that is, for $s \in F(U)$, we have $(s|_V)|_W = s|_W$.

Altogether, with these two pieces of data,

$$U \mapsto F(U), \qquad \{V \overset{\subseteq}{\hookrightarrow} U\} \mapsto \{F(U) \to F(V) \text{ via } s \mapsto s|_V\}$$

we are just reiterating that we have a functor from $\mathcal{O}(X)^{op}$ to **Set**. Together with the notion of a cover, we are now ready to define a sheaf.

Definition 122 (*Definition of a sheaf*) Assume given X a topological space, with $\mathcal{O}(X)$ its partial order category of open sets, and $F : \mathcal{O}(X)^{op} \to \mathbf{Set}$ a presheaf. Then, given an open set $U \subseteq X$ and a collection $\{U_i\}_{i \in I}$ of open sets covering $U = \bigcup_{i \in I} U_i$, we can define the following *sheaf condition*:

- Given a family of sections $s_1, \ldots, s_n$, where each $s_i \in F(U_i)$ is a value assignment (section) over U_i, whenever we have that for all i, j,

$$s_i|_{U_i \cap U_j} = s_j|_{U_i \cap U_j},$$

then there exists a *unique* value assignment (section) $s \in F(U)$ such that $s|_{U_i} = s_i$ for all i.

Whenever there exists such a unique $s \in F(U)$ for every such family, we say that F satisfies the sheaf condition for the cover $U = \bigcup_{i \in I} U_i$. The presheaf F is then a *sheaf* (full stop) on the space whenever it satisfies this sheaf condition for *every* cover.

Let us break this definition down into four, more easily digestible, steps. Given a presheaf on some space and a cover, the definition of a sheaf begins by making use of what is sometimes called a *matching family*:

Definition 123 A *matching family* $\{s_i\}_{i \in I}$ of sections over $\{U_i\}_{i \in I}$ consists of a section s_i in $F(U_i)$ for each i—chosen from the entire set $F(U_i)$ of all sections over U_i—such that for every i, j, we have

$$s_i|_{U_i \cap U_j} = s_j|_{U_i \cap U_j}.$$

In other words: given a data assignment s_i throughout or over region U_i and a data assignment s_j over region U_j, if there is agreement or consistency between the different data assignments when these are restricted to the subregion where U_i and U_j overlap, then together the data assignments s_i, s_j give a matching family. As the definition requires that it holds for every i, j, the idea is that we can build up larger matching families of sections via such pairwise checks for agreement.

Digesting the definition of matching family is the first step in grasping the definition of a sheaf. Next, the definition specifies what is sometimes called a *gluing* (or the *existence*

8; meanwhile, we may make use of the convenience of such language. For now, one thing to realize is that, for an arbitrary presheaf on a space, the set of global sections of the presheaf on the overall space may be different from the set of local sections given over all the open subsets; the "gluing" axiom (discussed below) is there precisely to make sure that this difference disappears.

condition). Given a matching family for our cover of the space U, we call a section over U itself a gluing if, whenever this data assignment over all of U is restricted back down to each of the subregions or pieces that make up the cover of the entire object, it is equal to the original local data assigned to each subregion. The definition stipulates that such a gluing $s \in F(U)$ exists.

Not only does such a gluing exist, but to have a sheaf, we require that there is a *uniqueness* condition. Specifically, there exists a *unique* section $s \in F(U)$ such that $s|_{U_i} = s_i$ for all i. In other words, if $t, s \in F(U)$ are two sections of $F(U)$ such that $s|_{U_i} = t|_{U_i}$ for all i, that is, they are equivalent along all their restrictions, then in fact $s = t$ (i.e., they must be the same, so we have *at most one* s with restrictions $s|_{U_i} = s_i$).

With the notion of a matching family and that of a unique gluing, we can form the notion of the sheaf condition, and the definition is basically complete. The idea is that if *for every matching family, there exists a unique gluing*, then we say that the presheaf F satisfies the *sheaf condition*. A presheaf will then be a sheaf whenever it satisfies this sheaf condition for every cover. That is all—we have defined what a sheaf is!

Stepping back, we can accordingly break down the definition of a sheaf on a topological space into a particular presheaf that moreover satisfies two conditions with respect to a cover: (1) existence (or gluing); and (2) uniqueness (or locality).

Definition 124 (*Definition of a sheaf (again)*) Given a presheaf $F : \mathcal{O}(X)^{op} \to \mathbf{Set}$, an open set U with open cover by $\{U_i\}_{i \in I}$, and an I-indexed family $s_i \in F(U_i)$, then F is a sheaf provided it satisfies both:

1. (*Existence/gluing*) If, for each i, there is a section $s_i \in F(U_i)$ satisfying that for each pair U_i and U_j the restrictions of s_i and s_j to the overlap $U_i \cap U_j$ match (or are "compatible")—in the sense that

$$s_i(x) = s_j(x)$$

for all $x \in U_i \cap U_j$ and all i, j—then there *exists* a section $s \in F(U)$ with restrictions $s|_{U_i} = s_i$ for all i. (Here, such an s is then called the gluing, and the s_i are said to be *compatible*.)

2. (*Uniqueness/locality*) If $s, t \in F(U)$ are such that

$$s|_{U_i} = t|_{U_i}$$

for all i, then $s = t$. (In other words, there is *at most one* s with restrictions $s|_{U_i} = s_i$.)

Together, these two axioms assert that compatible sections can be uniquely glued together.

If F and G are sheaves on a space X, then a morphism $f : F \to G$ will just be a natural transformation between the underlying presheaves. This lets us define $\mathbf{Sh}(X)$ the *category of sheaves* on X, which has sheaves for objects and natural transformations between them for morphisms. There are in general far more preshcaves on a space than there are sheaves on the space. This category of sheaves on X will be a (full) subcategory of the category of presheaves on X, giving the inclusion functor

$$\iota : \mathbf{Sh}(X) \to \mathbf{Set}^{\mathcal{O}(X)^{op}}.$$

5.1.1 A Sheaf as Restriction-Collation

Before launching into examples, let us briefly consider how the description of a sheaf (of sets) on a topological space can be motivated by simple observations concerning functions. We know that specifying a topology on a set X lets us define, in particular, which functions are continuous, so that we can consider all the continuous functions from a space X (or some open $U \subseteq X$) to the reals $\mathbb{R}$. Whether or not a function $f: U \to \mathbb{R}$ is continuous is something that can be *determined locally*. But what exactly does this mean? This fundamentally amounts to saying two things:[66]

1. *Restriction* (or *Identity*): If $f: U \to \mathbb{R}$ is continuous, and $V \subseteq U$ is open, then restricting f along V, that is, $f|_V : V \to \mathbb{R}$, yields a continuous function as well.
2. *Unique collatability* (or *Gluability*): If U is covered by open sets U_i, and the functions $f_i : U_i \to \mathbb{R}$ are continuous for all $i \in I$, then there will be *at most one* continuous $f: U \to \mathbb{R}$ with restrictions $f|_{U_i} = f_i$ for all i. Furthermore, this f will exist precisely when the given f_i *match* on all the overlaps $U_i \cap U_j$ for all i, j, that is, $f_i(x) = f_j(x)$ for all $x \in U_i \cap U_j$.

One might accordingly think about the "localness" of a given function's property (such as continuity) as involving two sorts of compatibility conditions or constraints tending in two different directions (the first "downward" and the second "upward"): (1) that which requires that information specified over a larger set is compatible whenever *restricted* to information over a smaller open set; (2) that which involves conditions on the *assembly* or *collation* of matching information on smaller opens into information given over larger open sets. One might also think of the first condition as the "localizing" part, and the second condition as the "globalizing" part.

While continuous functions provide a particularly natural example of these sorts of requirements, there is no need to restrict ourselves to *continuous* functions. Various properties such as differentiability, real analyticity, and other structures on a space X (including involving things that are not even functions, but are function-like) are in fact determined locally in the same sort of way. The underlying idea here is that certain functions (or things that behave like functions), thought of as having some property P, are defined on the open sets in such a way that one can check for this property in a neighborhood of every point of the space—this is fundamentally what makes it *local*—and then each inclusion $V \subseteq U$ of open sets in X will determine a function $\rho^U_V : P(U) \to P(V)$, for which we just write $t \mapsto t|_V$ for each $t \in P(U)$, treating it like the usual restriction of a function (which restriction is, moreover, transitive). Altogether, this just says that we have defined a functor $P: \mathcal{O}(X)^{op} \to \mathbf{Set}$; and saying that P is such a presheaf (functor) simply expresses the first (restriction) condition given above. The second condition mentioned above, unique collatability or gluability, can in turn be described category-theoretically in terms of an equalizer diagram for a corresponding covering.[67] Accordingly, the two requirements of restriction and unique collatability supply the model for how to define a sheaf more generally—as a functor for which the corresponding equalizer diagram, containing the information of the open sets and the cover, is an equalizer for all coverings. This description ultimately

66. This perspective of restriction-collation is derived from Mac Lane and Moerdijk (1994).
67. We will see how this works below.

enables the sheaf construction for a wide class of structures. But this motivation in terms of certain properties of classes of functions being checked locally is a particularly useful perspective to keep in mind as one thinks about the construction of sheaves in general.

Let us now dive right into some examples of sheaves. Over the course of the book, we will provide a multitude of examples, ranging from the intuitive to the more computationally explicit and involved. With the next examples, we start with a couple of simple sheaves, occasionally omitting some of the details, only to develop some initial *intuition* for the sheaf concept and to leave the reader with a number of suggestive pictures and guiding examples. In the sections and chapters that follow, more elaborate and complicated examples are given.

5.2 Examples

Example 125 The presheaf of continuous real-valued functions on a topological space X, first introduced in example 55 (chapter 2), is a sheaf, specifically a sheaf of real algebras associating to each open $U \subseteq X$ the algebra $F(U)$ of real-valued continuous functions defined there. Not only can we restrict functions down to any open subset, but we can also glue together local assignments whenever they agree on overlapping regions, producing a global assignment, that is, a consistent assignment *over the entire region* that will agree with the local assignments when restricted back down to each subregion. Uniqueness in this case is automatic from the fact that we are dealing with *functions*.

Overall, this process of the pairwise compatibility checks and the subsequent gluing is nicely captured by an image of the following sort (where, for each piece, we just depict the choice made from the overall set of all continuous functions over that region):

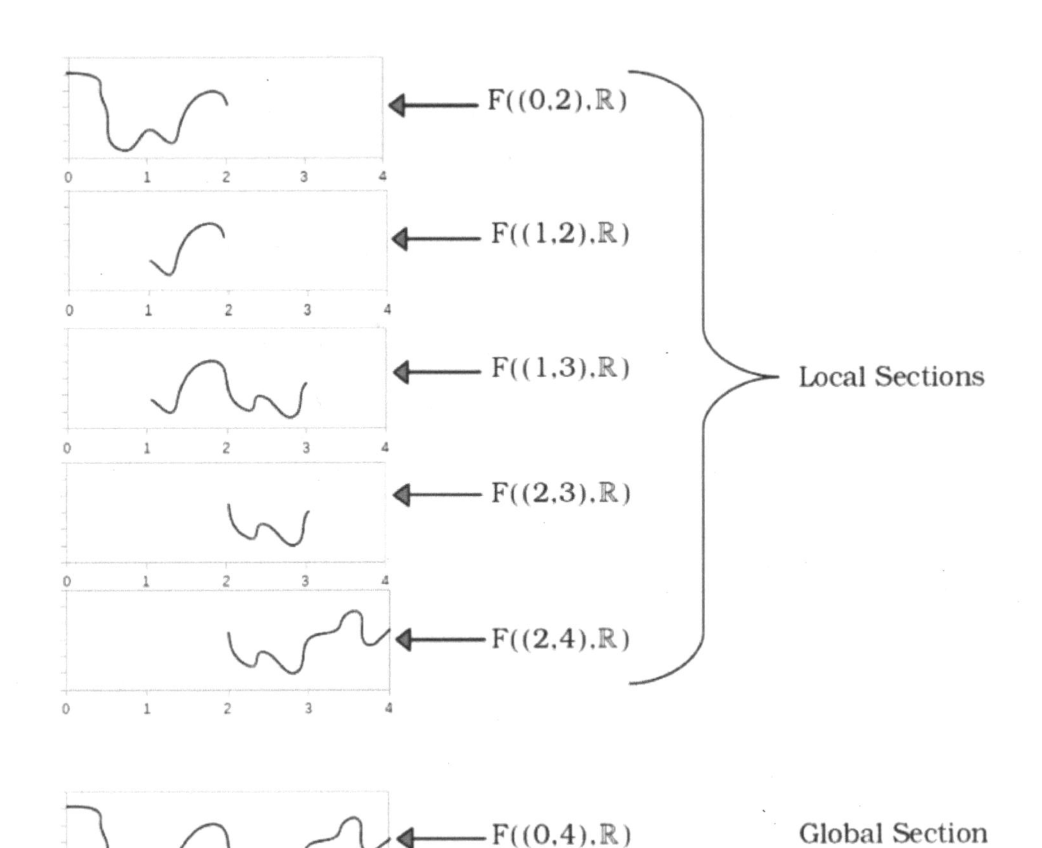

Example 126 Let us revisit the example of the presheaf of laws being respected throughout a jurisdiction (a geographic area over which some legal authority extends), again first introduced in example 55. For X the entire world, to each jurisdiction $U \subseteq X$ we assigned the set $R(U)$ of laws being respected throughout the region U. Is this presheaf R a sheaf? Well, we can check: Given some law respected throughout U and another law respected throughout W, do they amount to the same law on the subregion where U and W overlap?[68] (If there is no overlap, then this is trivially satisfied.) Now repeat this check for each such pair of overlapping regions.

For instance, on U there might be a law that stipulates "no construction near sources of potable water," while on W a law might stipulate "no construction in public parks." If it turns out that on the overlapping subregion $U \cap W$ all public parks are near sources of potable water (and vice versa), then the laws agree on that overlapping region, and thus can be glued together to form a *single* law about construction that holds throughout the union $U \cup W$ of the two.

68. In fact, as we will see later on, we do not strictly need them to be exactly the same law—we just need there to be a consistent system of translation between the sets of laws, that is, a set of isomorphisms translating between each such pairs of sets of laws.

This might seem like a rather harmless or trivial construction, but consider that the global sections of such a sheaf R would tell you exactly those laws that are respected by everyone throughout the planet. This would be a useful piece of information! (For instance, it might reveal the sorts of shared values that are ultimately respected, in one form or another, by every society.) The process of checking for agreement on overlapping regions is straightforward, but the resulting observations or data assignments one can now make concerning the entire space, via the global sections, can be very powerful and far-reaching.

Incidentally, the alert reader might wonder whether we have any right to speak of such a construction in terms of a presheaf and sheaf on open sets, since one might plausibly question whether the underlying jurisdictions even form open sets. One might assume, for instance, that states are in fact best thought of as closed sets with measure 0 overlaps; thus, laws would not stitch together in nontrivial ways, as the probability of observing a violation in that boundary territory is zero.

However, it can be argued that jurisdictions are indeed rightly thought of in terms of open sets. To see this, consider the following rough line of reasoning: Suppose a car is speeding across state lines, from state A into state B. Now suppose a cop from state A has a radar gun—think open ball—homing in on the boundary itself. But no matter how much the police department spends on refining its radar gun to arbitrary precision (error of ϵ), there is no way for the cop to verify (via a measurement device of some nonzero error) that the car clocked as speeding right on the boundary is in fact speeding, without also extending its observation into state B. Legally, states have jurisdiction (power to prosecute) any crime that occurs within that state, but not necessarily in another state. Moreover, the laws might not even be the same across state lines—in particular, the speeding car might not in fact be speeding in state B.

The moral is this: If a crime occurring in state A is to be prosecuted by that jurisdiction, then state A needs to be able to witness/verify that the crime occurred within state A. But if the boundary is to be included in state A itself, making state A a closed set, then any attempt to verify the crime (with a neighborhood or approximation window centered on the point on the boundary) will also require that state A assumes jurisdiction over happenings in state B. But, by definition, a state does not have such jurisdiction. It seems that state A should *not* be regarded as closed.

By contrast, it is entirely plausible to regard the extent of A's jurisdiction as an open set. In general, following the story developed in the appendix, we can interpret the interior of a region U as the set of states in which properties concerning U are verifiable, so that open sets are construed in terms of

> sets U such that U is equal to $\mathbf{int}(U)$ = observations/propositions whose truth is equivalent to their verifiability.

Moreover, whenever an observable property (such as speeding) can be verified as true, it is verified by an open set (think a radar gun), and any more refined approximation must also verify that property. This captures the idea of openness. Following the story developed in the appendix, we can construe open sets in terms of a verificationist paradigm—and since states have to verify crimes in order to prosecute and so have meaningful jurisdiction, it makes sense to regard state jurisdictions in terms of open sets.

Example 127 For the next example, we consider a satellite making passes over portions of a region of earth, collecting data as it goes, or various satellites doing the same thing. For concreteness, consider some specific portion of the earth, say Alaska, or the part of Alaska where the Bering Glacier resides, as a topological space X. Then given an open subset $U \subseteq X$, we can let $S(U)$ denote the set of functions from U to C, where C might be the set interval of wavelengths in the light spectrum, or some geo-referenced (perhaps time-stamped) intensity-valued image data, or some other data corresponding to the data feed of the satellites (or the processing thereof). This presheaf S is in fact a sheaf, since we can indeed fuse together the different data given over the open sets of X, forming a larger patched-together image of the glacier. For concreteness, assume we are given the following selection of three satellite images of the Bering Glacier, chosen from among the (possibly very large) sets of images assigned to each region:[69]

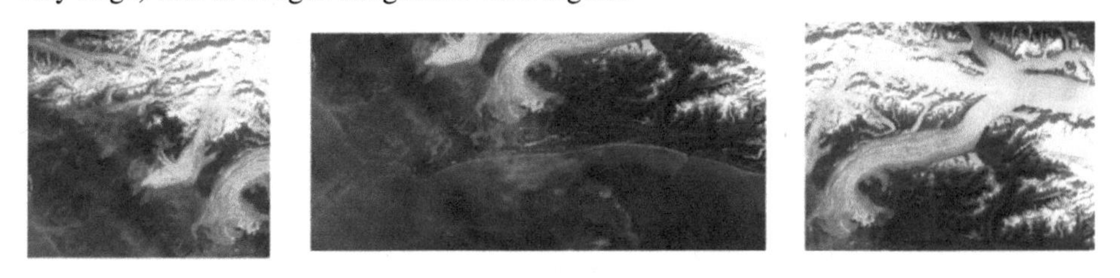

v_1 v_2 v_3

Each of the $v_i \in S(U_i)$ correspond to value assignments throughout or over certain subsets, U_1, U_2, U_3 of X, which together cover some subset $U \subseteq X$—say, the underlying region of the earth corresponding to the Alaskan glacier. In terms of the data sitting over each of these regions, as provided by each of the satellites in the form of individual images, we can check that the restriction of the value assignment v_1 to the underlying region $U_1 \cap U_2$ is equal to the restriction of the value assignment v_2 to the same subset $U_1 \cap U_2$, and so on, all the way down to their common restriction to $U_1 \cap U_2 \cap U_3$:

69. The images come from Landsat 8: https://earthobservatory.nasa.gov/IOTD/view.php?id=4710.

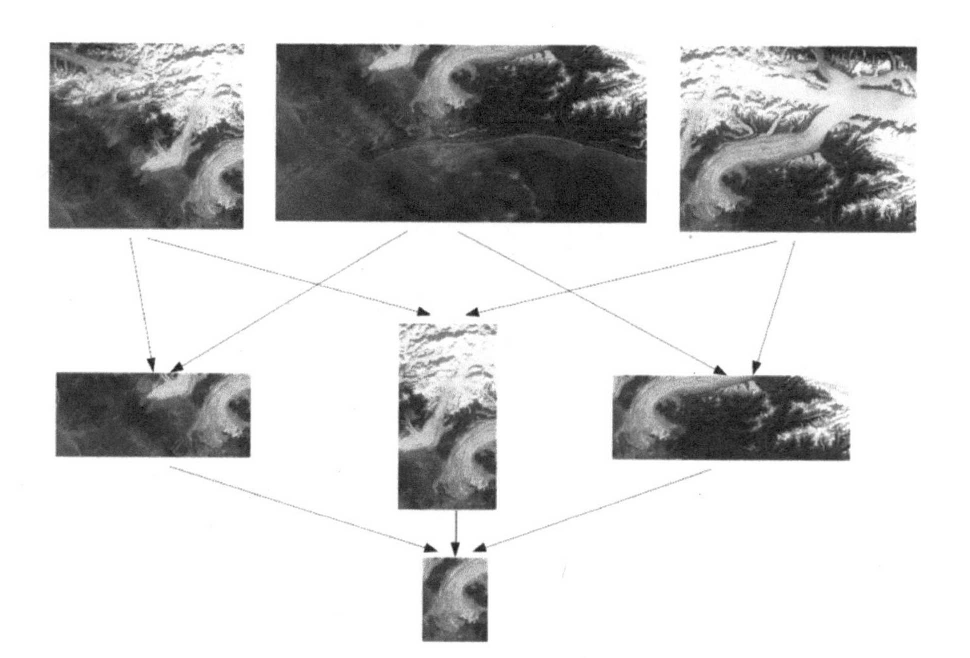

One can thus immediately *see* that the sheaf condition is met, and we can in fact patch together the given local pieces or sections over the members of the open covering of U to obtain a unique section specified over all of $U = U_1 \cup U_2 \cup U_3$. In summary, we have the following inclusion diagram (on the left) describing the underlying structure of the open sets of the topology, paired with the sheaf diagram (on the right) with its corresponding restriction maps (notice the change in direction):

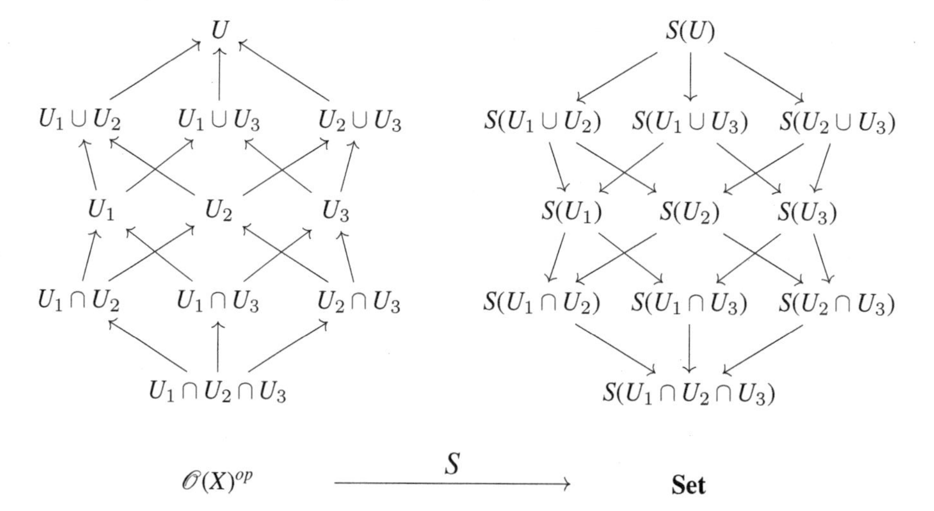

In terms of the given value assignments (actual images), the sheaf diagram on the right is pictured below, where we can think of the restriction maps as performing a sort of cropping operation, corresponding to a reduction in the size of the domain of the sensor, while the gluing operation corresponds to gluing or collating the images together along their overlaps all the way up to the topmost image (which corresponds to the section or assignment over all of U).

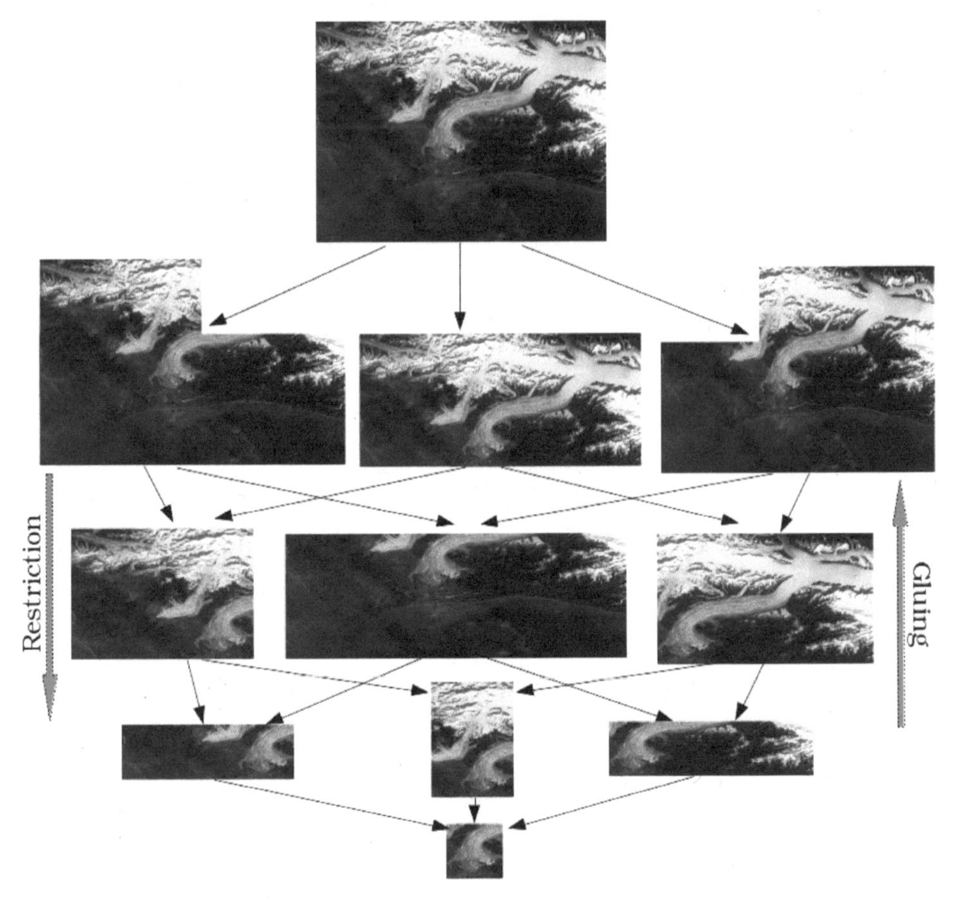

This mosaic example gives a particularly concrete illustration and motivation for another definition of a sheaf, namely as a functor $F : \mathcal{O}(X)^{op} \to \mathbf{Set}$ that *preserves limits*. Because we use the opposite category for domain in defining the presheaf-functor, this means that *colimits* in the domain category of open sets get sent to limits in **Set**. In the partial order (in fact, lattice) of open subsets of X, for an I-indexed family of open subsets $U_i \subseteq U$ (in the particular case described above, $I = 3$) that *covers* U—in the exact sense that the entire diagram comprising the sets U_i and the inclusions of their pairwise intersections $U_i \cap U_j$ has U for its colimit—the contravariant functor S given above *preserves* this colimit in the sense that it sends it to a *limit* in **Set**. Referring back to the universal characterization of these notions introduced in chapter 3, with this example one can basically immediately see that while all possible cocones will have to pass through the universal cocone given by U, that is, U is initial among cocones; $S(U)$, to which U is mapped by the functor, will be terminal among its associated cones. More formally,

Definition 128 (*Yet another definition of a sheaf on a topological space*) Given a presheaf $F : \mathcal{O}(X)^{op} \to \mathbf{Set}$ from the poset of open sets of the space X to **Set**, and defining an I-indexed family of open subsets $U_i \subseteq U$ as a *cover* for U when the entire diagram consisting of all the U_i together with the inclusions of their pairwise intersections $U_i \cap U_j$—that is, $U_j \hookleftarrow U_i \cap U_j \hookrightarrow U_i$ —has U for its colimit, then such a presheaf F is a *sheaf* (of sets) provided it preserves these colimits, sending them to limits in **Set**.

This means, in effect, that for any open cover $\{U_i\}_{i \in I}$ of U (colimit), the following is an equalizer diagram

$$F(U) \overset{\rho^U_{U_i}}{\rightarrowtail} \prod_{i \in I} F(U_i) \underset{q}{\overset{p}{\rightrightarrows}} \prod_{i,j \in I} F(U_i \cap U_j)$$

in **Set** (recall that an equalizer diagram is a *limit* diagram). Here, for $t \in F(U)$, and letting the equalizer map $\rho^U_{U_i}$ (i.e., $F(U_i \hookrightarrow U)$) be denoted by e, we will have that

$$e(t) = \{t|_{U_i} \mid i \in I\},$$

and for a family $t_i \in F(U_i)$, we will have

$$p(t_i) = \{t_i|_{(U_i \cap U_j)}\}, \qquad q(t_i) = \{t_j|_{(U_i \cap U_j)}\},$$

the map p being $\rho^{U_i}_{U_i \cap U_j}$ (i.e., $F(U_i \cap U_j \hookrightarrow U_i)$) composed with the appropriate projection map, while q is $\rho^{U_j}_{U_i \cap U_j}$ together with its projection map.

An arrow *into* a product is entirely determined by the components, namely its composition with the projections of the product. Thus, the maps e, p, and q of the equalizer diagram above are precisely the unique maps making the "unfolded" diagrams below commute for all $i, j \in I$ (where the vertical maps are the relevant projections of the products):

$$\begin{array}{ccccc}
 & & F(U_i) & \xrightarrow{\rho^{U_i}_{U_i \cap U_j}} & F(U_i \cap U_j) \\
 & \nearrow & \uparrow & & \uparrow \\
F(U) & \overset{e}{\dashrightarrow} & \prod_i F(U_i) & \underset{q}{\overset{p}{\dashrightarrow}} & \prod_{i,j} F(U_i \cap U_j) \\
 & \searrow & \downarrow & & \downarrow \\
 & & F(U_j) & \xrightarrow[\rho^{U_j}_{U_i \cap U_j}]{} & F(U_i \cap U_j).
\end{array}$$

The utility of this alternate description is that it furnishes us with a completely *categorical* description of the equalizer diagram, which means that the above definition of a sheaf will work even when we replace **Set** with other suitable categories (specifically, those with all small products). In other words, we might just as well have provided a definition of sheaves $F : \mathcal{O}(X)^{op} \to \mathbf{D}$ of **D**-objects on a space X; there are many prominent candidates for **D** in this more general definition, some of which we will meet in later chapters, for example giving rise to sheaves of abelian groups, vector spaces, rings, R-modules.

Before considering further examples, the reader should test their appreciation of the sheaf definition by verifying the following.

Theorem 129 If X is a topological space and F a sheaf (of sets) on X, then $F(\emptyset)$ will be a singleton set $\{*\}$.

Exercise 17 Verify this statement.

Solution Recall how the open set $\emptyset$ can be covered with the empty cover, that is, taking the index $I=\emptyset$. Thus, the gluing property will ensure that $F(\emptyset)\neq\emptyset$. The uniqueness property guarantees that any two sections given over $\emptyset$ will agree, since any cover of $\emptyset$ will be by empty sets. One can also appreciate this fact by considering how a product $\prod_i$ over an empty index set is just the singleton set $\{*\}$—as such, the above equalizer diagram is just

$$F(\emptyset) \rightarrowtail \{*\} \rightrightarrows \{*\},$$

from which we must have $F(\emptyset)=\{*\}$.

Here is another important result.

Example 130 Observe that on any topological space X, each of the open sets U of $\mathcal{O}(X)$ will determine a hom-functor presheaf Hom$(-, U)$, a fact that may be more readily appreciated after chapter 6. This is defined, for each open set V, as $\{*\}=1$ (the singleton set) if $V\subseteq U$, and $\emptyset$ otherwise. Recall also, from definition 41, the notion of a subpresheaf (subfunctor). Subfunctors are ordered by inclusion, and so form a partial order. We can further define a *subsheaf* of a sheaf F as just a subpresheaf (subfunctor) of F that is also a sheaf. More explicitly,

Definition 131 If F is a sheaf on X, then a subfunctor $S\subseteq F$ is a *subsheaf* iff, for every open set U, every element $f\in F(U)$, and every open covering $U=\bigcup U_i$, we have that $f\in S(U)$ iff $f|_{U_i}\in S(U_i)$ for all i.

Now observe that the terminal object in the sheaf category $\mathbf{Sh}(X)$ will be given by $Y_X=$ Hom$(-,X)$, called the terminal sheaf, denoted 1.

We can put these notions together to show that the open sets of a topological space are fundamentally the same as the subsheaves of the terminal sheaf.

Proposition 132 Given any topological space X, we have an isomorphism (of orders)

$$\mathcal{O}(X)\cong \text{Sub}_{\mathbf{Sh}(X)}(1).$$

Proof. There are a variety of more or less advanced ways of showing this, the others using notions we have not yet introduced. Let's instead show it in the most direct way.[70]

Suppose given any open set $V\subseteq X$. We can define a functor S_V on the open sets U by

$$S_V(U)=\begin{cases}1 & \text{if } U\subseteq V\\ \emptyset & \text{otherwise}.\end{cases}$$

S_V is itself a sheaf, and thus automatically gives a subsheaf of the terminal sheaf.

On the other hand, suppose given a subsheaf S of the terminal sheaf 1. Then, for each object, $S(U)$ will either be 1 or $\emptyset$. Being a functor, $S(U)=1$ and $W\subseteq U$ implies that $S(W)=1$ as well. Being a sheaf, and using the equalizer condition in the definition of sheaf, if $\{U_i\}_{i\in I}$ is an open cover of U and $S(U_i)=1$ for all i, then we must have $S(U)=1$. So we can let V

70. This proof follows Mac Lane and Moerdijk (1994, II. 2).

be the union of all the open sets U for which $S(U)=1$, from which it will follow that for all U, $S(U)=1$ iff $U \subseteq V$. But this just says that S is equal to the functor S_V defined a moment ago.

Altogether, with $V \mapsto S_V$ we thus have a bijection. This is moreover order-preserving, meaning that it is an isomorphism of orders (introduced properly in chapter 6). □

Remark 133 This seemingly unassuming result in fact tells us something rather significant—that the category of sheaves on a space X can be used to determine the topology on X. Put otherwise, the decisive data of the poset of open subsets of a topological space X can be recovered by just looking at all the sheaves on X. This is in part what motivated the remarks of Grothendieck (cited in the Introduction) to the effect that sheaves incorporate what is most essential about a space, and that the space can even be dropped and replaced by the category of sheaves on the space (and we know how to "reconstitute" the original space, as needed).

Example 134 Recall the functor $nColor : \mathbf{SmpGrph}^{op} \to \mathbf{Set}$, first introduced in example 37 (chapter 2). A graph is said to be *connected* if there is a path between every pair of vertices. Clearly, if a graph is not connected, then we can just color each connected component independently. As we are interested in coloring graphs with at most n colors, let us restrict attention to connected graphs.

We are already familiar with the notion of a subgraph of a graph; ultimately, a subgraph G' of a graph G corresponds to an inclusion arrow $G' \hookrightarrow G$ in **SmpGrph**. Then, a *cover* of a graph G is a family of subgraphs $\{G_i \hookrightarrow G \mid i \in I\}$ satisfying the condition that

$$\bigcup_{i\in I} \text{Nodes}(G_i) = \text{Nodes}(G) \text{ and } \bigcup_{i\in I} \text{Edges}(G_i) = \text{Edges}(G).$$

In the case of (undirected) connected graphs, we can define a subgraph G' of a graph G as a graph such that $\text{Edges}(G') \subseteq \text{Edges}(G)$, and the further fact that $\text{Nodes}(G') \subseteq \text{Nodes}(G)$ follows automatically since G' is assumed to be connected. In this setting, to define a cover of a graph G it thus suffices to specify a family of subgraphs $\{G_i \hookrightarrow G \mid i \in I\}$ satisfying the condition that

$$\bigcup_{i\in I} \text{Edges}(G_i) = \text{Edges}(G).$$

Using the subcategory $\mathbf{SmpGrph}_{\hookrightarrow}$ having connected undirected graphs for objects and just inclusion arrows for morphisms, and using the above notion of covers of a given connected graph by subgraphs, we can in fact form a sheaf out of the presheaf $nColor$: $\mathbf{SmpGrph}^{op}_{\hookrightarrow} \to \mathbf{Set}$. To appreciate this, observe that if graph G has an n-coloring, then clearly each of its subgraphs will have an n-coloring, so for any graph homomorphism (inclusion) $f : G' \hookrightarrow G$, $nColor(f)$ will be the function restricting the n-colorings of G to G'.[71]

For concreteness, we exhibit this in the case of a 3-coloring of the complete graph K_3 (together with its subgraphs, ordered in the natural way).[72] We first display what this functor assignment looks like over a particular subgraph of the graph K_3, then we display the

71. Srinivas (1991) appears to be the first to have described such a sheaf.

72. A *complete graph* on n nodes is just a (special sort of connected) graph in which every pair of distinct vertices is connected by a unique edge.

full diagram conveying the sheaf over the space of subgraphs. The pictures are very explicit and take care of all the details; by attending to the pictures, readers should be able to see for themselves how there are actually two distinct 3-coloring sheaves here, each got by selecting one of the two colorings (solutions) on all of K_3 and then restricting that down all the way through the inclusions.[73]

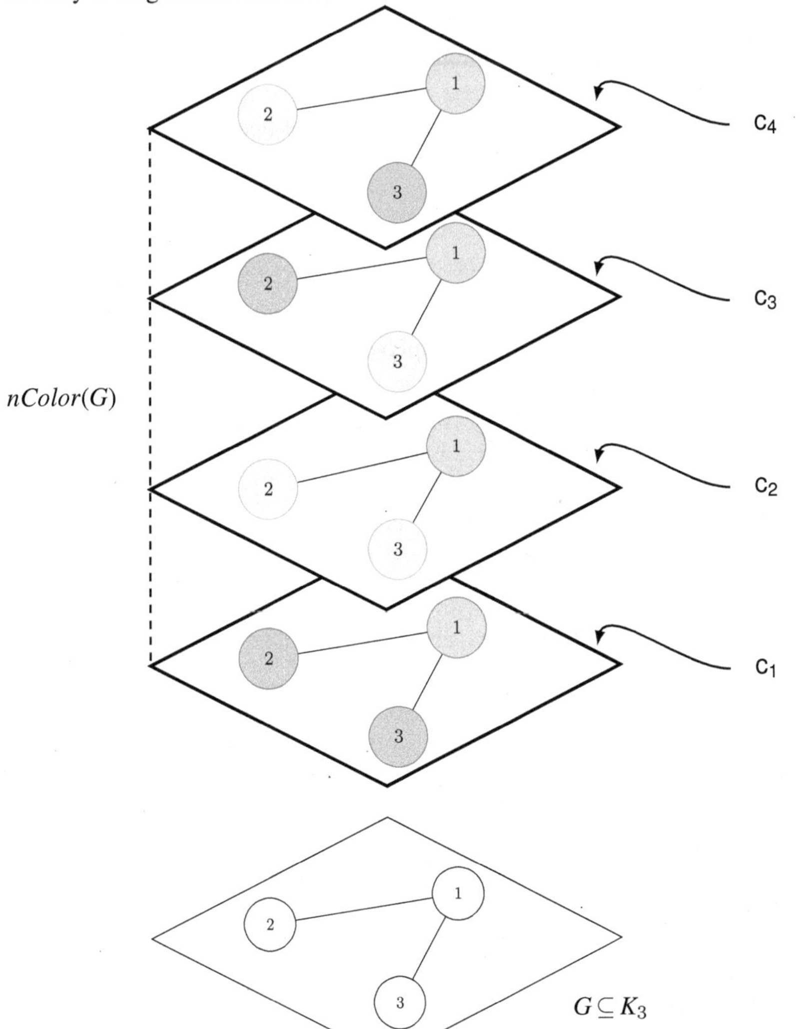

The next two suggestive examples are more "for fun," meant to emphasize or reinforce certain aspects of the *idea* of a sheaf, with less focus on the more mechanical or technical development of the construction.

73. The reader will note, however, that we do not represent all possible colorings, but only those colorings that have already fixed the coloring of the vertex 1 as blue. The rest, however, display colorings that are ultimately isomorphic to these two, reflecting the fact that of the six 3-colorings of K_3, there are only two nonisomorphic ones.

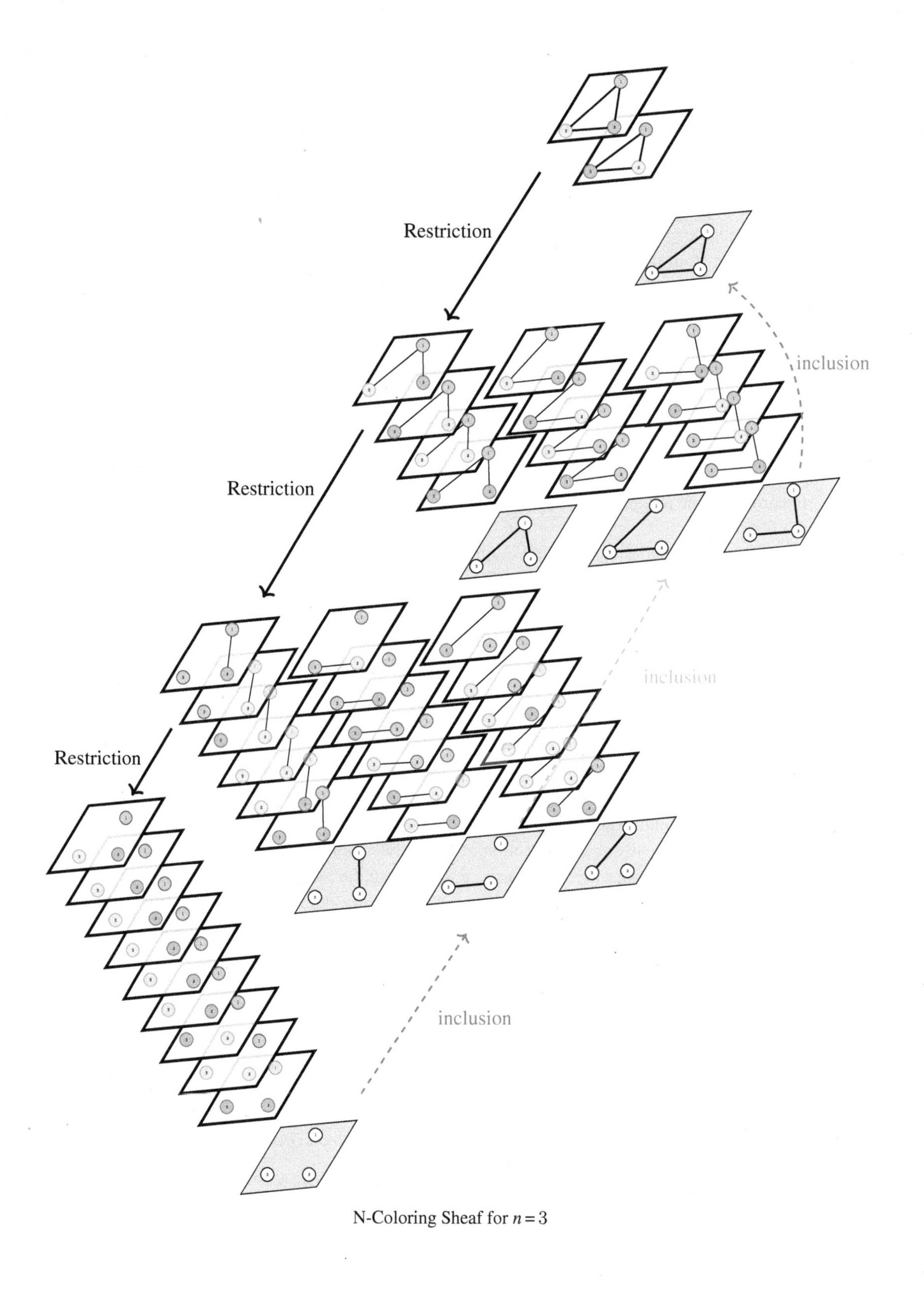

N-Coloring Sheaf for $n = 3$

Example 135 The twentieth century pianist Glenn Gould was one of the first to ardently defend the merits of studio recording, and use of the *tape splice* in the creative process, against those who held fast to the supposedly more "pure" tradition of the live concert performance and who accordingly thought that the only purpose of the splice would be to rectify performance mishaps or to alleviate the pressure of the "one-take" approach demanded by the concert form.[74]

Gould challenged the view that the only legitimate continuity of a unified interpretation could come from the one-takeness of traditional performance, proposing instead that the performer's newfound editorial control in the recording studio would bestow upon creators an even more demanding ethic concerning matters of architecture and integrity of vision. Gould defended the idea that new, explicit, and more demanding forms of continuity were to be found in this montage-based approach: "splicing builds good lines, and it shouldn't much matter if one uses a splice every two seconds or none for an hour so long as the result *appears* to be a coherent whole." Just as one does not demand or expect that the filmmaker shoot a film in one shot, Gould believed that one should not expect that the coherence or continuity of an interpretation of a musical piece can be secured only by the inexorable linearity of time and the single take—the musician has just as much a right to montage as the filmmaker.

Gould went as far as to test, with a controlled experiment involving eighteen participants, whether listeners (including laymen and recording experts) could detect the "in point" of any splice in certain selections of recordings, each of which selection had drastically different splice densities (in some cases, none).[75] What he found, in short, was that "the tape does lie and nearly always gets away with it." While originally (with analog magnetic tape splicing) the tape splice involved careful (and literal) cutting of the physical tape with scissors or a blade and (literal) gluing or taping of it to another section of tape (possibly from an entirely different recording session)—whenever relevant qualities, such as tempo and dynamics, of the two recordings could be made compatible enough on their overlapping measures to permit a seamless joining—Gould foresaw the power inherent in the more general *idea* of splicing and montage. He saw in this new approach the opportunity to provide a more analytically acute dissection of the minute connections ultimately defining the coherence of a particular piece of music, encouraging the performer to display the architectural coherence of a piece of music less dogmatically than if they had to rely on the "in-built continuity" alleged to belong to the one-take concert ideal—focusing instead on breaking a score down into parts, recording many distinct takes of sections of the score, and then splicing together the results of certain of those distinct performances whenever they could be made compatible on their overlap (all with the aim of producing a single, unified performance of the entire score). Gould's insistence on the virtues of tape splicing and montage in recording practice nicely captures something akin to the fundamental spirit of the sheaf construction.[76]

74. A "splice" is an edit point representing the confluence or merging of distinct takes or inserts (i.e., the recorded performance of a portion of a musical score).

75. The results can be found in his essay "The Grass Is always Greener in the Outtakes," in Gould (1990, 357-367).

76. Incidentally, Gould's closeness to the "sheaf philosophy" is evidenced in a number of aspects of his life, not just his approach to his art; for instance, via his insistence on forever integrating as many disparate planes and partial pieces of information as possible into a single coherent experience, like his alleged habit of simultaneously

Moreover, while the one-take approach to recording and unifying the musical idea is essentially *deductive*, reducing the individual (voice, line, note) to its participation in a prefabricated idea of totality and relying on a dubious notion of some immediate continuity, montage/splicing (like the sheaf construction) is fundamentally *inductive*, allowing the individual component materials of a work to create their own formal structure "from the bottom up" via insistence on the transparent and explicit unfolding of the principles by which the component parts can be patched together locally. According to Gould, it is precisely through the initial discontinuity induced by the decomposition into parts and the cutting process in montage/splicing that the task of making explicit the principle of their reorganization/patching into a unified totality is allowed to emerge, and the unity of the whole no longer regarded as something to be taken for granted. Just as in the sheaf construction, this approach essentially involves both cutting (decomposition/discontinuity) of the space (the score) and local patching or gluing (recomposition/continuity) of data or happenings associated to that space (distinct partial recordings), gradually building up to a unique data assignment over the entire space (recording of the entire score).

For a more concrete, if rough and simplistic, idea of how the splicing or montage approach to recording might be seen as akin to the construction of a sheaf, consider a score consisting of thirty-two measures. We might then consider that the "space" of the score has been decomposed into three principal parts or pieces: (A) spanning from measure 1 to the end of measure 16; (B) spanning from the beginning of measure 8 until the end of measure 24; and (C) spanning from the beginning of measure 16 until the final measure. Together, these portions collectively cover the entire thirty-two measure score, and there are the obvious pairwise overlapping measures. We can now imagine that to each section (A)–(C), there corresponds a (possibly very large) set of distinct recordings. If, for some selection of individual recordings from each of the three regions (A)–(C), the select recordings can be made to agree on their overlap—via some system of translation functions, such as slowing down one partial recording to match the tempo of another partial recording, adjusting the dynamics of chords bridging sections, and so on—then these can be spliced together into a unique single recording of the entire work.

Example 136 Detectives collect certain information pertaining to a crime that purportedly occurred in some area during a certain time interval. This information will most likely be heterogeneous in nature, that is, they may have camera footage of some part of the scene, some eyewitness testimony, some roughly time-stamped physical data, and so on. These

---

listening to all the conversations going on in a café. This mentality is perhaps most famously illustrated by the various accounts of him purposely dividing his concentration across multiple channels in order to better understand something. For instance, he claimed that he discovered he could best understand Schoenberg's Opus 23 if he listened to it while simultaneously playing the news on the radio, and he mastered a demanding section of a Beethoven sonata only after placing a radio and television next to the piano and turning them up as loud as they would go as he worked through the passage. This embodies something like the sheaf philosophy: Integration and coherence come about not through an enforced isolationism or homogeneity but precisely through immersion in the dense texture that arises by decomposition into pieces, by careful choices made locally, and by the resulting appreciation of the need to make explicit the most minute links between parts of a whole, as one gradually, piece by piece, assembles a more global or unified perspective. In this connection, we could also mention one of Gould's descriptions of his famous "contrapuntal radio" programs from the seventies, in which he claimed to try "to have situations arise cogently from within the framework of the program in which two or three voices could be overlapped, in which they would be heard talking—simultaneously, but from different points of view—about the same subject" (Payzant 2008).

various pieces of data are all considered to be local in the sense that they are assumed to concern (or be valid throughout) a delimited region of space-time, such as time-stamped camera footage of one of the parking lot's exits or an eyewitness testimony claiming to have heard a scream coming from the southern end of the parking lot sometime between 8:00 p.m. and 8:30 p.m. The various pieces of information may very well agree or be compatible with respect to what they say happened on the overlapping portion of their underlying space-time regions (to which these pieces of information are attached); for example, an eyewitness's testimony with respect to a particular half-hour interval and location might be checked against the camera feed concerning that same time interval and area. In general, the various pieces of data over the same interval may corroborate one another or contradict one another, either entirely or in some particular respect or with respect to some subregion of their overlap. It is not always as simple as verifying whether or not they provide the *same* information. It may happen, for instance, that the parking lot is constructed in such a way that certain barriers acoustically account for why the witness heard the scream coming from the southern end of the parking lot, when in fact it could only have come from the western end (which is where the camera shows the victim in conflict during that time). It is the job of the detective to find the appropriate "translation functions" making sense of these at first (potentially) conflicting local pieces of data and then use these functions to glue together, piece by piece, the data that can be made to locally cohere into a coherent and self-consistent account of what occurred over the entire spatiotemporal interval in question. In a rough sense, then, given a presheaf assigning information (such as camera data or propositions) locally over some collection of space-time regions, the detective is looking to build a sheaf over the entire space-time interval covered by all those regions.

This example also invites us to consider, in a very preliminary fashion, just one of the many interesting features of a sheaf: namely, that in addition to the fact that a sheaf lets us determine global unknowns or solutions from data given merely locally, whenever we do indeed have a sheaf, say F, this may enable the prediction of missing or underspecified data (or at least the specification of what data will be *possible*) with respect to some subregion, given some selection of data over another region.[77] For instance, given some time interval and some particular area, say, $U = (8{:}00, 9{:}00) \subseteq X = (7{:}00, 9{:}40)$ concerning the parking lot in question, we might consider some subset A of all the data we have over that interval, that is, $F(U)$. Now consider another region $V = (7{:}30, 8{:}30) \subseteq X = (7{:}00, 9{:}40)$. Diagrammatically, we thus have the following:

$$\begin{array}{ccc} & F(X) & \xrightarrow{\rho^X_V} & F(V) \\ & \downarrow{\scriptstyle \rho^X_U} & & \\ A \xrightarrow[i]{} & F(U) & & \end{array}$$

We know that because we are dealing with *sets*, we can form the pullback or fiber product, that is, $A \times_{F(U)} F(X) := \{(a, u, x) \mid i(a) = u = \rho^X_U(x)\}$:

77. The presentation of this observation in the next paragraph closely follows Spivak (2014, 429–430).

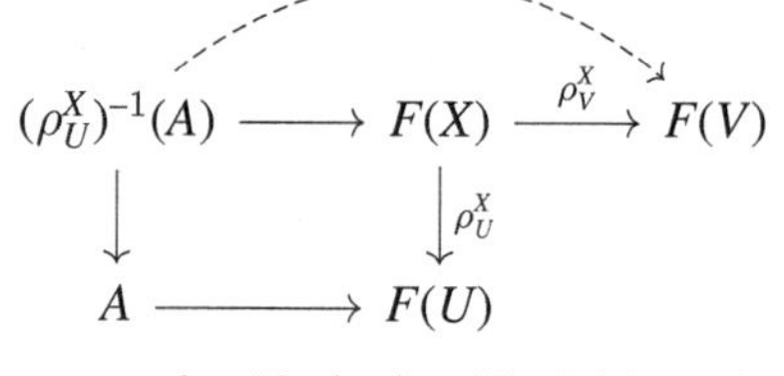

Then the image of the top composite (dashed) will yield a subset of $F(V)$, which informs us about the possible value assignments throughout V, *given* what we know to be the case (namely A) throughout U. Moreover, we could further consider maps into A and continue forming pullbacks; since the leftmost square will be a pullback iff the composite large rectangle forms a pullback, we can paste together pullbacks and further refine or constrain these predictions in a controlled way.

Example 137 Returning to the presheaf $S : \mathcal{T}^{op} \to \mathbf{Set}$ of a company's stockpile of products, we can describe a sheaf here. We might have described such a sheaf in terms of open sets, but using closed sets instead lets us consider a sheaf built on a closed set topology, and hint at some interesting nuances here, explored further in later chapters. While closed set sheaves exhibit interesting differences compared to the usual open set sheaves—in particular, manifesting the paraconsistent logic of closed set topologies of their base space, matters explored more in the appendix and parts of chapter 7—sheaves defined over the closed sets of a closed topology can be defined in fundamentally the same way as we did when using open sets of an open set topology.

If $\{[t_i, u_i] \mid i \in I\}$ covers the entire interval $[t, u]$—in our particular case, the interval from December 1 until July 1, say with cover given by two-month intervals—it is practically immediate that if a product is present in the company's stockpile throughout each of the pieces $[t_i, u_i]$ of the cover, then it will have to be present throughout all of $[t, u]$; and for any inclusion of intervals, S of that inclusion is the restriction function mapping each product onto itself (any product found throughout the larger interval must clearly be present throughout a subinterval). Global sections will then be given by those products that are consistently present throughout the entire time period, such as product B, as depicted below:

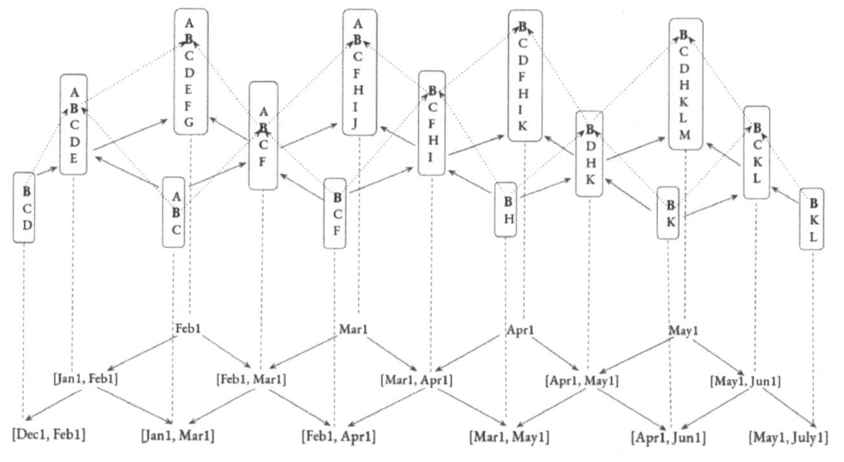

This may not appear to be a very exciting sheaf, but its simplicity can be useful in helping one achieve an initial working understanding of the difference between local sections that can extend to global sections and local sections that are purely local and satisfy certain local compatibility checks but that cannot be glued together into a global section. To see this, suppose, for instance, we had instead selected the product C, which indeed appears to be present throughout much of the overall time period. It is certainly present throughout all of $[Dec1, Apr1] = [Dec1, Feb1] \cup [Jan1, Mar1] \cup [Feb1, Apr1]$, and it is also present $[May1, Jun1]$. As can be seen by inspection,

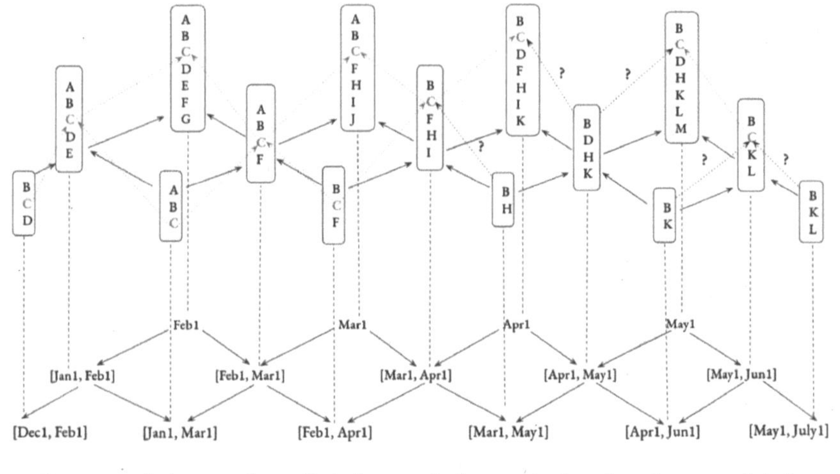

the persistence of the product C (witnessed, in particular, by the associated arrows) throughout all of $[Dec1, Apr1]$ is "proven" by the existence of local maps between all the subintervals covering this region. However, some (non)maps are indicated with question marks because there is not in fact any selection from the set of products given over (for example) $S([Mar1, May1] = \{B, H\}$ that could get mapped, under the prescribed action of S, to C in $S([Mar1, Apr1])$. Since there is, however, a map from $S([Mar1, Apr1])$ to $S([Apr1])$ that lands in C, the nonexistence of the previous map tells us, in particular, that at some point in the period after $Apr1$ through $May1$, the product C ceased to be present in the company's stockpile of products. Additionally, as there is a map from $S([May1, Jun1])$ to $S([May1])$ sending C to itself, we know that the product C could have been removed from the stockpile only in the period strictly between $Apr1$ and $May1$.

The point is that there is no way of gluing together the combined local sections of C "on the left" (from $Dec1$ through $Apr1$) to the local section "on the right" (from $May1$ to $Jun1$). This gap in the presence of C in the stockpile at some point in the time period between $Apr1$ and $May1$ is witnessed by the nonexistence of any maps at the presheaf level involving C, that would have let us pass from one side to the other. This makes the local section corresponding to the selection of C with its associated maps *strictly local*, since they cannot assemble into a global section, that is, be glued together into a section specified over *all* the diagram and covering the time period from December 1 until July 1.

Notice also how, among the strictly local sections, some sections can be "more local" than others, in the sense of how, for instance, the product D is only present throughout $[Dec1, Feb1]$ and $[Apr1, May1]$, as witnessed by the following restriction maps:

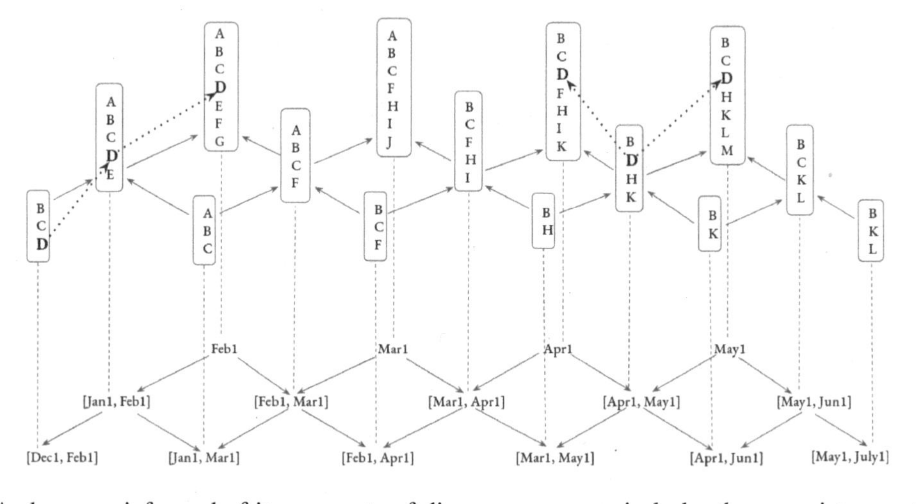

And we are informed of its moments of disappearance precisely by the nonexistence of certain (D-valued) maps.

A final nuance is worth noting before moving on: observe how, for instance, the product I shows up in the set $S([Mar1]) = \{A, B, C, F, H, I, J\}$, indicating that it is present in the company's stockpile throughout the instant $[Mar1, Mar1]$ at midnight. One can also see that it is in fact present throughout all of $[Mar1, Apr1]$. However, I is not present throughout $[Feb1, Mar1]$. In general, viewing such sheaves "internally" (i.e., in terms of the contents of the variable sets assigned to each piece of the covers), the sets can be thought of as being described by their behavior on small intervals or neighborhoods. On any given interval, the products present in the stockpile over that period either will or will not include I. However, intervals containing $[Mar1]$ will include a subinterval on the left, like $[Feb1, Mar1]$, during which I is not present and a subinterval on the right, like $[Mar1, Apr1]$, during which it is present. Thus, for such a closed set sheaf, the product I witnesses the failure of the law of noncontradiction: over arbitrary intervals of a cover containing March 1 at 0:00, it is incorrect to say that the product I cannot be both present and not present. In terms of the closed set topology of the base space, $Mar1$ acts as a boundary, and the algebra of closed sets in general is that of a co-Heyting algebra, which model paraconsistent logics (as explored in greater detail in the appendix and chapter 7). Paraconsistency is the blanket term for a logical system where the principle of noncontradiction can fail without the theory becoming trivial, that is, where a formula and its (paraconsistent) negation intersect at the boundary of their extensions. In the sheaf construction, the base topology is of special importance—for one thing, as we will see later in the book, the algebra of the base space topology can be seen in terms of the algebra of the sheaf section structure. Thus, by considering sheaves with respect to closed set topologies, we can introduce the valuable notion of boundary and paraconsistent negations—something that does not exist for the associated open set sheaf construction.[78]

78. Again, more extensive discussion of (co-)Heyting algebras, paraconsistency, the role of boundaries, and the connections to sheaves, appears in the appendix and parts of chapters 7 and 10. James (1992, 1995) are also good resources for closed set sheaves.

5.3 Philosophical Pass: Sheaf as Local-Global Passage

Box 5.1
The Idea of a Sheaf as Local-Global Passage

A sheaf is not to be situated in either the local (restriction) or the global (collation) registers, but rather is to be located in the passage forged between these two, in the translation system or glue that mediates between the two registers. Conceptually, the transit from the local to the global secured via the sheaf gluing (collatability) condition provides a deep but also precisely controllable connection between continuity (via the emerging system of translation functions guaranteeing coherence or compatibility between the local sections) and generality (global sections). One could make the case that sheaves are best seen as what happens when topology meets the category-theoretic notion of limit (which, as we saw, is bound up with both universality and a generalized notion of continuity).

By separating something into parts, that is, by specifying information locally, considering coverings of the relevant region, and enabling the decomposition or refinement of value assignments into assignments over restricted parts of the overall region (restriction condition), we are presented with a problem, a problem that in a sense can first appear only with such a downward movement towards greater refinement. Without having separated something into parts, we may appear to have a sort of trivial or default cohesion of parts—where, without being recognized in their separation, the parts yet remain implicit, and so the glue binding them together or the rule allowing one to transit from one part to another in a controlled fashion is simply not visible. However, having decomposed or discretized something into parts, we are at once presented with this separation of parts and the problem of finding and making explicit the glue that will serve to bind them together. A sheaf is a way of taking information that is locally defined or assigned and decomposing those assignments in a controlled fashion into assignments over smaller regions so as to draw out the specific manner of translating or gluing those particular assignments along their overlapping regions, and then using this now explicit system of gluings to build up a unique and comprehensive value assignment over the entire network of regions. In this sense, a sheaf equally involves both (1) controlled decomposition (discreteness), and (2) the recomposition (continuity) of what is partial into an architecture that makes explicit the special form of cooperation and harmony that exists between the decomposed items, items that may have previously been detached or that may have only *appeared* to stick together because we had not bothered to look closely enough.

Via the restriction/localization step, sheaf theory teaches us that we do not command a more global or integrated vision by renouncing the local nature of information or distinct planes and textures of reality or by glossing over the minute passages between things. Instead, it forces us to first become masters of the smallest link and, precisely through that control of the passages between the local parts, forge a coherent (collatable) vision of the largest scope.

Phenomenologically speaking, data or observations are frequently presented to us in "zones," fragmented or isolated in some way. These items can be thought of as various light-beams (perhaps of specific hues or brightness) cast over (and covering only parts of) a vast landscape, some of which may overlap. Even if this data clearly emerges as evolving over some region, it remains indexed or determined in some way by a particular zone or context. One interpretation of this initial particularity would be to suggest that the very fact that certain information initially presents itself in this local and bounded fashion is an indication that we are dealing with various discrete approximations, presented piecemeal, to phenomena that may in fact *really be* continuous. Whether or not that is the case, it is not difficult to accept that in its presentation to us in fragmented form, this step in the process is closely allied with the discrete (in a very general sense of the word). For centuries, the modes of restoring continuity to such partial information have been more or less haphazard. A sheaf removes this aspect of haphazardness. Significant is the at once *progressive* and *necessary* nature of the sheaf concept: how by gradually (progressively) covering fragments of reality and then systematically gluing them together into unique global solutions (necessary), the construction of sheaves encourages us to shift away from our standard ontologies or descriptions of reality as anchored in some absolute toward a more "contrapuntal and synthetic" perspective capable of registering "relative universals." (Zalamea (2013) contains an insightful discussion of precisely this latter perspective on sheaves.)

With the sheaf construction, a global vision is not *imposed* on the local pieces, obliterating the local nature of the presented information via some "sham" generalization, but emerges progressively, step-by-step, through the unfolding of precise translation systems guaranteeing the compatibility of the various components. A sheaf does not attempt to suppress the richness and polyphony of data in its particularity and relative autonomy, coercing a kind of standardized agreement as so many past models of generality have done. A sheaf is like a master composer who is not content to have her harmony prefabricated for her by habitual associations or who does not wish to achieve harmony only at the expense of suppressing all contrapuntal impulses and polyphony, imposing it *from above*, or restraining the local freedom of each voice to roam with some independence from the constraints that bind it in the name of some prefabricated schema. Rather, the sheaf-like composer achieves harmony only progressively, first by letting each component part unfold, in its relative autonomy, its own laws, then by insisting on making explicit even the most minute links and transits between the laws of movement of each of the parts, securing locally smooth passages for each transition, and from the glue or constraints that emerge out of this process, begins to build up a larger ensemble, step by step. It is not a compromise between the local and the global in the name of some idealogical preference for the more global or the universal. Sheaves earn their place as true mediators by virtue of their complete realization of the idea that—to paraphrase Hegel—true mediation comes about only from preserving the extremes as such, and true universality comes about only by sinking as deeply as possible into the particular.

6 There's a Yoneda Lemma for That

In which we cover what is perhaps the most important idea in category theory—the Yoneda results—by first focusing on what these results look like "in the miniature" (in order theory), then covering representability, and finally moving into the results in their full generality and reflecting on the associated "Yoneda philosophy."

Before continuing to develop the story of sheaves, the next two chapters take a step back to complete the account of category theoretic fundamentals. This chapter is devoted to what is perhaps the most important idea in category theory, the Yoneda results. Chapter 7 turns to consideration of adjunctions, and is focused on developing these ideas through a number of examples. In the coming sections of the present chapter, we will motivate the main ideas through a simplified special case, its analogue for posets (in fancier language, its "**2**-enriched" analogue). This motivation requires that one first understand *enrichment*, the introduction of which also gives us a chance to refine our understanding of categories in general.

6.1 First, Enrichment!

Not all categories were created equal. For instance, for certain categories, there may be a natural way of combining elements of the category, that is, of making use of an operation that takes two elements and "adds" or "multiplies" them together. Not all categories admit such a thing. Those that do are called *symmetric monoidal*.

Definition 138 A *symmetric monoidal structure* on a category $\mathcal{V}$ consists of the following data:

1. a bifunctor[79] $-\otimes-: \mathcal{V} \times \mathcal{V} \to \mathcal{V}$, called the *monoidal product*;
2. a unit object $I \in \text{Ob}(\mathcal{V})$, called the *monoidal unit*,

subject to the following specified natural isomorphisms:

$$v \otimes w \cong_\gamma w \otimes v \qquad u \otimes (v \otimes w) \cong_\alpha (u \otimes v) \otimes w \qquad I \otimes v \cong_\lambda v \cong_\rho v \otimes I$$

that witness symmetry, associativity, and unit conditions on the monoidal product. There are then standard "coherence conditions" that these natural transformations are expected to obey.

79. A *bifunctor* is just a functor whose domain is the product of two categories.

A category equipped with such a symmetric monoidal structure is then called a *symmetric monoidal category*, denoted, for example, $(\mathcal{V}, \otimes, I)$. A *monoidal category* is similarly defined, except one leaves out the symmetry natural isomorphism displayed above on the far left. If the natural isomorphisms involving associativity and the unit are replaced by *equalities*, then the monoidal structure is said to be *strict*.

This is defined on categories in general, but an especially simple special case comes from restricting the definition to preorders (as categories).

Definition 139 A *symmetric monoidal structure* on a preorder $(X, \leq)$ consists of

- an element $I \in X$ called the monoidal unit, and
- a function $\otimes : X \times X \to X$, called the monoidal product.

These must further satisfy the following, for all $x_1, x_2, y_1, y_2, x, y, z \in X$, where we use infix notation, that is, $\otimes(x_1, x_2)$ is written $x_1 \otimes x_2$:

- *monotonicity*: if $x_1 \leq y_1$ and $x_2 \leq y_2$, then $x_1 \otimes x_2 \leq y_1 \otimes y_2$;
- *unitality*: $I \otimes x = x$ and $x \otimes I = x$;
- *associativity*: $(x \otimes y) \otimes z = x \otimes (y \otimes z)$;
- *symmetry*: $x \otimes y = y \otimes x$.

Then a preorder equipped with a symmetric monoidal structure, $(X, \leq, I, \otimes)$, is called a *symmetric monoidal preorder*.

Monoidal units may be given by, for example, $0, 1$, false, true, $\{*\}$, etc. Monoidal "products" include the likes of $\otimes, +, *, \wedge, \vee, \times$, and so on.

Example 140 The simplest nontrivial preorder is $\mathbf{2} = \{0 \xrightarrow{\leq} 1\}$. Alternatively, you might think of this as $\mathbf{2} = \{\mathit{false}, \mathit{true}\}$ with the single nontrivial arrow $\mathit{false} \leq \mathit{true}$. There are two different symmetric monoidal structures on it. To consider one of these: let the monoidal unit be *true* and the monoidal product be $\wedge$ (AND), leaving us with a monoidal preorder $(\mathbf{2}, \leq, \mathit{true}, \wedge)$.[80]

Example 141 For a set S, the powerset $\mathbb{P}(S)$ of all subsets of S, equipped with the natural order $A \leq B$ given by subset relation $A \subseteq B$, in fact has a symmetric monoidal structure on it: $(\mathbb{P}(S), \leq, S, \cap)$ is a symmetric monoidal preorder.

In particular, taking S a two-element set, this is isomorphic to $(A_4, \leq_k, B, \otimes)$, Belnap's four-valued "knowledge lattice" (or "approximation lattice") $A_4 = (\{\bot, t, f, \top\}, \leq_k)$, often used by relevance and paraconsistent logicians, where the values are the various subsets of $\{\mathit{true}, \mathit{false}\}$ (i.e., $\{t, f\}$). Here, $\top$ (or sometimes B) is "*both* true and false"; $\bot$ is "*neither* true nor false"; $\otimes$ is a "consensus" connective corresponding to meet; and $(A_4, \leq_k)$ is the (complete) lattice corresponding to an ordering on epistemic states ("how much information/knowledge").

80. Fong and Spivak (2020) sensibly calls this **Bool**, but we will just stick with calling it **2**, after its carrier preorder. The reader who desires a more in-depth treatment of enrichment, or who is intrigued by any of these matters, will surely enjoy the recent Fong and Spivak (2020). Readers with a higher tolerance for abstraction might also find Kelly (2005) useful.

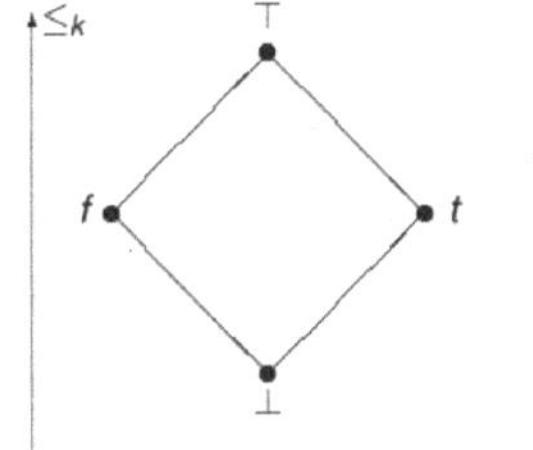

This structure accordingly has four "truth values": the classical ones (t and f); a truth value $\bot$ that intuitively captures the notion of a *lack of information* ("neither t nor f"); and a truth value $\top$ that can be deployed to represent *contradictions* or *inconsistency* ("*both* t and f"). The underlying partial order of the lattice has t and f as its intermediate truth values, $\bot$ as the $\leq_k$-minimal element, and $\top$ as the $\leq_k$-maximal element. Overall, the partial order $(\{\bot, t, f, \top\}, \leq_k)$ of the lattice is often regarded as serving to rank the "amount of knowledge or information," where $\leq$ captures the notion of "approximates the information in"—that is, if $x \leq_k y$, then y gives us *at least as much* information as x (possibly more). A move up in the lattice represents an increase in the amount of information, with $\otimes$ taking the uppermost element below both x and y.[81]

Example 142 Let $[0, \infty]$ be the set of nonnegative real quantities, together with ∞. Consider the preorder $([0, \infty], \geq)$, with the natural order $\geq$, for example, $\pi \geq 0.8$, $14.\overline{33} \geq 11$, and of course $\infty \geq x$ for all $x \in [0, \infty]$. There is a symmetric monoidal structure here, with monoidal unit 0 and monoidal product + (where in particular $x + \infty = \infty$ for any $x \in [0, \infty]$). After Fong and Spivak (2020), we can call this symmetric monoidal preorder $\textbf{Cost} := ([0, \infty], \geq, 0, +)$, since we think of the elements of $[0, \infty]$ as costs.

In the standard definitions of a category that we have seen thus far in this book, the hom-sets are *sets*, that is, objects of the category **Set**. On this approach, with such categories, that the hom-sets are specifically sets effectively means that the task or question of getting from (or relating) one object to another has a *set* of approaches or answers or names. But what if we generalized this story and let the hom-sets of a category come from some category other than **Set**?

Symmetric monoidal categories are important, in large part, because of something we can *do* with them: we can *enrich* an (arbitrary) category in them! What does that mean? Fong and Spivak (2020) suggest a very nice intuitive way of thinking of this: Enriching in, say, a monoidal preorder $\mathcal{V} = (V, \leq, I, \otimes)$ just means "letting $\mathcal{V}$ structure the question of the relations or paths between the objects of the underlying category." In this general context, enriching in different monoidal categories often recovers (while generalizing) important entities in math. For instance, it emerges that categories "enriched in **Cost**," or **Cost**-categories, provide a powerful generalization of the notion of metric space.

Definition 143 Let $\mathcal{V} = (V, \leq, I, \otimes)$ be a symmetric monoidal preorder. A $\mathcal{V}$-category $\mathcal{X}$ consists of

- specification of a set $\text{Ob}(\mathcal{X})$, elements of which are objects,

81. For more on this lattice, see Belnap (1992).

- for every two objects x, y, specification of an element $\mathcal{X}(x,y) \in V$, called the hom-object,

and where these satisfy the two properties

- for every object $x \in \mathrm{Ob}(\mathcal{X})$, we have $I \leq \mathcal{X}(x,x)$, and
- for every three objects $x, y, z \in \mathrm{Ob}(\mathcal{X})$, we have $\mathcal{X}(x,y) \otimes \mathcal{X}(y,z) \leq \mathcal{X}(x,z)$.

In this case, we call $\mathcal{V}$ the *base of the enrichment* for $\mathcal{X}$, or just say $\mathcal{X}$ is *enriched in* $\mathcal{V}$.

Example 144 What happens if we enrich in **Cost** $=([0,\infty], \geq, 0, +)$? Following the definition: a **Cost**-category $\mathcal{X}$ consists of

1. a collection $\mathrm{Ob}(\mathcal{X})$, and
2. for every $x, y \in \mathrm{Ob}(\mathcal{X})$ an element $\mathcal{X}(x,y) \in [0,\infty]$.

The idea here is that $\mathrm{Ob}(\mathcal{X})$ provides the "points," while $\mathcal{X}(x,y) \in [0,\infty]$ plays the role of supplying the "distances." Still just following the definition, the properties of a category enriched in **Cost** are given by:

- $0 \geq \mathcal{X}(x,x)$ for all $x \in \mathrm{Ob}(\mathcal{X})$, and
- $\mathcal{X}(x,y) + \mathcal{X}(y,z) \geq \mathcal{X}(x,z)$ for all $x, y, z \in \mathrm{Ob}(\mathcal{X})$.

Note that since $\mathcal{X}(x,x) \in [0,\infty]$, the property $0 \geq \mathcal{X}(x,x)$ implies that $\mathcal{X}(x,x) = 0$. So this is in fact equivalent to the first condition $d(x,x) = 0$ describing a metric. And the second condition here is clearly the usual triangle inequality! We have thus defined, with the notion of a **Cost**-category, an extended (Lawvere) metric space.

Recall the usual definition of a metric space from definition 73 (chapter 4). If we instead take a function $d: X \times X \to [0,\infty] = \mathbb{R}_{\geq 0} \cup \{\infty\}$, then we have an *extended metric space*. From the categorical viewpoint, the generalized construction of a **Cost**-category recovers the notion of a metric space, while already suggesting that the usual definition's conditions

- (2) if $d(x,y) = 0$, then $x = y$;
- (3) $d(x,y) = d(y,x)$

are somehow not as natural or primitive as the other two conditions (triangle inequality and that points are at "zero distance" from themselves). Indeed, there are contexts in which (2) is not satisfied, yet we would still like to have a metric. Also, requiring (3) or symmetry prevents us from regarding a number of constructions we would like to regard as metrics as legitimate metrics, so relaxing this condition also seems desirable.

Example 145 Now take the symmetric monoidal preorder $\mathbf{2} = (\{\mathit{false}, \mathit{true}\}, \leq, \mathit{true}, \wedge)$. Enriching in **2** recovers the notion of a preorder, since for any $x, y \in \mathcal{P}$, with $\mathcal{P}$ a preorder, there is either 0 ("false") or 1 ("true") arrow from x to y. Accordingly, the "homs" here will be objects of **2**, not **Set**. More formally, a **2**-category[82] consists of

- a specification of a set of objects
- for every x, y, an element $\mathcal{X}(x,y) \in \mathbf{2}$

where this data satisfies

82. Not to be confused with the notion of a 2-category, mentioned in remark 49 (chapter 2).

1. for every element $x \in \text{Ob}(\mathcal{X})$, $\mathit{true} \xrightarrow{\leq} \mathcal{X}(x, x)$, so $\mathcal{X}(x, x) = \mathit{true}$
2. for every x, y, z, $\mathcal{X}(x, y) \wedge \mathcal{X}(y, z) \xrightarrow{\leq} \mathcal{X}(x, z)$

The first condition above just amounts to reflexivity and the second to transitivity, used to define a preorder; understanding $\mathcal{X}(x, y) = \mathit{true}$ to just mean that $x \leq y$, clearly this just recovers the notion of a preorder. Thus, the theory of **2**-enriched categories just recovers precisely the theory of ordered sets and the monotone maps between them.

Example 146 Returning to A_4, we can understand t as "told True," f as "told False," $\bot$ as "told nothing" (i.e., *neither* told True nor told False), $\top$ as "*both* told True and told False." $\bot$ is at the bottom of the lattice as it gives no information at all, while $\top$ is at the top since it gives "too much" (or inconsistent) information.

When we enrich in A_4, the resulting A_4-category $\mathcal{X}$ will describe, for any two objects x, y of $\mathcal{X}$, all the (true, false, null, inconsistent) information that has been received or inputed (perhaps from several independent sources) about whether you can "get from" x to y.

Enriching in A_4 implies that the issue of passing from x to y is structured by how much information/knowledge we (or some system, like a computer, prepared to receive and reason about inconsistent information) might have about the question. For instance, "I have been told that 'yes' ('no') one can (cannot) pass from x to y"; or "I have been told both that you can and that you cannot pass from x to y"; or "I have not been told anything about whether or not you can pass from x to y."

In the next few sections, we will make use of this notion of enrichment to build towards a particularly simple presentation of the abstract Yoneda results.

6.2 Downsets and Yoneda in the Miniature

Given a poset $\mathcal{P} = (P, \leq)$, we have seen how we can regard $\mathcal{P}$ as a category. The following notion will be of use to us.

Definition 147 Let $\mathcal{P}$ be a poset, and $A \subseteq \mathcal{P}$ a subset. Then, we call the subset A a *downset* if for each $p \in A$ and $q \in \mathcal{P}$, we have that $p \in A$ and $q \leq p$ implies that $q \in A$. Dually (i.e., reversing all the arrows), a subset $U \subseteq \mathcal{P}$ is an *upper set* (or *up-set*) provided: if $p \in U$ and $p \leq q$, then $q \in U$.

We can further define, for each element $p \in \mathcal{P}$, the downset generated by p—called its *principal downset*, denoted D_p (or just $\downarrow p$)—as

$$\downarrow p := \{q \in \mathcal{P} \mid q \leq p\}.$$

Dually, as one might expect, we can also define, for each point p, its *principal upper set* U_p (or just $\uparrow p$) as

$$\uparrow p := \{q \in \mathcal{P} \mid p \leq q\}.$$

For instance, consider the following poset $\mathcal{P}$, built from $P = \{a, b, c, d\}$, and given by $a \leq c, b \leq c, b \leq d$, and the obvious identity (reflexivity) $x \leq x$ for all $x \in P$. The data of this poset is perhaps more helpfully displayed in the picture:

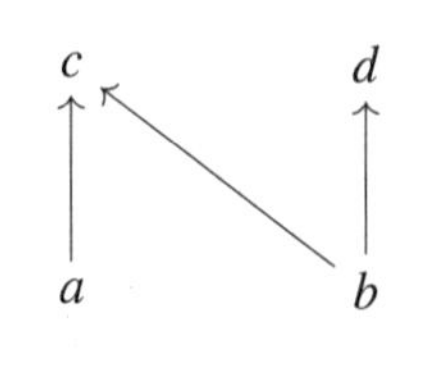

Exercise 18 *Is $\{a,b,c\}$ a downset? How about $\{a,b\}$? And $\{a,c,d\}$?*

Solution Yes, $\{a,b,c\}$ is a downset; same with $\{a,b\}$. But $N=\{a,c,d\}$ is not a downset; for, in particular $d \in N$, yet considering $b \in P$, since $b \leq d$, we should have that $b \in N$, in order for N to be a downset. Yet $b \notin N$.

In general, we denote by $\mathcal{D}(\mathcal{P})$ the collection of all downsets of the poset $\mathcal{P}$. Observe that $\mathcal{D}(\mathcal{P})$ has a natural order on it—namely, $U \leq V$ provided U contained in V. Then, $(\mathcal{D}(\mathcal{P}), \subseteq)$ is itself an order under inclusion, one that we will sometimes just denote by $\mathcal{D}(\mathcal{P})$, or **Down**$(\mathcal{P})$ when we want to emphasize that we are regarding this order as a *category*.[83] This resulting poset consisting of the collection of all downsets of $\mathcal{P}$, ordered by inclusion, is sometimes called the *downset completion*.

The following diagram displays the information of all the downsets of our given $\mathcal{P}$, ordered by inclusion:

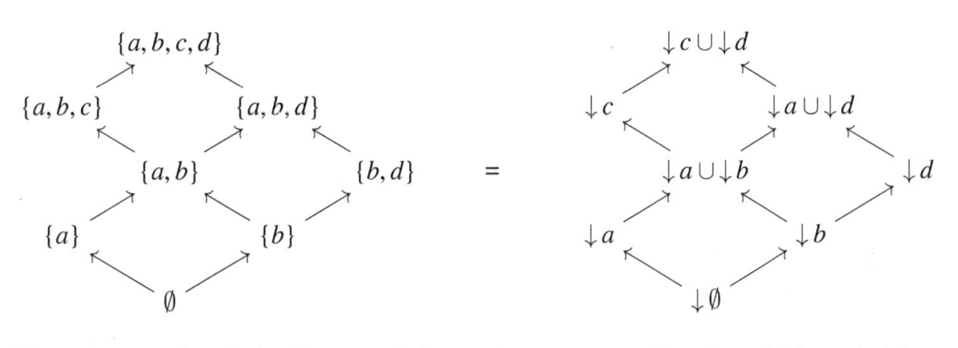

There are a couple of valuable general observations to note at this point, which can be illustrated via this particular example. The first observation will allow us to construe downsets in terms of monotone maps (functors) from $\mathcal{P}^{op}$ to the order **2**.

First, consider that any given element A of $\mathcal{D}(\mathcal{P})$ represents something like a "choice" of elements from the underlying set P, with the further requirement that, as a downset, whenever $x \in A$, then any $y \in P$ such that $y \leq x$ in $\mathcal{P}$ is also in A. But this requirement is the same as saying that for any x, y such that $y \leq x$ in $\mathcal{P}$, if we have that "it is *true* that $x \in A$," then we must also have that "it is *true* that $y \in A$." And this is just to say that

$$y \leq x \text{ implies } \phi(y) \geq \phi(x),$$

where ϕ is an antitone map from the order $\mathcal{P}$ to order **2**; or, equivalently, it is a monotone map from the opposite order $\mathcal{P}^{op}$ to **2**. Such maps are themselves ordered under the pointwise inclusion ordering. If we designate such a poset of monotone maps, ordered by

83. Dually, we write $\mathcal{U}(\mathcal{P})$ for the collection of all the upper sets of $\mathcal{P}$; this also has a natural order on it—namely, $U \leq V$ provided U is contained in V, making $(\mathcal{U}(\mathcal{P}), \subseteq)$ an order as well.

inclusion, by $\mathbf{2}^{\mathcal{P}^{op}}$ or Monot$(\mathcal{P}^{op}, \mathbf{2})$, then we can see that there is a map between the orders

$$\mathcal{D}(\mathcal{P}) \to \text{Monot}(\mathcal{P}^{op}, \mathbf{2})$$
$$D \mapsto \phi_D,$$

where ϕ_D acts as the characteristic (or indicator) function, mapping to 1 on D and 0 elsewhere. In other words, given a downset D of $\mathcal{P}$, we define $\phi_D : \mathcal{P}^{op} \to \mathbf{2}$ by setting $\phi_D(x) = 1$ precisely when $x \in D$ (i.e., assigns it to the characteristic function of D). Conversely, given a monotone map in Monot$(\mathcal{P}^{op}, \mathbf{2})$, we can send this to the inverse image $\phi^{-1}(1) \in \mathcal{D}(\mathcal{P})$, recovering a unique downset (you can verify for yourself that the subset $\phi^{-1}(1)$ is a downset). In order theory, in general, a map $F : P \to Q$, where P and Q are posets, is said to be an *order-embedding* provided

$$x \leq y \text{ in } P \text{ iff } F(x) \leq F(y) \text{ in } Q,$$

and then such an order-embedding yields an *order-isomorphism* between P and Q. But since, in our case, $A \subseteq B$ iff $\phi_A \leq \phi_B$, altogether we have thus described an order-embedding, giving us an order-isomorphism

$$\mathcal{D}(\mathcal{P}) \cong \text{Monot}(\mathcal{P}^{op}, \mathbf{2}).$$

Notice how, included among the maps $\mathcal{P}^{op} \to \mathbf{2}$ are those ϕ_p for any given element $p \in \mathcal{P}$. These send $q \mapsto 1$ iff $p \leq q$ in $\mathcal{P}$ (or, equivalently, but perhaps more clearly, iff $q \leq p$ in $\mathcal{P}^{op}$, which is our domain), for such a map generally acts as the indicator function of the set of all $x \leq y$, that is, the principal downset $\downarrow y$ of y.

Observe that the principal downset $\downarrow p := \{q \in P : q \leq p\}$ is itself a downset (i.e., it will belong to $\mathcal{D}(\mathcal{P})$), *for any* p in our set $\mathcal{P}$. (This follows from the transitivity of $\mathcal{P}$.) The principal downsets are actually rather special objects among the downsets. To see this, consider again the diagram of the downsets of our particular $\mathcal{P}$. You can see that every object in $\mathcal{D}(\mathcal{P})$ is a principal downset or a union of principal downsets, and that the principal downsets $\downarrow x$ run through all $x \in \mathcal{P}$. Observe also that for $A \subseteq \mathcal{P}$, $\downarrow A$ will be the *smallest* downset that contains A, and moreover $A = \downarrow A$ iff A is a downset. We will return to these facts, and make better sense of them, shortly.

For now, we need to realize that $\downarrow$ can actually be regarded as a monotone map from $\mathcal{P}$ to $\mathcal{D}(\mathcal{P})$, which in our particular case may be pictured as

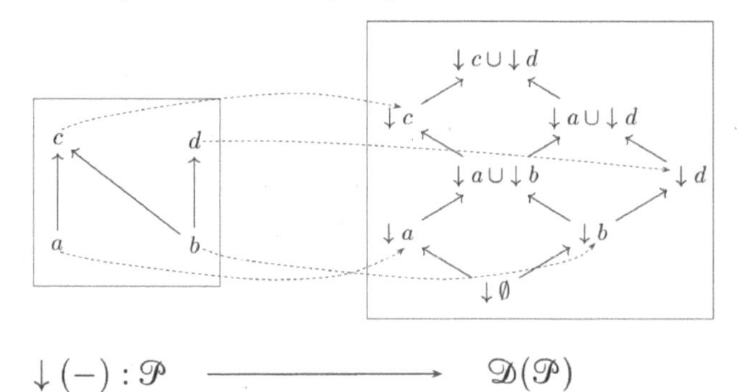

That this $\downarrow$ really just defines a monotone map—that is, $x \leq y$ implies $\downarrow x \subseteq \downarrow y$—is easy to see in the general case as well. For, let $x \leq y$ in $\mathcal{P}$. We want $\downarrow x \subseteq \downarrow y$ in $\mathcal{D}(\mathcal{P})$. Take any

$x' \in \downarrow x$. Then we must have that $x' \leq x$. By transitivity of the order, $x' \leq x$ and the assumed $x \leq y$ yield $x' \leq y$. Thus, $x' \in \downarrow y$. Altogether, this shows that $\downarrow x \subseteq \downarrow y$.

We have the other direction as well, namely $\downarrow x \subseteq \downarrow y$ in $\mathcal{D}(\mathcal{P})$ implies $x \leq y$ in $\mathcal{P}$. Altogether, then, we actually have another *order-embedding*, embedding any poset into its downset completion:

$$\downarrow(-) : \mathcal{P} \to \mathcal{D}(\mathcal{P}) \quad p \mapsto \downarrow p.$$

This is all part of a much bigger story, so the main results are set aside for emphasis.

Proposition 148 (*Yoneda lemma for posets*) Given $\mathcal{P}$ a poset, $x \in \mathcal{P}$, and $A \in \mathcal{D}(\mathcal{P})$, then

$$x \in A \text{ iff } \downarrow x \subseteq A.$$

Proof. ($\Rightarrow$) Let $x \in A$. Then all $y \leq x$ is also in A, as A is a downset, and in particular $x \in \downarrow x$. Thus, $\downarrow x \subseteq A$.

($\Leftarrow$) Take $y \in \downarrow x$. But then $y \leq x$, and so $y \in A$ since A is a downset. □

Applying the Yoneda lemma to two principal downsets, we have that $y \leq z$ iff for all x[84]

$$x \leq y \Rightarrow x \leq z.$$

The most important corollary, or application, of the lemma is the following:

Proposition 149 (*Yoneda embedding for posets*) This $\downarrow$ defines an order-embedding (embedding any poset into its downset completion):

$$\downarrow(-) : \mathcal{P} \to \mathcal{D}(\mathcal{P}) \quad p \mapsto \downarrow p.$$

Proof. Let $x \leq y$. Then $x \in \downarrow y$ (conversely, if $x \in \downarrow y$, clearly we must have $x \leq y$). Applying the previous lemma (taking $A = \downarrow y$), $x \in \downarrow y$ holds precisely when $\downarrow x \subseteq \downarrow y$. On the other hand, the converse holds as well, that is, $\downarrow x \subseteq \downarrow y$ implies $x \in \downarrow y$, which implies $x \leq y$. Altogether then,

$$x \leq y \text{ iff } (\downarrow x) \subseteq (\downarrow y).$$

□

The Yoneda results thus assure us, in a slogan, that

To know everything "below" an element is just to know that element.

Let us start to generalize this story. Given a poset $\mathcal{P}$, by considering $\mathcal{P}$ as a category, we might have looked at presheaves $\mathcal{P}^{op} \to$ **Set**. But since preorders are the same thing as **2**-enriched categories, supposing we want not *sets* for our hom-sets, but rather "truth values," it is natural instead to consider **2**-enriched "presheaves" on $\mathcal{P}$. Instead of arbitrary set-valued data, then, such a **2**-presheaf assigns to each $x \in \mathcal{P}$ a *truth-value* in **2**; and this will just recover the monotone maps (functors) $\mathcal{P}^{op} \to \mathbf{2}$ (or, equivalently, an antitone map, or contravariant functor (presheaf), from $\mathcal{P}$ to **2**).

We saw how such a monotone map (from the opposite order) effectively acts as the characteristic (or indicator) function ϕ_A of a downset $A \subseteq P$, forming part of the important

84. Readers familiar with real analysis may recognize in this the construction of *Dedekind cuts*!

order-isomorphism

$$\mathcal{D}(\mathcal{P}) \cong \text{Monot}(\mathcal{P}^{op}, \mathbf{2}).$$

We know, moreover, how to convert any poset into a category. Thus, renaming $\mathcal{D}(\mathcal{P}) := \mathbf{Down}(\mathcal{P})$ and $\text{Monot}(\mathcal{P}^{op}, \mathbf{2}) := \mathbf{Monot}(\mathcal{P}^{op}, \mathbf{2})$ to emphasize that we are now dealing with categories, the above actually describes

$$\mathbf{2\text{-}PreSh}(\mathcal{P}) := \mathbf{Monot}(\mathcal{P}^{op}, \mathbf{2}) \cong \mathbf{Down}(\mathcal{P}).$$

In the order setting, via the principal downsets, we were able to construct an embedding $\mathcal{P} \to \mathbf{Down}(\mathcal{P})$. Similarly, but in much greater generality, we will see that there is an embedding $\mathbf{C} \to \mathbf{PreSh}(\mathbf{C})$, taking a general category $\mathbf{C}$ to its category of presheaves. Before defining the Yoneda lemma and embedding in the general case, we need to take a step back for a moment and discuss *representability*.

To motivate this, consider that in a poset, regarded as a category, a principal downset on an element $p \in \mathcal{P}$ is just all the arrows into p, where these amount to all elements that p "looks down on" (or all elements that "look up to" p). This identification of "arrows into p" and "*elements* below p" can be made, since a poset is precisely a category for which there is *at most one* arrow $q \to p$ for any p, q, allowing us to *identify* $\text{Hom}_{\mathcal{P}}(q, p)$ with the element q. In other words,

$$\text{Hom}_{\mathcal{P}}(-, p) = \downarrow p.$$

Before cashing in on the power of this statement, we will discuss the "representability" operative here by considering a sort of miniaturized version of this phenomenon.

6.3 Representability Simplified

Consider a general map

$$T \times X \to Y$$

that has a product for its domain (you can think of this as involving sets and functions for now). We cannot typically expect to be able to reduce this to a specification of what is happening on T and X separately, as the interaction of the two factors of the product is essentially involved in supplying the values of the mapping itself. Thus, in considering a map that has for domain a product (all three objects different, for the most general case),

$$T \times X \xrightarrow{f} Y,$$

we can ask (nontrivial) questions about the relations between any of the separate objects involved in the product and the codomain object.

Observe that if we use the terminal object 1 to pick out "points" of X, via $1 \xrightarrow{x} X$, any such point will give rise, via f, to the map f_x

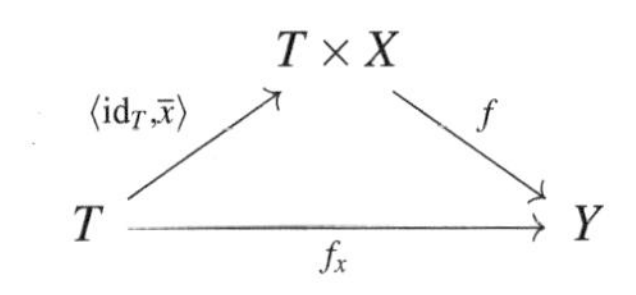

where $\bar{x}$ is defined as the composite constant map $T \to 1 \xrightarrow{x} X$. Thus,

$$f_x(t) = f(t, x)$$

for all t. In this way, we are now regarding the map f as an X-parameterized family of maps $T \to Y$, one for each of the points of X. In this setting, one possible question would be to consider, for a pair of sets T, Y, whether there is a set X *large enough* for its points to supply, via the maps $f(-, x)$, *all* maps $T \to Y$. As a very simple illustration of this, in the simple case of sets described by their cardinality (number of elements), if T is a set with four elements and Y a set with three elements, then X would need to have

$$3^4 = 81$$

elements, for that is the number of maps $T \to Y$.

This is in fact part of a much more general story—one that gets at representability—than the above discussion might suggest. For, we do not need to restrict attention to sets and their cardinal number properties or even make this a matter of *size*. For a function $g : T \to Y$, whenever there is at least one $1 \xrightarrow{x_0} X$ such that

$$g(-) = f(-, x_0),$$

that is, for all $t \in T$,

$$g(t) = f(t, x_0),$$

we say that g is *representable* by x_0, or *f-represented by* x_0. This might be seen as a toy instance of the much more general notion of representable functors. Recall the hom-functor $\text{Hom}_{\mathbf{C}}(-, c)$ (and its dual $\text{Hom}_{\mathbf{C}}(c, -)$), where this may be given by

$$\text{Hom}_{\mathbf{C}}(-, -) : \mathbf{C}^{op} \times \mathbf{C} \to \mathbf{Set}.$$

For any categories $\mathbf{C}, \mathbf{D}, \mathbf{E}$, an isomorphism can be demonstrated between

$$\mathbf{E}^{\mathbf{C} \times \mathbf{D}} \cong (\mathbf{E}^{\mathbf{D}})^{\mathbf{C}},$$

allowing us to move freely between functors $\mathbf{C} \times \mathbf{D} \to \mathbf{E}$ and $\mathbf{C} \to \mathbf{E}^{\mathbf{D}}$. Thus, if we fix one of the variables of this $\text{Hom}_{\mathbf{C}}$, then we get the important *representable functors*:

$$\begin{aligned} \text{Hom}_{\mathbf{C}}(a, -) : \mathbf{C} &\to \mathbf{Set} \\ b &\mapsto \text{Hom}_{\mathbf{C}}(a, b) \\ f : b \to c &\mapsto \text{Hom}_{\mathbf{C}}(a, f) : \text{Hom}_{\mathbf{C}}(a, b) \to \text{Hom}_{\mathbf{C}}(a, c) \\ & \qquad\qquad g \mapsto f \circ g, \end{aligned}$$

and

$$\begin{aligned} \text{Hom}_{\mathbf{C}}(-, a) : \mathbf{C}^{op} &\to \mathbf{Set} \\ b &\mapsto \text{Hom}_{\mathbf{C}}(b, a) \\ f : b \to c &\mapsto \text{Hom}_{\mathbf{C}}(f, a) : \text{Hom}_{\mathbf{C}}(c, a) \to \text{Hom}_{\mathbf{C}}(b, a) \\ & \qquad\qquad h \mapsto h \circ f. \end{aligned}$$

But this is just to describe the Yoneda-embedding functors taking, for instance in the contravariant case,

$$\begin{aligned} \mathbf{C} &\to \mathbf{Set}^{\mathbf{C}^{op}} \\ x &\mapsto \text{Hom}_{\mathbf{C}}(-, x). \end{aligned}$$

Functors (presheaves) of this form are then said to be *representable*. More formally,

Definition 150 For a locally small category $\mathbf{C}$, we say that a functor $F : \mathbf{C} \to \mathbf{Set}$ is a *representable functor* if there exists an object $c \in \mathbf{C}$ (sometimes called the *representing object*) together with a natural isomorphism $\mathrm{Hom}_{\mathbf{C}}(c, -) \cong F$; or, equivalently, one speaks of a *representation* for a (covariant) functor F as an object $c \in \mathbf{C}$ together with a specified natural isomorphism $\mathrm{Hom}_{\mathbf{C}}(c, -) \cong F$.[85]

If F is a contravariant functor, then the desired natural isomorphism is given between $\mathrm{Hom}_{\mathbf{C}}(-, c) \cong F$.

In the covariant case, the representable functor can be thought of, intuitively, as encoding how a category "is seen" or "is acted on" by a certain object; in the contravariant case, how the category "sees" or "acts on" the chosen object. For instance, in the category of topological spaces **Top**, if we regard all the maps from $\mathbf{1}$ (the one-point space) to a space X, this just produces the points of X, that is, "$\mathbf{1}$ sees points."[86]

It is worth lingering a bit with this notion of representability. It might be useful to mention, moreover, that most functors (valued in **Set**) are *not* representable. If you were to pick a functor randomly, the odds are it would not be representable. Thus, examples of *non*representable functors abound; but perhaps a few concrete nonexamples are in order.

Example 151 The *covariant* powerset functor $\mathbb{P} : \mathbf{Set} \to \mathbf{Set}$ is not representable. This functor $\mathbb{P}$ is such that $\mathbb{P}(X)$ is just the power set of X, and for any function $X \to Y$, the map $\mathbb{P}(X) \to \mathbb{P}(Y)$ takes $A \subseteq \mathbb{P}(X)$ to the image under f, that is, to $f(A)$.

To see that it is not representable, suppose we have a representing object X, that is, $X \in \mathbf{Set}$ represents $\mathbb{P}$. Then, in particular, we will need that

$$|\mathrm{Hom}_{\mathbf{Set}}(X, -)| = |\mathbb{P}(-)|$$

for all sets, that is,

$$|\mathrm{Hom}_{\mathbf{Set}}(X, Y)| = |\mathbb{P}(Y)|$$

for all $Y \in \mathbf{Set}$. But then we can take $Y = \{*\}$, a singleton set. For any nontrivial X, there can be only one map to the singleton set, so $|\mathrm{Hom}_{\mathbf{Set}}(X, Y)| = 1$. Yet the powerset of a singleton set is, of course, of cardinality 2. Thus

$$|\mathrm{Hom}_{\mathbf{Set}}(X, Y)| \neq |\mathbb{P}(Y)|$$

for our given $Y = \{*\}$. This contradiction tells us that there can be no such representing object X in **Set** for the covariant powerset functor.

On the other hand, the contravariant powerset functor (presheaf) is representable![87] Specifically, $\mathbb{P}$ (contravariant now) is representable by the two-element set $2 = \{0, 1\}$, so that for each set Y, we have the isomorphism

$$\mathrm{Hom}_{\mathbf{Set}}(Y, 2) \cong \mathbb{P}(Y)$$
$$f \mapsto f^{-1}(\{1\}).$$

85. Ordinarily, one requires that the domain $\mathbf{C}$ of a representable functor be locally small so that the hom-functors $\mathrm{Hom}_{\mathbf{C}}(c, -)$ and $\mathrm{Hom}_{\mathbf{C}}(-, c)$ are valued in the category of sets.

86. This example is lifted from Leinster (2014).

87. In general, it seems to be easier to find representables among *contravariant* functors.

Effectively, this says that 2 is a set that contains a universal subset $\{1\}$ that pulls back to any other subset, via the characteristic function of that subset.

The great importance of representable functors is in part due to the fact that representable functors can encode a universal property of its representing object. For instance, a category $\mathbf{C}$ will have an *initial object* precisely when the constant functor $* : \mathbf{C} \to \mathbf{Set}$ is representable, that is, an object $c \in \mathbf{C}$ will be initial iff the hom-functor Y^c is naturally isomorphic to the constant functor sending every object to the singleton set. Dually, an object $c \in \mathbf{C}$ will be *terminal* iff the functor Y_c is naturally isomorphic to the constant functor $* : \mathbf{C}^{op} \to \mathbf{Set}$. Put otherwise: an object $c \in \mathbf{C}$ is *initial* if, for all objects $d \in \mathbf{C}$, there exists a unique morphism $c \to d$; while an object $c \in \mathbf{C}$ is *terminal* if, for all objects $d \in \mathbf{C}$, there exists a unique morphism $d \to c$.

The absence of such universal properties can be used, as we effectively did in dealing with the (covariant) powerset functor above, to show that a candidate nonrepresentable functor is in fact not representable. The general idea here—which method you might use to convince yourself of the nonrepresentability of the functors described in the coming examples—is to (1) assume the functor is representable; (2) consider a possible "universal" element for the functor; and then (3) produce a contradiction by showing that this element cannot actually have the special universal property that it needs to have.

Recall that in **Set**, every one-element (singleton) set is a terminal object (a special object with a special universality property). Thus, in our discussion of the covariant powerset functor, another way of saying that

$$|\mathrm{Hom}_{\mathbf{Set}}(X, \{*\})| \neq |\mathbb{P}(\{*\})|$$

would accordingly have been to say that $\mathbb{P}$ does not preserve the terminal object.

Example 152 The (covariant) functor $\mathbf{Group} \to \mathbf{Set}$ that takes a group to its set of subgroups is not representable.

Example 153 The (covariant) functor $\mathbf{Rng} \to \mathbf{Set}$ that takes a (nonunital) ring R to its set of squares, that is, $\{r^2 : r \in R\}$, is not representable.

It turns out that all universal properties themselves can be captured by the fact that certain data defines an initial or terminal object in an appropriate category, specifically the category of elements of the representable functor, a fact that can be rather useful (but that we simply record, without proof, before giving examples).

Proposition 154 A covariant (contravariant) set-valued functor is representable iff its category of elements has an initial (respectively, terminal) object.

Example 155 Applied to a poset $\mathcal{P}$, consider $\mathcal{P}$ as a category. For an arbitrary element $p \in \mathcal{P}$, first check that the slice category $\mathcal{P}/p$ (also denoted $(\mathcal{P} \downarrow p)$)[88] is just the principal downset generated by p; dually, the co-slice category $(p/\mathcal{P})$ (also denoted $(p \downarrow \mathcal{P})$) is the principal upper set of p. Recall that we can construct the category of elements in terms

88. I hope this latter notation is not too confusing in this context, given that we are also talking about principal downsets!

of the slice category. Thus, a **2**-presheaf (i.e., downset $A \subseteq P$) is representable iff it has a greatest element.

Example 156 Recall the discussion of graph coloring and the functor *nColor*, first described in example 37 (chapter 2). Recall also that a complete graph is just a graph in which every pair of distinct vertices is connected by a unique edge. In graph theory, by K_n we mean the complete graph on n nodes. In terms of graph colorings, it should be obvious that this graph will be the graph with the fewest vertices and edges that needs at least n colors to be colored. This fact actually gets at a universal property of the graph K_n. Leaving the details to the reader, we indicate that the functor *nColor* is represented by K_n. This basically says that if we want to know about the set $nColor(G)$ of n-colorings of a graph G, we can just look at the set of graph homomorphisms from G to the complete graph K_n on n vertices. To appreciate this, notice how, given a morphism $f : G \to K_n$, the condition (by definition of being a graph homomorphism) that $f(e)$ be in the edge-set of K_n for any edge e in the edge-set of G will force $f(x) \neq f(y)$ whenever x and y are adjacent, recapturing the notion of an n-coloring of G.

Incidentally, the reader might wish to ponder which concept from graph theory is captured by morphisms going in the other direction, that is, by homomorphisms $f : K_n \to G$.

In a moment, we will see the most important category-theoretic result, which morally shows how an object is defined completely by its functorial (relational) properties. But as representability often seems to baffle the newcomer, the next (optional) section offers an elaboration on the phenomenon of (non)representability. The reader eager to press on to the main Yoneda results can skip ahead a few pages.

6.4 More on Representability, Fixed Points, and a Paradox

Above, when we were thinking of general morphisms $T \times X \to Y$ as a family of morphisms $T \to Y$ parameterized or indexed by the elements of X—in which setting we were considering a sort of "miniature" version of representability—this was effectively to look at arbitrary maps

$$\hat{f} : X \to Y^T,$$

which led to the question of when (and which) X can "parameterize" all the maps from T itself to some Y. This is effectively the same as asking when such $\hat{f}$ are surjective. In particular, though, we can ask this for $X = T$, so that we are considering

$$\hat{f} : T \to Y^T$$

or

$$f : T \times T \to Y,$$

which are effectively "Y-valued" *relations* or *predicates* on T (or Y-*attributes* of type T). Via the object (function) Y^T, Y-valued predicates can be thought of as "talking about" T. A special circumstance would be where *all* the ways of "talking about itself" can be said by T itself! This is captured by the surjectivity of the map, that is, when *every* element $f : T \to Y$ of Y^T is representable in T. The next result concerns when this can occur.[89]

89. The interested reader may wish to consult Lawvere (2006) and Lawvere and Schanuel (2009) for more extensive discussion of this and the following theorem.

Theorem 157 (*Lawvere's Fixed-Point Theorem*) If

$$\hat{f}: X \to Y^X$$

is surjective (i.e., every $g: X \to Y$ is representable, in the above sense), then Y will have the *fixed-point property*, that is, *every* endomap $\tau: Y \to Y$ has at least one *fixed point*, where this of course means some $y \in Y$ such that

$$\tau(y) = y.$$

Proof. Consider $p: X \to Y$, an arbitrary "predicate" (i.e., element of Y^X). Since any endomap $\alpha: Y \to Y$ just "shuffles around" the elements of Y, we can define p as the composite of the diagonal map, the function f (got from $\hat{f}$ via the standard exponential conversion), and an endomap,

$$\begin{array}{ccc} X \times X & \xrightarrow{f} & Y \\ \delta \uparrow & & \downarrow \alpha \\ X & \xrightarrow[p]{} & Y. \end{array}$$

By assumption, moreover, there will be an $x \in X$ that *represents* p. Thus,

$$p(x) = \alpha(f(\delta(x))) = \alpha(f(x, x)) = \alpha(p(x)),$$

making $p(x)$ a fixed point of α. □

Notice that Y has the fixed-point property provided *every* endofunction on Y has a fixed point. But any set with more than one element clearly has an endofunction on it that does not have a fixed point (hint: the simplest example is a two-point set, where the points are "true" and "false"; then, an endomap without fixed points is given by the familiar negation map); thus, no set with more than one element will have the fixed-point property. In order to appreciate the importance of the theorem in **Set**, we can present the theorem in another light, namely via the contrapositive.

Theorem 158 (*Cantor's Theorem*) If Y has at least one endomap τ that has no fixed points (i.e., for all $y \in Y, \tau(y) \neq y$), then for every object X and for every

$$X \xrightarrow{\hat{f}} Y^X$$

$\hat{f}$ is not surjective.

In other words, $\hat{f}$ not being surjective means that for every attempt $\phi: X \times X \to Y$ to parameterize maps $X \to Y$ by the points of X, there must be at least one map $g: X \to Y$ that gets left out, that is, it is not representable by ϕ (meaning, does not occur as $\phi(-, x)$ for any point x in X).

Proof. Again, define g as the composite of the diagonal map, the function f (got from $\hat{f}$ via the exponential conversion), and an endomap,

$$\begin{array}{ccc} X \times X & \xrightarrow{f} & Y \\ \delta \uparrow & & \downarrow \alpha \\ X & \xrightarrow[g]{} & Y. \end{array}$$

In other words,

$$g(x) = \alpha(f(x,x)).$$

Then, for all $x \in X$,

$$g(-) \neq f(-,x)$$

as functions of one variable. For, if we *did* have $g(-) = f(-,x_0)$ for some $x_0 \in X$, then by evaluation at x_0,

$$f(x_0,x_0) = g(x_0) = \alpha(f(x_0,x_0)),$$

where the leftmost equality follows from g being representable, and the second equality is by definition. But then, α has a fixed point. This is a contradiction. □

Cantor's famous result that there is no surjective map from a set to its powerset

$$X \to 2^X$$

is a special case of the above.

It is best, though, to see how this special result is part of something more general. Given our way of thinking about maps Y^X as providing a particular way (or name for how) X "speaks about" or describes itself, the generalized version of the above can be regarded as saying that, provided the truth-values or properties of X are nontrivial, there will be no way that the elements of object X can "talk about" themselves (in the sense of talking about their own truthfulness or their own properties). The result appeals to an observation concerning the fundamental limitations in how an object X can address its own properties. Many apparent paradoxes of the past seem to play off this. For instance, the Liar paradox was an ancient way of exhibiting the trouble one can get into when natural languages attempt to construct self-referential statements that speak about their own truthfulness—if one permits this, it seems one must open the door to certain inconsistencies in natural language. Russell's famous paradox was basically a simplified version of something Cantor himself already found, one that did not involve the notion of *size*: namely that if we take T as the set of *all* sets, then by Cantor's theorem, there is a set larger than T, namely the powerset of T, yet T is assumed to contain all sets, so we are saying that T contains a subset that is larger than itself. Gödel's famous incompleteness results revealed limitations in formal systems and provability statements within those systems. Brandenburger-Keisler's paradox (a sort of two-person or interactive version of Russell's paradox) concerns the description of a belief situation in which "Ann believes that Bob believes that Ann believes that Bob believes something false about Ann." The paradox is: does Ann believe that Bob has a false belief about Ann? This suggests that not every description of beliefs can be "represented." There are a variety of other results[90] that one could enumerate as further examples of what are arguably all variations on the same theme:

Letting things address their own properties, without limitations, can lead to problems.

The phenomenon of (non)representability is really at the core of such problems. The following (apparently paradoxical) example is mostly meant to get the reader thinking more about some of the subtleties in issues of representability.

90. See Yanofsky (2003) for more; Abramsky (2014) is also of interest, in this connection.

Example 159 The issue underlying the following example sometimes goes under the name of "Grelling's paradox."[91] Consider the set of all English words. Some of these words *describe themselves*, while others (most) do not. Adjectives, perhaps more than any other type of word, are used to describe things. So let us restrict attention to the set of adjectives, which we may denote *Adj*. Certain adjectives *describe themselves*, while others (most) do not. Those that describe themselves are said to be *autological* (or *homological*). For instance, the following adjectives are homological: "English" (is English!); "polysyllabic" (is polysyllabic); "Hellenic" (is of Greek origin); "unhyphenated." Those adjectives, by contrast, that *do not describe themselves* are said to be *heterological*. For instance, the following are heterological: "Spanish" (not a Spanish word!); "misspelled" (is spelled correctly!); "long" (is hardly long); "monosyllabic"; "hyphenated."

It seems plausible that all adjectives will be either homological or heterological. However, consider the adjective "heterological." Is it heterological? Suppose it is not. Then, it might naturally be assumed, it will be homological. So it describes itself. Thus "heterological" (which says that it does not describe itself) must be heterological after all. So if "heterological" is not heterological, then it is heterological. On the other hand, then, we suppose that the answer to the question is affirmative, that is, that "heterological" is heterological. Then "heterological," being heterological, does not describe itself. But this implies that it is *not* heterological after all—since "heterological" says that it is of the sort that does not describe itself, and we just said that "heterological" does not describe itself, so it is not described by the description "does not describe itself"!

We might formalize this seemingly paradoxical situation by first considering that we are dealing with a function

$$f : Adj \times Adj \to 2$$

defined on all adjectives a_1, a_2 by

$$f(a_1, a_2) = \begin{cases} 1 & \text{if } a_2 \text{ describes } a_1 \\ 0 & \text{if } a_2 \text{ does not describe } a_1. \end{cases}$$

Then we know there is a predicate (a map $Adj \to 2$) that can be defined on *Adj* that is not representable by any element of *Adj*. We get this by applying the fixed-point theorem, with α the negation map $\neg : 2 \to 2$, setting $\alpha(0) = 1$ and $\alpha(1) = 0$. More explicitly, using the idea from before,

$$\begin{array}{ccc} Adj \times Adj & \xrightarrow{f} & 2 \\ {\scriptstyle\delta}\uparrow & & \downarrow{\scriptstyle\alpha} \\ Adj & \xrightarrow[g]{} & 2, \end{array}$$

we know how to construct g as a (nonrepresentable) function naming a particular property of adjectives, namely as the characteristic function of a subset of adjectives that cannot be described by any adjectives. In particular, the adjective "heterological" will be in this subset. In terms of the above, that g is such a characteristic function just says that we must have that

$$g(-) \neq f(-, a)$$

91. Yanofsky (2003) has a very nice discussion of this and a number of other such "paradoxes."

for all adjectives a, since if there *were* an adjective a_0 that satisfied $g(-)=f(-,a_0)$, evaluating at a_0 would give

$$f(a_0,a_0)=g(a_0)=\alpha(f(a_0,a_0)),$$

the first equality from the (assumed) representability of g and the second by definition of g. But this is certainly false, due to the nature of the map α.

Observe that the hypothetical

$$f(a_0,a_0)=g(a_0)=\alpha(f(a_0,a_0)),$$

which yields a contradiction (whether we choose $f(a_0,a_0)=1$, when "a_0 describes itself," or $f(a_0,a_0)=0$, when "a_0 does not describe itself"), makes precise exactly the "paradox" described at the beginning.

Altogether, it is perhaps more telling to consider such a situation in terms of the non-representability of the g given above, where this just means that the property of "not being described by" an adjective (which applies to "heterological" in particular) is *not representable*, for there is no adjective that might represent itself via f.

6.5 Yoneda in the General

Let us now make good use of the notions of representability and the model of the Yoneda results for posets. In the special case of posets, we saw that we can identify a principal downset $\downarrow p$ with a representable functor $\mathrm{Hom}_{\mathcal{P}}(-,p)$. For any element $p\in\mathcal{P}$, there will be a representable **2**-presheaf (think: it is *represented by p*)

$$\phi_p:\mathcal{P}^{op}\to\mathbf{2}$$

that takes $q\mapsto 1$ iff $p\leq q$. In this way, the representable presheaves act as the "characteristic maps" of the principal downsets of $\mathcal{P}$, and the **2**-enriched version of the Yoneda embedding taking each $p\mapsto\phi_p$ is the same as the inclusion of the elements of the poset into its downsets (which is, in turn, the same as considering the **2**-enriched presheaves on $\mathcal{P}$)

$$\mathcal{P}\hookrightarrow\mathbf{Down}(\mathcal{P})\simeq\mathbf{2\text{-}PreSh}(\mathcal{P}).$$

The Yoneda results in the case of categories more generally, that is, in the **Set**-enriched setting, are effectively a far-reaching generalization of this idea, and supply perhaps the most important and well-utilized results in category theory.

Proposition 160 (*Yoneda lemma*) For any functor $F:\mathbf{C}\to\mathbf{Set}$, where $\mathbf{C}$ is a locally small category, and for any object $c\in\mathbf{C}$, the natural transformations $Y^c\Rightarrow F$ are in bijection with elements of the set $F(c)$, that is,[92]

$$\mathrm{Nat}(Y^c,F)\cong F(c). \tag{6.1}$$

Moreover, this correspondence is natural in both F and c. In the contravariant case, that is, for $F:\mathbf{C}^{op}\to\mathbf{Set}$, things are as above, except we have

$$\mathrm{Nat}(Y_c,F)\cong F(c). \tag{6.2}$$

92. Recall that by Y^c we just mean $\mathrm{Hom}_{\mathbf{C}}(c,-)$, while Y_c is used for $\mathrm{Hom}_{\mathbf{C}}(-,c)$.

We are not going to prove this (it is a good exercise to actually attempt to prove this yourself!), but instead will unpack it and then discuss its significance at a more general level. We will confine attention to the contravariant version in what follows (but dual statements can be made for the covariant version).

The idea is that for a fixed category $\mathbf{C}$, given an object $c \in \mathbf{C}$ and a (contravariant) functor $F : \mathbf{C}^{op} \to \mathbf{Set}$, we know that the object c gives rise to another special (representable) functor $Y_c : \mathbf{C}^{op} \to \mathbf{Set}$. A very natural question to ask, then, is about the maps $Y_c \Rightarrow F$,

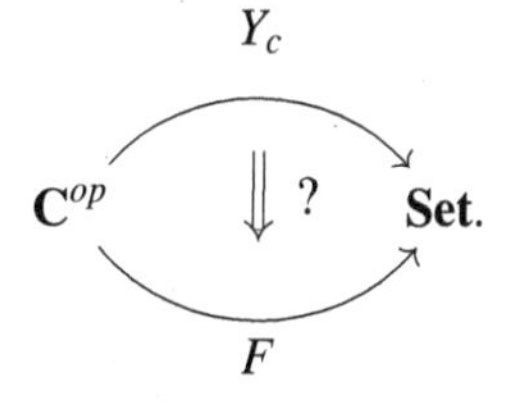

The functors we are comparing both live in $\mathbf{Set}^{\mathbf{C}^{op}}$, so the collection of maps from Y_c to F are just the natural transformations that belong to $\mathrm{Hom}_{\mathbf{Set}^{\mathbf{C}^{op}}}(Y_c, F)$. But what is this set? Notice that from the input data F and c we were given ("given an object c and a functor F"), we could have also constructed the set $F(c)$, by simply applying F on the given object c. The Yoneda lemma just assures us that these two sets are the same! Moreover, all the generality of natural transformations is encoded in the particular case of identity maps (used in the proof of the lemma).

The "naturality" in F mentioned in the lemma just means that, given any $\upsilon : F \to G$, the following diagram commutes:[93]

$$\begin{array}{ccc} \mathrm{Hom}(Y_c, F) & \xrightarrow{\cong} & F(c) \\ {\scriptstyle Hom(Y_c,\upsilon)}\downarrow & & \downarrow{\scriptstyle \upsilon_c} \\ \mathrm{Hom}(Y_c, G) & \xrightarrow[\cong]{} & G(c). \end{array}$$

On the other hand, naturality in c means that, given any $h : c \to c'$, the following diagram commutes:

$$\begin{array}{ccc} \mathrm{Hom}(Y_c, F) & \xrightarrow{\cong} & F(c) \\ {\scriptstyle Hom(Y_h,F)}\uparrow & & \uparrow{\scriptstyle F(h)} \\ \mathrm{Hom}(Y_{c'}, F) & \xrightarrow[\cong]{} & F(c'). \end{array}$$

The most significant application of the Yoneda lemma is given by the *Yoneda embedding*, which tells us that any (locally small) $\mathbf{C}$ will be isomorphic to the full subcategory of $\mathbf{Set}^{\mathbf{C}^{op}}$ spanned by the contravariant representable functors, while $\mathbf{C}^{op}$ will be isomorphic to the full subcategory of $\mathbf{Set}^{\mathbf{C}}$ spanned by the covariant representable functors. We have seen that for each $c \in \mathbf{C}$, we have the covariant functor Y^c going from $\mathbf{C}$ to $\mathbf{Set}$ and the contravariant functor Y_c going from $\mathbf{C}^{op}$ to $\mathbf{Set}$. If we let this functor vary over all the objects of $\mathbf{C}$, the resulting functors can be gathered together into the (for example, covariant) functor $Y^\bullet : \mathbf{C}^{op} \to \mathrm{Hom}(\mathbf{C}, \mathbf{Set})$. Dually, we have the contravariant functor Y_c going from $\mathbf{C}^{op}$ to

93. Note: all the "Homs" are $\mathrm{Hom}_{\mathbf{Set}^{\mathbf{C}^{op}}}$.

Set, and collecting these functors together as we let c vary will give a functor $Y_\bullet : \mathbf{C} \to \mathrm{Hom}(\mathbf{C}^{op}, \mathbf{Set})$.[94]

Definition 161 The *Yoneda embedding* of $\mathbf{C}$, a locally small category, supplies functors

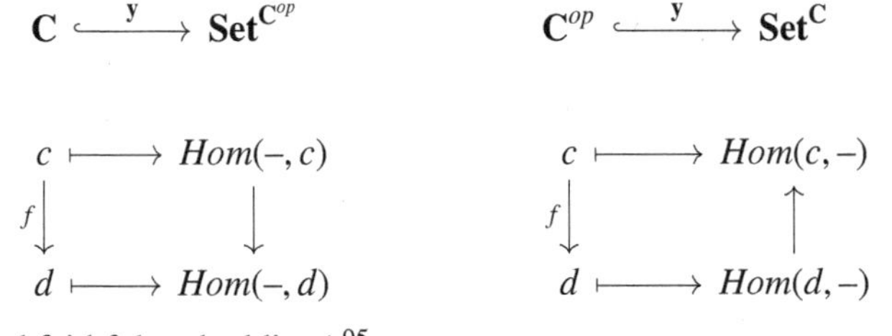

defining full and faithful embeddings.[95]

The Yoneda embedding $\mathbf{y}$ gives us a representation of $\mathbf{C}$ in a category of set-valued functors and natural transformations. An important consequence of the embedding is that any pair of isomorphic objects $a \cong b$ in $\mathbf{C}$ are representably isomorphic, that is, $Y^a \cong Y^b$. The Yoneda lemma supplies the converse, namely if either the (co- or contravariant) functors represented by a and b are naturally isomorphic, then a and b will be isomorphic; so in particular, if a and b represent the same functor, then $a \cong b$. In many cases, it will be easier or more revealing to give such an arrow $Y^a \to Y^b$ or $Y_a \to Y_b$ than to supply $a \to b$, for the category $\mathbf{Set}^{\mathbf{C}^{op}}$ in general has more structure than does $\mathbf{C}$—namely, it is complete, cocomplete, and "Cartesian closed" (basically, any morphism defined on a product of two objects can be identified with a morphism defined on one of the factors). Thus we can use the more advanced tools and universal properties (like the existence of limits) that come with the presheaf category, and be sure that an arrow of the form $Y_a \to Y_b$, for instance, comes from a unique $a \to b$ even if $\mathbf{C}$ on its own may not allow the advanced constructions. Analogously, representing a rational number in terms of downward (upward) closed sets under the standard ordering results in a Dedekind cut, and altogether this embeds the rationals into the reals, allowing for solutions to more equations. Passing from a category $\mathbf{C}$ to its presheaf category can also be regarded as adjoining colimits (think generalized sums) to $\mathbf{C}$, and doing so in the most "free" way.[96] In general, in passing to the presheaf category, many nonrepresentable presheaves will show up as well. However, the representables have a very special role to play.

Before concluding our discussion of these matters, let us record a final result. We have seen how an object is defined completely by its functorial (relational) properties. The

94. It is not unusual to rename these functors, as we do in the following definition, with a lowercase (bold) $\mathbf{y}$ in both cases, leaving the appropriate variance to context.

95. In general, a functor $F : \mathbf{C} \to \mathbf{D}$ induces, for every pair of objects c, c' in $\mathbf{C}$, a function on the hom-sets—that is, $F_{c,c'} : \mathrm{Hom}_{\mathbf{C}}(c, c') \to \mathrm{Hom}_{\mathbf{D}}(F(c), F(c'))$. The functor F is called *faithful* provided this function $F_{c,c'}$ is injective for every c, c' in $\mathbf{C}$, and called *full* provided $F_{c,c'}$ is surjective for every c, c' in $\mathbf{C}$. A full and faithful functor is thus bijective on hom-sets. An *embedding* in the categorical sense is a faithful functor that is also injective on objects (up to isomorphism)—that is, if $F(c) \cong F(c')$, then $c \cong c'$. An embedding in this sense reveals the domain category to be a subcategory of the codomain category. A full and faithful functor that is injective on objects thus gives us a *full embedding*, identifying the domain category as a full subcategory of the codomain category. The proof of the main result can be found in any text on category theory.

96. This is a powerful and general idea, but the reader who desires a more concrete way of thinking about the previous statement, might consider the unions (*colimits*) that showed up in the downset poset, after we embedded $\mathcal{P}$ into $\mathcal{D}(\mathcal{P})$, where these were not present in $\mathcal{P}$ itself.

next proposition tells us that even if a functorial definition does not correspond to an object—that is, if the particular functor is not representable—it is still "built out of" the representables (in particular, it is the colimit of a diagram of representables).

Proposition 162 Every object P in the presheaf category $\mathbf{Set}^{\mathbf{C}^{op}}$ (i.e., every contravariant functor on $\mathbf{C}$) is a colimit of a diagram of representable objects, in a canonical way, that is,

$$P \cong \operatorname{colim}\left(\int P \xrightarrow{\pi_P} \mathbf{C} \xrightarrow{\mathbf{y}} \mathbf{Set}^{\mathbf{C}^{op}}\right),$$

where π is the projection functor and $\mathbf{y}$ is the Yoneda embedding.

This proposition states that given a functor $P : \mathbf{C}^{op} \to \mathbf{Set}$, there will be a canonical way of constructing an indexing category $\mathbf{J}$ and a corresponding diagram $A : \mathbf{J} \to \mathbf{C}$ of shape $\mathbf{J}$ such that P is isomorphic to the colimit of A composed with the Yoneda embedding. The indexing category that serves to prove the proposition is the category of elements of P.[97]

Specializing to posets gives another way to think about this: taking joins (unions) of downsets is the same thing as taking colimits in $\mathbf{Down}(\mathcal{P})$ regarded as a category; so any element in $\mathbf{Down}(\mathcal{P})$ can be regarded as the colimit (join) of representable functors (principal downsets).

6.6 Philosophical Pass: Yoneda and Relationality

Occasionally, an idea is powerful enough that it seems, almost effortlessly, to transcend its local and native context of application and speak articulately to many other contexts. The idea underlying the Yoneda results is like this. In the most general sense, it might be regarded as saying something like

To understand an object it suffices to understand all its relationships with other things.

Or, to say it in another way,

If you want to know whether two objects A and B are the same, just look at whether all the ways of probing A with other things (or, dually, probing other things with A) is the same as all the ways of probing B with other things (or, dually, probing other things with B).

The previous two slogans are an attempt to convey what is sometimes called the "Yoneda philosophy." It tells us, fundamentally, that if you want to understand what something *is*, there's no need to chase after some "object *in itself*"; instead, just consider all the ways the (candidate) object transforms, perturbs, and constrains other things (or, dually, all the ways it is transformed, perturbed, and constrained by others)—this will tell you what it is!

This idea that an object is determined by its totality of behaviors in relation to other entities capable of affecting it (or being affected by it), in addition to appealing to many of our intuitions, is an idea that one can find versions of throughout a number of different

97. A proof of this fact can be found in Riehl (2016). That *every presheaf is a colimit of representable presheaves* is closely related to another construction, namely the *Cauchy completion* (or *Karoubi envelope*) of a category, in which the fact that representable presheaves are *continuous* in a precise sense is exploited. The main idea here is that while we have the powerful Yoneda (full and faithful) embedding sending a category $\mathbf{C}$ to the category of presheaves $\mathbf{Set}^{\mathbf{C}^{op}}$, in general, a category $\mathbf{C}$ cannot be recovered from $\mathbf{Set}^{\mathbf{C}^{op}}$, so a natural question to ask is how or to what extent, given $\mathbf{Set}^{\mathbf{C}^{op}}$, it can be said to determine $\mathbf{C}$. Basically, if a category $\mathbf{C}$ (or $\mathbf{C}^{op}$) can be shown to be Cauchy complete, then not only can it be recovered (up to equivalence) from the presheaf category (or covariant functor category of variable sets $\mathbf{Set}^{\mathbf{C}}$), but it can be shown to *generate* the original presheaf (variable set) category.

contexts, for instance in the seventeenth-century philosopher Spinoza's idea that what a body *is* (its "essence") is inseparable from all the ways that the body can affect (causally influence) and be affected (causally influenced) by *other bodies*. In this general approach, what an object *is* can be entirely encapsulated by regarding *all at once* (generically) all of its interrelations or possible interactions with the other objects of its world. In the covariant case, we do this by regarding its ways of affecting other things; dually, in the contravariant case, by its particular ways of being affected by the other objects that inhabit its world. To use another metaphor: if you want to understand if two "destination points" are the same, just inspect whether, ranging over all the addresses with which they can communicate, the networks of routes connecting them to the addresses are the same. For a given object c, the representable functor just captures, all at once, the most generic and universal picture of that object, supplying a placeholder for each of the possible attributes of that object (and, following Spivak [2014], then Yoneda's lemma can be thought of as saying that to specify an actual object of type c, it suffices to fill in all the placeholders for every attribute found in the generic thing of type c).

If one regards $\mathrm{Hom}_{\mathbf{C}}(-,A)(U)=\mathrm{Hom}_{\mathbf{C}}(U,A)$ as telling us about "A viewed from the perspective of U," then the fact (from the Yoneda embedding) that

$$\mathrm{Hom}_{\mathbf{C}}(-,A)\cong\mathrm{Hom}_{\mathbf{C}}(-,B) \text{ iff } A\cong B \text{ iff } \mathrm{Hom}_{\mathbf{C}}(B,-)\cong\mathrm{Hom}_{\mathbf{C}}(A,-)$$

can be glossed as saying that two objects A and B will be the same precisely when they "look the same" from all perspectives U.[98] This paradigm seems especially natural in many contexts, beyond mathematics. It seems especially appropriate to an adequate description of *learning*, wherein an object comes to be known and recognized through probing it with other things, varying the perspectives on it. For instance, suppose you are tasked with having a robot learn how to identify objects that it has never been exposed to before, without relying on much training data or manual supervision, with the aim of having it come away with an ability to correctly discriminate between (what you naively take to be) different objects and readily recognize other instances of objects of the same type in other contexts.

You might place the robot in a room with a number of other objects, say, a ceramic cup, some red rubber balls, a steel cable, a small plant, a plastic bottle filled with water, a worm, and a thermostat. If the robot cannot interact with the objects in any way, and the observable interactions and changes that would unfold without its intervention are rather uneventful or slow to unfold, it is not clear how the robot could learn anything at all. On the other hand, suppose you have enabled the robot to inspect, manipulate, or otherwise instigate or probe the objects in a number of ways, and observe the outcomes. At first, these action attempts might be more or less randomized. The robot might simply locally perform simple action sequences or gestures such as

Grasp, Release, Put, Pull, Push, Rotate, Twist, Throw, Squeeze, Bend, Stack, See, Hear, Locate.

98. It is not uncommon to hear such interpretations, namely that we can retrieve the object itself via all the "perspectives on it." In case there are any unscrupulous listeners ready to confuse this with a kind of relativism, the relationism of Yoneda has nothing to do with this. Thankfully, such misunderstandings can be blocked by Yoneda itself. For the (untenable) relativist interpretation would need to assume, among other questionable assumptions, that there is an object (namely of the type "human being") that can represent *any* functor whatsoever, that thereby itself mediates all possible exchanges between objects. But Yoneda tells us no such thing.

It may grasp a rubber ball, twist the plastic water bottle, attempt to bend the steel cable, see something move (without its having performed any other action that might explain this motion) or hear it wriggling. Such actions can provide the robot with much information about the objects populating the room, and sometimes even the mere successful implementation of a certain isolated action can confirm additional information, such as about the location ("within reach radius") of an object-candidate that is engaged by *Grasp*. And once the robot has a decent working sense of some possible object-candidates, it can use one to probe others and learn even more. It may *Push* on a number of objects, with little to no effect, and then push on (what we know to be) the thermostat, altering the room's temperature, on which change it may observe different effects throughout the room (e.g., the water evaporates, the plant withers, the worm moves more, other things remain unchanged in certain relevant respects). In this way, our robot goes around the room and probes "possible object" regions in different ways, and observes the effects of these variations, giving them their own name. Via such probings of the objects of the room, and composite action-sequences, such as *Grasp, then Release, then Hear*, it seems that our robot will have a chance at learning "what is what."

Later, if we take our robot and place it in a new room, one that has in it just (what we know to be) a red balloon and a (similarly shaped) blue rubber ball; and if, in the previous room, all that the robot came to know of (what we know to be) the red rubber ball was the visual information (gathered through *See*) of its color and shape; then it may assume that the red balloon *is* the object it knew as the rubber ball (which will go under the name of a mapping from some A_i to color data), while it may assume the blue rubber ball is some entirely new thing. On the other hand, if our robot *had* probed the red rubber ball in more ways—say, having subjected it to *Bounce, Twist, Throw, Hear, Grasp*—you can be sure that it would take much less for it to come to recognize that the blue rubber ball was something like the object it knew in the previous room, while the red balloon was something very different.

The idea of Yoneda is that we can be assured that if the robot wants to learn whether some object A is the same thing as object B, it will suffice for it learn whether

$$\mathrm{Hom}_{\mathbf{C}}(-, A) \cong \mathrm{Hom}_{\mathbf{C}}(-, B)$$

or, dually,

$$\mathrm{Hom}_{\mathbf{C}}(B, -) \cong \mathrm{Hom}_{\mathbf{C}}(A, -).$$

In terms of the discussion above, this means having the robot explore whether

> all the ways of probing A with objects of its environment amount to the same as all the ways of probing B with objects of its environment.

This is a fascinating idea, philosophically, and one that we think has much in its favor even beyond the narrower context of mathematics. In Japanese, the word for human being, *ningen*, is made up of two characters, the first of which means something like a human or person, while the second is a representation of the doors of a gate and means something like "betweenness," so that the literal meaning of the word "human" is "the relation between persons." This—rather than the narrative of atomistic individualism, that often ignores or glosses over the immense load of relational constraints and determinations that come with differential obligations, pressures, and opportunities—seems more attuned to the Yoneda way of thinking.

The fundamental intuition behind the Yoneda philosophy, then, is that to know or access an object it suffices to know or access how it can be transformed by different objects, or how other objects transform into it. More exactly, Yoneda tells us that if there is a natural way of passing an object c's vision of its world (or how it is seen by its world) on to a functor F on that same category, then to recover this vision it suffices to ask F how it acts on c. While, mathematically speaking, the usefulness of the lemma often boils down to the fact that we are able to reduce the computation of natural transformations (which can be unwieldy) to the simple evaluation of a (set-valued) functor on an object, in a sense the full philosophical significance of the lemma points in the other direction. Given a category and an object in that category, rather than regard the object "on its own" (moreover, treating the entire category in a detached manner, as delimiting the outer boundaries of our consideration), via Yoneda we can regard that object as entirely characterized by its perspective or action on its world (or its world's perspective or action on it), and moreover place the category in which it lives in the wider category of all presheaves or sets varying over that category. Via Yoneda, we can perform this sort of passage from the detached consideration of a given object to the consideration of all its interrelations with the other objects of its world *for every object of a given category*. In doing so, we can think of ourselves as taking an entire category **C** that previously was itself being regarded in a detached manner, and placing it in the more "continuous" (in a loose sense) context of the category of all the presheaves over **C**. The category of presheaves over **C** into which **C** is embedded not only has certain desirable properties that the original category may lack, like possessing all categorical limits, but it can be understood (in both intuitive and in various technical ways) as providing the continuous counterpart to the detached consideration of the original category.

7 Adjunctions

In which we use detailed examples and constructions from a variety of areas to illustrate the last of the really fundamental notions in category theory—that of adjunctions, or adjoint functors—and explore their key features.

In chapter 2, we saw how the notion of an isomorphism of categories—using a pair of functors $F:\mathbf{C}\rightarrow\mathbf{D}$ and $G:\mathbf{D}\rightarrow\mathbf{C}$ inverse to each other in the sense that $G\circ F=\text{id}_{\mathbf{C}}$ and $F\circ G=\text{id}_{\mathbf{D}}$—was too restrictive to be useful, and that a far better category-theoretic notion of "sameness" of categories was supplied by the notion of an equivalence of categories, using natural transformations to relax the equalities and leave us with functors inverse "up to natural isomorphism," in the sense that $G\circ F\cong\text{id}_{\mathbf{C}}$ and $F\circ G\cong\text{id}_{\mathbf{D}}$. The notion of an adjunction, or adjoint functors, is in a sense a further step in generalization, weakening the notion of an equivalence of categories, so that we are interested less in a relation (of sameness) between two categories and more in a relation between specific functors moving between those categories. Another perspective would be to say that adjunctions represent something like a further broadening of the notion of inverse, involving *unique* "reversal attempts" that supply the "closest thing" to inverses and capture important relations that exist even when inverses in the strict sense do not.

Like so many other important notions in category theory, these arise in an especially simple form in the special context of orders, so that adjoint relations between orders allow one to display the features of adjointness in a particularly accessible form. Adjoint functors between orders first appeared under the name of *Galois connections*. The next few examples will illustrate and motivate, via explicit examples in the context of orders, some of the many fundamental general features and properties of adjunctions.

7.1 Adjunctions through Morphology

Example 163 Suppose you receive a dark photocopy of some text, where the pen or marker appears to be bleeding:

With the help of your favorite programming language, you might perform what the image processing community would call an "erosion" of the image. After doing this (perhaps a few times), you would be left with something like

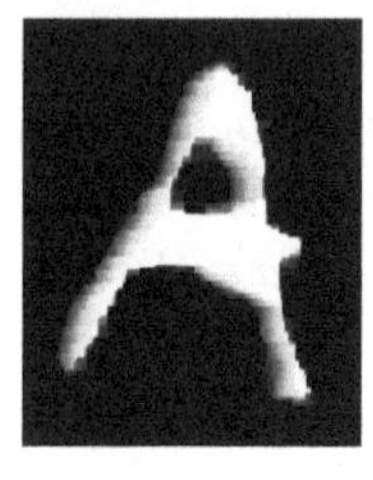

As one can see, erosion effectively acts to make thicker lines skinnier and detects, or enhances, the holes inside the letter A.

Suppose, instead, that you attempted a *dual* operation, called "dilation," the effect of which is to thicken the image, so that lightly drawn figures are presented as if written with a thicker pen, and holes are (gradually) filled. In the case of dilating the original image, you would be left with something like

Now suppose that, for instance, after eroding the image you received, certain things have become harder to read. You decide that you would like to undo what you have done, perhaps because you have lost some important information. It seems sensible to hope you might undo it, and get back to the original image, by dilating the result of your erosion. But, in general, erosion and dilation do not admit inverses—in particular, they are not one another's inverse—and there is no operation that would recover the exact original image from a dilated (or eroded) image. If an image is eroded and then dilated (or conversely), the resulting image will not be the original image. These operations *discard* information, so perhaps it is not so surprising that one would not get back to the original by "undoing" an erosion, for instance, by dilating the result.

However, in the failed search for an inverse to each operation, you will very quickly alight on something new, the basic properties of which appear to be useful and interesting in their own right. Erosion and dilation, while not inverses of each other, do seem to be related in a rather special way. In particular, eroding after we have dilated an image yields a very different result than dilating after eroding, even though neither composite gives back the original image. However, for an arbitrary image I, you will notice that it is "bounded" in both directions, in the sense of containing the result of one of the two composite operations while being contained by the result of the other, that is,

Dilating after eroding $I \subseteq I \subseteq$ Eroding after dilating I.

If we erode an image and then dilate the eroded image (making use of the same "structuring element" through both operations, on which more below), we arrive at a subset of the original image, in a process sometimes called *opening* the image by the image processing community. For instance, if we start with the following image

then opening it yields

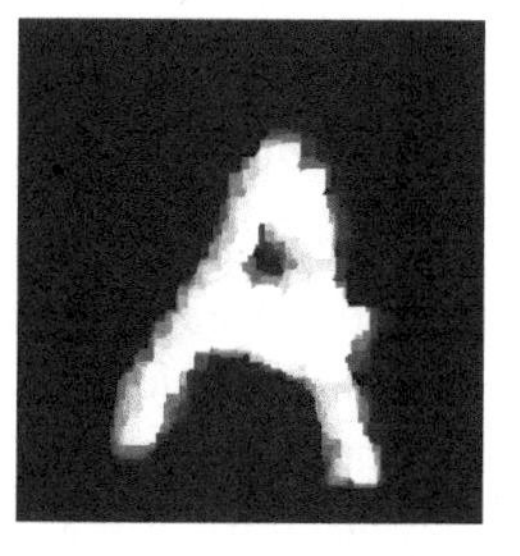

Opening an image will leave an image that is generally smaller than the original, as it removes noise and protrusions and other small objects from an image, while preserving the shape and size of the more substantial objects in the image. On the other hand, dilating an image and then eroding the dilated image (with the same structuring element throughout)—sometimes called *closing* the image—leaves one with an image that is generally larger than the original. Starting with the following image

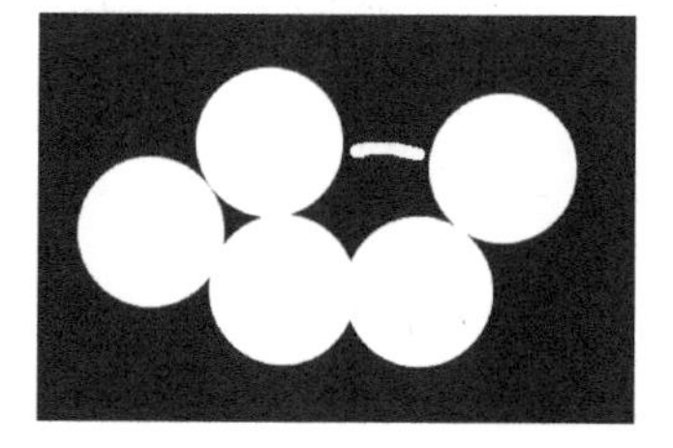

closing it will get rid of small holes, fill gaps in contours, smooth sections of contours, and fuse thin gulfs or breaks between figures. The result of closing the above yields

However, you may quickly learn that opening (or closing) an image twice leaves one with the same image as opening (or closing) it once. In other words, opening and closing are *idempotent* operations.

Altogether, the two basic operations of dilation and erosion are not quite inverses of one another, yet, as may already be evident from the discussion of the ways their idempotent composites "bound" the original above and below, they are nevertheless related in a special way, and are, in a sense, the "closest thing" to inverses (when these do not exist). We will explore that notion more closely and formally now.

Mathematical morphology is a field that deals with the processing of binary, gray-level, and other signals, and has proven useful in image processing. The majority of its tools are built on the two fundamental operators of dilation and erosion, and combinations thereof. It has a variety of applications involving image processing and feature extraction and recognition, including applications in x-ray angiography, biometrics, text restoration, among others.[99]

A fundamental idea, in this setting, is that of "probing" an image with a basic, predefined shape and then examining how this fixed shape relates to the shapes comprising the image. One calls the probe the *structuring element*, which is itself a subset of the space, so that, for example, in the simple case of binary images, such a structuring element is itself just a binary image; these are, moreover, taken to have a defined origin. For instance, in the digital space $E = \mathbb{Z}^2$ (imagine a grid of squares), one might take for structuring element a 3×3 square, that is, the set

$$\{(-1,-1),(-1,0),(-1,1),(0,-1),(0,0),(0,1),(1,-1),(1,0),(1,1)\},$$

or perhaps a 3×1 rectangle with a designated central square, or a diamond or disk-shaped element, and so on. Going back to our earlier example of the blurry binary image letter A, we may take 0s to represent background and 1 for foreground, so that you basically have something like:

0 0 0 0 1 0 0 0 0
0 0 0 1 1 1 0 0 0
0 0 0 1 0 1 0 0 0
0 0 1 1 0 1 1 0 0
0 0 1 1 1 1 1 0 0
0 1 1 1 0 0 1 1 0
0 1 1 0 0 0 1 1 0
1 1 0 0 0 0 0 1 1

Then, we might take for structuring element $B \subseteq E$ a "diamond" (with origin boxed off)

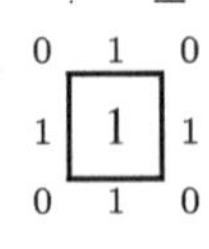

99. For classic introductions to mathematical morphology and filtering, see Serra (1986) and Serra and Vincent (1992).

In the simple case of our running example of a binary image in a bounded region, we took the structuring element B, placing B's origin at each pixel of our image X as we scan over all of X, and compute at each pixel the dilation and erosion by taking the maximum or minimum value, respectively, of all pixels within the "window" or neighborhood covered by the structuring element (so that, e.g., in the case of dilation, a pixel is set to 1 if any of its neighboring pixels has the value 1).

More specifically, assuming we have fixed an origin in E, to each point p of E there will correspond the translation map that takes the origin to p; such a map will then take B in particular onto B_p, the translate of B by p. In general, translation by p is a map $E \to E$ that takes x to $x+p$; thus, it takes any subset X of E to its *translate* by p,

$$X_p = \{x+p \mid x \in X\}.$$

For a structuring element B, then, we can consider all its translates B_p. Given a subset (image) X of E, we can examine how the translates B_p of a given structuring element B interact with X. In the simple case of Boolean images (as subsets of a Euclidean or digital space), we carry out this examination via two operations:

$$X \oplus B = \{x+b \mid x \in X, b \in B\} = \bigcup_{x \in X} B_x = \bigcup_{b \in B} X_b,$$

called *Minkowski addition*, and its dual,

$$X \ominus B = \{p \in E \mid B_p \subseteq X\} = \bigcap_{b \in b} X_{-b}.$$

The former transformation (taking X into $X \oplus B$) is in fact what gives us a *dilation*, the basic property of which is that it distributes over union, while the latter (taking X into $X \ominus B$) is an *erosion*, the basic property of which is that it distributes over intersection. In the simple set-theoretical binary image case, dilations coincide with Minkowski addition; yet erosion of an image is the intersection of all translations by the points $-b$. In short,

Dilation of X by B is computed as the union of translations of X by the elements of B,

while

Erosion of Y by B is the intersection of translations of Y by the reflected elements of B.

While we will see that we can give more general definitions of these operations, the particular behaviors of these operations in fact already follow from a general relationship underlying these operations.

Proposition 164 For every subset X, Y, B of our space E, where B is any structuring element, we have

$$X \oplus B \subseteq Y \text{ iff } X \subseteq Y \ominus B.$$

Proof. ($\Rightarrow$) Suppose $X \oplus B \subseteq Y$ and let $z \in X$ and $b \in B$. Then $z+b \in X \oplus B$, and thus $z+b \subseteq Y$. And $z+b \subseteq Y$ for any $b \in B$ implies that $z \in Y \ominus B$.
($\Leftarrow$) Suppose $X \subseteq Y \ominus B$ and let $z \in X \oplus B$. Then there exists $x \in X$ and $b \in B$ such that $z = x+b$. But $x \in X$ and $X \subseteq Y \ominus B$ entails that $x \in Y \ominus B$. Thus, for every $b' \in Y$, we have $x+b' \in Y$, and in particular, $b \in B$, so $x+b \in Y$. But $z = x+b$, so $z \in Y$. □

Before discussing the significance of this more generally, it is worth noting that there is no need to restrict attention, as we have thus far, to consideration of dilation and erosion in the simple case of Boolean images in digital space. We can also define the dilation and erosion of a *function* by a structuring element (itself regarded as a *structuring function*). For instance, given a function $f : E \to T$ from a space E to a set T of, for instance, gray-levels (i.e., a complete lattice that comes from a subset of $\overline{\mathbb{R}} = \mathbb{R} \cup \{-\infty, +\infty\}$), and given a point $p \in E$, the *translate* of f by p is the function f_p whose graph is obtained by translating the graph $\{(x, f(x)) \mid x \in E\}$ by p in the first coordinate, that is, $\{(x+p, f(x)) \mid x \in E\}$, so that for all $y \in E$, we have

$$f_p(y) = f(y-p).$$

This defines the translation of a function by a point. Of course, if we extend this to the translation by a pair (p, t), we get that for all $y \in E$,

$$f_{(p,t)}(y) = f(y-p) + t.$$

This approach allows us to define Minkowski addition and the dual operation for two *functions* $E \to T$. Where f plays the role of a (gray-level) image, and b is the functional analogue of a structuring element (i.e., a *structuring function*), for all $p \in E$, these operations will take on values

$$(f \oplus b)(p) = \sup_{y \in E}(f(y) + b(p-y))$$

and

$$(f \ominus b)(p) = \inf_{y \in E}(f(y) - b(y-p)).$$

Then the operator $\delta_g : T^E \to T^E$ taking $f \mapsto f \oplus g$ is *dilation by* g, and $\epsilon_g : T^E \to T^E$ taking $f \mapsto f \ominus g$ is *erosion by* g. In a low-dimensional case, using a portion of a disk for structuring function, these operators might do something like:[100]

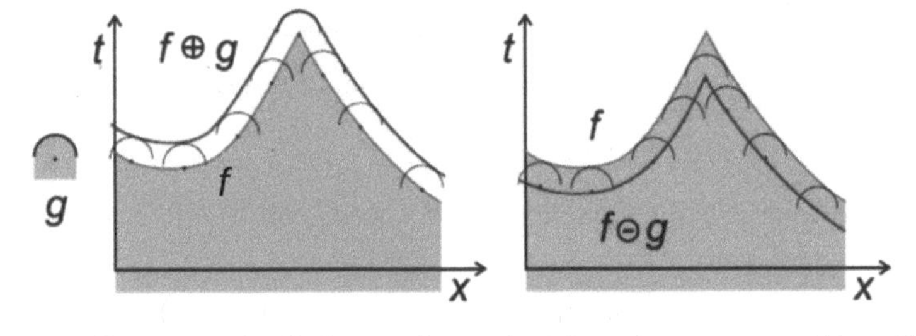

When we use a "flat structuring" element instead, such as that represented by a line, things are simplified even further, and we are effectively applying max and min filters, that is, for dilation

$$(f \oplus g)(x) = \sup_{y \in E, x-y \in B} f(y) = \sup_{y \in B_x} f(y)$$

and for erosion

$$(f \ominus g)(x) = \inf_{y \in E, x-y \in B} f(y) = \inf_{y \in \breve{B}_x} f(y),$$

where $\breve{B}$ is the transpose or symmetrical of B, that is, $\{-b \mid b \in B\}$. Erosion by a flat structuring function acts to shrink peaks and flatten valleys, while dilation acts in a dual fashion,

100. This image (and the one on the following page) is taken, with slight modifications, from Hlavac (2020).

flattening or rounding peaks and accentuating valleys. Taking, for instance, some price data, we might then have something like

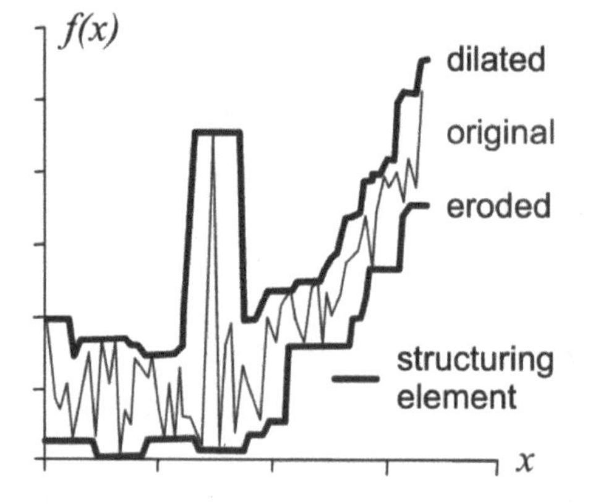

Whether in its functional treatment, or in terms of the special relation $X \oplus B \subseteq Y$ iff $X \subseteq Y \ominus B$, the relation between dilation and erosion is part of a much more general and powerful story, exemplifying the notion of a Galois connection, itself an instance of the more general notion of an adjunction. We first supply the relevant definitions, and then further explore some of the powerful abstract features of this notion through the particular case of the operations of dilation and erosion.

Definition 165 Let $\mathcal{P} = (P, \leq_P)$ be a preordered set, and $\mathbb{Q} = (Q, \leq_Q)$ another preorder. Suppose we have a pair of monotone maps $F : P \to Q$ and $G : Q \to P$,

$$P \underset{G}{\overset{F}{\rightleftarrows}} Q$$

such that for all $p \in P$ and $q \in Q$, we have the two-way rule

$$\frac{F(p) \leq_Q q}{p \leq_P G(q)}$$

where the bar indicates "iff." If such a condition obtains, the pair (F, G) is said to form a monotone *Galois connection* between $\mathcal{P}$ and $\mathbb{Q}$.

When such a connection obtains, we also say that F is the *left* (or *lower*) *adjoint* and G the *right* (or *upper*) *adjoint* of the pair, and write (for reasons we will see in a moment) $F \dashv G$ to indicate the relation.

Such a situation can be expressed in terms of the behavior of certain special arrows associated to each object of P and Q. In particular, let p be an object of P, and set $q = F(p)$. Then

$$\frac{F(p) \leq F(p)}{p \leq G(F(p))},$$

where the top is the identity arrow on $F(p)$ in Q, indicating reflexivity (which holds in any order). If we use θ to designate the bijection that realizes the "iff" situation, then $\theta(\mathrm{id}_{F(p)})$ is a special arrow of Q, called the *unit* of p, where this arrow enjoys a certain universality property. There is a corresponding dual notion of a *counit*. In short, for each $p \in P$, we call the *unit* an element $p \leq GF(p)$ that is least among all x with $p \leq G(x)$; dually, for each $q \in Q$, the *counit* is an element $FG(q) \leq q$ that is greatest among all y with $F(y) \leq q$. It can be shown that, given order-preserving maps $F : P \to Q$ and $G : Q \to P$, it is in fact *equivalent* to say (1) $F \dashv G$ and (2) $p \leq GF(p)$ and $FG(q) \leq q$.

Changing the variance of the functors involved gives us a slightly different notion.

Definition 166 Let $\mathcal{P} = (P, \leq_P)$ and $\mathbb{Q} = (Q, \leq_Q)$ be orders. Suppose we have a pair of antitone (order-reversing) maps $F : P \to Q$ and $G : Q \to P$,

$$P \underset{G}{\overset{F}{\rightleftarrows}} Q$$

such that for all $p \in P$ and $q \in Q$, we have the two-way rule

$$\frac{q \leq_Q F(p)}{p \leq_P G(q)}.$$

If such a condition obtains, the pair (F, G) is said to form an *antitone Galois connection* between $\mathcal{P}$ and $\mathbb{Q}$.

We have seen many times now how any order $\mathcal{P}$ can be regarded as a category by taking

$$x \leq_P y \text{ iff there exists an arrow } x \to y.$$

And in this setting, we further know that covariant functors between such categories are just monotone (order-preserving) functions, and that contravariant functors are antitone (order-reversing) functions. Thus, it is entirely natural to attempt to regard Galois connections in a more general, categorical guise. Doing so gives us the notion of an adjunction, which can accordingly be seen as a straightforward categorical generalization of the notion of a (monotone) Galois connection.

Definition 167 An *adjunction* is a pair of functors $F : \mathbf{C} \to \mathbf{D}$ and $G : \mathbf{D} \to \mathbf{C}$ such that there is an isomorphism

$$\mathrm{Hom}_{\mathbf{D}}(F(c), d) \cong \mathrm{Hom}_{\mathbf{C}}(c, G(d)),$$

for all $c \in \mathbf{C}, d \in \mathbf{D}$, which is moreover natural in both variables. When this obtains, we say F is left adjoint to G, or equivalently G is right adjoint to F, denoted $F \dashv G$.[101]

In saying that the isomorphism is "natural in both variables," we mean that for any morphisms with domain and codomain as below, the square on the left commutes (in $\mathbf{D}$) iff the square on the right commutes (in $\mathbf{C}$):

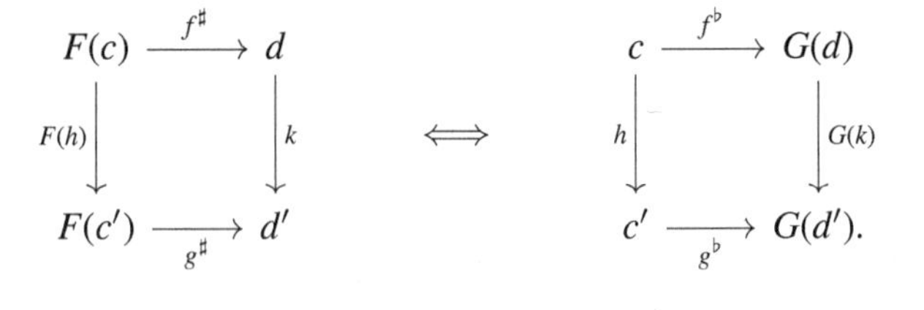

101. Sometimes the morphisms $F(c) \xrightarrow{f^\sharp} d$ and $c \xrightarrow{f^\flat} G(d)$ of the bijection given above are said to be *adjunct* or *transposes* of each other.

As one might expect, by considering functors of different variance, corresponding to *antitone* Galois connections, there is another notion, namely that of *mutually right adjoints* (and further, mutually left adjoints).

Definition 168 Given a pair of functors $F: \mathbf{C}^{op} \to \mathbf{D}$ and $G: \mathbf{D}^{op} \to \mathbf{C}$, if there exists a natural isomorphism

$$\mathrm{Hom}_{\mathbf{D}}(F(c), d) \cong \mathrm{Hom}_{\mathbf{C}}(G(d), c),$$

then we say that F and G are *mutually left adjoint*. Given the same functors, if there exists a natural isomorphism

$$\mathrm{Hom}_{\mathbf{D}}(d, F(c)) \cong \mathrm{Hom}_{\mathbf{C}}(c, G(d)),$$

then we say that F and G are *mutually right adjoint*.

Observe that an antitone Galois connection just names a mutual right adjoint situation between preorders (posets). Still following the example of the Galois connection definition, we can also recover the notions of the unit and counit of an adjunction.

Definition 169 Given an adjunction $F \dashv G$, there is a natural transformation

$$\eta : \mathrm{id}_{\mathbf{C}} \Rightarrow GF,$$

called the *unit* of the adjunction. Its component

$$\eta_c : c \to GF(c)$$

at c is the transpose of the identity morphism $\mathrm{id}_{F(c)}$.

Dually, there is a natural transformation $\mu : FG \Rightarrow \mathrm{id}_{\mathbf{D}}$, called the *counit* of the adjunction, with component

$$\mu_d : FG(d) \to d$$

at d defined as the transpose of the identity morphism $\mathrm{id}_{G(d)}$.

Any adjunction comes with a unit and a counit. In fact, conversely, given opposing functors $F: \mathbf{C} \to \mathbf{D}$ and $G: \mathbf{D} \to \mathbf{C}$, supposing they are equipped with natural transformations $\eta : \mathrm{id}_{\mathbf{C}} \Rightarrow GF$ and $\mu : FG \Rightarrow \mathrm{id}_{\mathbf{D}}$ satisfying a pair of conditions, then this data can be used to exhibit F and G as adjoint functors. In other words, we can use the natural transformations exemplifying the counit and unit maps, together with some conditions on these, to actually define an adjunction.

Definition 170 (*Adjunction, again*) An *adjunction* consists of a pair of functors $F: \mathbf{C} \to \mathbf{D}$ and $G: \mathbf{D} \to \mathbf{C}$, equipped with further natural transformations $\eta : \mathrm{id}_{\mathbf{C}} \Rightarrow GF$ and $\mu : FG \Rightarrow \mathrm{id}_{\mathbf{D}}$ satisfying what are sometimes called the *triangle identities*:

Then, the isomorphism $\text{Hom}_{\mathbf{D}}(F(c), d) \cong \text{Hom}_{\mathbf{C}}(c, G(d))$ realizing F and G as an adjoint pair, will exist precisely where there exists a pair of natural transformations, as above, satisfying the triangle identities.

Let us now return to the dilation and erosion of images and breathe some life into these ideas. The special relation that related Minkowski addition to its dual did not really depend on the particular form given to it in the translation-invariant case of a binary image, but exemplifies a more general notion of dilation and erosion on an arbitrary complete lattice, using the operations of supremum and infimum, where we have the adjunction

$$\text{dilate} \dashv \text{erode}.$$

To see how this works, we can first observe that both operations, dilation and erosion, are order-preserving (monotone), in the sense that $X \subseteq Y$ implies $X \oplus Z \subseteq Y \oplus Z$ and also $X \ominus Z \subseteq Y \ominus Z$. Moreover, while the order of an image intersection (union) and a dilation (erosion) cannot be interchanged freely, the dilation of the union of two images is indeed equal to the union of the dilations of the images, so the order can be interchanged; likewise, erosion of the intersection of two images yields the intersection of their erosions.

In dealing with the pair (dilate, erode), there is no need to be restricted to the poset of subsets of a digital space. There are clearly many choices for the underlying object space, where the images in question are held to reside. It is common to consider $\mathbb{P}(E)$ the space of all subsets of E (where E is the d-dimensional Euclidean space $\mathbb{R}^d$ or the digital space $\mathbb{Z}^d$). But we might also consider Conv(E) the space of all convex subsets of E; or $\mathcal{P}^E$ the space of "image functions" from E a discrete space to $\mathcal{P}$ a pixel lattice; or the space $(T^3)^E$ of RGB color images (where RGB colors are triples (r, g, b) of numerical values, T^3 the lattice of RGB colors under componentwise order, and an RGB image is a function $E \to T^3$ taking each point $p \in E$ to a triple $(r(p), g(p), b(p))$ representing the RGB coloration of p); and so on. The takeaway, though, is that despite the differences in these underlying spaces, all such spaces form *complete lattices* (where this means that the underlying poset has all joins and meets). So if we consider a complete lattice $\mathcal{L}$, with the order $\leq$, supremum $\bigvee$, infimum $\bigwedge$, least element 0, and greatest element I, such a lattice can be thought of as our "image lattice," corresponding to a particular set of images we are working with. Traditionally, one then defines dilations and erosions as follows.

Definition 171 Let $\mathcal{L}$ and $\mathcal{M}$ denote complete lattices. For $\delta : \mathcal{L} \to \mathcal{M}$ and $\epsilon : \mathcal{M} \to \mathcal{L}$, we say that

- δ is a *dilation* provided for every $S \subseteq \mathcal{L}$,

$$\delta\left(\bigvee S\right) = \bigvee_{X \in S} \delta(X).$$

- ϵ is an *erosion* provided for every $T \subseteq \mathcal{M}$,

$$\epsilon\left(\bigwedge T\right) = \bigwedge_{Y \in T} \epsilon(Y).$$

Note that this also applies in the case of S, T empty, in which case a dilation is held to preserve 0, while an erosion preserves I.

Our dilate-erode pair is actually an antitone Galois connection, where $\mathcal{M} = \mathcal{L}^{op}$, which just means that for all S, T in $\mathcal{L}$

$$\frac{T \leq^{op} \delta(S)}{S \leq \epsilon(T)},$$

which is, of course, the same as

$$\frac{T \geq \delta(S)}{S \leq \epsilon(T)}$$

or, equivalently,

$$\frac{\delta(S) \leq T}{S \leq \epsilon(T)},$$

where the order here is now the same, that given on $\mathcal{L}$, above and below the line. Thus, we have recovered the usual notion of an adjunction with $\delta : \mathcal{L} \to \mathcal{L}$ order-preserving and $\epsilon : \mathcal{L} \to \mathcal{L}$ order-preserving! Dilations and erosions are then precisely just the order-preserving (monotone) transformations on a complete lattice that moreover commute with the supremum and infimum, respectively.[102]

Morphological operators are thereby given a unified treatment in the general framework of an adjoint pair on complete lattices. A number of well-established properties concerning the interaction of these operators then fall out immediately from the general framework of adjunctions. Conversely, we can illustrate such general facts via the present operators on images.

Suppose we have an adjunction $\delta \dashv \epsilon$ on a complete lattice $\mathcal{L}$.[103] Then a number of morphologically significant facts come "for free" as corollaries of general categorical truths about an adjoint pair. Even the fact that δ is a dilation and ϵ is an erosion in the first place can be derived from the existence of this special adjoint relationship. In what follows, we explore some of these general truths through the lens of some notable particular truths about dilations and erosions.

7.1.1 Uniqueness of Adjoints

Proposition 172 To each dilation δ there corresponds a unique erosion ϵ, namely

$$\epsilon(X) = \bigvee \{S \in \mathcal{L} \mid \delta(S) \leq X\},$$

and to each erosion ϵ there corresponds a unique dilation,

$$\delta(X) \bigwedge \{S \in \mathcal{L} \mid \epsilon(S) \geq X\}.$$

102. Note that, if we were to regard the pair as comprising an antitone Galois connection, then we would be saying that the operators exchanged suprema and infima, in the sense that, for example, $\delta\left(\bigvee_i x_i\right) = \bigwedge \delta(x_i)$.

103. Looking ahead to what will have to be true of such functors, they have been given the names δ and ϵ, suggesting dilations and erosions; however, at this point, we do not yet require or assume anything about the maps δ and ϵ, except that they form an adjoint pair moving between $\mathcal{L}$ and itself.

This ultimately derives from a general result that assures us that, like inverses, adjoints are unique (well, actually "unique up to unique isomorphism," but we can ignore this in our special case):

Proposition 173 Adjoint maps are unique.

In the case of orders, with order-preserving maps between them, this just means

1. if F_1 and F_2 are left adjoints of G, then $F_1 = F_2$;
2. if G_1 and G_2 are right adjoints of F, then $G_1 = G_2$.

Proof. (We focus on the simple case of orders, and prove (1); (2) follows by duality) From the adjointness assumptions, we have both

$$\frac{F_1(p) \leq q}{p \leq G(q)}$$

and

$$\frac{p \leq G(q)}{F_2(p) \leq q},$$

so immediately we have that $F_1(p) \leq q$ iff $F_2(p) \leq q$. Set $q = F_1(p)$, making $F_1(p) \leq q$ trivially true, forcing $F_2(p) \leq F_1(p)$ to be true as well. Similarly, set $q = F_2(p)$ and use the trivial truth $F_2(p) \leq F_2(p)$ to force $F_1(p) \leq F_2(p)$. In a poset, this entails that $F_1(p) = F_2(p)$, p arbitrary. □

The adjunction then gives rise to the formulas

$$G(q) = \bigvee \{p \mid F(p) \leq q\}$$

and

$$F(p) = \bigwedge \{q \mid p \leq G(q)\},$$

which displays the uniqueness of the adjoints, and so explains the unique erosion (dilation) corresponding to each dilation (erosion), as written above.

In general, a given map may or may not have a left (or right) adjoint; the map may have one without the other, neither, or both (where these may be the same or different). But if it does have a left (or right) adjoint, we can be confident that, even though they are not quite inverses, the adjoint is unique up to isomorphism.

Adjoint functors also interact in particularly interesting and useful ways with the limit and colimit constructions, a connection we now explore.

7.1.2 Limit and Colimit Preservation

Proposition 174 δ is a dilation and ϵ is an erosion, and both are order-preserving.

This follows immediately from a very important category-theoretic result, namely that

Proposition 175 Right adjoints preserve limits (*RAPL*); left adjoints preserve colimits (*LAPC*).[104]

104. Terminologically, recall that a general functor that is limit-preserving is said to be a *continuous* functor, while a colimit-preserving functor is a *cocontinuous* functor. Another related concept we will make use of later in the book is the following: a functor is said to be *left exact* if it preserves *finite* limits, and *right exact* if it preserves *finite* colimits. Speaking of (co)limits, it is worthwhile noting that entities exhibiting universality, like colimits and limits, can themselves be phrased entirely in terms of adjoint functors. Then, one of the advantages of this

Instead of proving this in the general case, we will show how it obtains in our special case of maps between orders, in which setting limits are infima (meets) and colimits are suprema (joins).

Proposition 176 *RAPL and LAPC in the special case of orders:*

1. If $f:\mathbb{Q}\to\mathcal{P}$ has a right adjoint (i.e., is a left adjoint), then it preserves the suprema that exist in $\mathbb{Q}$.
2. If $g:\mathcal{P}\to\mathbb{Q}$ has a left adjoint (i.e., is a right adjoint), then it preserves the infima that exist in $\mathcal{P}$.

Proof. (Of (1), since (2) follows by duality) Assume $S=\{q_i\}_{i\in I}$ is a family of elements of $\mathbb{Q}$ with a supremum $\bigvee S$ in $\mathbb{Q}$. Claim: $f(\bigvee S)$ is the supremum in $\mathcal{P}$ of the family $\{f(q_i)\}_{i\in I}$, that is,

$$f\left(\bigvee S\right)=\bigvee f(S).$$

But $f\left(\bigvee S\right)\leq p$ iff $\bigvee S\leq g(p)$ (since, by assumption, f has a right adjoint, call it g). And this latter inequality holds iff for all $q_i\in S$, we have $q_i\leq g(p)$. But then we can again use the assumed adjoint relation $f\dashv g$, and see that this latter inequality will hold iff $f(q_i)\leq p$ for all $q_i\in S$, and this in turn will hold iff for all $t\in f(S)$, we have $t\leq a$. In sum, then, we have that $f\left(\bigvee_i q_i\right)\leq p$ if and only if $\bigvee_i f(q_i)\leq p$, or that f preserves any suprema that exist in $\mathbb{Q}$. □

But a dilation (erosion) was just *defined* as an order-preserving map that commutes with colimits (limits). So δ being a left adjoint suffices to tell us that δ must be a dilation (the dual situation holding for an erosion).

7.1.3 Adjoints Compose

Proposition 177 Given two dilations $\delta:\mathcal{L}\to\mathcal{M},\delta':\mathcal{M}\to\mathcal{N}$ and two erosions $\epsilon:\mathcal{M}\to\mathcal{L},\epsilon':\mathcal{N}\to\mathcal{M}$ such that $\delta\dashv\epsilon$ and $\delta'\dashv\epsilon'$, then their composition forms an adjunction $\delta'\circ\delta\dashv\epsilon\circ\epsilon'$.

This exemplifies a general result in category theory, namely:

Proposition 178 Left (right) adjoints are closed under composition, that is, given the adjunctions

$$\mathbf{C}\underset{G}{\overset{F}{\underset{\longleftarrow}{\overset{\longrightarrow}{\bot}}}}\mathbf{D}\underset{G'}{\overset{F'}{\underset{\longleftarrow}{\overset{\longrightarrow}{\bot}}}}\mathbf{E},$$

the composite $F'\circ F$ is left adjoint to the composite $G\circ G'$:

$$\mathbf{C}\underset{G\circ G'}{\overset{F'\circ F}{\underset{\longleftarrow}{\overset{\longrightarrow}{\bot}}}}\mathbf{E}.$$

In this way, arbitrarily long strings of adjoints can be produced.

---

adjunction perspective is that the (co)limit of *every* **J**-shaped diagram in **C** can be defined all at once, rather than just taking the (co)limit of a particular **J**-shaped diagram $X:\mathbf{J}\to\mathbf{C}$. We will exhibit the relevant adjunction in the final section of this chapter.

Moreover, another fact from morphology follows from the facts that adjoints compose (and using LAPC and RAPL), namely that for dilations and erosions on the same complete lattice, if $\delta_j \dashv \epsilon_j$ forms an adjoint pair for every $j \in J$, then $\left(\bigvee_j \delta_j, \bigwedge_j \epsilon_j\right)$ is an adjunction.

7.1.4 Units and Counits

The "opening" operator bounds an image on the left, while its "closing" bounds it on the right, that is,

Proposition 179

$$\delta\epsilon \leq \mathrm{id} \leq \epsilon\delta.$$

This is immediate from the unit natural transformation, $\mathrm{id} \leq \epsilon\delta$, and counit $\delta\epsilon \leq \mathrm{id}$.

7.1.5 Fixed Point Formulae

Proposition 180 $\delta\epsilon\delta = \delta$ and $\epsilon\delta\epsilon = \epsilon$.

The unit and counit maps, satisfying the triangle identities, give the following general "fixed point formulae" result underlying the above:

Proposition 181 If $\mathcal{P}$ and $\mathbb{Q}$ are posets and $F : \mathcal{P} \to \mathbb{Q}$ and $G : \mathbb{Q} \to \mathcal{P}$ form a (monotone) Galois connection (adjunction), with $F \dashv G$, then the following *fixed point formulae* will hold for F and G:

$$FGF = F \text{ and } GFG = G.$$

Proof. The triangle identities give $F(p) \leq FGF(p) \leq F(p)$ for all $p \in \mathcal{P}$, so $F = FGF$. The second formula follows similarly. □

7.1.6 Returning to Main Discussion

In the last two items, we saw how the unit and counit maps determine two important endomaps, namely $\delta \circ \epsilon$ ("opening") and $\epsilon \circ \delta$ ("closing"). The presence of unit and counit further give us the fixed point formulae, which translates to the morphologically significant fact,

$$\delta\epsilon\delta = \delta \qquad \text{and} \qquad \epsilon\delta\epsilon = \epsilon.$$

This "stability" property of openings and closings means, in terms of the interpretation of such operations as filters, that they effectively "complete their task" (unlike many other filters, where repeated applications can involve further modifications of the image, with no guarantee of the outcome after a finite number of iterations). In general, the above fixed point formula further entails, in particular, that $\epsilon\delta$ and $\delta\epsilon$ are each idempotent. Thus, altogether, the composite monotone map $\epsilon\delta$, for its part, has the properties that

- $p \leq \epsilon\delta(p)$, and
- $\epsilon\delta\epsilon\delta(p) = \epsilon\delta(p)$.

But this is exactly to say that $\epsilon\delta$ is a *closure* operator, in the following general sense.

Definition 182 A *closure operator* on a poset $\mathcal{P}$ (typically some poset of subobjects, for example, the powerset poset)[105] is an endomap $K : \mathcal{P} \to \mathcal{P}$ such that

1. for each $p \leq p' \in \mathcal{P}$, $K(p) \leq K(p')$ (monotonicity);
2. for each $p \in \mathcal{P}$, $p \leq K(p)$ (extensivity);
3. for each $p \in \mathcal{P}$, $K(K(p)) = K(p)$ (idempotence).

There is an important dual notion to closure, called the kernel operator (or dual closure), where this is an endomap that, like K, arises from a Galois connection, and is both monotone and idempotent, yet satisfies the dual of the extensivity property.

Definition 183 A *kernel operator* (or *dual closure*) is an endomap L satisfying

1. for each $p \leq p' \in \mathcal{P}$, $L(p) \leq L(p')$ (monotonicity);
2. for each $p \in \mathcal{P}$, $L(p) \leq p$ (contractivity);
3. for each $p \in \mathcal{P}$, $L(L(p)) = L(p)$ (idempotence).

In short, these notions are all part of a much more general story, namely that for a Galois connection or adjunction on posets such as $\delta \dashv \epsilon$ as above, the composite $\epsilon \circ \delta$ will automatically be monotone, extensive, and idempotent, that is, a closure operator on the underlying poset (or lattice) $\mathcal{P}$; dually, $\delta \circ \epsilon$ will be monotone, contracting, and idempotent, that is, a kernel operator on $\mathbb{Q}$.

Before leaving this example, we will explore a few last notions via morphology. We will let the induced kernel operator—which is precisely what the morphology community calls by the name "opening"—be denoted $\phi = \delta\epsilon$, while $\kappa = \epsilon\delta$ will denote the induced closure (or "closing" operator, to be consistent with the mathematical morphology literature). In the binary case, opening and closing are typically defined, respectively, as

$$X o B = (X \ominus B) \oplus B = \bigcup \{B_p \mid p \in E, B_p \subseteq X\}$$

$$X \bullet B = (X \oplus B) \ominus B.$$

Morphological closing is just dilation (by some B) followed by erosion of the result by B, while morphological opening is the erosion (by some B) followed by dilation of the resulting image by B. Closing acts to fill out narrow holes. In terms of translations with the structuring element, the opening of an image A by B is the complement of the union of all translations of B that fall outside (do not overlap) A. As extensive (i.e., larger than the identity mapping), closings of an image are generally "larger" than the original image. Opening, for its part, acts to remove noise, narrow connections between regions, and parts of objects, generally attenuating peaks and other small protrusions or components. If you have a note where the writing appears to be growing tiny roots from its edges, opening effectively acts to remove these outer leaks at the boundary, rounding the edges. In terms of translations with the structuring element, the opening of A by B is the union of all translations of B that fit completely within A. As antiextensive (contracting), openings of an image are generally "smaller" than the original.

105. "Subobjects" are formally introduced in chapter 11. For now, you can just think of them as categorical generalizations of the notion of a *subset*.

Exercise 19 Composing dilations and erosions, we found the composite operations of opening ($\phi = \delta\epsilon$) and closing ($\kappa = \epsilon\delta$), which were moreover idempotent. Further composing openings and closings with one another (e.g., $\kappa \circ \phi$), how many more distinct operations can we produce? Describe, in terms of their effect on images, at least one of these "image filters." Finally, consider how the composite operators must be related to one another.

Solution By (alternately) composing openings $\phi(= \delta\epsilon)$ and closings $\kappa(= \epsilon\delta)$, we can obtain four new filters in total, each four of which are idempotent.

1. closing-after-opening: $\kappa\phi$
2. opening-after-closing: $\phi\kappa$
3. opening-after-closing-after-opening: $\phi\kappa\phi$
4. closing-after-opening-after-closing: $\kappa\phi\kappa$.

You can easily convince yourself that no other nontrivial operator can be obtained by further composition with any combination of ϕs and κs. Any attempt to produce further new operations by pre- or postcomposing the above four with κ or ϕ will just reduce back to one of those four, by the idempotence of these operators (together with the idempotence of κ and ϕ themselves).

These composites are used in the course of various image processing tasks, such as "smoothing" an image or performing image segmentation. An example of $\kappa\phi$ is given by the following:[106]

In terms of the relations between these four (together with the original opening and closing operators as well), it is easy to show that

$$\phi \leq \phi\kappa\phi \leq \{ {}^{\kappa\phi}_{\phi\kappa} \} \leq \kappa\phi\kappa \leq \kappa,$$

and moreover $\phi\kappa\phi$ will be the greatest filter smaller than $\phi\kappa \wedge \kappa\phi$, while $\kappa\phi\kappa$ will be the smallest filter greater than $\phi\kappa \vee \kappa\phi$.

7.2 Adjunctions through Modalities

We can get an even better handle on adjunctions by looking at further applications of such notions, specifically some applications involving *modalities*. At least since Aristotle's attempt to understand certain statements containing the words "necessary" and "possible," philosophers and logicians have been interested in the logic of different operators

106. This image is taken from Bobick (2014).

describing different *ways of being true*. Modal logic began as the study of *necessary* and *possible* truths, but over the course of at least the last 100 years it has been recognized that modalities abound in both natural and formal languages. These days modal logic is more commonly regarded as the much broader study of a variety of constructions that modify the truth conditions of statements (which includes, most notably, statements concerning knowledge, belief, ethics, temporal happenings, how computer programs behave).[107] Moreover, one can construe modal operators in terms of adjunctions—and, in this way, modalities can be shown to arise in a number of other more unexpected settings. In particular, there are in fact a number of close connections between erosion and dilation, and the modal operators $\Box$ of necessity and $\Diamond$ of possibility, respectively.

After a brief subsection devoted to establishing some background on the more traditional logical treatment of modalities, the subsequent four subsections realize all these ideas in four different settings: (1) in connection with the treatment of *negation*, (2) in the **Qua** category, (3) in graphs, and (4) in topology.

7.2.1 Some Background on Modalities

The reader is likely already familiar with classical propositional logic (PL). Classical PL is ultimately built from propositional variables $p, q, r, \ldots$, where these are variables (representing *propositions*) that can be assigned a truth value, and certain symbols called logical connectives, where this includes $\neg$ ("not"), $\wedge$ ("and"), $\vee$ ("or"), $\rightarrow$ ("implies"), $\leftrightarrow$ ("if and only if"). The connectives enable compound propositions or formulas to be constructed from simpler propositional variables, and in such a way that the truth-value of the compound formula is defined as a function of the truth values of the simpler propositions. We define a *formula* by specifying that it is any expression constructed according to the following rules:

1. every propositional variable is a formula;
2. given any formulas ϕ and ψ already constructed, the expressions $\neg\phi$ (negation), the conjunction $\phi \wedge \psi$, the disjunction $\phi \vee \psi$, the implication $\phi \rightarrow \psi$ ("ϕ implies ψ"), and the biconditional $\phi \leftrightarrow \psi$ ("ϕ if and only if ψ") are formulas.

Because one can combine simple propositions into compound ones in many ways, additional symbols like parentheses are needed to avoid ambiguities. In classical PL, an *interpretation* of a formula is a functional assignment of truth-values to its constituent variables. It is of course possible that a formula will have different truth-values under different interpretations—but a formula is said to be *valid* if it is true under every interpretation.

Intuitionistic PL is defined by the following axiom schemata (meaning one can substitute arbitrary formulae, obtaining *instances* of the axioms):

1. $\phi \rightarrow (\psi \rightarrow \phi)$
2. $(\phi \rightarrow (\psi \rightarrow \rho)) \rightarrow ((\phi \rightarrow \psi) \rightarrow (\phi \rightarrow \rho))$
3. $(\phi \wedge \psi) \rightarrow \phi$
4. $(\phi \wedge \psi) \rightarrow \psi$
5. $\phi \rightarrow (\psi \rightarrow (\phi \wedge \psi))$

107. For a nice history of formal approaches to modal logic, see Goldblatt (2003).

6. $\phi \rightarrow (\phi \vee \psi)$
7. $\psi \rightarrow (\phi \vee \psi)$
8. $(\phi \rightarrow \rho) \rightarrow ((\psi \rightarrow \rho) \rightarrow (\phi \vee \psi) \rightarrow \rho))$
9. $(\phi \rightarrow \psi) \rightarrow ((\phi \rightarrow \neg\psi) \rightarrow \neg\phi)$

together with one inference rule (*modus ponens*)

$$\frac{\phi \qquad \phi \rightarrow \psi}{\psi}$$

where this means that if ϕ and $\phi \rightarrow \psi$ are derivable from the axioms, then so too is ψ.

We then use the axioms of the system, together with the inference rule, to produce proofs or derivations—where a proof is just a finite sequence of formulas, usually displayed vertically, such that each line of the proof is either an axiom instance of or is inferable from earlier lines using the inference rule.

When we add to this list of axioms a tenth axiom,

$$\phi \vee \neg\phi,$$

the law of excluded middle (or $\neg\neg\phi \rightarrow \phi$), we get classical PL.

In general, an axiom system is said to be *sound* if every formula that is derivable from the axiom system is valid. An axiom system is said to be *complete* if every valid formula can be derived from this axiom system. The ten axioms we gave above supply one of the sound and complete axiom systems for classical PL.

One can then "upgrade" PL by augmenting it with quantifiers "for all" ($\forall$) and "there exists/there is at least one" ($\exists$), where the role of these operators is of course to tell us *of how many* a proposition is true. The resulting logic—predicate logic—supplies us with a way of proving the validity of a number of perfectly valid arguments that are simply invalid when symbolized in the somewhat limited notation of PL. For instance, the venerated syllogism

1. All humans are mortal
2. Socrates is a human

∴ Socrates is mortal

can only be symbolized in PL in a manner that makes it invalid—for instance, as

1. ϕ
2. ψ

∴ ρ.

Yet, of course, the syllogism ultimately represents a valid argument, and by augmenting PL with quantifiers, we can show this. PL, and its extension to predicate or quantifier logic (by adding the quantifiers $\forall$ and $\exists$), is of course rather useful in the formalization and analysis of many arguments. However, in the same way that the syllogism above could not be shown valid without a way of reasoning with quantifications of statements, there are further valid arguments that cannot be shown to be valid using just propositional (or predicate) logic. Take, for instance, the following argument:

> *If a new course on sheaves is to be offered next year, then proposal submissions must be made to the department before September. If proposal submissions are to be made to the department before September, then a departmental meeting must be had. A month's notice must be given if a*

departmental meeting is to be had. But it is already August. Because it is not possible to give a month's notice, it follows that it is not possible to offer a new course on sheaves next year.

Surely we would like to be able to show that such an argument is valid. But in order to show that such an argument is indeed valid, we need a way of reasoning with what is expressed by these various ways of *qualifying* how a proposition is true—in particular, capturing the notions of *must* and *possible* used above. The idea of reasoning with such qualifications of "ways of being true"—where these are called *modalities*, the most conspicuous and extensively studied of which have been "*it is necessarily the case that...*" and "*it is possibly the case that...*"—has been studied by philosophers for millennia, since at least the time of Aristotle, and reasoning with such qualifications was extensively discussed and debated in the Middle Ages. Roughly, a modality is just a phrase or concept that is applied to a given statement (ϕ) to create a new statement, where this latter statement now makes an assertion of the mode of truth of ϕ—where or how ϕ is true, when ϕ is true, under which circumstances ϕ may be true.

As mentioned, modal logic began as the study of *necessary* and *possible* truths. But even confining our attention to logicians, these days modal logic is more commonly regarded as the much broader study of a variety of constructions that modify the truth conditions of statements. Here is a partial sample of some prominent modalities considered by modal logicians:

- epistemic logic: it is known (to agent X) that
- doxastic logic: it is believed that
- deontic logic: it is obligatory that
- dynamic logic: after the program/computation terminates, the program enables that
- metalogic/provability logic: it is provable that
- tense logic: at all future times it is true that

In the formal analysis of modalities a key feature to emerge was that many modal operators came in dual pairs, similar to how the universal quantifier $\forall$ and the existential quantifier $\exists$ are dual and so interdefinable using negation

$$\forall x(\ldots x \ldots) := \neg \exists x \neg (\ldots x \ldots)$$

or

$$\exists x(\ldots x \ldots) := \neg \forall x \neg (\ldots x \ldots).$$

In a similar fashion, necessity (symbolized using $\Box$) acts as a sort of universal counterpart to the notion of possibility (symbolized using $\Diamond$), which plays the role of its existential dual, in the sense that

$$\Box := \neg \Diamond \neg$$

$$\Diamond := \neg \Box \neg,$$

reflecting the idea that to hold that "it is necessarily the case that ..." is equivalent to maintaining that "it is not possible that it is not the case that ...." As these operators are given different readings, and put to different use, key dualities are captured—for instance,

- "it is obligatory that..." vs. "it is permissible that...";
- "at all future times it is true that..." vs. "at some future time it is true that...";
- "it is provable that..." vs. "it is consistent (not provable that it is not the case) that...."

Realizing such matters as a case of an adjoint relationship allows us to connect the story of modalities and such dualities to an even broader class of structures, beyond the confines of logic. The next four sections offer a glimpse into some of those further connections.

7.2.2 On What Is Not

Since at least the time of one of the first Western philosophical texts, attributed to Parmenides (around 500 BCE), the nature of *negation* has been on people's minds. This includes a number of issues, such as the following:

- Would a complete description of *what is* need to include any description of *what is not*? In other words, what is the "ontological status" of the negated entities or negative states of affairs?[108]
- When is the negation of a negation (negated entity) the identity (the original entity)?

One might further motivate such concerns as follows. One might try to argue that, philosophically, holes, shadows, fissures, boundaries—and other such "negative" or derivative entities—seem somehow less real or fundamental than (or at least not to be on the same footing as) the "positive" objects that produce or surround or support them. At the very least, this sort of observation seems to have some validity in that it does seem somehow more difficult to supply identity criteria for holes, for instance, compared to ordinary material objects (for holes appear to be *made of nothing*), or even to speak of what holes are (what are the *parts* of a hole?).

Today, we are most accustomed to thinking of negation as a linguistic or logical operator on a language, where the operation leads from an expression to the contradictory expression. Typically, the expression in question is a proposition or a part of a proposition. But we might attempt to regard negation, more broadly, as an operation that can also take place on larger wholes, on entire structures or theories. Moreover, one might argue that, however one approached it, the "right" understanding (and description) of negation would need to capture, above all, the relation or dependence between what (the structure) is being negated and the result of this operation of negation.

Such a perspective on negation is arguably exemplified in the facts that (1) negation is a contravariant functor on a particular category (to itself) and (2) this functor has special relations to itself, in that it is adjoint—in fact, *self-adjoint*. This perspective, developed formally in the following discussion, might even suggest, informally, that one think of the action of certain contravariant functors as a generalized sort of negation of structures. To develop these ideas, let us first recall some notions.

In general, regarding a poset as a category, recall that the colimit recovers the notion of supremum, while limit recovers that of infimum. In the case of the (co)limit over a diagram consisting of just two objects, we get the (co)product: the coproduct of objects x and y is the join (least upper bound) $x \vee y$, and the product is the meet (greatest lower bound) $x \wedge y$.

108. One position, in this context, might articulate the view that everything is what it is—as the individual thing it is—only on account of how it is *not* some other things, and accordingly try to take very seriously the idea that "all determination arises from a negation" (to paraphrase the old principle *omnis determinatio est negatio*, "every determination is a negation"). An opposing position might argue that negations always just describe *privations*, and a complete and accurate description of reality would not need to involve mention of any "negative entities."

Definition 184 A poset $\mathcal{P}$ is called a *join-semilattice* if each two-element subset $\{x, y\} \subseteq P$ has a join, denoted $x \vee y$. A poset $\mathcal{P}$ is called a *meet-semilattice* if each two-element subset $\{x, y\} \subseteq P$ has a meet, denoted $x \wedge y$.

Definition 185 A poset is a *lattice* if it is both a join- and a meet-semilattice.

By induction, the binary case of joins and meets of a lattice can be extended so that every nonempty finite subset of a lattice has a join and a meet. In general, supposing we have $\mathcal{P}$ a poset, then for all $x, y \in \mathcal{P}$,

$$x = \bigwedge \{x, y\} \iff x \leq y \iff y = \bigvee \{x, y\}.$$

For such a poset $\mathcal{P}$, for all $y \in P$, we can define the following:

Definition 186 (*A top as empty meet; bottom as empty join.*)

1. y is the empty meet, that is, $y = \bigwedge \emptyset$, iff it is a top (greatest) element; and
2. y is the empty join, that is, $y = \bigvee \emptyset$, iff it is a bottom (least) element.

Note that empty meets and joins need not exist. We could have two minimal elements, for instance, but no least element. A least element would have to be less than everything. An empty meet (top) is often written as $\top$ or 1 while an empty join (bottom) is written as $\perp$ or 0.

If a lattice $\mathcal{L}$ has a greatest element (top) 1 and a least element (bottom) 0, which further satisfy that

$$0 \leq x \leq 1 \text{ for every } x \in \mathcal{L},$$

then we call the lattice a *bounded lattice*. Category-theoretically, this makes 0 and 1 the (unique) initial and terminal objects of the lattice, considered as a category. Thus, altogether, a lattice with 0 and 1 is a poset that, regarded as a category, has all finite limits and all finite colimits.

Posets are not guaranteed to have anything except the order $\leq$. Lattices, on the other hand, have all finite meets and joins. Further properties of lattices may obtain—for instance, distributive lattices obey an additional distributive law that brings them closer to logic.

Definition 187 A *distributive lattice* $\mathcal{L}$ is a lattice in which the identity

$$x \wedge (y \vee z) = (x \wedge y) \vee (x \wedge z)$$

holds for all x, y, z. This identity gives the dual distributive law

$$x \vee (y \wedge z) = (x \vee y) \wedge (x \vee z).$$

We can also define the following notions:

Definition 188 For $\mathcal{L}$ a bounded lattice, with least element 0 and greatest element 1, we define a *complement* for an element x of the lattice as an element $a \in L$ such that $x \wedge a = 0$ and $x \vee a = 1$.

If a lattice is distributive, then a complement a, provided it exists, will be unique. In general, when it exists, the unique complement a of an element x is denoted by $a = \neg x$, meant to evoke a lattice-theoretic analogue of logical negation. This lets us define the following:

Definition 189 A *Boolean algebra* is a distributive lattice with 0 and 1 for which every element x has a complement $\neg x$.

Heyting algebras are examples of a still more general notion, namely of distributive lattices for which some members may lack complements. More explicitly,

Definition 190 A *Heyting algebra* H is a poset with all finite products and coproducts, and that is moreover Cartesian closed. Another way of describing such an H is as a distributive lattice with a least element 0 and a greatest element 1, expanded with an "implication" operation $\Rightarrow$, meaning that for any two elements p, q of the lattice, there exists an exponential q^p, usually written

$$p \Rightarrow q.$$

This operation is characterized by an adjunction, specifically

$$r \leq (p \Rightarrow q) \text{ iff } r \wedge p \leq q.$$

In other words, $\Rightarrow$ is a binary operation on a lattice with a least element, such that for any two elements p, q of the lattice, $\max\{r \mid r \wedge p \leq q\}$ exists (where this latter set contains an element greater than or equal to every one of its elements, and such a least upper bound for all those elements r where $r \wedge p \leq q$ is what is denoted by $p \Rightarrow q$).[109]

In the general case, in any Heyting algebra, we can define the "negation" of an element p as

$$\neg p := (p \Rightarrow 0).$$

Heyting algebras serve as models for intuitionistic propositional calculus—in which setting, variables are regarded as propositions, $\wedge$ as "and," $\vee$ as "or," and $\Rightarrow$ as implication—and in that connection, it is sensible to think of "not p" as effectively saying that "p implies *false*."

Note, moreover, how on account of the way $\Rightarrow$ is defined, we can rewrite this as

$$q \leq \neg p \text{ iff } q \wedge p = 0,$$

revealing $\neg p$ to be the join of all those q whose meet with p in the lattice is 0, the least element. In any Heyting algebra H, we will have not just that

$$p \leq \neg\neg p,$$

but also that

$$p \leq q \text{ implies } \neg q \leq \neg p.$$

But this reveals how negation is just a contravariant functor from the Heyting algebra to itself! More explicitly,

Proposition 191 $\neg$ is a functor $\neg : H \to H^{op}$ (and also $\neg : H^{op} \to H$). This functor is, moreover, adjoint to itself, since $p \leq \neg q$ iff $q \leq \neg p$.

Let us spell out this self-adjointness more explicitly. The first inequality, $p \leq \neg\neg p$, is immediate from $q \leq \neg p$ iff $q \wedge p = 0$, using the further fact that, for any Heyting algebra, $p \wedge \neg p =$

109. Another way to think of this $p \Rightarrow q$ is in the setting of the propositional calculus, where it is the weakest condition needed for the inference rule of modus ponens to hold, that is, to enforce that from $p \Rightarrow q$ and p we can infer q.

0 (this follows from the adjunctive definition of $\Rightarrow$ together with the definition of negation as $\neg p = (p \Rightarrow 0)$). For the second, suppose $p \leq q$ in H. Then, $p \wedge \neg q \leq q \wedge \neg q$ and the right-hand side of this inequality is 0. So $p \wedge \neg q = \neg q \wedge p = 0$, and so by $q \leq \neg p$ iff $q \wedge p = 0$, we have that $\neg q \leq \neg p$. Incidentally, this moreover shows that

$$\neg p = \neg\neg\neg p$$

which we might call the "$1=3$" fact. This is a result of the contravariant functoriality of $\neg$ and that $p \leq \neg\neg p$. For, suppose $p \leq \neg\neg p$. Then, by the contravariant functoriality inequality, we also have that $\neg\neg\neg p \leq \neg p$. And since $p \leq \neg\neg p$ holds for all p in H, it holds in particular for $\neg p$. Thus, $\neg p \leq \neg\neg\neg p$, giving the other side of the equality, and so, altogether, $1=3$.

That $\neg$, as a functor $H \to H^{op}$ and also $H^{op} \to H$, is adjoint to itself means that for all $p \in H$ and $q \in H^{op} (= H)$, we have the two-way rule

$$\frac{\neg p \leq^{op} q}{p \leq \neg q}$$

or

$$\frac{\neg p \geq q}{p \leq \neg q,}$$

where the top (holding in H^{op}) holds if and only if the bottom (holding in H) does. For another way of seeing the truth of the fact that for all $x \in H$, $x \leq \neg\neg x$, notice that as an adjoint, letting $q = \neg p$, we must have

$$\frac{\neg p \geq \neg p}{p \leq \neg\neg p,}$$

and since the top is always true, the bottom must be as well.

An adjoint is a kind of generalized inverse, and as such, an adjunction describes a kind of relaxation or weakening of the notion of equivalence. In the present situation, asking when it is in fact the case that $p = \neg\neg p$ (when the "do nothing" functor is *equal* to applying the negation functor twice) is like asking when the above adjunction happens to be an equivalence. If the relations $\leq$ are replaced by =, then we get isomorphisms. This distinction captures the following well-known relation between Heyting algebras and Boolean algebras:

Proposition 192 A Heyting algebra H is Boolean (i.e., $\neg\neg x = x$ for all $x \in H$) if and only if the above adjunction is an equivalence.

This is a stricter requirement, and in general we need not have $x = \neg\neg x$. This requirement is something one might not always want to impose, and this is in fact one of the merits or utilities of working with the more general Heyting algebras. For any topological space X, the set $\mathcal{O}(X)$ of open sets of X forms a Heyting algebra; for instance, the opens in the real line accordingly form a Heyting algebra, but one that is not Boolean, since the complement of an open set is not necessarily open.

We could go on to dualize things and describe a dual notion, namely that of *co-Heyting algebras*, which support a corresponding but different notion of negation. The utility of considering such things can be motivated with another problem related to natural language. In many natural languages, for instance in English, one often has recourse to forms of negation that do not seem to be captured by, or behave as, the single negation operator of classical logic. Suppose someone is described to you as "not honest," after they act in a particular way in a particular situation. This is not necessarily to say that they are "dishonest." In natural language, we can deny that a person is honest in at least two distinct ways: (1) by asserting that someone is not honest (negating the predicable "to be honest"); or (2) by asserting that they are dishonest (negating the adjective "honest"). It is easy to appreciate, intuitively, how the second ("dishonest") is a stronger form of negation than the first ("not honest"). Moreover, suppose your friend Abe is someone you would be willing to describe as "not dishonest." This does not seem to convey the same thing as describing Abe as "honest." Finally, while we expect it to be the case that Abe is either honest or not honest, it seems plausible to assert, as well, that he is neither honest nor dishonest. We would like to know how to capture these observations more formally, and describe formal relations between the different forms of negation, as applied to natural language. The story that follows begins to address this.

A little more generally, compare the sentence

It is false that not p,

with the sentence

It is not false that p.

These sentences clearly do not say the same thing. The first indicates the *necessity* of p, while the second indicates its *possibility*. We can make sense of this in the context of a particular algebra called a bi-Heyting algebra, by interpreting the "it is false" in such sentences as the Heyting negation and the "not" as the corresponding "co-Heyting" negation. This setting will also allow us to define modal operators in terms of pairings of both negations.

Definition 193 A *co-Heyting algebra* is a poset whose dual is a Heyting algebra.

Unpacking this, we can equivalently observe that a co-Heyting algebra will be a bounded lattice expanded with a binary operation $\setminus$ such that for every p, q, r, we have the adjunction rule

$$(p \setminus q) \leq r \text{ iff } p \leq q \vee r.$$

In other words, $p \setminus q = \bigwedge\{r \mid p \leq q \vee r\}$.

A corresponding unary negation operation $\sim$ can then be defined by

$$\sim p := (1 \setminus p).$$

It follows that we have the following adjunction rule for this "negation" $\sim$:

$$\frac{\sim p \leq q}{1 = p \vee q}$$

Similar to how we have $p \wedge \neg p = 0$ in a Heyting algebra (though not necessarily $p \vee \neg p = 1$), notice how in a co-Heyting algebra we will have

$$p \vee \sim p = 1.$$

Likewise, similar to how the negation $\neg$ is order-reversing and satisfies $x \leq \neg\neg x$ in a Heyting algebra, in a co-Heyting algebra the negation $\sim$ is also order-reversing and it satisfies $\sim\sim x \leq x$. Moreover, just as in a Heyting algebra $p \vee \neg p$ is not necessarily the top (*true*), so in a co-Heyting algebra $p \wedge \sim p$ is not necessarily the bottom (*false*). In particular, then, in a co-Heyting algebra, we can thus define the generally nontrivial notion of the *boundary* of p, as

$$\partial p := p \wedge \sim p.$$

Incidentally, this recovers, in purely algebraic terms, the spatial and geometric notion of *boundary*, thus suggesting certain deep connections between logic and geometry and the study of space (connections explored in greater detail in the appendix). Bi-Heyting algebras—a bounded distributive lattice that is both a Heyting and a co-Heyting algebra—can accordingly be deployed to shed further light into some of these deep connections between logic and geometry.

We can work in the context of a bi-Heyting algebra and combine these negations to form our modal operators. In particular, we will let

$$\Diamond p := \sim \neg p,$$

read as "possibly p." This begins to suggest how we might formalize the fact that, for instance, John not being dishonest does not let us conclude, in general, that John is honest, but rather only that he is *possibly* honest. Similarly, we will let

$$\Box p := \neg \sim p,$$

read as "necessarily p."

Given such definitions, we can show that

Proposition 194 We have the adjunction $\Diamond \dashv \Box$.

Proof. To show that $\Diamond \dashv \Box$, we need only verify some equivalences, each of which more or less follows automatically from definitions. By definition, $\Diamond = \sim \neg$ and $\Box = \neg \sim$, so the adjunction just says that $\sim \neg \dashv \neg \sim$. But

$$\sim \neg p \leq q$$

just says that

$$1 \leq \neg p \vee q$$

which is of course equivalent to

$$1 \leq q \vee \neg p,$$

which itself is just to say that

$$\sim q \leq \neg p.$$

Using this last inequality, and the definition of $\neg$, we have

$$\sim q \wedge p \leq 0$$

or, equivalently,

$$p \wedge \sim q \leq 0.$$

This last line can be written

$$p \leq (\sim q \Rightarrow 0),$$

which is the same as saying that

$$p \leq \neg \sim q.$$

Altogether, this string of equivalences shows that

$$\sim \neg p \leq q \text{ iff } p \leq \neg \sim q,$$

which is precisely what is needed to show that $\sim\neg \dashv \neg\sim$ (or $\Diamond \dashv \Box$). □

Working in a bi-Heyting algebra and using the above definitions of $\Box$ and $\Diamond$, we can moreover consider repeated applications of these operators. As a matter of simplifying computations involving repeated applications of the operators, we will define $\Box_0 = \Diamond_0 = \mathrm{id}$, and

$$\Box_{n+1} := \neg \sim \Box_n, \qquad \Diamond_{n+1} := \sim \neg \Diamond_n,$$

where $\Box_n$ is of course just the result of iterating (n times) the composition of $\neg\sim$, and $\Diamond_n$ by iterating (n times) the composition of $\sim\neg$. $\Box_n$ and $\Diamond_n$ are clearly both order-preserving, for all n, as is evident from the double fact that, in a Heyting algebra, $\neg$ is order-reversing and satisfies $p \leq \neg\neg p$ for all p, and that, in a co-Heyting algebra, $\sim$ is order-reversing and $\sim\sim p \leq p$. Moreover, we have

1. $\Box_{n+1} \leq \Box_n \leq \mathrm{id} \leq \Diamond_n \leq \Diamond_{n+1}$ for all n; and
2. $\Diamond_n \dashv \Box_n$ for all n.

Proof. 1. First, we know that, for any p in the bi-Heyting algebra, we have that $\neg p \leq \sim p$. From this, taking $p = \sim p$, we have that

$$\neg \sim p \leq \sim\sim p,$$

and since $\sim\sim p \leq p$, this gives that

$$\neg \sim p \leq p.$$

Moreover, $p \leq \neg\neg p$, and, using $\neg p \leq \sim p$ again, applied now to $p = \neg p$, we get that

$$\neg\neg p \leq \sim \neg p.$$

Altogether, this gives that

$$\neg \sim p \leq p \leq \sim \neg p.$$

By definition, then, this reads as

$$\Box p \leq \mathrm{id}(p) \leq \Diamond p,$$

and further iterating this, letting $p = \Box_n p$, and then $\Diamond_n p$, gives the main result. □

Proof. 2. We want that $\Diamond_n \dashv \Box_n$ for all n. But since adjoints compose, this follows from the preceding result, by iterating. □

In a bi-Heyting algebra where countable suprema and infimi exist satisfying

$$\frac{b \leq \bigwedge_n a_n}{b \leq a_n \text{ for all } n}$$

and

$$\frac{\bigvee_n a_n \leq b}{a_n \leq b \text{ for all } n,}$$

we can further define

Definition 195

$$\Box p := \bigwedge_n \Box_n p,$$

$$\Diamond p := \bigvee_n \Diamond_n p.$$

Then, for a bi-Heyting algebra that has countable suprema and infima satisfying the above rules, the modal operators $\Box$ and $\Diamond$ are such that $\Box p$ will be the largest complemented x such that $x \leq p$, and $\Diamond p$ will be the smallest complemented x such that $p \leq x$. This moreover realizes the fact that $\Box$ and $\Diamond$ are both order-preserving and that $\Box p \leq p \leq \Diamond p$.

More generally, for any bounded distributive lattice and operators $\Box$ and $\Diamond$ defined just as above, the same properties will hold of $\Box$ and $\Diamond$. In particular, it can be shown that $\Diamond \dashv \Box$, and thus

1. $\Box \leq \text{id} \leq \Diamond$,
2. $\Box\Box = \Box, \Diamond\Diamond = \Diamond$,
3. $\text{id} \leq \Box\Diamond$,
4. $\Diamond\Box \leq \text{id}$,
5. $\Diamond(\phi \wedge \Box\psi) = \Diamond\phi \wedge \Box\psi$.

7.2.3 Adjoint Modalities in the Qua Category

Example 196 For another realization of some of the ideas explored in the previous section, recall the *interpretation* functor on the *qua* category **Qua** from example 38 (chapter 2).[110] Applying all this to the particular subcategory **A** from earlier, an interpretation amounts to a set X together with a set of predicates of X, where by "predicate of X" is just meant a family $\{\phi_A\}_{A \in \mathrm{Ob}(\mathbf{A})}$ of subsets of X with the functorial property

$$\text{if } x \in \phi_A(\text{ i.e., } x \in_A \phi) \text{ and } A' \to A \in \mathbf{A}, \text{ then } x \in \phi_{A'}.$$

And since **A** was just the comma or slice category $\boxed{\text{a scf}} \downarrow \mathbf{CN}$, so that all aspects are of the form $\boxed{\text{a scf}}$ *qua* $\boxed{\text{B}}$, an interpretation of **A** just associates to every aspect the same set X. Notice that we can collect together such families of predicates (as subfunctors of X) into the set $\mathcal{P}(X)$ of all predicates of X. These predicates in fact form a bounded distributive lattice with two negations—in fact, a bi-Heyting algebra—as

$$(\mathcal{P}(X), \leq, \vee, \wedge, 1, 0, \neg, \sim),$$

110. Again, this material on the *qua* category is ultimately derived from Reyes et al. (1999); the reader who desires to pursue these matters further than the discussion here can find many more interesting details in that paper.

where there is the natural ordering

$$\phi \leq \psi \text{ iff } \forall A \in \mathbf{A}, \forall x \in X, x \in_A \phi \Rightarrow x \in_A \psi.$$

Also, as expected, we have

$$x \in_A (\phi \vee \psi) \text{ iff } x \in_A \phi \text{ or } x \in_A \psi$$

and

$$x \in_A (\phi \wedge \psi) \text{ iff } x \in_A \phi \text{ and } x \in_A \psi.$$

0 (or $\perp$) is the predicate "false," the bottom element of the order, while 1 (or $\top$) is the predicate "true," the top element of the order. Given a predicate ϕ, we define two negations, $\neg\phi$ and $\sim \phi$, which are in fact two new predicates (i.e., have the requisite functorial property):

$$x \in_A \neg\phi \text{ iff } \forall A' \to A \in \mathbf{A} \;\; x \notin_{A'} \phi$$

and

$$x \in_A \sim \phi \text{ iff } \exists A \to A' \in \mathbf{A} \;\; x \notin_{A'} \phi.$$

Applied to the predicate "honest," for instance, we have the natural reading: "$\neg$ honest" as "dishonest," and "$\sim$ honest" as "not honest."

As we saw in our earlier discussion of these same negations, we have the following adjunctions for our negations (which hold for arbitrary properties ϕ, ψ):

$$\frac{\psi \leq \neg\phi}{\psi \wedge \phi = 0,}$$

$$\frac{\sim \phi \leq \psi}{1 = \psi \vee \phi.}$$

Notice also that for every property ϕ, it is a consequence of the above that $\phi \wedge \neg\phi = 0$ and $\phi \vee \sim \phi = 1$. However, $\phi \wedge \sim \phi$ need not be 0, and similarly, $\phi \vee \neg\phi$ is not necessarily 1.

Applying all this, suppose we return to modeling a discussion and decision regarding your friend Abe's honesty. We can assume the aspects considered relevant to Abe's honesty have been agreed upon, and likewise, agreement has been achieved on his honesty under each of the relevant aspects, that is, for every aspect $A \in \mathbf{A}$, we assume it is known whether $Abe \in_A$ *honest* or $Abe \notin_A$ *honest*. If Abe fails to be honest under every subaspect of one of the aspects, say $F =$ family man, then we would say that "Abe is dishonest" under that aspect, that is, *qua* family man. By contrast, we would say that "Abe is not honest" under the aspect F precisely when he fails to be honest with respect to one of the superaspects of F. The ultimate judgment regarding Abe's honesty is then obtained by restricting to the global aspect, G (or "*qua* scf"), where this means that Abe is "honest," "not honest," or "dishonest" precisely when $Abe \in_G$ *honest*, $Abe \in_G \sim$ *honest*, $Abe \in_G \neg$*honest*, respectively. In more detail,

$Abe \in_G$ *honest* iff $\forall A$ Abe $\in_A$ *honest*, that is, "Abe is honest iff Abe is honest under any aspect."

$Abe \in_G \sim$ *honest* iff $\exists A$ Abe $\notin_A$ *honest*, that is, "Abe is not honest iff Abe fails to be honest under at least one of the aspects."

$Abe \in_G \neg$*honest* iff $\forall A$ Abe $\notin_A$ *honest*, that is, "Abe is dishonest iff Abe fails to be honest under every one of the aspects."

If Abe is honest, then he cannot be dishonest (and conversely), that is, $\phi \wedge \neg\phi = 0$. However, that $\phi \vee \neg\phi$ is not necessarily 1 means, of course, that it is not always the case that Abe is either honest or dishonest. One scenario in which this might occur would be where Abe is honest under the aspect S (Abe is honest *qua* student) but fails to be honest in all the other aspects. This reveals how the negation is not, in general, Boolean. Yet we can note that $\sim$ is Boolean globally. This means that for the "global aspect," we will have that Abe is either honest or not honest, but not both, that is, $honest \vee \sim honest = 1$ and $honest \wedge \sim honest = 0$. However, it may occur that Abe is both honest and not honest under the very same aspect (as long as this is not the global aspect).

Thus, notice that even though Abe is "not honest" under a particular aspect precisely when he is not honest under every aspect, for all aspects other than the global one, there is a difference between "not honest" under an aspect and *failing* to be honest under that aspect, for the former has the functoriality property: if Abe is not honest under aspect A, then he is not honest under any subaspect $A' \to A$. Failing to be honest under a given aspect, by contrast, is simply the absence of Abe's honesty under that aspect. Absence of honesty under an aspect is not functorial, so there may be an absence of Abe's honesty under an aspect and the presence of Abe's honesty under another.

Suppose Abe is not dishonest under an aspect. What can we conclude from this? Not that Abe is honest; rather, only that he is *possibly* honest. Abe is not dishonest under a given aspect precisely when he is honest under at least one aspect, as the following string of equivalences reveal:

$$\begin{array}{c} Abe \in_A \sim\neg honest \\ \hline \exists A \to A' \; Abe \notin_{A'} \neg honest \\ \hline \exists A \to A' \exists A'' \to A' \; Abe \in_{A''} honest \\ \hline \exists A' \; Abe \in_{A'} honest \end{array}$$

Restricting to the global level $A = G$, gives

$$\frac{Abe \in \sim \neg h}{\exists A' Abe \in_{A'} h,}$$

or "Abe is not dishonest iff Abe is honest under at least one aspect." Since this works for any predicate, this suggests we define our modal operator

$$\Diamond\phi := \sim \neg\phi,$$

read as "possibly ϕ."

Similarly, we can calculate

$$\begin{array}{c} Abe \in_A \neg\sim honest \\ \hline \forall A' \to A \; Abe \notin_{A'} \sim honest \\ \hline \forall A' \to A \exists A' \to A'' \; Abe \in_{A''} honest \\ \hline \forall A' \; Abe \in_{A'} honest. \end{array}$$

Again, letting $A = G$ the global aspect, this becomes

$$\frac{Abe \in \neg \sim honest}{Abe \in honest,}$$

that is, Abe is necessarily honest under a given aspect iff he is honest under every aspect. In other words, $x \in_A \Box\phi$ iff for all aspects A', $x \in_{A'} \phi$ iff $x \in_G \phi$. This suggests we define a further modal operator,

$$\Box\phi := \neg \sim \phi,$$

read as "necessarily ϕ." Observe that $\Diamond\phi$ and $\Box\phi$, defined thus, are clearly themselves predicates of X. Moreover, they are themselves adjoint, $\Diamond \dashv \Box$, and as such satisfy the (in)equalities we isolated at the end of section 7.2.2.

7.2.4 Adjoint Modalities in Graphs

Example 197 The modalities discussed in the last few sections are particularly well illustrated and tangible in the context of graphs and their subgraphs. If we have a directed multigraph G, the lattice of subgraphs of G constitutes a bi-Heyting algebra,[111] where a subgraph X of G is a directed multigraph (i.e., consisting of a subset X_0 of the vertices of G and a subset X_1 of the edges of G) such that every edge in X_1 has both its source and target in the vertex set X_0. It is clear that we can take unions and intersections of subgraphs, but what it means to take "complements" is not as evident. The set-theoretical complement X^c of a subgraph X will not work, since in general it will not even be a graph, as we might wind up with edges whose source or target is missing in the set X^c. There are two obvious ways to address the insufficiencies with this complement operation, though: we can either discard such problem edges or, on the other hand, we can retain them and "complete" them by adding their sources and targets in the underlying graph. The first option in fact gives rise to our Heyting negation $\neg X$, and the second to the co-Heyting $\sim X$.

It is instructive to see these notions "at work" and to make explicit computations with them. So observe that given the graph G together with its subgraph Y, as displayed below, then for $\sim Y$ we will have

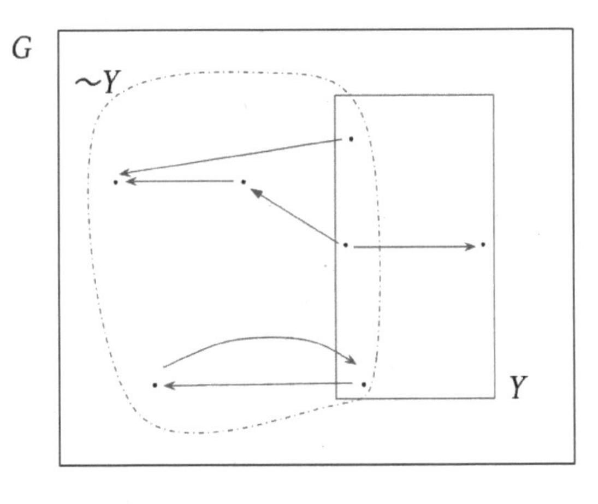

Running this again, we compute $\sim\sim Y$ (which is, importantly, not identical to Y!),

111. This is actually a special case of something that will be discussed further in chapter 10, namely that *any presheaf topos is bi-Heyting*. We know that the category of (multi)graphs can be represented as $\mathbf{Set}^{\mathbf{C}^{op}}$ for $\mathbf{C}$ the index category of two objects and two nontrivial morphisms between those objects; moreover, it can be shown that the lattice of "subobjects" of any object of a bi-Heyting topos is itself a (complete) bi-Heyting algebra.

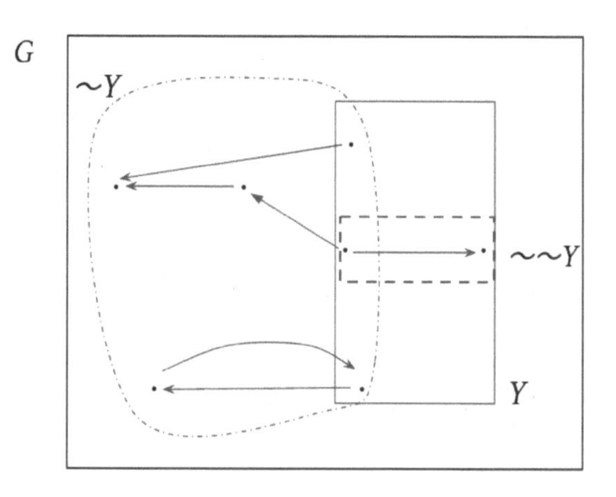

Readers can also verify for themselves that for such a G and Y, $\neg Y \neq \sim Y$.[112]

Let us now look closely at an example that computes modalities in graphs. Suppose we have the following graph G, and the subgraph X of G, as follows:

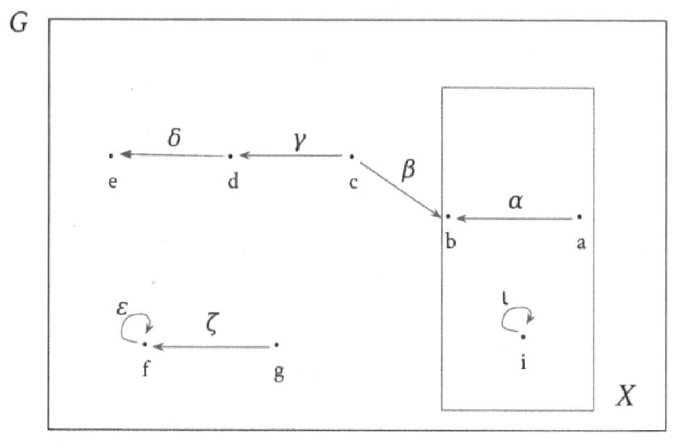

Then, for $\neg X$ we will get the largest subgraph disjoint from X:

112. In general, though, a bi-Heyting algebra for which the two sorts of negation or complementation collapse—that is, $\neg x = \sim x$ for all x—will necessarily be a Boolean algebra.

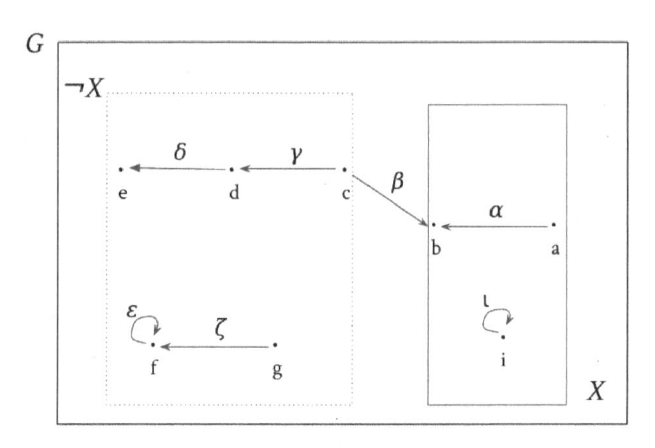

$\sim X$, for its part, is here different from $\neg X$, and yields the smallest subgraph whose union with X gives all of G:

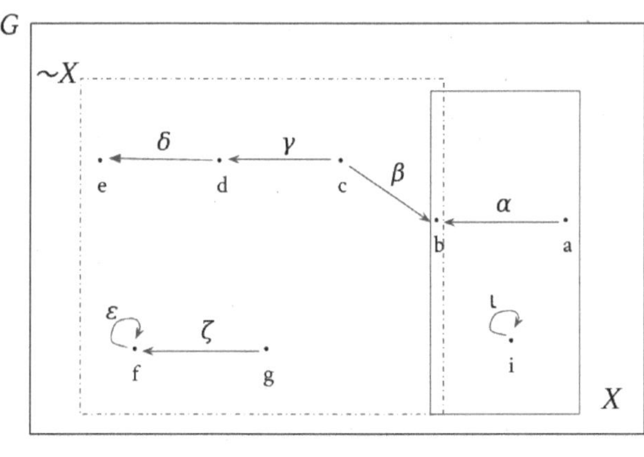

Incidentally, notice that the boundary of X, $\partial X = \sim X \wedge X$, is not the empty subgraph (which functions as 0), but is the sole vertex b. Intuitively, it makes sense to think of the vertex b as the "boundary" of X, since it liaisons between the "inside" of X and the "outside," as there is an arrow, namely β coming in to X from the outside of X via the target b.

Let us now compute $\Diamond_1 X$, that is, $\sim \neg X$:

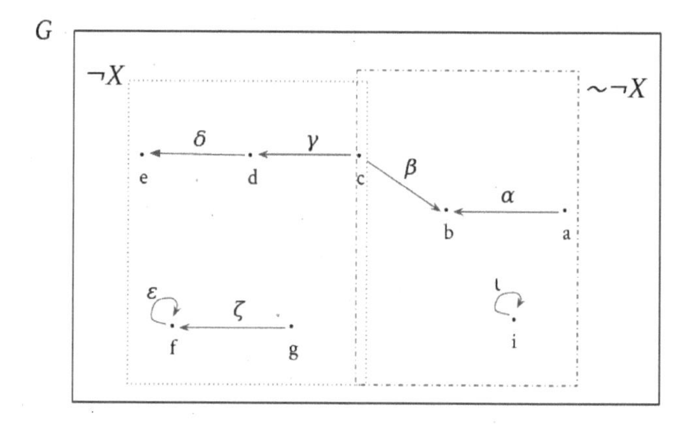

To compute $\Diamond_2 X$, we must first compute $\neg\Diamond_1 X$

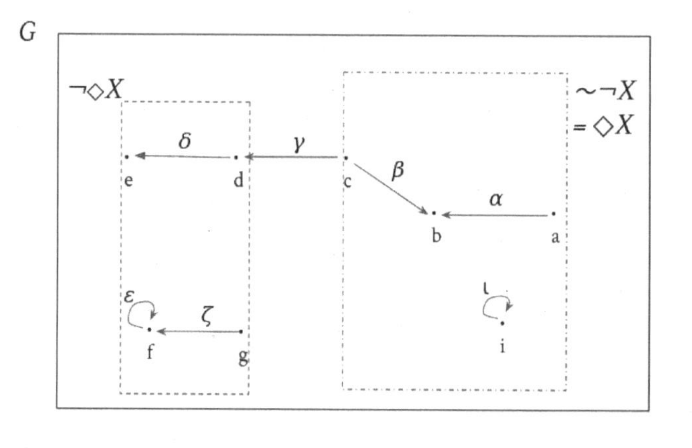

and then $\sim\neg\Diamond_1 X$, that is, $\Diamond_2 X$,

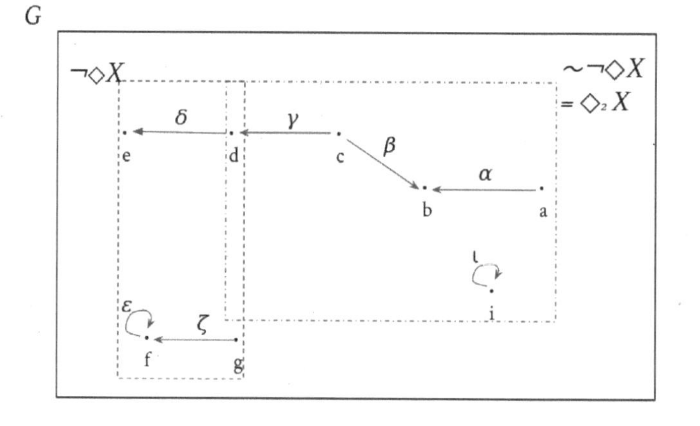

Running this one more time, we first get $\neg\Diamond_2 X$,

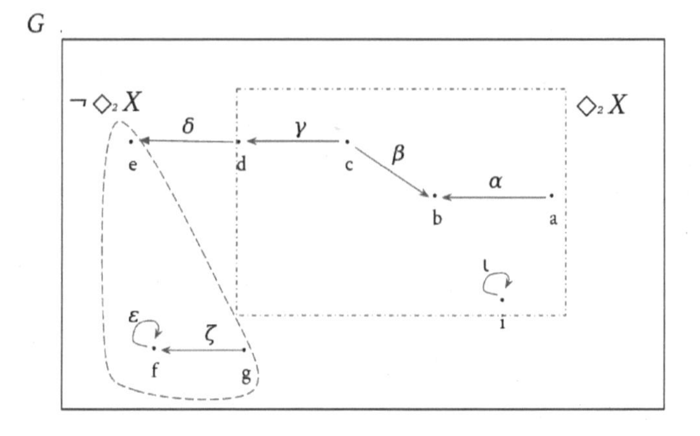

and then $\sim$ of this, that is, $\sim \neg \Diamond_2 X = \Diamond_3 X$,

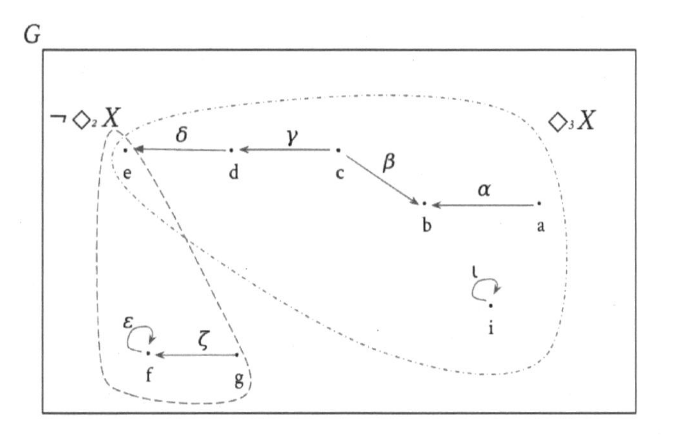

At this point, something interesting happens. If we run $\neg$ on $\Diamond_3 X$, we get

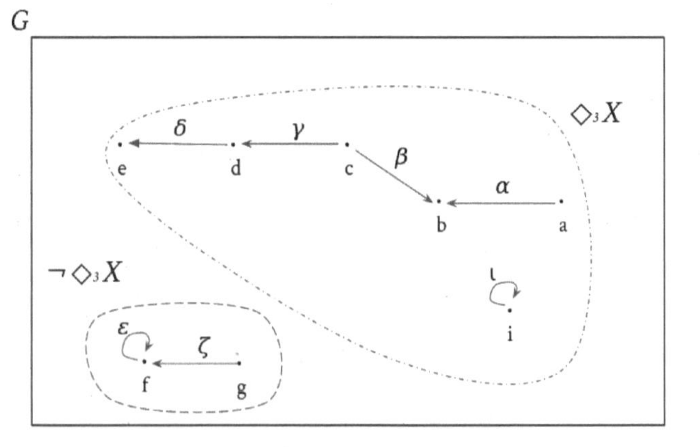

and then $\sim$ of the above, which gives us $\Diamond_4 X$, is revealed to be the same as $\Diamond_3 X$. Thus, this iterative operation stabilizes at $\Diamond_3 = \Diamond_4$, and we have that $\Diamond X = \Diamond_3 X$, having captured, in a subgraph, all those elements of the graph G that can be reached from X through some path, which is clearly something of a picture of the *possibility* of X (or, perhaps more accurately stated, of what is *possible for* X). This illustrates a more general feature as well, namely that every arrow or vertex that is connected to X via some path will end up in $\Diamond X$ after a finite number of steps. In general, applying the operator $\Diamond_n$ to the subgraph should be thought of as capturing those elements connected with X within n paths. Notice also that taking $\neg$ of $\Diamond_3$ is the same as taking $\sim$ of $\Diamond_3$. As such, $\Diamond X$ has no boundary (and no edge going out of it), as $\partial \Diamond X = \Diamond X \wedge \sim \Diamond X$ is the empty subgraph. In general, in the land of graphs, taking the boundary of a subgraph X yields the subgraph whose elements are connected to the outside of X.

We can perform similar computations, in reverse order, to compute $\Box X$, which will supply us with a subgraph whose elements are those that are *not* connected to the outside. First, we compute $\neg$ of $\sim X$, to get $\neg \sim X = \Box_1 X$,

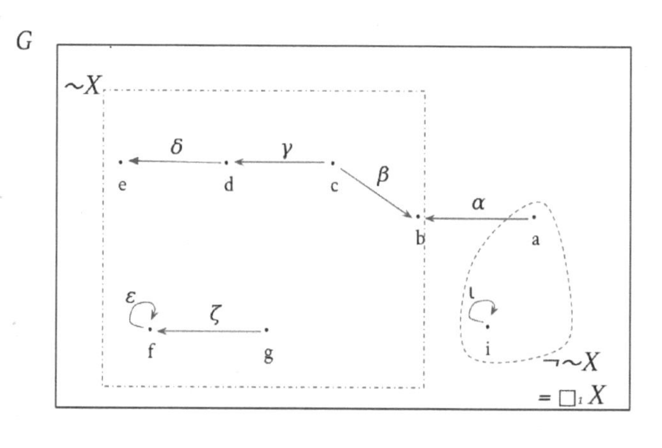

Then, taking $\sim$ of this, we get $\sim \Box_1 X$, and stripping away one more layer, by taking $\neg$ of the result, we end up with $\neg \sim \Box_1 X = \Box_2 X$:

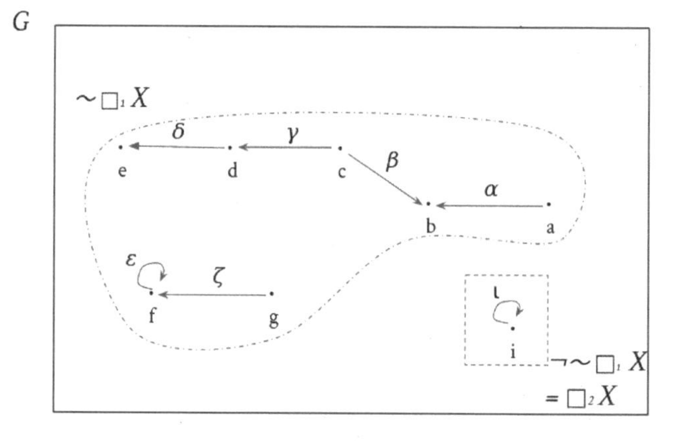

and we have stability, as any further iterations $\Box_{2+n}X$ will just reduce back to $\Box_2 X$. Notice that what is left, $\Box X$, just consists of those elements of X that are not connected to the outside (of X in G), which seems to align with some intuitions we have about the notion of "necessity" for X.

Altogether then, and for a general graph, $\Diamond X$ supplies the elements of the ambient graph G that can be reached from X via some path, while $\Box X$ has those elements in X not connected, via any path, to the outside. Note, finally, that both the subgraphs $\Diamond X$ and $\Box X$ are complemented sums of connected components.

Exercise 20 Consider the following graph G of routes, with subgraph X corresponding to some region of the northeast (including the nodes Boston, New York, and Long Island, together with the indicated routes between them):

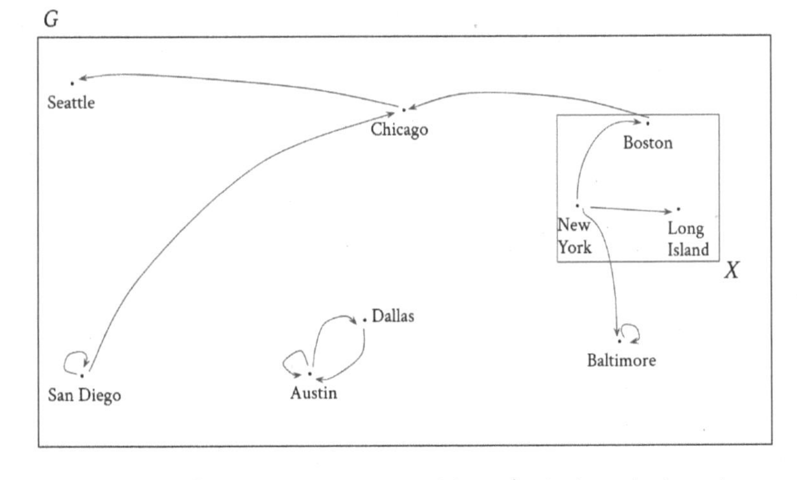

Compute $\Diamond X$, $\Box X$, and ∂X. Then consider, via this example, how the boundary operator ∂ interacts with the modal operators.

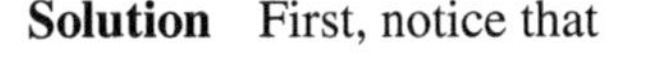

Solution First, notice that

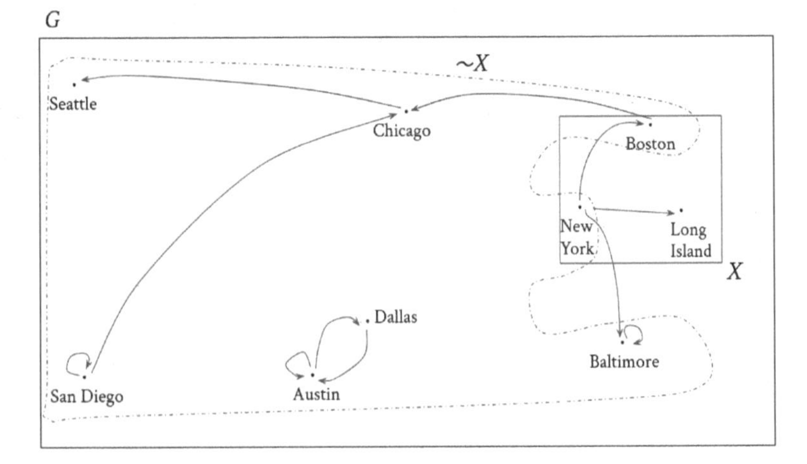

and

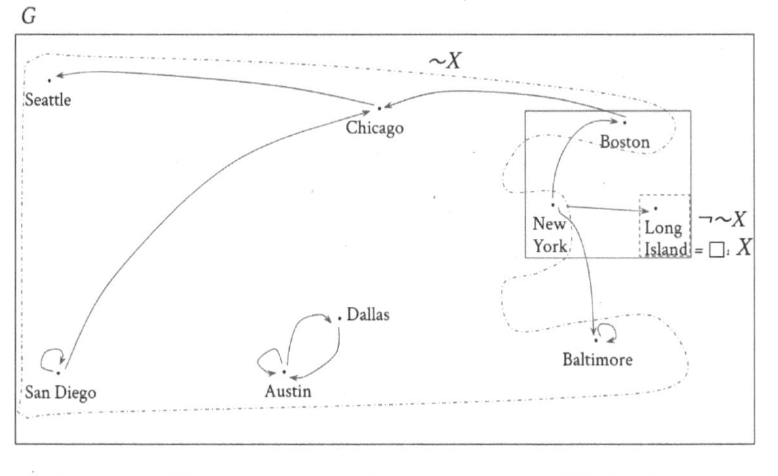

Things stabilize here, with $\square_1 = \square$. $\square X$ includes those parts of X that have no connection to the outside of X. The meaning of $\square X = \{$ Long Island $\}$ is this: if one is in X and ends up in Long Island, one will never get out of X—having arrived there, one is *necessarily* in X.

What about $\Diamond X$, or "possibly" X?

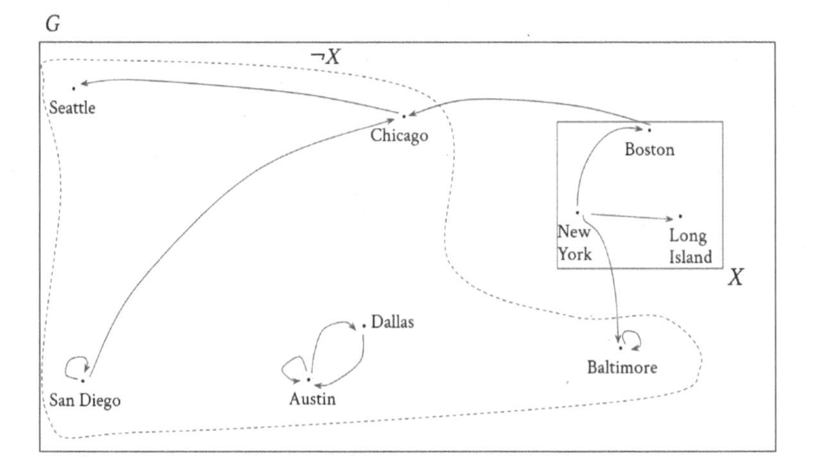

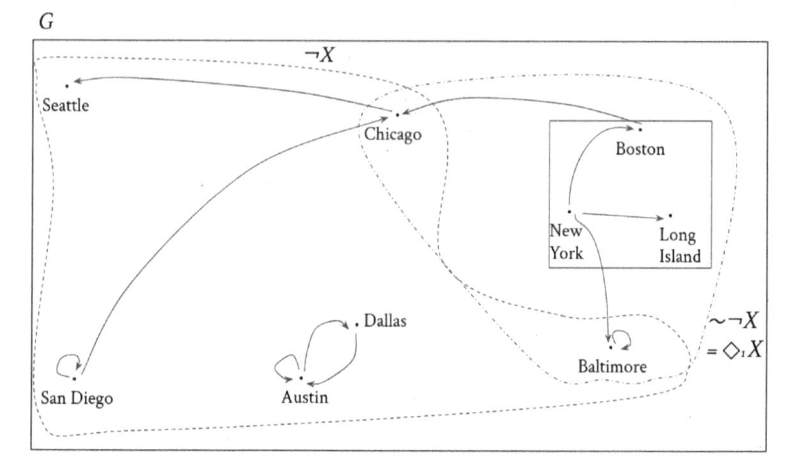
G
¬X
Seattle
Chicago
Boston
New
York
Long
Island
Dallas
~¬X
= ◇₁X
San Diego
Austin
Baltimore

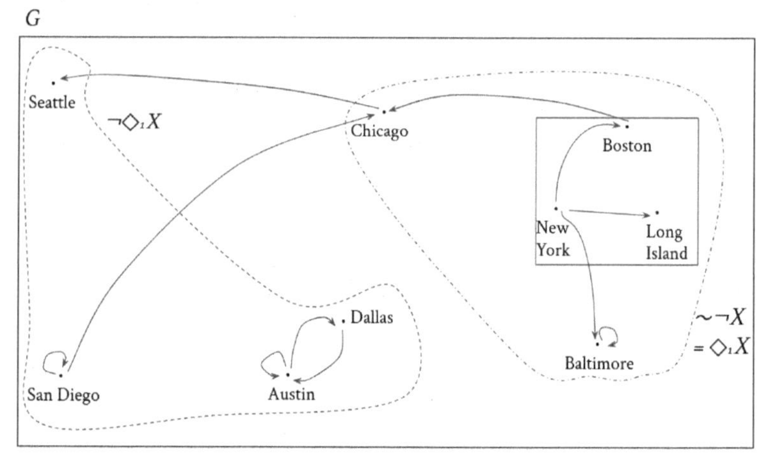
G
Seattle
¬◇₁X
Chicago
Boston
New
York
Long
Island
Dallas
~¬X
= ◇₁X
San Diego
Austin
Baltimore

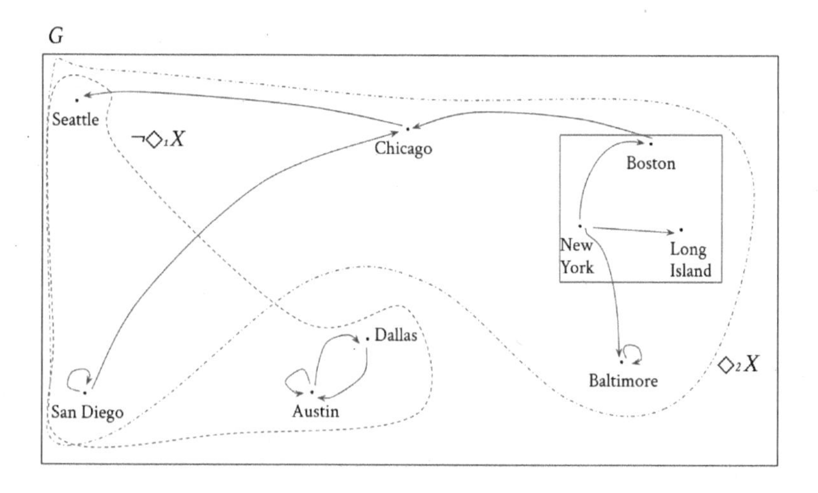
G
Seattle
¬◇₁X
Chicago
Boston
New
York
Long
Island
Dallas
◇₂X
San Diego
Austin
Baltimore

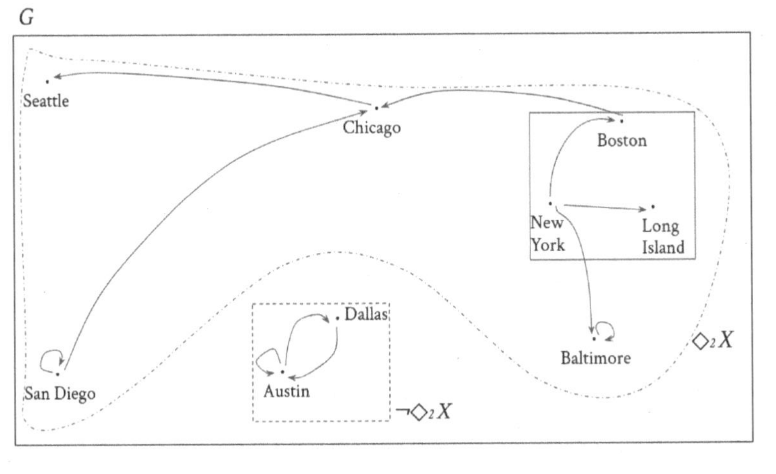

Taking $\sim$ of $\neg\Diamond_2 X$ just returns $\Diamond_2 X$, so we achieve stability at $\Diamond_3 = \Diamond_2$, and so $\Diamond_2 X$ gives us $\Diamond X$, a picture of the "possibility" of X. Intuitively, this makes sense, as $\Diamond X$ supplies those parts of G that can be connected, directly or indirectly, with some part of X. For instance, then, even though San Diego is not directly reachable from any city in X or any city reachable by X, anyone in San Diego can get to Chicago, and Chicago *is* reachable from X. In other words, a person from X might meet someone from San Diego in Chicago or Seattle—and so, for someone from X, San Diego may form part of their picture of reality. Dallas and Austin, by contrast, are inaccessible to X—given the graph above, someone from anywhere in X could never meet anyone from Dallas or Austin.

Finally, consider $\partial X = \sim X \wedge X$. As one can see, this will give the "vertices" Boston and New York. Intuitively, this makes sense, as these cities are those parts of X that mediate between the "inside" of X (as parts of X) and the "'outside" in G. Seeing how the boundary operator ∂ interacts with the modal operators can further solidify the intuitiveness of the reading of $\Diamond$ as "possibility" and $\Box$ as "necessity," even in contexts like that of graphs. As one can easily verify,

$$\partial\Box X = \sim \Box X \wedge \Box X = \Box X$$

which confirms the intuition that the boundary of what is necessarily X—that is, those parts of X that have no connection to the outside of X—is just trivially $\Box X$ itself. Also,

$$\partial\Diamond X = \sim \Diamond X \wedge \Diamond X = \emptyset$$

or, more accurately, the empty subgraph. Intuitively, this realizes the idea that the "world" of what is *not possible* for X has empty overlap with what's *possible* for X.

7.2.5 Adjoint Modalities in Topology

Example 198 Recall from chapter 4 (and see the appendix for) the extended discussion of open and closed sets of a topology, and the associated operations of taking the interior and closure. If we let $\mathscr{O}(X)$ denote the open sets, and $\mathscr{C}(X)$ the closed sets, of some space X, then **int** can be regarded as an inclusion-preserving map from $\mathscr{C}(X)$ to $\mathscr{O}(X)$, and **cl** as an inclusion-preserving map from $\mathscr{O}(X)$ to $\mathscr{C}(X)$. Building on the work of chapter 4 and the appendix, we can leave it to the reader to verify that these maps satisfy that, for any U

open and A closed,

$$\mathbf{cl}(U) \subseteq A$$

iff

$$U \subseteq \mathbf{int}(A).$$

This situation in fact just describes, in category theoretic terms, that **cl** is left adjoint to **int**

$$\mathscr{O}(X) \underset{\mathbf{int}}{\overset{\mathbf{cl}}{\underset{\longleftarrow}{\overset{\longrightarrow}{\bot}}}} \mathscr{C}(X).$$

Moreover, suppose given the region R

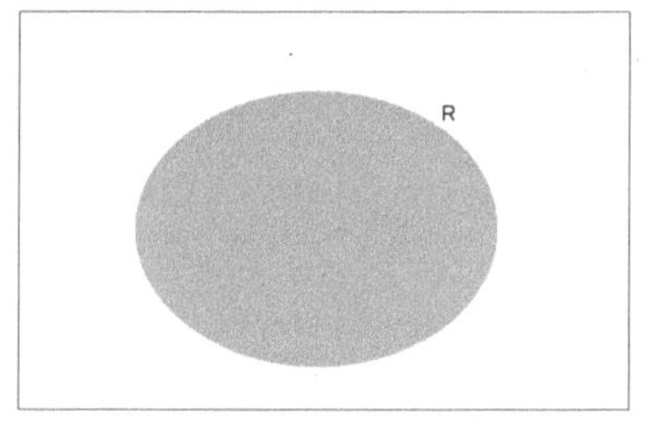

There are two ways of negating (taking the complement of) a part R of a space, where we may assume we initially know nothing of whether or not R includes its boundary. Namely:

1. do not include the boundary, that is, take $\neg R$;
2. do include the boundary, that is, take $\sim R$.

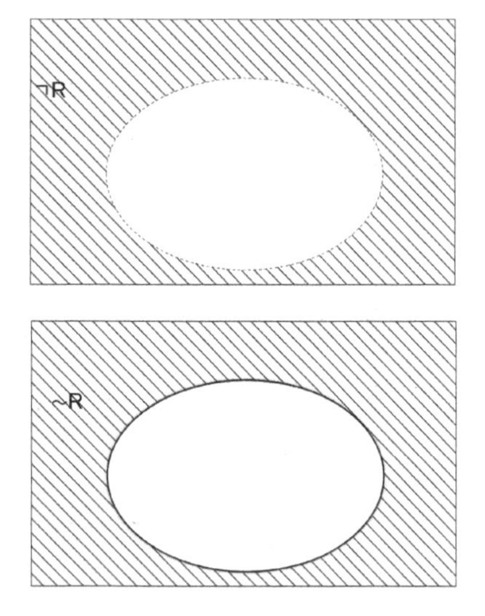

The second operation acts as the closure of the complement. By carrying out these operations again on the results, we can in particular get

1. $\sim\neg R$ (closure of R), or $\Diamond R$;
2. $\neg\sim R$ (interior of R), or $\Box R$.

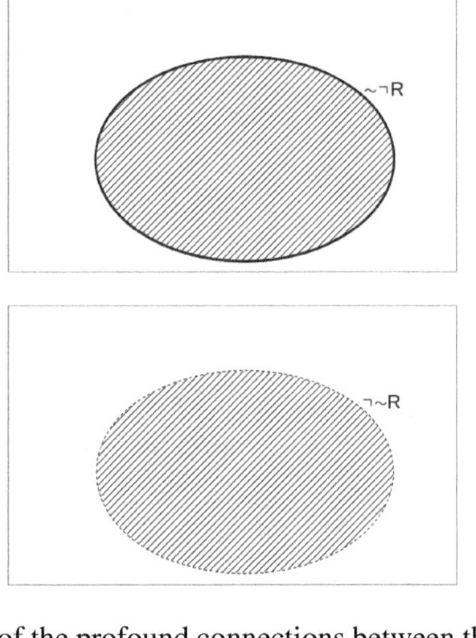

This starts to hint at some of the profound connections between the modal operators $\Box$ and $\Diamond$, on the one hand, and the topological operators **int** and **cl**, on the other—connections explored in the appendix. The more general story in terms of the adjointness of both pairs further allows us to exhibit and unify otherwise seemingly arbitrary results from each special area. The following displays one such case.

In modal logic, a *modality* is any sequence of zero or more monadic operators that involves $\neg$, $\Box$, and $\Diamond$. Iterated modalities include more than one such operator. One is sometimes asked to prove that the logic **S4** has (up to equivalence) exactly fourteen modalities, specifically: (1) id; (2) $\Box$; (3) $\Diamond$; (4) $\Box\Diamond$; (5) $\Diamond\Box$; (6) $\Box\Diamond\Box$; (7) $\Diamond\Box\Diamond$; together with the negations of each (i.e., adding $\neg$ to each), making for fourteen in total. In the affirmative case (unnegated), the seven modalities are related by implication as displayed in the following diagram (where the arrow should be read "implies"):

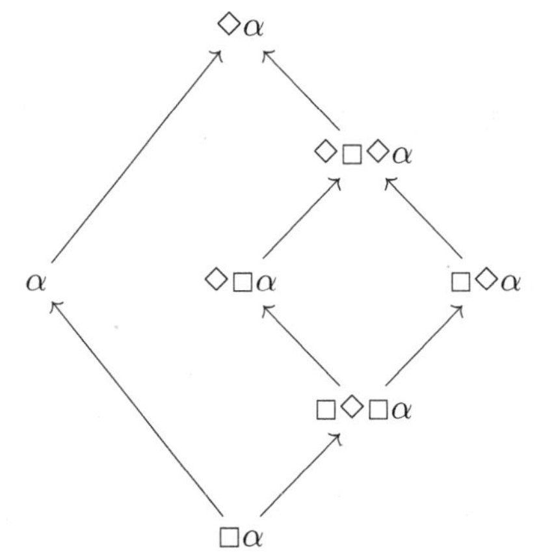

In the context of topology, one can verify that

$$\mathbf{cl}^2 = \mathbf{cl},\ ((-)^c)^c = \text{id},\ \mathbf{int} = (\mathbf{cl}((-)^c))^c,\ \mathbf{int}^2 = \mathbf{int},\ \mathbf{int}((-)^c) = (\mathbf{cl}(-))^c,\ \mathbf{cl}((-)^c) = (\mathbf{int}(-))^c,$$

and there is the following diagram relating the closure and interior operators, and combinations thereof,

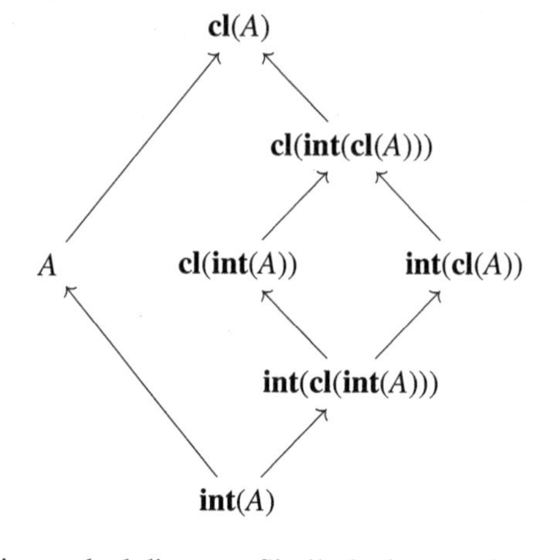

Taking complements gives a dual diagram. Similarly, by negating each of the modalities in the first diagram of modalities, and accordingly reversing the order of implication, we can display the other seven modalities and their mutual relations:

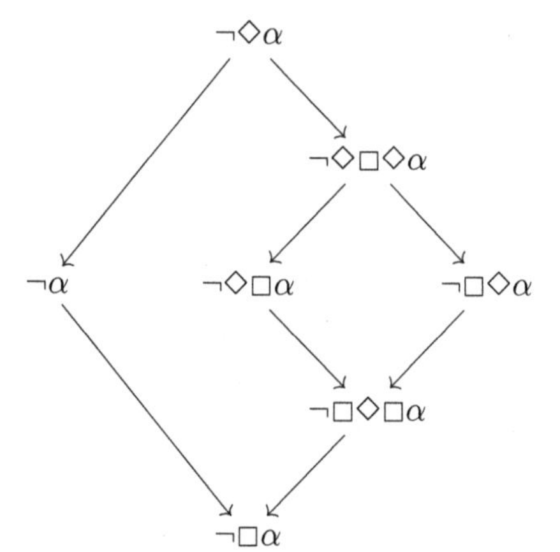

This can be compared to the diagram governing the topological operators, where $\neg$ becomes complement $(-)^c$, and $\Box$ is replaced with **int** and $\Diamond$ with **cl**.

In topology, there is a somewhat mysterious result—called Kuratowski's 14-set theorem—which says that in a topological space, fourteen is the maximum possible number of distinct sets that can be generated from a fixed set by taking closures, interiors, and complements (or just by taking interiors and complements, or closures and complements). The above treatment of these matters, together with the connections explored in the appendix, gives a unified framework for understanding why this, and the fourteen modalities results, should be true, and begins to suggest how they form part of a single story. As it turns out, while this number fourteen is generally a *maximum*, a topological space X will actually contain fourteen such sets when it is "sufficiently rich," in particular when it contains a copy of the Euclidean line—which fact has close connections to the fact (explored in the appendix) that the logic **S4** models the Euclidean line.[113]

7.3 Some Additional Adjunctions and Final Thoughts

More examples of important adjunctions will arise organically throughout the book, and the facts learned in this chapter will be put to use. At this point, there are a number of more advanced category-theoretic results and constructions that we could pursue. But with a good working understanding of categories, functors, natural transformations, and adjunctions—built on a number of carefully selected examples and applications—the reader is already well equipped to delve deeper into sheaves.

For now, this chapter ends by sketching a few more examples, the last two of which are left deliberately somewhat vague, while being correct "in spirit" (and, in fact, they can be developed to be formally correct). These are meant to provide what I hope are some engaging examples of adjunctions, while encouraging more fastidious readers to work out the unspecified details on their own. The chapter ends with a brief Philosophical Pass.

Example 199 Recall how every (directed) graph gives rise to a category and how every category is itself a directed graph with some extra data concerning composition. There is in fact a "free-forgetful" adjunction $F \dashv U$ here

$$\mathbf{Cat} \underset{F}{\overset{U}{\rightleftarrows}} \mathbf{Grph}.$$

The functor F gives the free category on a graph first mentioned in example 34 (chapter 2), which takes vertices for its objects, edges for its morphisms, adds identity arrows at each vertex, and lets the set of morphisms between two vertices be the set of all finite paths between the two. The functor U, for its part, just assigns to a category its underlying graph—it does this by forgetting all the data and structure of the category except for the objects (sent to vertices) and morphisms (sent to edges), leaving out any information or conditions pertaining to identities and composition, as these things are not native to graphs.

Example 200 Universal concepts like colimits and limits, initial objects and terminal objects, can be phrased entirely in terms of adjoint functors. As mentioned already, one of the advantages of this adjunction perspective is that the (co)limit of *every* **J**-shaped diagram in **C** can be defined all at once, rather than just taking the (co)limit of a particular **J**-shaped diagram $X : \mathbf{J} \to \mathbf{C}$.

Recall the notion of diagrams of shape **J** and the constant diagram functor $\Delta : \mathbf{C} \to \mathbf{C}^{\mathbf{J}}$, which sends each object $c \in \mathbf{C}$ to the associated constant functor, defined as the composite functor $c \circ t : \mathbf{J} \to \mathbf{C}$, where $t : \mathbf{J} \to \mathbf{1}$ is the unique functor to the terminal category and c is the object represented by the functor $c : \mathbf{1} \to \mathbf{C}$. Then, given an object X of $\mathbf{C}^{\mathbf{J}}$, we want an object of **C**. Indeed, it turns out that a category **C** admits all limits of diagrams indexed by a small category **J** iff the constant diagram functor Δ admits a right adjoint, and that **C** admits all colimits of diagrams indexed by **J** iff Δ admits a left adjoint:

113. For more on Kuratowski's 14-set theorem, see Sherman (2004). For more on the connections between the notions of modal logic and features of topology, see the appendix.

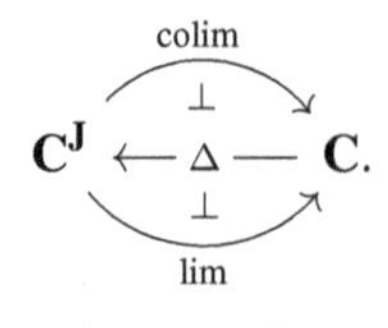

Example 201 There is a connection between how the world *appears* to an agent and what that agent *believes* to hold of their world. But "appears" and "believes" are not quite inverses of one another. Instead, we might conjecture that

$$\text{appearance} \dashv \text{belief},$$

in the sense that there is an adjunction (in a slogan, "belief as the right adjoint of appearance")

$$\frac{A_\alpha(m) \leq m'}{m \leq B_\alpha(m')}$$

realizing, effectively, how all that appears to an agent to hold at, or given, state m entails state m' if and only if whenever m holds in the "real world," this entails that all that the agent *believes* to hold on the assumption that m' holds does in fact hold. Thus, in general, $B_\alpha(m)$ would stand for agent α's belief at m and will consist of those propositions that agent α *believes to hold* whenever m holds.

Example 202 Both small-scale and large-scale projects, such as those in research or development, require resources. Resource allocation (through grants, investment funding, contracts, etc.) requires a detailed plan (for how those resources are to be spent), especially as the project increases in scale. If **Rsrc** is a category consisting of relevant resources, so that objects are resources (e.g., for simplicity, different-sized checks or bags of money) and morphisms are given by a natural relation between those resources (e.g., $\leq$ in the case of a uniform money-valuation of the different resource objects), and if **ProjPlan** is a category consisting of project tasks, given some natural ordering (e.g., by order of priority in the carrying out of the plan), then we might consider the functor

$$V : \mathbf{Rsrc} \to \mathbf{ProjPlan}$$

that maps a resource r to the collection of plans p_i that are *viable* given that resource, and the functor

$$N : \mathbf{ProjPlan} \to \mathbf{Rsrc}$$

taking a project task p to all those resources that are necessary to complete the task (which, depending on how **Rsrc** is structured, say in a simple case of "costs," might just amount to returning an interval bounded by the *least* cost for which the task could be carried out, and including all other more ample amounts).

We would probably not expect V and N to construct strict *inverses* to one another, for we do not expect that, for any given resource r, a list of necessary resources for those plans that are deemed viable given r would be *equal* to r. Though we might expect that, among the resources, $r \leq NV(r)$. Similarly, we would not expect that, for a given project task p, the result of applying N to p and then V to $N(p)$, would always *equal* the same task p. Yet we would expect $VN(p) \leq p$ in **ProjPlan**. This suggests that we have an adjunction,

$$\mathbf{Rsrc} \underset{N}{\overset{V}{\underset{\longleftarrow}{\overset{\longrightarrow}{\bot}}}} \mathbf{ProjPlan}.$$

At this point, having seen a variety of examples and explored some of the basic facts of the general theory of adjoint functors (e.g., adjoints are unique up to unique natural isomorphism, adjunctions can be composed, and so on), readers should feel more comfortable with the idea of adjunctions. This simple concept—fundamentally consisting of an opposing pair of functors with a special relationship to one another—is incredibly powerful and useful. Adjoint functors really are ubiquitous throughout mathematics,[114] and the examples considered in this chapter are but a very small fraction of panoply of adjunctions that interest mathematicians. Adjoints generally tell us very important things. For instance, if a given functor has a left adjoint, then (being a right adjoint of the adjoint pair) it will commute with limits (i.e., it is continuous); if it has a right adjoint, then (being a left adjoint of the pair) it commutes with colimits (i.e., it is cocontinuous). Moreover, we know that the left (or right) adjoint to a functor, provided it exists, will be unique up to natural isomorphism. But adjoint functors need not exist. Given a functor, it need not admit an adjoint. Readers who may very well appreciate the utility and interest of adjoint functors might still be wondering: how do we find such things and how can we know if they even exist? In searching for adjoint pairs, it can often be useful to keep in mind those general situations where adjoints do exist—for instance, how when dealing with categories that consist of algebras (groups, vectors, etc.), forgetful functors will have left adjoints—and consider if they apply. More abstractly, we can *use* some of the general results to help us with the question. In particular, if your given functor T can be shown not to preserve either limits or colimits (i.e., not to be continuous or cocontinuous), then it cannot be either the left or right adjoint of an adjoint functor pair. In other words, (co)continuity gives us a necessary condition for a functor to have a left (right) adjoint. What about sufficient conditions? As it turns out, it can be shown that a functor $T : \mathbf{A} \to \mathbf{S}$ admits a left adjoint iff for each $s \in \mathbf{S}$ the comma category $s \downarrow T$ has an initial object, which effectively means that the problem of finding a left adjoint to a continuous functor can be treated as a problem of finding an initial object in the associated comma category given for each object of $\mathbf{S}$.[115] There are additional conditions—supplied by the so-called "adjoint functor theorems"—that will apply in certain contexts and give sufficient conditions for an adjoint to exist.

In later chapters—especially chapter 11—we will explore important adjunctions in greater depth, in the course of which we will see how, in situations where we can interpret constructions in terms of presheaf (or sheaf) categories, there will automatically be certain pairs of adjoint functors.

7.4 Philosophical Pass: The Idea of Adjointness

Box 7.1
The Idea of Adjointness

114. See Mac Lane's oft-cited slogan, "Adjoint functors arise everywhere" (Mac Lane 1998).

115. For a proof and discussion of this, as well as the other adjoint functor theorems referred to in the following sentence, see Riehl (2016, Lemma 4.6.1 and Epilogue).

Adjoint functors are perhaps the most decisive concept in category theory. In addition to being found all over mathematics, adjoint functors frequently arise from constructions that have a certain universal property, where the constituent functors of an adjoint pair give the "most efficient" solution to a problem. To put it another way: many constructions with universal properties can be translated into statements of adjointness.

Beyond the connections to universal properties, adjunctions exhibit what category theory is really all about. To somewhat overstate the point: categories are not what is important. And the purpose of category theory is not to *categorize*, if that means what it usually means—creating an inventory of sorts that we can consult to ensure that things are put where they belong. What *is* important—the reason why we care about categories in nearly all cases—is supplied by the functors between them and the relations between those functors. Many mathematical structures will have all sorts of important relations to other structures, relations that cannot be reduced to questions of whether the underlying structures they are relating are the "same" (in the sense of isomorphism or equivalence). As just the right formal substitute for the stricter notion of inverse, adjoint functors supply a powerful tool that can capture relations too subtle for the blunt instrument of inverses (which really only "care" about whether the underlying structures are the *same*).

Conceptually, the role adjunctions play in the *theory* of category theory is not dissimilar to what can be seen in the philosophical idea—developed in very different contexts by a variety of different thinkers, each with their own different aims and terminology—of *dialectics*. Proponents of dialectics essentially start out by maintaining that—very abstractly speaking—any notion or model of the "space of concepts" as a sort of static inventory of individual concepts, each set off on its own and capable of "determining itself," is not to be taken at face value. Unpacking an individual concept by using certain abstract transformations or determinations that lead it to be rearticulated against the backdrop of a concept other than it, then doing the same for the latter concept, pushing it into the setting of the former—these transformations will help to extract previously unarticulated antagonisms in the concepts and uncover ways the original concepts were not as capable of "determining themselves" as they first appeared. Finally, combining the two determinations into a sort of conceptual "unity," gives a new and more dynamic understanding of the original concepts, each revealed in some fashion as being *what it is* by the way it achieves this unity of distinct determinations in relation to a concept other than it. Using this very schematic idea, one might think of adjunctions as similarly yielding generalized (conceptual) "unities," where this weakened notion is meant to name the "closest thing" to an *equivalence*.

Philosophers (like Hegel) who are most associated with the development of a concept of dialectics were largely concerned with one particular determination—namely, *negation*. To the extent that we could compare the heuristic supplied by the concept of dialectics and the precise notion of adjunction, we might thus say such philosophers were largely bound by consideration of (proto)adjunctions involving the negation functor and categories in relation to their "opposing" categories. But these situations form a very small, and extreme, slice of the sorts of adjoint relations we can find. In this way, one might say that the notion of adjunction is not just a way of making precise the imprecise idea of dialectics but also a way of releasing it from its arbitrary restriction to involving only functors of the "negation" type.

In short, you could think of the notion of an adjunction as representing something like the wedding of the category-theoretic notion of *universality* with (a much improved version of) the conceptual notion of a *dialectic* or back-and-forth between self-determination and determination-by-another. In chapter 11, we will return to this idea and exhibit adjoint triples that make the connections with the notion of dialectics more precise.

8 Sheaves Revisited

In which we return to sheaves, first taking a closer look at three historically significant examples, using these to further unpack important general features of sheaves, then considering when and why we can fail to have a sheaf, and finally looking at a result concerning the relationship between presheaves and sheaves when dealing with orders.

This chapter returns to sheaves. We first take a closer look at three examples that were historically rather significant in the early development of sheaf theory—sheaves in the context of manifolds, analytic continuation, and "cross sections" of a bundle. In the course of developing these examples, we take the opportunity to dwell further on important features of sheaves. We then include a section dedicated to what is *not* a sheaf, or when and why we can fail to have a sheaf. Finally, we look at an important result concerning the relationship between presheaves and sheaves in the case of working with orders.

8.1 Three Historically Significant Examples

Example 203 Euclidean space, the space of classical geometry, is comparatively "well behaved." In many areas of geometry and physics, one has to deal with fairly complicated structures, and it is desirable to have a description of these things in terms of simpler properties found in Euclidean space. This is where the concept of a *manifold* comes in. Manifolds are, roughly, topological spaces that look locally like $\mathbb{R}^n$, where *local* here means that every point of a manifold will have an open neighborhood that admits a one-to-one map onto an open set of $\mathbb{R}^n$. You can think of a manifold as a space that is "curved" in certain ways, but such that locally—in a small enough neighborhood of any of its points—it "looks like" a part of the usual Euclidean space. In other words, even if globally a manifold does not look like a Euclidean space, locally it will resemble Euclidean space near each of its points.

Manifolds may be regarded as being seamlessly glued together from various different patches, so that overlapping patches form "sheets" that make up the entire space. One of the principal reasons for moving to manifolds is that there are a number of useful instruments available to us in the context of $\mathbb{R}^n$—such as those of integral and differential calculus—that it is desirable to import to the study of other (more complicated) spaces. Sometimes we have a topological space on which we would like to employ the instruments of calculus, for example, and we find that such spaces are locally like open subsets of the Euclidean space $\mathbb{R}^n$ even while they do not provide coordinates valid everywhere. The basic idea with manifolds is that we can cope with such spaces by transferring the instruments that are available in $\mathbb{R}^n$ to small open sets and then patching those sets together, in a sense "recovering" important aspects of the original topological space, while taking advantage of the usual tools.

In chapter 5 (5.1.1), we discussed how functions of various sorts—such as continuous, infinitely differentiable, real analytic, holomorphic functions—are not only all continuous, but the condition for each continuous function to belong to its given class is in fact a *local* condition, meaning that the question of it having a property is in fact equivalent to it having that property in the neighborhood of every point in its domain of definition. In general, when piecing together a manifold, one must just consider what the local structure is that needs to be preserved from one piece to another. In short, manifolds and the functions on them are distinguished by being constructed by the pasting together of pieces that have a particular "nice" property locally.

All of this—the issue of locality and the notion of gluing or patching together local patches—should suggest that sheaves are perhaps lurking somewhere in the background. Indeed, sheaves arise in a particularly natural way in the context of manifolds, and were accordingly pivotal in the early historical development of sheaves.

Formally, a manifold is defined as follows:

Definition 204 An n-dimensional topological *manifold* M is a second countable (its topology has a countable base) Hausdorff space[116] such that each point $q \in M$ has an open neighborhood U homeomorphic to an open set $W \subseteq \mathbb{R}^n$, that is, for each point in $q \in M$, there is an open neighborhood of this point such that there exists a map ϕ from this open neighborhood into $\mathbb{R}^n$, which must further satisfy: (1) ϕ is invertible (i.e., we have $\phi^{-1} : \phi(U) \to U$); (2) ϕ is continuous; (3) ϕ^{-1} is continuous.[117]

If the reader has never worked with manifolds, this definition may be difficult to parse at first. If the reader is intimidated by this definition, it is fine to just think of a sphere or the surface of a globe, or a finite cylinder.

The homeomorphism $\phi : U \to W(\subseteq \mathbb{R}^n)$ given in the definition is usually called the *chart map* for the cover (on which more below). This map, together with the open set U, gives the pair (U, ϕ), called a *chart* for M. Note that ϕ is just a map $\phi(q) = (\phi^1(q), \phi^2(q), \ldots, \phi^n(q))$ (with n entries), where each ϕ^j is just a map $\phi^j : U \to \mathbb{R}$ for $j = 1, 2, \ldots, n$. In other words, the result is n many real numbers, $x_1, \ldots, x_n$, the result of n component maps acting on the point. These component maps are called the *coordinate maps*, and provide the *local coordinates*.

All we require is that, for every point of the manifold, there exists a chart that contains the point. The globe cannot be represented by any one single flat map, so in using flat maps or charts to navigate the Earth's surface, one must collect many (occasionally overlapping) charts together into an atlas. Similarly, in general it is not possible to describe a manifold via one chart alone. But the whole manifold may be covered by a collection of charts. A

116. A Hausdorff space is one in which distinct points are contained in disjoint open neighborhoods.

117. In other words, a homeomorphism is a continuous function between topological spaces that has a continuous inverse; as such, a homeomorphism is what an isomorphism looks like within the category of topological spaces.

collection of charts $\{(U_i, \phi_i) \mid i \in I\}$ is called an *atlas* of the manifold M if $M = \bigcup_{i \in I} U_i$. More formally,

Definition 205 An *atlas* for a manifold M is an indexed set $\{\phi_i : U_i \to W_i\}$ of charts such that together all the domains U_i cover M.

As an example of this, the torus (3-D) is a 2-D manifold, since it can be covered by charts to $\mathbb{R}^2$. Something like a cross, or branching lines, on the other hand, does not yield a manifold, for at the branch points it is not possible to find an invertible continuous map on an open neighborhood of those points.

The proper understanding of manifolds is nicely suggested by the case of the two-sphere manifold S^2, in which context we can provide an atlas consisting of two charts via *stereographic projection*. Of course, this sphere can be described as the subset $\{(x, y, z) \mid x^2 + y^2 + z^2 = 1\}$ in $\mathbb{R}^3$. But we can also describe it *intrinsically*, meaning without reference to the ambient space $\mathbb{R}^3$, in terms of the points via some parameterization. It is not possible to cover the sphere with a single chart, since the sphere is a compact space, and the image of a compact space under a continuous map is compact, while $\mathbb{R}^n$ is noncompact—so there cannot be a homeomorphism between S^n and $\mathbb{R}^n$.[118] However, it *is* possible to cover the sphere by two charts, as will be seen.

In order to develop this further, we will first need to review the relevant notions involved in stereographic projection. The basic idea is that we want to "picture" the sphere as a plane by projecting it onto the plane. We could do this by first isolating the south pole of the sphere, $S = [0, 0, -1]$. To parameterize the sphere we consider a point in the equatorial plane given by $z = 0$, namely $P = [r, s, 0]$, and then draw a line from the south pole through this point. Such a line will intersect the sphere somewhere, say Q, and the resulting line from S to Q will be unique and will intersect the plane in exactly one point. We can describe this line in vector form as $Q = [0, 0, -1] + \lambda(S - P) = [0, 0, -1] + \lambda(r, s, 1)$ (where round brackets denote a vector and square brackets a point). This is of course equal to a point $[\lambda r, \lambda s, \lambda - 1]$. We are looking for what values of λ land us on the sphere, that is, whenever

$$
\begin{aligned}
&(\lambda r)^2 + (\lambda s)^2 + (\lambda - 1)^2 = 1 \\
&\iff \lambda^2(r^2 + s^2 + 1) - 2\lambda = 0 \\
&\iff \lambda = 0 \text{ or } \lambda = \frac{2}{(1 + r^2 + s^2)}.
\end{aligned}
$$

Substituting the nontrivial value of λ in to our equation for Q, we get the point where the line from S to P meets the sphere

$$Q = \left[\frac{2r}{1 + r^2 + s^2}, \frac{2s}{1 + r^2 + s^2}, \frac{1 - r^2 - s^2}{1 + r^2 + s^2}\right]$$

and where for every value r, s, we get a point on the sphere, and every point on the sphere, with one exception, is obtained in this way. The sole exception that prevents this from

118. This argument is not quite complete, as is. A chart is a homeomorphism onto an open subset of $\mathbb{R}^n$, which need not be $\mathbb{R}^n$ itself. Thus, one must use the fact that compact subsets of Euclidean space are closed and bounded. In sum, this means that any chart that is defined on the entire sphere is a homeomorphism to both a closed and open subset of $\mathbb{R}^n$, which by the properties of connectedness and nonemptiness must be $\mathbb{R}^n$ itself.

giving a bijection is obviously the point $S=[0,0,-1]$, corresponding to the line tangent to the south pole.

In other words, then, for every point on the plane (i.e., every r, s from $[r,s,0]$), there is an associated point on the sphere (with the exception of $S=[0,0,-1]$). We thus have a map $\phi_s^{-1}: \mathbb{R}^2 \to S^2 - \{S\}$, the inverse projection map, or the rational parameterization of the sphere. We will denote the sphere with the south pole removed $S^2-\{S\}$ by U_s. But now note that S, P, and $[x,y,z]$ all belong to the same line. Thus, $[r,s,0]=\lambda[x,y,z]+(1-\lambda)[0,0,-1]$, which gives that $\lambda=\frac{1}{z+1}$, and thus that $r=\frac{x}{1+z}, s=\frac{y}{1+z}$. This is in fact the *projection map*, $\phi_s: U_s \to \mathbb{R}^2$. On all the points for which they are defined, this map and the inverse map are clearly continuous; thus we have homeomorphisms. This moreover means that we have produced a coordinate chart for U_s. We can do exactly the same sort of thing starting with the north pole, with the expected result that all points are covered except for a sole exception, the *north* pole itself. Similar to before, we can let U_n denote $S^2-\{N\}$. Then we have another homeomorphism pair, $\phi_n^{-1}: \mathbb{R}^2 \to U_n$ and $\phi_n: U_n \to \mathbb{R}^2$.

Note that together, U_s and U_n provide a *cover* of S^2, and that together with their respective charts, they give us an atlas for all of S^2. Moreover, on the intersection $U_n \cap U_s$ (= $S^2-\{S,N\}$, i.e., the sphere without its poles), we have that $\phi_i(U_n \cap U_s)=\mathbb{R}^2-\{(0,0)\}$. The idea is that we can obtain *all of* S^2 by taking these two homeomorphic copies of $\mathbb{R}^2$—namely U_s and U_n, or rather $\phi_s(U_s) \cong W_s \subseteq \mathbb{R}^2$ and $\phi_n(U_n) \cong W_n \subseteq \mathbb{R}^2$—and pasting together the $\phi_s(U_s \cap U_n) \cong W_{sn} \subseteq W_s$ to $\phi_n(U_s \cap U_n) \cong W_{ns} \cong W_n$ via appropriate transition functions (discussed below). It is straightforward to define the transition function between the two projections which enables such gluing. In this way, we will have thus constructed the atlas $\{(U_s,\phi_s),(U_n,\phi_n)\}$, and an atlas determines $M=S^2$ as a topological space.

Stepping back, a little more generally now, assume given two charts (U_i,ϕ_i) and (U_j,ϕ_j). The general idea here is captured in something like the following figure:

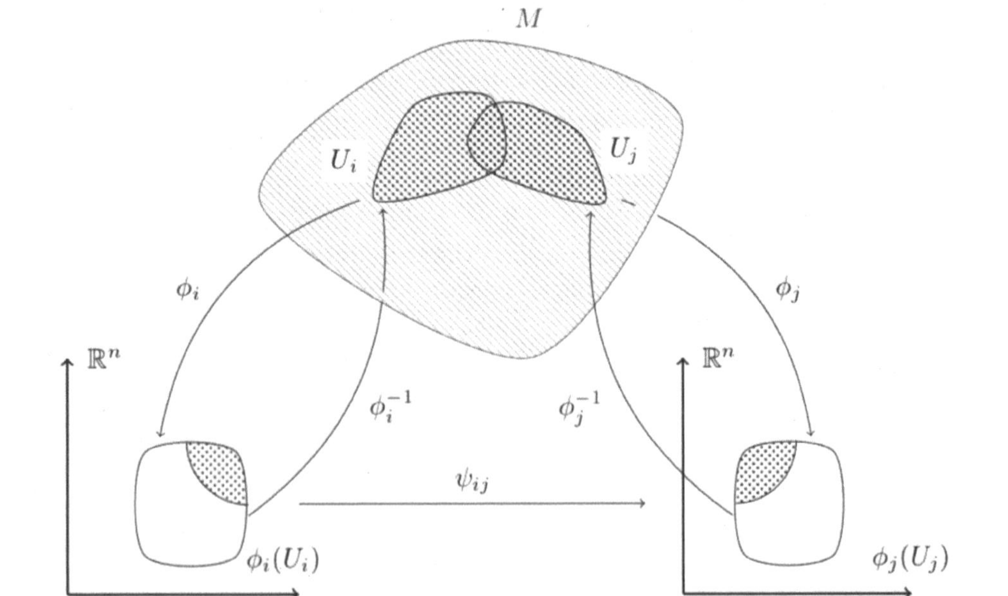

Two charts ϕ_i and ϕ_j of an atlas might overlap on the set $U_i \cap U_j$. But then by composition of ϕ_i with the inclusion $U_i \cap U_j \subseteq U_i \to W_i$, we get a homeomorphism $\phi_{ij} : U_i \cap U_j \xrightarrow{\cong} W_{ij}$ that goes from the overlap to an open set $W_{ij} \subseteq W_i \subseteq \mathbb{R}^n$. However, while a point in the intersection is mapped to one set of coordinates via ϕ_i, it is in general mapped to another set of coordinates in ϕ_j. Accordingly, ϕ_j gives an in principle different homeomorphism $\phi_{ji} : U_i \cap U_j \xrightarrow{\cong} W_{ji}$ from the intersection to a different open set $W_{ji} \subseteq W_j \subseteq \mathbb{R}^n$. In the above figure, the dotted regions on the bottom left and right figures in $\mathbb{R}^n$ represent $\phi_{ij}(U_i \cap U_j) \cong W_{ij}$ and $\phi_{ji}(U_i \cap U_j) \cong W_{ji}$, respectively. Notice that, as homeomorphisms, these are not just *maps*, but maps with inverses.

The question then is: how do we translate or transition between the two homeomorphisms on the overlap? We are looking for a map $\psi_{ij} : W_{ij} \to W_{ji}$, a rule that enables us to get from a point in one set of coordinates to that same point in terms of the other coordinates. In our example of the manifold S^2, such a transition rule is evident and easy to construct explicitly. To find this more generally, notice that the map ϕ_{ij} is invertible (it describes a homeomorphism). So we can take a point $q \in \phi_{ij}(U_i \cap U_j) \subseteq \mathbb{R}^n$ and apply the inverse map to it to get back to the manifold. We then apply ϕ_{ji}, landing us back in $\mathbb{R}^n$, but this time with the coordinates specified by ϕ_j. In other words, we have constructed the composite map $(\phi_{ji} \circ \phi_{ij}^{-1})(q)$, producing coordinates of q in the other chart, and we can of course do this for any such q in the overlapping region. Thus we are really mapping one open set of $\mathbb{R}^n$ to another, and doing so homeomorphically (the composite of homeomorphisms is a homeomorphism).

With such a *chart transition map* we thus have the following:

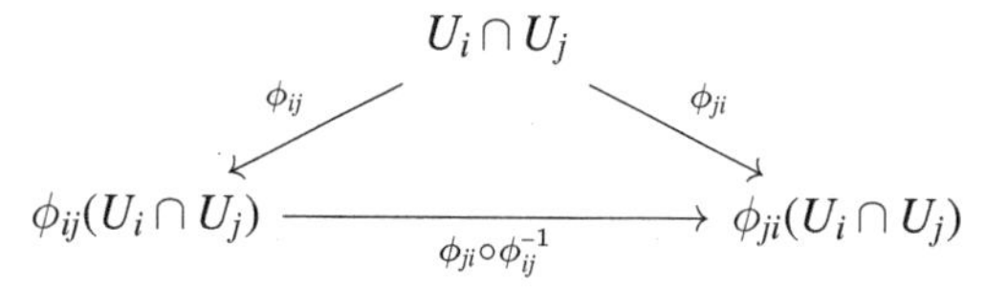

where, as in the following, the two bottom objects are contained in $\mathbb{R}^n$.

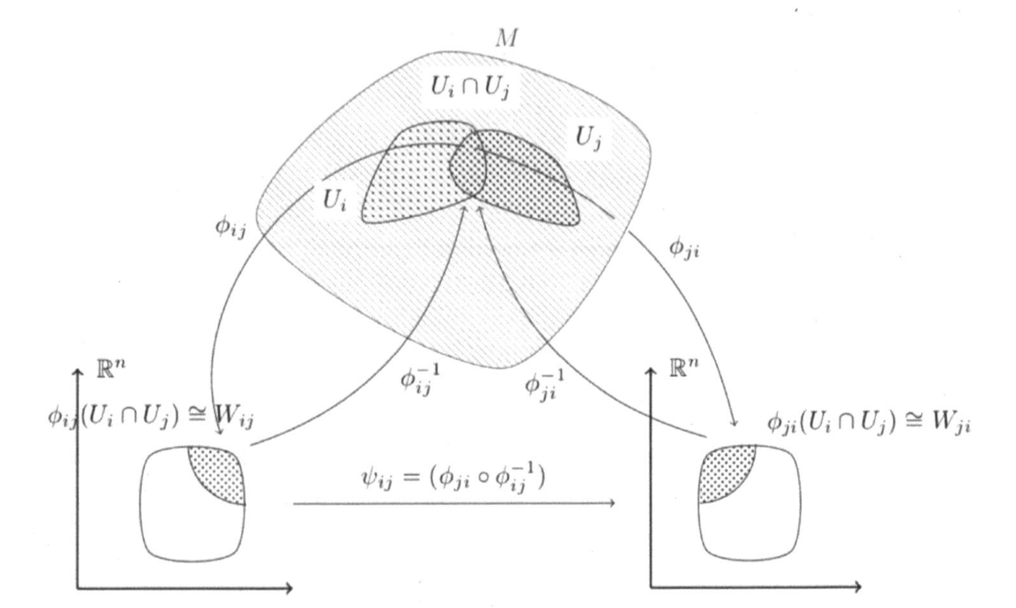

This transition map ψ_{ij} gives us our desired *change of coordinates*, a map from an open part of $\mathbb{R}^n$ to another open part of $\mathbb{R}^n$. Since we already had continuous inverses ϕ_i^{-1} and ϕ_j^{-1}, the transition map comprising the restricted maps is guaranteed to be continuous. Note also that in changing coordinates, from the "perspective" of $p \in U_i \cap U_j$ (think of this as the "real world"), this p "does not care" what coordinates it is in, or that you have changed coordinates.

The really important point here is that in general the chart transition maps contain the instructions for *how to glue together the various charts of an atlas*. In other words, with these maps, you have all the *global* information you need. In the case of our example with S^2, there were only two charts that needed gluing. But in the general case, the conditions detailed above hold for all i, j in some index set I, ultimately allowing an entire manifold M to be covered. This process of obtaining the entire manifold by taking all the $W_i \subseteq \mathbb{R}^n$ and pasting the $\phi_i(U_i \cap U_j) \cong W_{ij} \subseteq W_i$ to $\phi_j(U_i \cap U_j) \cong W_{ji} \subseteq W_j$ by the transition functions can be formally described with the following coequalizer diagram:

$$\coprod_{i,j} U_i \cap U_j \underset{\beta}{\overset{\alpha}{\rightrightarrows}} \coprod_i U_i \xrightarrow{\gamma} M$$

Here $\coprod_i$ denotes the coproduct in the category **Top**, which is the disjoint union; γ takes each $x \in U_i$ and sends it to the same point $x \in M$; the maps α takes each point x_{ij} in the intersection $U_i \cap U_j$ to the same x_{ij} in U_i, while β takes each point x_{ij} in the intersection $U_i \cap U_j$ to the same x_{ij} in U_j. M is the coequalizer of these maps α and β in the category **Top**. Given how consideration of locally defined functions over open sets ordered by inclusion reverses the direction of the arrows, and given the definition of a sheaf in terms of an *equalizer diagram*, the close connection to sheaves should be apparent.

Returning to S^2 again, a further important feature of all this is that we can now test a function on S^2 for certain properties, like continuity or differentiability, by trying it on each of the two parts (each of the two charts) separately. Moreover, a function will be continuous on $U \subseteq M$ when its composite with ϕ^{-1} is continuous on $W \subseteq \mathbb{R}^n$. Via the following important *direct image sheaf* construction, we will be able to appreciate how the chart will determine a particular sheaf of continuous functions on U.

Definition 206 If $f : X \to Y$ is a continuous map of spaces, then each sheaf F on X yields a sheaf $f_* F$ on Y defined, for V open in Y, by $(f_* F)(V) = F(f^{-1}(V))$. In other words, since f is continuous, it induces an order-preserving map, f^{-1}. Composition with f^{-1} then gives $f_* F$ defined as the composite functor

$$\mathcal{O}(Y)^{op} \xrightarrow{f^{-1}} \mathcal{O}(X)^{op} \xrightarrow{F} \mathbf{Set}$$

This sheaf is usually called the *direct image* of F under f.[119]

Notice that the map f_* is in fact a functor

$$f_* : \mathbf{Sh}(X) \to \mathbf{Sh}(Y)$$

and that $(f \circ g)_* = f_* g_*$, which means that by defining $\mathbf{Sh}(f) = f_*$, we have that **Sh** becomes a functor on the category of all (small) topological spaces. And, in particular, if $f : X \to Y$ is a homeomorphism, then f_* gives an isomorphism of categories between sheaves on X and sheaves on Y.[120]

Returning to our example, then, our chart (U, ϕ) will in fact determine a particular sheaf C_U of continuous functions on U, specifically as the direct image $C_U = (\phi^{-1})_* C_W$. The coordinate projections $\mathbb{R}^n \to \mathbb{R}$, composed with ϕ after being restricted to W, give us the local coordinates for the chart ϕ, that is, the n-coordinate functions $x_1, \ldots, x_n : U \to \mathbb{R}$. Going the other way, these coordinate functions determine the chart as the continuous map $U \to \mathbb{R}^n$ with components $x_i : U \to \mathbb{R}$, the map to $\mathbb{R}^n$ being restricted to its image $W \subseteq \mathbb{R}^n$.

We now make a key general observation, one that can be applied to our present situation. Consider how if we let U be an open set in a general space X, then any sheaf F on X, when restricted to open subsets of U, clearly gives us a sheaf $F|_U$ on U. And so, in this way, $U \mapsto \mathbf{Sh}(U)$ and $V \subseteq U \mapsto (F|_U \mapsto F|_V)$ defines a contravariant functor on $\mathcal{O}(X)$. Moreover, since the notion of a sheaf is "local," this suggests that this functor *itself* might be a sheaf.

Theorem 207 If $X = \bigcup W_k$ is an open covering of the space X, and if, for each k, F_k is a sheaf of sets on W_k such that

$$F_k|_{(W_k \cap W_l)} = F_l|_{(W_k \cap W_l)}$$

for all indices k and l, then there will exist a sheaf F on X, unique up to isomorphism, with isomorphisms $F|_{W_k} \cong F_k$ for all indices k, which match on the equation shown above.[121]

119. This direct image sheaf is also sometimes called the *pushforward* of a sheaf.

120. As you might expect, a continuous map of spaces f will induce another functor going in the other direction on the associated categories of sheaves. Performing the reverse operation yields the *inverse image* sheaf f^*, or the *pullback* of a sheaf along a map. Pushforwards and pullbacks are especially useful in providing us with a canonical way to register the effects of changing base spaces. We look more closely at these two functors in later chapters.

121. A proof of this can be found in Mac Lane and Moerdijk (1994). Instead of going through this, we note, following Serre, that more generally we could replace the equality in the above theorem with *isomorphisms* of sheaves, showing that sheaves are in fact *collatable up to isomorphism*. Explicitly (since this is actually the version that most concerns us in the present case): If X again has an open cover with the W_k, and if, for each index

Many sheaves can be built up in this way from the local pieces F_k.

It is basically immediate from this theorem how we can produce the sheaf of continuous functions on S^2, or the sheaf of differentiable functions on S^2, and so on, and thus ultimately sheaves of smooth structures on S^2.

An important point that emerges in this discussion of sheaves on manifolds is that in the standard definition of sheaf the reader might be misled into thinking that what we need on overlapping regions is *equality* "on the nose"; but this is not quite right. In fact, all that is actually needed in patching together different information on the same overlapping region is just a consistent *translation system* that can take us from one data piece to the other and back—that is, we need a set of isomorphisms allowing us to translate between them, not simple equality. This system of translations typically leads to what is called *descent data*, and can be thought of, intuitively, as providing some sort of generalized matching.

Before moving on, it is worth noting that the "meaning" of the manifold only really emerges once we have glued together the charts according to these transition maps. The idea is to study manifolds by "pushing them down" onto the charts in $\mathbb{R}^n$ with their transitions, thus allowing us to work with far simpler maps from some portion of $\mathbb{R}^n$ to another. Often it is desirable to define properties, like continuity or differentiability, of a "real-world object" (a map from $\mathbb{R}$ to M) by judging suitable conditions on a chart representative of that real-world object. Say we start with a path or curve $\gamma : \mathbb{R} \to M$ in a manifold. We then take an open set U enclosing it, yielding a map $\gamma : \mathbb{R} \to U$. We can now push down the real-world curve (γ, in bold) to a chart via some chart map ϕ, to get its image in $\mathbb{R}^n$:

$$\begin{array}{ccc} \mathbb{R} & \xrightarrow{\gamma} & U \\ & & \downarrow{\scriptstyle\phi} \\ & & \phi(U) \subseteq \mathbb{R}^n \end{array}$$

But then, of course, instead of looking at the curve in the real world, we can focus just on the map $\phi \circ \gamma$:

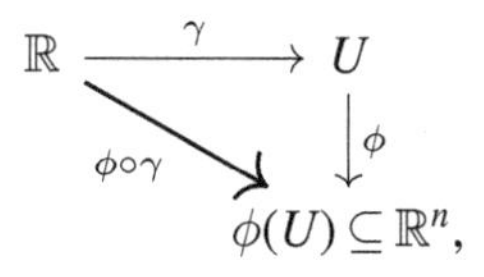

which represents the original curve as some curve in $\mathbb{R}^n$ down in the chart. If we want to know about the continuity of the real-world curve, we instead look at continuity in this chart representative of it; but one should not confuse this chart representative for the real

---

k, F_k is a sheaf on W_k, so for all j and k

$$\theta_{jk} : F_j|_{(W_j \cap W_k)} \cong F_k|_{(W_j \cap W_k)}$$

is an isomorphism of sheaves, and for each i, j, k we have

$$\theta_{ik} = \theta_{jk} \circ \theta_{ij}$$

whenever this is defined; then it can be shown, in the same way as the less general case, that there will exist a sheaf F (unique up to isomorphism) on X and isomorphisms $\phi_k : F|_{W_k} \to F_k$ (unique up to isomorphism) such that

$$\phi_j = \theta_{ij}\phi_i$$

when this is defined.

world (for instance, one cannot define differentiability on the real-world trajectory given by γ), or forget that ϕ could be ill-defined. In order to overcome this latter issue, and ensure that the chart is not arbitrary, we make sure that a given property (e.g., continuity) does not change under change to another chart. Formally, what we mean is:

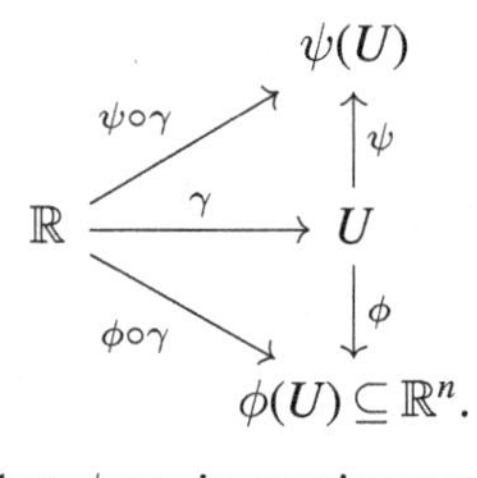

Then if we know, for instance, that $\phi \circ \gamma$ is continuous, we also want to know if we can say the same for $\psi \circ \gamma$. But this arrow is given by another path around the diagram, in particular,

$$\psi \circ \gamma = \psi \circ (\phi^{-1} \circ \phi) \circ \gamma = (\psi \circ \phi^{-1}) \circ \phi \circ \gamma.$$

But we know that $(\psi \circ \phi^{-1})$ is continuous, and we know that $\phi \circ \gamma$ is continuous, and the composition of continuous functions is continuous. So $\psi \circ \gamma$ must be continuous as well. The map $\phi \circ \gamma$ is thus well defined and continuous, and since the downward map ϕ is continuous, we can show that we can "lift" continuity up to γ. But notice that if we had said that $\phi \circ \gamma$ was differentiable, on the other hand, that does not automatically guarantee that $\psi \circ \gamma$ is differentiable. Everything depends, then, on the nature of the transition map.[122]

We can be more general. Two charts (U, ϕ) and (V, ψ) of a topological manifold are said to be *blank*-compatible if either $U \cap V = \emptyset$ or whenever $U \cap V \neq \emptyset$, we have that the following transition maps

$$\psi \circ \phi^{-1} : \phi(U \cap V) \to \psi(U \cap V)$$
$$\phi \circ \psi^{-1} : \psi(U \cap V) \to \phi(U \cap V)$$

have the *blank* property. We then take a restriction of the maximal atlas to get an atlas $\mathscr{A}_{blank}$, and say that this atlas is a *blank*-compatible atlas as long as any two charts in $\mathscr{A}_{blank}$ are *blank*-compatible. A *blank*-manifold is then a triple $(M, \mathcal{O}, \mathscr{A}_{blank})$.

Of course, *blank* could be, for instance, the property of differentiability. But it can be many other things as well. For instance, it could represent: C^0 (trivially, because *every* atlas is a C^0 atlas), C^1 (i.e., differentiable once, the result of which is continuous), C^k (i.e., k-times continuously differentiable), C^∞ (i.e., continuously differentiable arbitrarily many times), C^ω (i.e., real analytic, meaning there exists a [multi-dimensional] Taylor expansion), $\mathbb{C}^\infty$ (i.e., complex differentiable, meaning that each continuous map from $\mathbb{R}^n$ to $\mathbb{R}^n$ satisfies the Cauchy-Riemann equations—this gives a *complex manifold*). In this list, our atlases are becoming more and more restrictive, as we place stronger and stronger conditions on the transition functions. A well-known theorem informs us that as long as $k \geq 1$, any C^k-atlas $\mathscr{A}_{C^k}$ of a topological manifold will contain a C^∞-atlas. There are some topological manifolds where you simply cannot remove a chart or charts in such a way that

122. In terms of differentiability, the transition map might preserve continuity but introduce an edge, preventing differentiability. But if we just "rip out" all those charts in our atlas that are only continuous, but not differentiable, then we get a *restricted atlas* where all the transition functions are differentiable.

all that remains is still an atlas and has continuously differentiable transition functions. However, what the theorem tells us is that whenever you can achieve that the transition functions are at least once continuously differentiable for an atlas, you can be sure that by removing more and more charts you will eventually have a C^∞-atlas. This guarantees that we may always, without loss of generality, consider C^∞-manifolds, which are called *smooth manifolds* ("always" meaning as long as we guarantee C^1).

In terms of the general picture given above, we will be able to use the homeomorphism ϕ_i to transfer smoothness on $W_i \subseteq \mathbb{R}^n$ to smoothness on $U_i \subseteq M$, giving us the sheaf C_i^k of all smooth functions on the open subsets of U_i. Then the smoothness expected to hold of the transition functions guarantees that the sheaves C_i^k and C_j^k will agree when restricted to the overlap $U_i \cap U_j$. Thus, via the result from theorem 207, we know that these sheaves C_i^k themselves can be collated, yielding the sheaf C^k of *all* smooth functions on opens of M. Moreover, this sheaf provides an instance of the general notion of a subsheaf, the description of which in a sense nicely exhibits the local character of a sheaf. Our particular sheaf will be a subsheaf of the sheaf of continuous functions; obviously, being a sheaf, its restriction to each of the U_i gives a sheaf C_i^k. This is just to say that each smooth manifold supports what is sometimes called a *structure sheaf*, namely the sheaf C^k of smooth functions. This is, importantly, not just a sheaf of *sets*, but a sheaf of $\mathbb{R}$-algebras, a sheaf of rings.

For the Euclidean n-space $X = \mathbb{R}^n$, more generally, there are a number of examples of sheaves, which then give rise to a number of subsheaves. For U open in $\mathbb{R}^n$, we let $C^k(U)$ be the set of all $f : U \to \mathbb{R}$ having continuous partial derivatives of all orders up to (and including) order k. This gives a functor $C^k : \mathcal{O}(X)^{op} \to \mathbf{Set}$ with values in $\mathbf{Set}$ (or in $\mathbf{Mod}_{\mathbb{R}}$), which in fact yields a sheaf. Notice that *each* C^k will be a sheaf on $\mathbb{R}^n$, leading to a nested sequence of subsheaves on $\mathbb{R}^n$:

$$C^\infty \subseteq \cdots \subseteq C^k \subseteq C^{k-1} \subseteq \cdots \subseteq C^1 \subseteq C^0 = C.$$

There are other constructions found in the theory of manifolds that in fact lead to sheaves, for instance, via the notion of a tangent bundle for a smooth (C^∞) manifold M. The story is pretty similar for various types of manifolds. But perhaps the most important, historically, involves complex analytic manifolds, covered in the next example (after a brief interlude).

Box 8.1
Philosophical Pass: Manifolds and Sheaves

The construction of a manifold, as presented in the previous example, supplies us with further philosophical motivation for the relationist perspective, introduced in earlier chapters (especially chapter 6). The inability to describe the two-sphere in a single coordinate chart provides us with another example of the need for multiple perspectives and their fusion.

As we saw, we have objects that are locally like domains in Euclidean space, where the classical devices of calculus are available to us. Using such objects, we can transfer such devices to small open sets and then patch them together, piece by piece—the sheaf notion can be seen as the tool used in this patching. One of the main virtues of the sheaf notion is that it enables us to deal, in a systematic way, with objects that may not have a global definition but are only locally defined. In this way, via the need for multiple perspectives and their subsequent patching, the sheaf notion emerges as an important part of the relationist approach.

8.1.1 Returning to the Examples

Example 208 Historically, the idea of a sheaf largely stems from problems surrounding the *analytic continuation* of functions (together with the development of Riemann surfaces). Analytic continuation involves the attempts to extend the given domains of definition of functions to larger domains. Recall the following notion from analysis:

Definition 209 A function $f : U \to \mathbb{C}$, where U is open in $\mathbb{C}^n$, is *analytic* (or *holomorphic*) if it is described by a convergent power series in a neighborhood of each point $p \in U$.

Another way of saying this is that a function is analytic precisely when its Taylor series about a point, say x_0, converges to the function in some neighborhood for every x_0 in its domain, that is, locally it is given by a convergent power series.

Now, let V be an open subset of $\mathbb{C}^n$, and for each open subset $U \subseteq V$, let $A_V(U)$ be the set of all analytic functions on U. Then A_V will in fact give us a sheaf (specifically, of $\mathbb{C}$-algebras).

Let us explore the connection between analytic continuation and sheaves more closely. We begin with a pair (D,f), where $D \subseteq \mathbb{C}$ is a domain and $f : D \to \mathbb{C}$ is analytic. We say that another such pair (D_1,f_1) is a *direct analytic continuation* or an analytic extension of (D,f) if $D \cap D_1 \neq \emptyset$ and $f|_{D\cap D_1} = f_1|_{D\cap D_1}$. As one should now expect, this gives us a *gluing condition*. In such a case, we then define

$$g : D \cup D_1 \to \mathbb{C}$$

by

$$g|_D = f, \qquad g|_{D_1} = f_1.$$

Analyticity is locally defined, so we can observe that g must be analytic, and we say that g is obtained by *gluing f and* f_1. One of the interesting notions here is that f may be defined very differently than how f_1 is defined. In general, an analytic function could be given by a power series, by a formula, or by an integral. The main question that presents itself in this setting is then: what is the *largest* open set to which a function can be extended, and can we describe such an extended function in some way?

What follows is a simple, concrete illustration of these ideas. Take

$$D = \{z : |z| < 1\}$$

and $f(z) = 1 + z + z^2 + z^3 + \cdots$, the geometric series. Then $f(z)$ is a power series, centered at z. The radius of convergence is the unit disk, that is, inside this disk the function always represents an analytic function (if you write out the Taylor expansion at zero, you will get back this power series). Now consider

$$D_1 = \mathbb{C} \setminus \{1\}$$

and $f_1 = \frac{1}{1-z}$. This is a (reciprocal of a) polynomial, and so as long as the denominator does not vanish, it will be analytic everywhere, that is, an *entire* function. But writing out the Taylor expansion of this function at zero gives exactly the same thing as for f, and in fact

(D_1,f_1) is an analytic continuation of f. We can now observe that while f lived only strictly inside the unit disk, f_1 (to which f is equal!) lives *everywhere* except at points *on* the unit circle. The function thus extends to all the complex plane (except for the points on the unit circle)! The moral of the story is that just because the power series representation lives only in a certain region does not mean that the analytic function that it represents lives only there. It is worth noting, however, that analytic extension is not always possible.

Now, the gluing condition stated above already suggests the close connection with sheaves. We also have the required uniqueness. If (D_1,f_1) and (D_1,f_2) are two analytic extensions of (D,f), defined on the same domain D_1, then $f_1 = f_2$. Note that it is immediate from the fact that both f_1 and f_2 are analytic extensions of f that if you take $f_1|_{D \cap D_1}$, this is equal to $f|_{D \cap D_1}$ which is in turn equal to $f_2|_{D \cap D_1}$. To check that the two analytic functions are equal on a domain (specifically the restricted domain in this case), all you have to do is find one convergent sequence in the domain that converges to a point in the domain, and verify that for each point of the sequence the two analytic functions take the same value.

So far, we mentioned only a direct analytic extension from one domain to another, but we can extend this idea from a pair of domains to any finite number of domains. The idea is that you do not insist that they *all* intersect—only that there is an ordering such that every member intersects with the next. More formally: suppose we have pairs (D_α,f_α), where the D_α are open connected domains and the $f_\alpha : D_\alpha \to \mathbb{C}$ are analytic, as well as a total ordering of the indexing set consisting of the $\{\alpha\}$. For simplicity, let A be finite, that is, $A = \{\alpha_1 < \alpha_2 < \cdots < \alpha_m\}$. Now, for all i, let $D_{\alpha_i} \cap D_{\alpha_{i+1}} \neq \emptyset$ and

$$f_{\alpha_i}|_{D_{\alpha_i} \cap D_{\alpha_{i+1}}} = f_{\alpha_{i+1}}|_{D_{\alpha_i} \cap D_{\alpha_{i+1}}}.$$

In other words, you just have a chain of direct analytic continuations. In this case, we say that $(D_{\alpha_m},f_{\alpha_m})$ is the (indirect) analytic continuation of $(D_{\alpha_1},f_{\alpha_1})$. This process of generating a chain of direct analytic continuations—that is, interlocking or overlapping regions of extension of various locally defined functions whereby the domain of that function is extended step by step—might be pictured in the following way (where to each D_{α_i} is of course attached a function f_{α_i}):

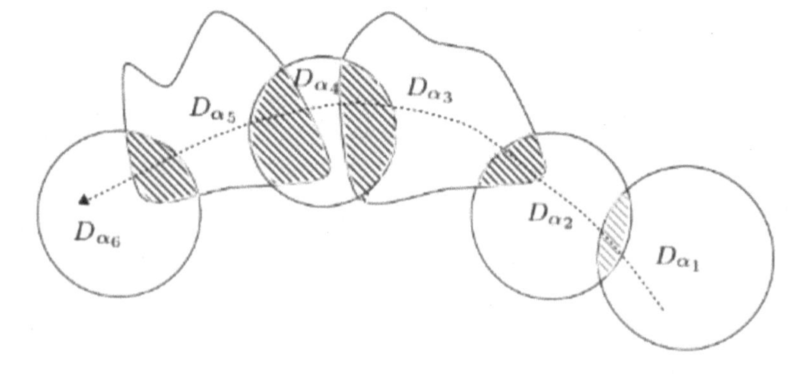

Now, something interesting can happen in such situations. We can have a chain of analytic continuations such that it happens that the final domain is the same as the first domain, but the function we get will be completely different. What is in fact happening in this case is that these functions are two so-called *branches* of an analytic function. (So this process of analytic continuations helps you find all possible branches of an analytic function.) Note also that this process is only nontrivial and useful if there are singularities. If the function were already entire, there would be nothing to extend.

Stepping back a bit, and recalling the previous example: smooth and complex analytic manifolds are particular examples of what are called *ringed spaces*, where a ringed space X is a topological space equipped with a fixed sheaf of rings, in which setting such a sheaf is referred to as the structure sheaf. In addition to leading to the notion of a Riemann surface, the consideration of analytic functions of one complex variable leads to the notion of a *bundle*. The next example provides a closer inspection of this notion, and covers a few notable abstract results that emerge in that context.

Example 210 In the previous example, we mentioned that the consideration of holomorphic functions of one complex variable can lead to the notion of a *bundle*. In what follows, we introduce the concept of a bundle, a notion that in a sense can be thought of as capturing the underlying set-theoretic structure of the sheaf idea. We will begin by considering a bundle of sets (over the discrete base space I).

Suppose we start with a coproduct $\coprod Y_i$ of sets, that is, the disjoint union $\bigsqcup_{i \in I} Y_i$ or $\bigcup_i Y_i \times \{i\}$, where the additional factor $\{i\}$ enforces disjointness. Then a bundle is the *entire* structure depicted in the following figure:

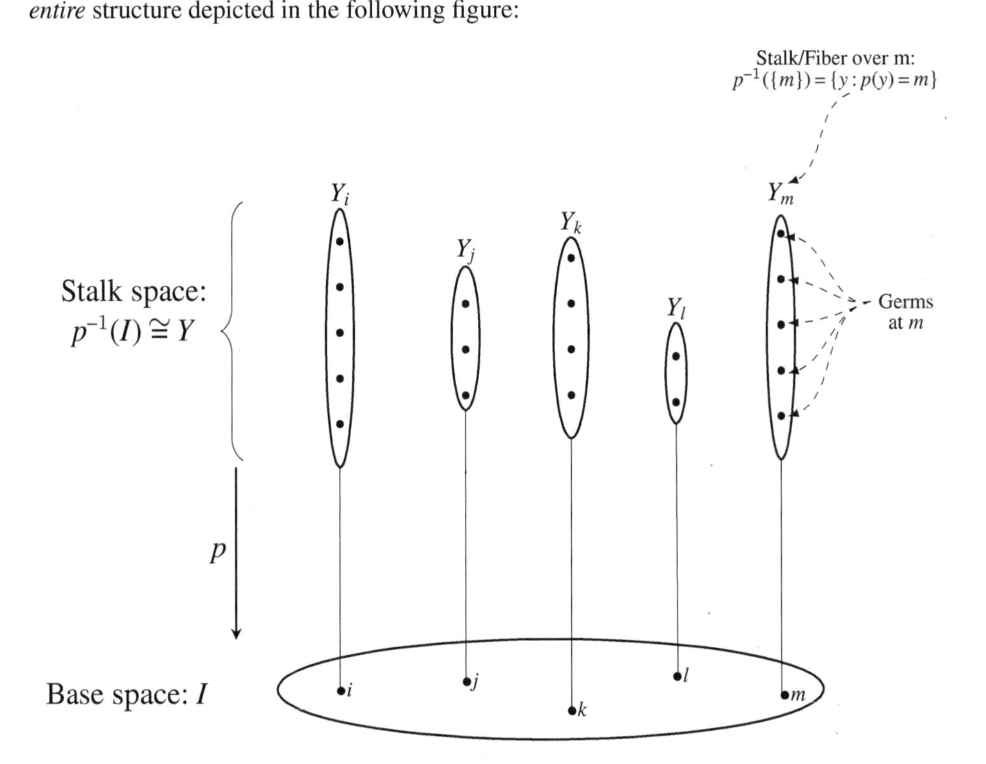

Let us unpack this figure a bit, and clarify some of the terminology. Y_i is called the *stalk*, or *fiber*, over i. The map $p : Y \to I$ is distinguished by the fact that if $y \in Y$, then there exists precisely one Y_i such that $y \in Y_i$, in which case we set $p(y) = i$. This just means that every member of the Y_i gets sent to i. Thus, the stalk or fiber Y_i can be got as the inverse image under the projection map p of $\{i\}$:

$$p^{-1}(\{i\}) = \{y \mid p(y) = i\} = Y_i.$$

A *stalk* over some $i \in I$ can also be seen as consisting of a collection of *germs*, where the germs at i are just the members of Y_i, represented by dots in the above picture. For simplicity, focusing on just a part of I, namely $U = \{j, k, l\}$, suppose we have

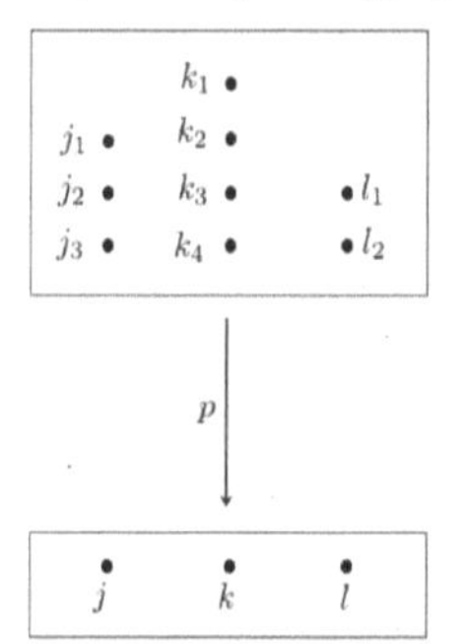

where, of course, each element of the space "above" is mapped via p to the element of I directly below it—for example, $p(j_1) = p(j_2) = p(j_3) = j, p(l_1) = p(l_2) = l$, and so on. Then the fiber or stalk over l, for instance, is of course just $\{l_1, l_2\}$, while the fiber over j is just $\{j_1, j_2, j_3\}$, and so on. Referring back to the main picture, the set $Y = \{y \mid y \in Y_i \text{ for some } i\}$ consisting of all the elements in *any* fiber over I, is then called the *stalk space* (or *l'espace étalé*) of the bundle. And the entire structure is then called a *bundle* (of sets) over the base space I.

This bundle construction is possible whenever there are functions. More specifically, given $p : Y \to I$ an arbitrary function from a set Y to I, we can define the bundle of sets over I with stalk space Y by first defining Y_i as the preimage $p^{-1}(\{i\})$ for each $i \in I$, and then taking

$$\{p^{-1}(\{i\}) \mid i \in I\} = \{Y_i \mid i \in I\}.$$

But then it should be obvious that a bundle of sets over I is basically just a function with codomain I, and so equivalent to set-valued functors defined on I, where I is a discrete category. In other words, this is nothing other than the slice (comma) category $(\mathbf{Set} \downarrow I)$ of functions with codomain I, though it is also common to denote the category of bundles over I by $\mathbf{Bn}(I)$. The reason for discussing such things is that a sheaf can be defined as a bundle with some extra topological structure. We first redefine a bundle in a topological context as follows:

Definition 211 For any topological space X, a *bundle over* X is a topological space Y equipped with a continuous map $p : Y \to X$.

As we did a moment ago, you can continue to imagine Y as "sitting above" X, and the map p as *projecting* the points of Y onto their "shadows" $p(y) \in X$. The bundle is all of this—that is, a triple (Y, p, X), where Y is the total space, X the base space, and p the projection map.

Notice from the definition that this is just to say that (topological) bundles are the objects of the slice category $(\mathbf{Top} \downarrow X)$, where an arrow $f : p \to p'$ is a continuous (in fact, open) map $f : Y \to Y'$ such that $p' \circ f = p$, where p' is a map from Y' to X. The close connection between the resulting category $(\mathbf{Top} \downarrow X)$ (or sometimes just $\mathbf{Top}(X)$) and the category of sheaves over X will emerge in what follows. It will turn out that *every* sheaf can be regarded as arising from a bundle; and, conversely, every sheaf gives rise to a bundle. To approach this important connection, we first introduce another important notion:

Definition 212 A *cross-section* (or just *section*) of a bundle $p : Y \to X$ is a continuous map $s : X \to Y$ such that $p \circ s = \mathrm{id}_X$.

In general, a section s for a map p can be regarded as a procedure that at once picks out an element from each of the fibers of p. In terms of the earlier case, where we were dealing with a set bundle over a discrete space, we can depict one such (cross) section s_1 as follows:

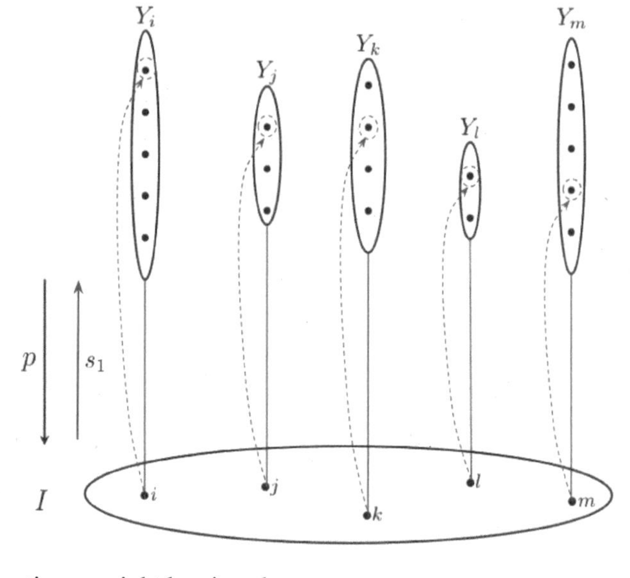

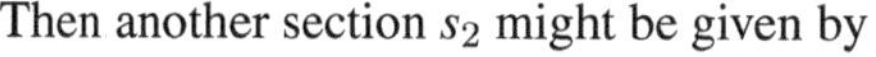
Then another section s_2 might be given by

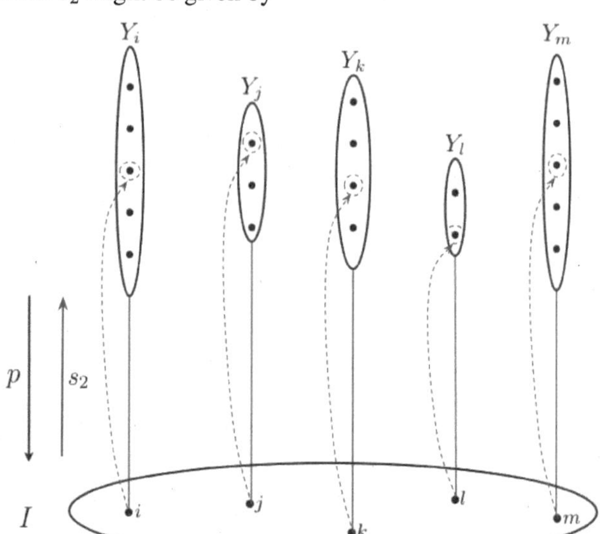

Still more evocatively, we might picture a particular section with something like

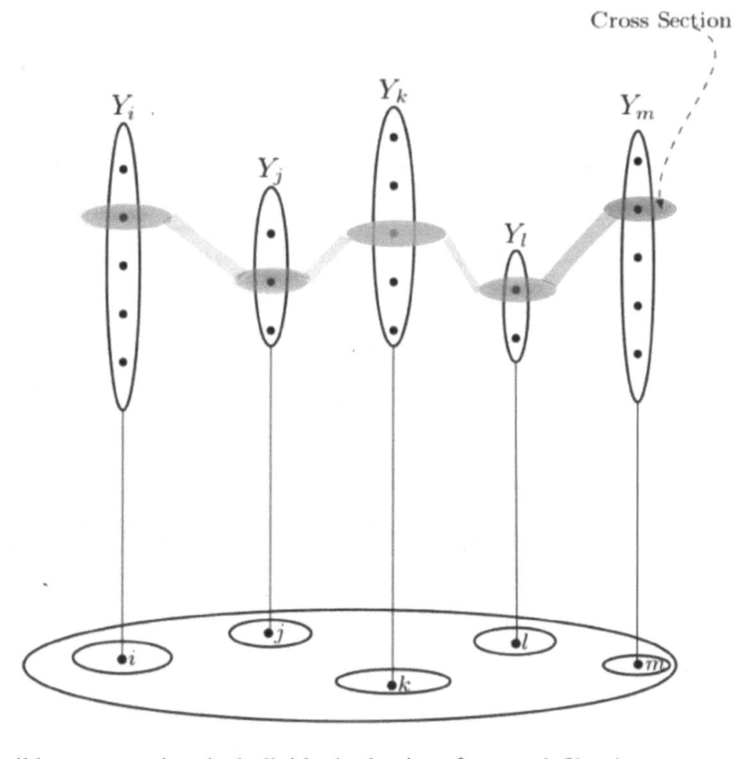

where the ribbon connecting the individual selections from each fiber is meant to anticipate the gluing process for the same sheaf, that is, where the various individual selections are glued together (via the topology of Y) to create a section of a larger open set (the entire ribbon representing a global section). Altogether, we think of a bundle as the indexed family of fibers $p^{-1}(x)$, one per point $x \in X$, glued together by the topology of Y.

8.1.2 Bundles to (Pre)Sheaves

A little more generally now, suppose we have a bundle (Y, p, X). If we consider some open subset U of the base space X for the bundle $p : Y \to X$, then p clearly restricts to a map $p_U : p^{-1}(U) \to U$, and this map will itself yield a bundle (now over U). Then, the following diagram, with horizontal arrows as inclusions, will be a pullback in **Top**, the "best" way of completing two given morphisms into a commutative square:

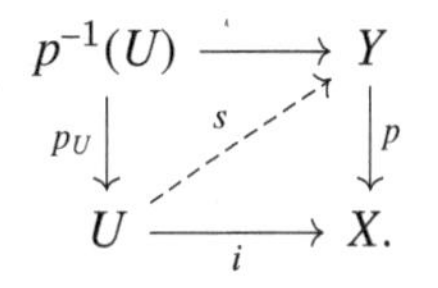

But this lets us define a cross-section (section) s of this bundle p_U: namely, a section of the bundle p over U is a continuous map $s : U \to Y$ such that the composite $p \circ s$ is equal to the *inclusion* $i : U \to X$, that is, $p \circ s = \mathrm{id}_U$.[123] In the case of discrete spaces—where every

123. Note that it may very well happen that a map admits a locally defined section over a subset $U \subseteq X$, but not a global section (i.e., a section over all of X).

subset is open, making p automatically continuous—things are especially easy to present. For instance, suppose we have p on $U = \{j, k, l\}$ as earlier

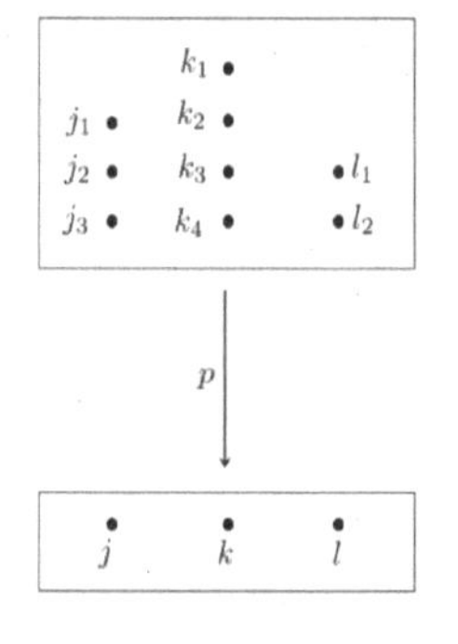

Then a section will of course just map each point of U to a point in Y that sits directly above it. For instance, one section s_1 for such a p over U will be given by

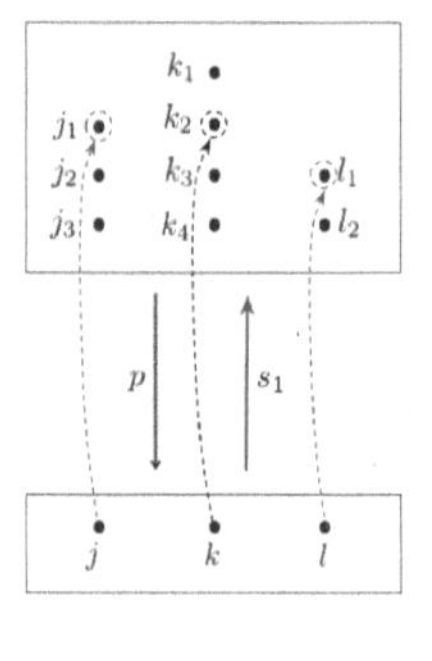

while another s_2 might be given by

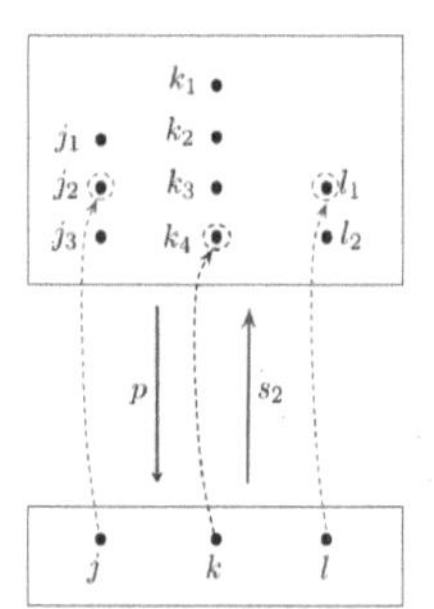

Observe that there would be a total of thirty-six distinct sections over $U = \{j, k, l\}$, given such a map p. Similarly, for more general spaces, here is a picture of sections over an open $U \subseteq X$

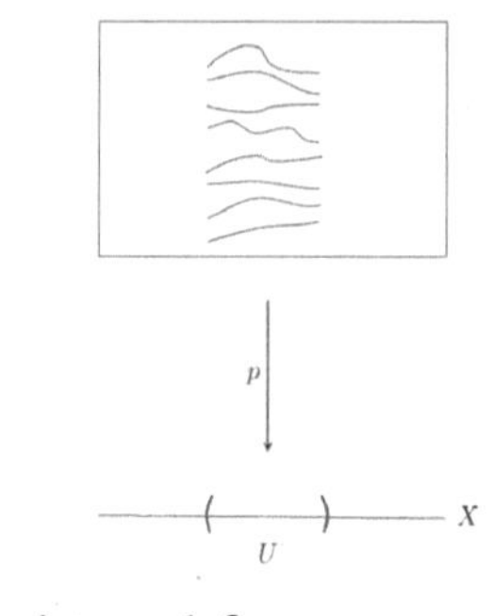

Collecting the sections together lets us define

$$\Gamma_p(U) = \{s \mid s : U \to Y \text{ and } p \circ s = i : U \subseteq X\}$$

the set of *all* cross-sections over U. But now observe that whenever $V \subseteq U$, we will have the induced (restriction) operation $\Gamma_p(U) \to \Gamma_p(V)$, restricting a function to a subset of its domain. Via this assignment on objects (open sets) U, and the induced restriction operation, altogether this just tells us that $\Gamma_p(-)$ defines a presheaf

$$\Gamma_p : \mathcal{O}(X)^{op} \to \mathbf{Set}.$$

On objects U, $\Gamma_p(U)$ supplies all the sections over U, and a given element $s \in \Gamma_p(U)$ will just be a choice of an element from each fiber over U. On arrows, the presheaf Γ_p acts by *restriction* (for every subset inclusion $V \subseteq U$). Given an inclusion $V \subseteq U$, and given a section s over U, we just restrict this to what s does on V,

$$\Gamma_p(U) \to \Gamma_p(V)$$
$$s \mapsto s|_V.$$

It should be clear that, given a function s on U, one can test locally whether or not s amounts to a section. But this locality suggests something more: namely that Γ_p is not just a presheaf, but in fact forms a sheaf on X, called the *sheaf of cross-sections* (or *sheaf of sections*) of the bundle p.

To see that Γ_p is in fact a sheaf, not just a presheaf, we need to see that it satisfies the sheaf condition for every cover. Let us explore how this works for the particularly simple case of the discrete sets we have been working with. Suppose given the sets $U_1 = \{j, k\}$ and $U_2 = \{k, l\}$, so that, together, we have a covering of the set $U = U_1 \cup U_2 = \{j, k, l\}$. We have a presheaf Γ_p and a cover. We can then define a matching family for the cover: this will, of course, just be a section t_1 given over U_1, such as

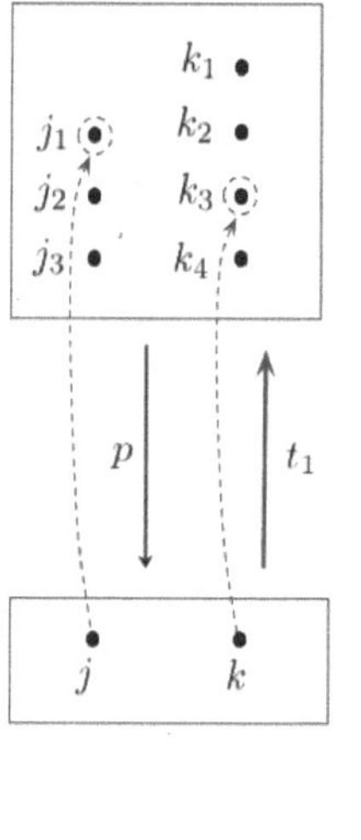

and a section t_2 over U_2, such as

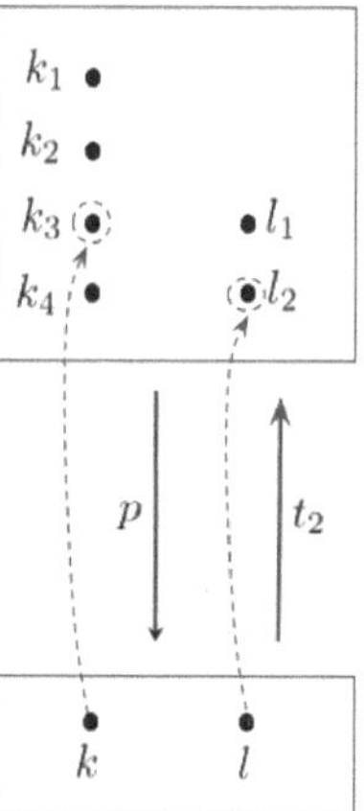

where these agree on the overlapping set $U_1 \cap U_2 = \{k\}$, as these particular sections t_1 and t_2 do (since they both map k to k_3). Pairs of sections that match on the overlap, such as t_1 and t_2, can then be *glued* together to yield a single section $t \in \Gamma_p(U_1 \cup U_2)$ over the entire set $U = U_1 \cup U_2$:

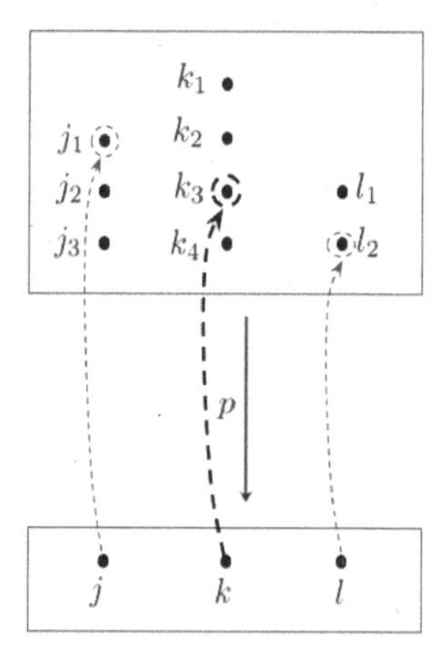

Moreover, observe how this section t is such that $t|_{U_1} = t_1$ and $t|_{U_2} = t_2$. In this way, we can build up sections over a space by gluing together sections given over local parts.

This example suggests how we can build a sheaf from the presheaf Γ_p. This is a general procedure: proceeding in fundamentally the same way as above, every bundle over X will give rise to a sheaf on X. Moreover, we know that in the category of bundles $\mathbf{Bn}(X)$ on X, given (objects) bundles $p: Y \to X$ and $p': Y' \to X$, a morphism $p \to p'$ from the first to the second is just a continuous map $f: Y \to Y'$ making the triangle $p' \circ f = p$ commute. But each such map $p \to p'$ of bundles over X (or maps in the slice category) will induce a map $\Gamma_p \to \Gamma_{p'}$ of (pre)sheaves on X, just as one would expect (a morphism of presheaves). This means we actually have a functor Γ going from the category of bundles on X to presheaves on X:

$$\Gamma : \mathbf{Bn}(X) \to \mathbf{Set}^{\mathcal{O}(X)^{op}}.$$

Actually, this really gives a functor from bundles to *sheaves*, a fact we will highlight in a moment. Earlier, though, we mentioned that we could also move in the other direction—that is, that *every* sheaf is in fact a sheaf of sections of a suitable bundle. The next section is devoted to seeing how every (pre)sheaf on X can be regarded as a (pre)sheaf of sections of some bundle.

8.1.3 (Pre)Sheaves to Bundles

Suppose we start with a presheaf

$$P : \mathcal{O}(X)^{op} \to \mathbf{Set}.$$

We want to use P to construct a bundle over X. We will want to construct a collection of sets, indexed by the points $x \in X$, take a union of these sets, and then place a topology on this, leaving us with a space from which we can then define a map down to X. This will give us our bundle.

To see how to construct the relevant sets in our desired collection, first observe that the presheaf P acts on *open sets*, so it does not yet give us sets for the *points* of X. This is where the notion of a *germ* comes in. The full demonstration that every sheaf is a sheaf of cross-sections of a suitable bundle relies heavily on this idea of a germ of a function. Earlier, with our base space discrete, we thought of germs as basically elements of a set. But the more general idea of germs is that functions that agree in a neighborhood of the given "germ point" are to be treated as equivalent. Two continuous functions f and g are said to have the same germ at a point x provided they agree in an open neighborhood of x. Intuitively, this language of "germs" at a point x can be thought of as naming what data (such functions) look like under a microscrope zeroing in on x. Notice that

$$(\text{germ}_x f = \text{germ}_x g) \Rightarrow (f(x) = g(x)),$$

that is,

> if two functions have the same germ at a point, they must have the same value at that point,

but the converse does not necessarily hold, that is,

> just because two functions have the same value *at* a point does not mean that they will agree *near* or *around* that point (that is, are *locally equal*).

This notion is suggested by the following picture of the functions $y=x^2, y=|x|, y=x$, and $y=-x^2$:

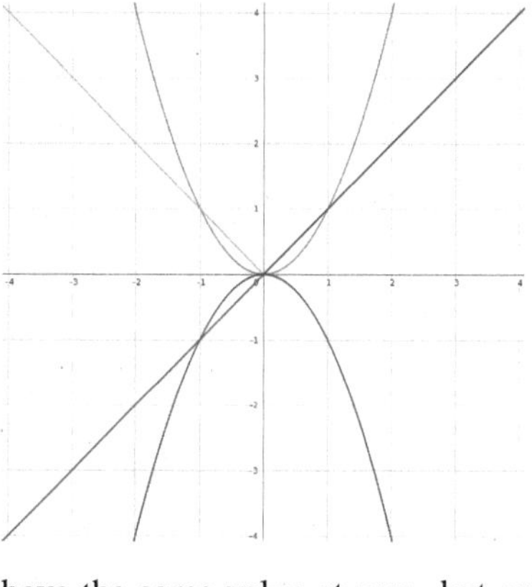

All of the functions above have the same value at zero, but only $y=x^2$ and $y=-x^2$ *agree in a small neighborhood* around 0—so of the four, only those two will have the *same germ* at zero. Two functions provide the same germ precisely when they become equal when we restrict down to some neighborhood of x. This explains why we suggested the germ of a function can accordingly be thought of as what you get when you focus your microscrope and increasingly magnify the point x. Relating this to the example involving analytic continuation, two holomorphic functions $h, k: U \to \mathbb{C}$ are similarly said to have the same germ at a point $a \in U$ if their power series expansions around a are the same, which is to say that h and k agree on some neighborhood of a.

Now suppose given a presheaf $P: \mathcal{O}(X)^{op} \to \mathbf{Set}$ on a space X, a point x, two open neighborhoods U and V of x, and two elements $s \in P(U)$ and $t \in P(V)$. Then we can define the following notion:

Definition 213 $s \in P(U)$ and $t \in P(V)$ *have the same germ at* x provided there exists some open set $W \subseteq U \cap V$ containing x and s and t agree with respect to this set, that is,

$$s|_W = t|_W \quad \in P(W).$$

The relation of "having the same germ" at a point is in fact an equivalence relation, and we call the equivalence class of any one such section s the *germ* of s at x, denoted $\text{germ}_x s$ or $[s]_x$, if we can forget about the neighborhood on which it is defined. Suppose, then, we denote by

$$P_x = \{\text{germ}_x s \mid s \in P(U), x \in U, U \text{ open in } X\},$$

the set of all germs at x. By taking all such functions that have the same germ (at a point) and identifying them, we just get back the *stalk* P_x of all germs at x. In a sense, the notion of a "stalk" P_x of a sheaf P is a generalization of the germ of a function, informing us about the properties of a sheaf near a point x. Recall that the presheaf P does not yield sets for the points of X, but rather for the open sets U of X. We will want to narrow in on smaller and smaller open neighborhoods $U' \subseteq U$ of a point x. But in looking at smaller and smaller neighborhoods $U' \subseteq U$ of a point x, it will not suffice to take any *single* neighborhood, since a smaller one can always be taken. So we need to take some sort of limit, putting some universal property in play. If $U' \subseteq U$, we know that we will have the induced presheaf restriction map $P(U) \to P(U')$. As we range over all the neighborhoods of x, then, we will want to take the *colimit* of all the sets $P(U)$. Accordingly, the set P_x of all germs at x is the stalk of P at x, where the stalk is defined—over all open $U \subseteq X$ that contain the given point x—as the colimit (or *direct limit*)

$$P_x = \varinjlim_{U \ni x} P(U).$$

An element of the stalk P_x will then be given by a germ of sections at x, that is, a section over a neighborhood of x, where two such sections will be regarded as equivalent provided their restrictions agree on a smaller neighborhood.

More explicitly, considering the restriction of the functor P along the open neighborhoods of x, the functions $\text{germ}_x : P(U) \to P_x$ just takes a section $s \in P(U)$ defined on an open neighborhood U of x to its germ at x, generalizing the usual notion of a germ (of functions), where the set P_x of all germs at x is the colimit, with germ_x the colimiting cone, of the functor P restricted to open neighborhoods of x.

Now, we also have that any morphism $h : P \to Q$ of presheaves (i.e., any natural transformation of functors) will induce at each point $x \in X$ a unique function $h_x : P_x \to Q_x$ such that the following diagram commutes for any open set U with $x \in U$:

$$\begin{array}{ccc} P(U) & \xrightarrow{h_U} & Q(U) \\ \downarrow{\scriptstyle germ_x} & & \downarrow{\scriptstyle germ_x} \\ P_x & \dashrightarrow_{h_x} & Q_x. \end{array}$$

But then notice that the assignments $P \mapsto P_x$, $h \mapsto h_x$ altogether just describe a functor $\mathbf{Set}^{\mathcal{O}(X)^{op}} \to \mathbf{Set}$, a functor you can think of as *taking the germ at x*.

Now that we have our sets P_x of germs, we can range over the $x \in X$ and further combine the various sets P_x of all germs in the disjoint union (which we will call Λ_P) over $x \in X$:

$$\Lambda_P = \coprod_x P_x = \{\text{all germ}_x s \mid x \in X, s \in P(U)\}.$$

Using this Λ_P, a disjoint union, there will be a unique projection function

$$\pi : \Lambda_P \to X$$

that projects each $\text{germ}_x s$ down the point x where it is taken. With such a map π and the set Λ_P, observe how we are making progress toward a description of a presheaf P as a bundle over X. However, it remains to put a topology on Λ_P, in order to speak about the continuity of π, and so to finish the construction. We can use the equivalence classes of sections—that is, germs—to form a basis for the relevant topology.

Notice that each $s \in P(U)$ determines a function from U to Λ_P taking all $x \in U$ to the $\text{germ}_x s$, a function that is in fact a section of π, in the sense that $\pi \circ s = \text{id}_U$. In this manner, each element s of the original presheaf can be replaced by an actual function to the set Λ_P of germs, in the following way: for any open set $U \subseteq X$, $s \in P(U)$, and $x \in U$, the element $s \in P(U)$ defines a section $\bar{s} : U \to \Lambda_P$ mapping x to $\text{germ}_x s \in P_x$, the germ of s at x. But then the collection

$$\{\bar{s}(U) \mid U \text{ open in } X, s \in P(U)\}$$

of subsets of Λ_P in fact forms a basis for a topology on Λ_P. For a basis for the topology, we first take open sets around each point in Λ_P. Recall that a point in Λ_P is just a germ, specifically a point $q \in \Lambda_P$ will lie in some set P_x for some x, making it the germ at x of some function $s \in P(U)$, where U is an open neighborhood of the point x. We thus get our base of open sets for Λ_P by taking all the image sets $\overline{s}(U) \subseteq \Lambda_P$. Then an open set of Λ_P will be a union of such sections $\overline{s}$.

With such a topology, Λ_P becomes a topological space, and $\pi : \Lambda_P \to X$ (and also every function $\overline{s}$) will be continuous (in fact, a homeomorphism). For any element $s \in P(U)$, moreover, the function $\overline{s} : U \to \Lambda_P$ will be a continuous section of Λ_P, where a section v of the space Λ_P is continuous precisely when every point $x \in X$ has a neighborhood U such that $v = \overline{s}$ on U for some $s \in P(U)$. To see this, let $s \in P(U)$ and $t \in P(V)$, and let these determine the sections $\overline{s}$ and $\overline{t}$, respectively, that agree at some point $x \in U \cap V$. Then, by the definition of a germ, we know that the set of all points $y \in U \cap V$ with $\overline{s}(y) = \overline{t}(y)$ will be an open set $W \subseteq U \cap V$ with $\overline{s}|_W = \overline{t}|_W$. Thus, each $\overline{s}$ will be continuous.

Moreover, if $h : P \to Q$ is a natural transformation between presheaves, the disjoint union of the functions $h_x : P_x \to Q_x$ yields a map $\Lambda_P \to \Lambda_Q$ of *bundles*, that is moreover continuous. Therefore, altogether, we have described a functor

$$\Lambda : \mathbf{Set}^{\mathcal{O}(X)^{op}} \to \mathbf{Bn}(X)$$

$$P \mapsto \Lambda_P$$

from presheaves to bundles.

We have thus sketched how we can turn presheaves into bundles. In the prior section, we saw how to turn bundles into (pre)sheaves. In the next section, we come to the important takeaway of all this, which involves what happens when we relate and then compose these functors.

8.1.4 The Bundle-Presheaf Adjunction

Theorem 214 For any space X, the bundle functor (assigning to each presheaf P the bundle of germs of P)

$$\Lambda : \mathbf{Set}^{\mathcal{O}(X)^{op}} \to \mathbf{Bn}(X)$$

is left adjoint to the sections functor (assigning to each bundle $p : Y \to X$ the sheaf of all sections of Y)

$$\Gamma : \mathbf{Bn}(X) \to \mathbf{Set}^{\mathcal{O}(X)^{op}}.$$

Recall that whenever you have an adjoint pair, with left adjoint $L : \mathbf{C} \to \mathbf{D}$ and right adjoint $R : \mathbf{D} \to \mathbf{C}$, this comes with a unit map

$$\eta : \mathrm{id} \Rightarrow RL$$

and a counit

$$\epsilon : LR \Rightarrow \mathrm{id}.$$

That $\Lambda \dashv \Gamma$ thus means that we will have unit and counit natural transformations

$$\eta_P : P \to \Gamma\Lambda(P),$$

for P a presheaf, and

$$\epsilon_Y : \Lambda\Gamma(Y) \to Y,$$

for Y a bundle.

We will not give a proper proof of this here,[124] but instead focus on these unit and counit maps. Consider for a given presheaf P on X the sheaf $\Gamma\Lambda_P$ of sections of the bundle $\Lambda_P \to X$, formed by first running the "take the germ" functor Λ and then following this with the "sections" functor Γ. For each open subset U of X, there will be a function $\eta_U : P(U) \to \Gamma\Lambda_P(U)$, taking $s \in P(U)$ to $\overline{s}$, and where restriction of s to the opens of U will agree with η, thus informing us that we have in fact just described a natural transformation

$$\eta : P \to \Gamma \circ \Lambda_P.$$

The next result is one of the main punch lines of all this.

Theorem 215 If the presheaf P is a sheaf, then the η shown above will be an *isomorphism* $P \cong \Gamma\Lambda_P$—that is, *every sheaf is a sheaf of sections.*

Proof. Suppose we have a presheaf $P : \mathcal{O}(X)^{op} \to \mathbf{Set}$. We know that for each open subset U of X, the function $\eta_U : P(U) \to \Gamma\Lambda_P(U)$, takes $s \in P(U)$ to $\overline{s}$, where, recall,

$$\overline{s} : U \to \Lambda_P, \qquad \overline{s}(x) = \text{germ}_x s, x \in U.$$

To show that η_U is an isomorphism, let us first show that it is injective, that is, that

$$\text{if } \overline{s} = \overline{t}, \text{ then } s = t,$$

for $s, t \in P(U)$. But $\overline{s} = \overline{t}$ just means that the germ of s and the germ of t agree on all points of U—that is, $\text{germ}_x s = \text{germ}_x t$ for each point $x \in U$. Thus, for each x, there will moreover be an open set $V_x \subseteq U$ such that $s|_{V_x} = t|_{V_x}$. But notice that the V_x supply a cover of U, which means that in

$$P(U) \to \prod_x P(V_x)$$

the elements s and t will have the same image. But then, supposing P is in fact a sheaf, we know that there can be at most one such element (i.e., $s = t$), which shows η_U injective.

To finish the proof, one must then show that $r : U \to \Lambda_P$ an arbitrary section of the bundle of germs over an open $U \subseteq X$ is in the image of η, altogether showing that η is an isomorphism. Details are left to the reader.[125] □

We can say even more about this special map η. Given a presheaf P, a sheaf F, and $\theta : P \to F$ any morphism of presheaves, then there will be a unique map $\sigma : \Gamma\Lambda_P \to F$ of sheaves making

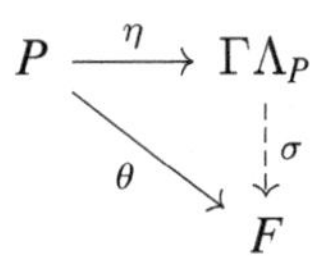

commute (i.e., $\sigma \circ \eta = \theta$). Another way of saying this is that the morphism η is *universal* from P to sheaves.[126]

124. A proof can be found in Mac Lane and Moerdijk (1994, II.6).

125. A proof can be found in Mac Lane and Moerdijk (1994, II.5).

126. See Mac Lane and Moerdijk (1994, II.5) for further details. All of this is in Grothendieck for the first time.

This important composite functor

$$\Gamma\Lambda : \mathbf{Set}^{\mathcal{O}(X)^{op}} \to \mathbf{Sh}(X) \tag{8.1}$$

is known as the associated sheaf functor, or the *sheafification functor*. The previous remark about the universality of η further entails that this sheafification functor $\Gamma\Lambda$ is in fact a left adjoint to the (full subcategory) inclusion functor

$$\mathbf{Sh}(X) \to \mathbf{Set}^{\mathcal{O}(X)^{op}}.$$

It is common to suggest that one think of $\Gamma\Lambda$ as taking each presheaf P to the "best approximation" $\Gamma\Lambda_P$ of P by a sheaf.[127]

Returning to the main theorem: constructing sheaves in this way as sheaves of cross-sections of a bundle suggests the further idea that a sheaf F on X can be replaced by the corresponding bundle $p : \Lambda F \to X$. But in constructing the topology on Λ_P, we mentioned that every function $\bar{s} : U \to \Lambda_P$ will not only be continuous, but will actually be a local homeomorphism. This basically means that each point of Λ_P will have an open neighborhood that is mapped via p homeomorphically onto an open subset of X. This leads to the following definition:

Definition 216 A bundle $p : E \to X$ is said to be *étalé* (or *étalé over* X) when p is a *local homeomorphism*, where this means that to each $e \in E$, there is an open set V, with $e \in V \subseteq E$, such that $p(V)$ is open in X and $p|_V$ is a homeomorphism (bicontinuous isomorphism) $V \to p(V)$.

When we have the étalé space given by taking $E = \Lambda_P$, this can be imagined as a space spread out over X, where its open sets "look like" the opens down in X. If we now let **Etale**(X), the category of étalé bundles over X, denote the (full) subcategory of the category **Bn**(X) of bundles over X, then we actually have the following powerful result:

Proposition 217 The adjoint functors Λ and Γ from before

$$\mathbf{Set}^{\mathcal{O}(X)^{op}} \underset{\Gamma}{\overset{\Lambda}{\rightleftarrows}} \mathbf{Bn}(X).$$

restrict—by restricting these functors to the subcategories **Sh**(X) and **Etale**(X)—to an equivalence $\mathbf{Sh}(X) \simeq \mathbf{Etale}(X)$, that is,

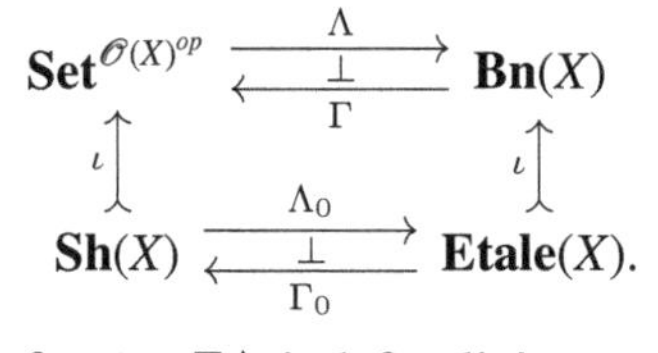

And while the sheafification functor $\Gamma\Lambda$ is left adjoint to the inclusion of sheaves into presheaves, $\Lambda_0\Gamma_0$ is right adjoint to the inclusion of étalé bundles into bundles.

The idea is that just as we saw how P is a sheaf precisely when η_P is an isomorphism, it can be shown that a bundle (Y, p, X) is étalé precisely when the counit morphism ϵ_Y is an isomorphism. The proof of the proposition above basically follows from general categorical reasoning regarding how the adjunction at the top restricts to consideration

127. We will see more of this functor in subsequent chapters.

of subcategories. One could also use the equivalence of categories between **Sh**(X) and **Etale**(X) to imply the fact we discussed earlier, namely that every sheaf can be viewed as a sheaf of cross-sections.[128]

8.1.5 Takeaways

In the last section, we described the basic adjunction involving presheaves and bundles. Just as sheaves are a special sort of "nice" presheaf, étalé bundles are a special sort of "nice" bundle. That there is an *equivalence* of the subcategories of sheaves (of sets) on a space X and the category of étalé bundles is like saying that "to be a nice presheaf is the same thing as being a nice bundle." One of the advantages of this perspective—allowing us to regard sheaves as étalé spaces (and the converse)—is that certain constructions may be simpler to define in one of the two settings. For example, pullbacks are very easily defined in the context of local homeomorphisms, so the *pullback sheaf* is more easily defined in this setting (conversely, the *direct image sheaf* is simpler to define in the context of sheaves seen as a set-valued functor).

The relationships explored in the previous section also allow us to take three equivalent ways of viewing morphisms between sheaves. Specifically, a morphism $h: F \to G$ of sheaves F, G can be described (equivalently) in terms of: (1) just a natural transformation $h: F \to G$ of functors; (2) a continuous map $h: \Lambda F \to \Lambda G$ of bundles over X; and (3) as a family $h_x : F_x \to G_x$ of functions on the fibers over each $x \in X$ such that, for each open set U and each $s \in F(U)$, the function $x \mapsto h_x(s(x))$ is a continuous $U \to \Lambda G$.

The point of view given by the equivalence of (1) and (3) in particular—namely of sheaf maps $h: F \to G$ in terms of stalk maps $h_x : F_x \to G_x$—can be rather useful, for it allows many facts about sheaves to be checked "at the level of stalks," something that cannot be done in general for presheaves (which also reflects, conceptually, the local nature of sheaves). But the more general definition of sheaves in terms of functors satisfying certain properties is arguably superior in that it allows us to consider many nontopological cases where there is no notion of an étalé space. Thus, while it is valuable to see the important close conceptual (and historical) connection between sheaves and étalé bundles, we will usually just insist on the more general approach, where a morphism $F \to G$ of sheaves will just be a natural transformation of functors, so that with **Sh**(X) we will just think of the category that has for objects all sheaves F (in this case, of sets) on X, and for morphisms the natural transformations between them.

Thus far, we have confined our attention to sheaves F on a topological space X. For such sheaves on spaces, we have been exploring basically two important candidate descriptions:

1. the *restriction-collation* description, and
2. the *section* description.

The first description was motivated by structures, such as classes of functions with certain "nice" properties (like continuity), that are defined locally on a space. The previous example introduced and developed the importance of the second of these two perspectives. The

128. We refer the reader to Mac Lane and Moerdijk (1994, II.5–6) for more details about the equivalence of these categories. The reader should note, however, that throughout Mac Lane and Moerdijk (1994), the authors write étale, when they mean étalé (the former being something else entirely).

idea here was that we took a sheaf to be some sort of principled way of assigning to each point x of the underlying space a set P_x consisting of all the germs at x of the functions being considered (where these germs were just equivalence classes identifying sets that look locally the same in neighborhoods of x), after which these sets P_x then get patched together by a topology to form a space (or bundle) projected onto X (a suitable function for this sheaf then being a cross-section of the projection of this bundle). According to this perspective, the sheaf P can ultimately be thought of as a set P_x that varies with the points x of the space.

After learning a new concept and seeing some descriptions and preliminary examples, it is important to think a bit about when and how this can go wrong (and acquire a store of nonexamples). We thus take the opportunity to consider what is *not* a sheaf—that is, when and why a construction fails to satisfy the sheaf conditions. After that, this chapter ends with a brief but important discussion of a general result allowing us to blur the distinction between presheaves and sheaves in the special case of posets.

8.2 What Is *Not* a Sheaf?

Even when structures are determined locally, sometimes local properties alone do not suffice to determine global properties. In such cases, we will not have a sheaf. A common example given to illustrate this is the set $B(U)$ of all *bounded* functions on U to $\mathbb{R}$—this will give a presheaf (functor) on U, but not a sheaf. The reason for this is that while the collation of functions that are bounded does indeed define a unique function on U, such a function may be unbounded. Presheaves are meant to carry local information, so they can fail to behave when the local information fails to transfer to global information, as boundedness does. Regarding the two sheaf conditions presented in definition 124 (chapter 5), notice how this construction satisfies the uniqueness condition, while the gluing condition is where the problem arises. This is somewhat typical—of the two defining conditions, the gluing condition is the one that seems to fail most often in practice.

Exercise 21 Make sense of the previous sentences by giving an example of a collection of bounded functions on subintervals whose collation is not bounded.

If a structure is not even determined locally, specifically in the sense that it does not even obey the first (restriction) condition, then it certainly cannot be a sheaf. An intuitive example of this might be given by the game of Scrabble™, where one thinks of this as follows: the 15×15 board with its squares labeled in some sensible way (with x and y coordinates of a tessellation, so that $(1, 1)$ would indicate the leftmost top corner square), may be regarded as a topological space, with a notion of covers. Then one might attempt to regard the assignment of sets of legal (English) word-forming letter combinations to subsets of the grid of squares (satisfying a further constraint capturing how words are to be read down and to the right) as a functor. However, while each inclusion of "opens" in the underlying grid of squares would have to induce a function restricting the word-forming letters assigned over a bigger region of the board to the word-forming letter assignments over a subregion, in general not every subword of a word is a word, so it is not clear how to make this work. Even if we agreed to treat individual letters as (legal) words, it is evident that the inclusions of open subsets will sometimes determine a "restriction" to a particular part of a word that does indeed form a word (now a *different* word), but on other occasions such a process will not result in a word at all. For instance, confining our attention for simplicity to a small 3×3 region of the board, and displaying a portion of the opens ordered by inclusion

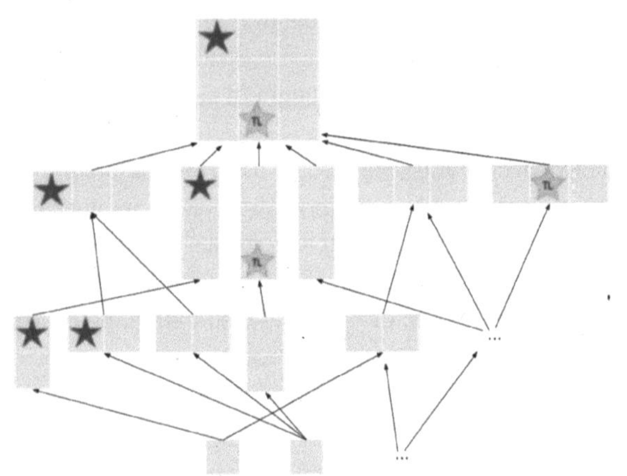

we might then regard a particular selection of possible letter assignments as follows

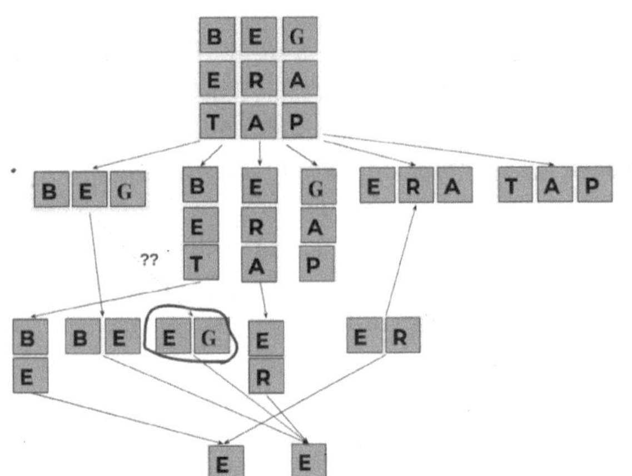

Here, with such an assignment, even though the letter assignments over the entire 3×3 portion result in valid words (in every possible three-letter combination, e.g., "beg," "bet," "era," "gap," "tap"), some of whose parts even themselves form words (like "be" in "beg"), it is not clear what to do with the (failed) restriction from the word "beg" down to "eg," which is not a word.

Even when we do have a presheaf, in general a presheaf can itself fail to be a sheaf in two (fundamentally independent) ways:

- **Nonlocality:** If a presheaf has a section $s \in F(U)$ that cannot be constructed from sections over smaller open sets in U—via a cover, for instance—then F fails to be a sheaf.
- **Inconsistency:** If a presheaf has a pair of sections $s \neq t \in F(U)$ such that when restricted to every smaller open set they define the same section, then F fails to be a sheaf. In other

words, informally, the presheaf has local sections that "ought to" patch together to give a unique global section, but do not.

It is often thought that the second sort of failure is somehow easier to understand. But that does not mean that there are not examples of the first sort of failure. A standard illustration of nonlocality is given by the following example. Consider X the topological space consisting of two points p, q, endowed with the discrete topology (i.e., every set is open). Then X consists of the open sets $\{p,q\}, \{p\}, \{q\}, \emptyset$, ordered by inclusion. We can form the *constant presheaf* P on X which assigns a set (or abelian group) to each of the four open sets and the identity map to each of the nine restriction maps (five plus the four trivial self-maps). For concreteness, let this presheaf assign $\mathbb{Z}$ to each of the sets.

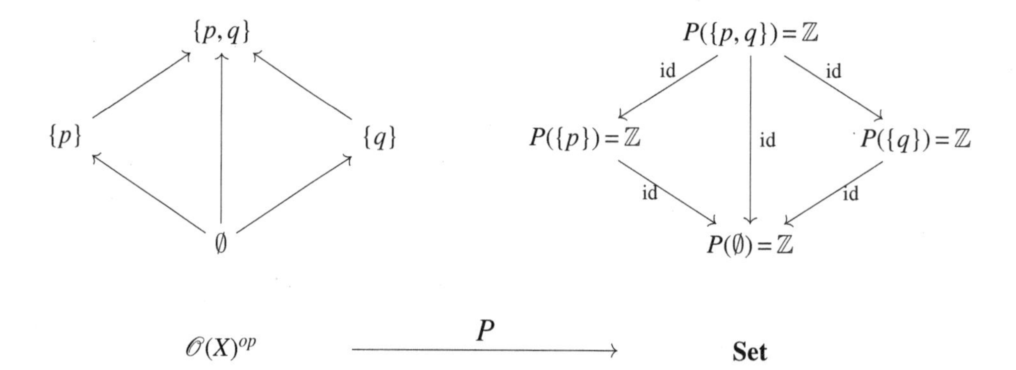

This presheaf P does indeed satisfy the gluing axiom. However, it fails to satisfy the locality/identity axiom, specifically with respect to the assignment on the empty set. The empty set is covered by the empty family of sets; but clearly any two local sections of P are equal when restricted to their common intersection in the empty family. If the locality axiom were satisfied, then any two sections of P over the empty set would be equal—however, this need not be true.

Another example of nonlocality is given by the following: take an open set $X \subseteq \mathbb{C}^n$, and for open $U \subseteq X$, let

$$S(U) := \{f : U \to \mathbb{C} \mid f \text{ is holomorphic}\}.$$

We then define the restriction maps by stipulating that for $V \subsetneq U$, we set the restriction $\rho^U_V = 0$, and set $\rho^U_U = \text{id}$. This S is a presheaf but it is not a sheaf precisely because it obviously has a section in $S(U)$ that cannot be built from sections over smaller open sets in U (for which f must be the 0 map).

To illustrate the second type of failure, inconsistency, consider again the constant presheaf P from before. We construct a new presheaf G over the same X with the same discrete topology, which is just like P except that we now let $G(\emptyset) = \{*\}$, where $\{*\}$ is a one-element set (the terminal object). We retain $\mathbb{Z}$ as our value assignment for the remaining nonempty sets. Now, however, for each inclusion of opens that has the empty set for domain, G will assign the unique map 0; otherwise, it just assigns the identity map as before.

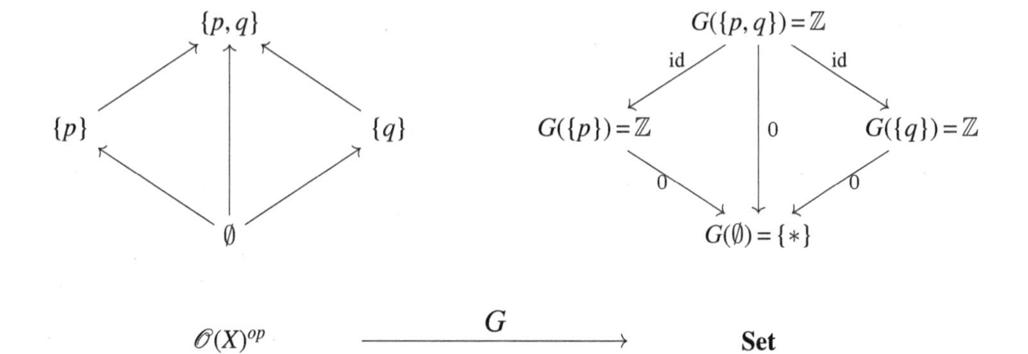

This presheaf G satisfies the locality/identity axiom, but now it fails to be a sheaf on account of not satisfying the gluing axiom. Let us unpack this. The entire set $X=\{p,q\}$ is covered by $\{p\}$ and $\{q\}$, which individual sets obviously have empty intersection. By definition, the sections on $\{p\}$ and $\{q\}$ will just be an element of $\mathbb{Z}$, that is, an integer. By selecting a section m (an integer) for our section over $\{p\}$ and n (another integer) for our section over $\{q\}$, such that $m \neq n$, we can easily see this violation. m and n must restrict to the same element over $\emptyset$ on account of the action of the trivial 0 restriction map; but then, if the gluing axiom were satisfied, because m and n restrict to the same element over their (trivial) intersection, we would need the existence of a unique section s over the union of the two sets (i.e., in $G(\{p,q\})$), which moreover restricts back to m on $\{p\}$ and to n on $\{q\}$. But the restriction maps from $G(\{p,q\})$ along $\{p\}$ and $\{q\}$, being the identity map in both cases, forces that $s=m$ and $s=n$, from which $m=n$, contradicting the assumption that the sections (integers) m and n were different.

Before moving on, we take the opportunity to briefly mention a common issue that may arise in the construction of sheaves in practice (construed as sheaves of sections), but one that is somewhat distinct from the failures of the previous two sorts. It concerns a situation where we may in fact be dealing with a sheaf, but *given* certain selections from the stalks or local sections, we may find that we simply cannot extend those assignments to produce a global section, since no matter what we assign to the remaining open set(s), we will run into inconsistency with respect to the other sections. This may simply be a problem with failing to select the "right" elements from the sets of possible values assigned to each underlying open. In the next chapter, we will see a number of explicit instances of this.

Finally, let us briefly look at a particularly interesting example (due to Goguen 1992) of a presheaf that is not a sheaf, for reasons distinct from the two discussed above.

Example 218 Taking our indexing (domain) category to be some base for some data assignment or observations, where this base is a poset, then the particular base consisting of intervals of natural numbers beginning with an "initial time" 0—where the various intervals beginning from 0 may represent periods of continuous observation of the system—can be described as

$$\mathscr{I}_0(\omega)=\{\emptyset, \{0\}, \{0,1\}, \{0,1,2\}, \dots\} \cup \{\omega\},$$

with ω representing the domain for observations or data assignment over an infinite time. Using such a base, we might consider a "fair scheduler" F for events a, b—that is, if a

occurs, then at some point b must occur, and vice versa. More formally, this means

$$F(\omega) = (a^+b^+ + b^+a^+)^\omega,$$

that is, we have a concatenation of strings where the components consist of some number of as followed by some number of bs or some number of bs followed by some number of as, on to infinity. For each natural number n, we will have that $F(\{0, 1, \ldots, n-1\}) = \{a, b\}^n$, which is just to say that we might have any combination of as and bs (including all as or all bs) for some finite interval. The point, however, is that while F is indeed a presheaf, if we had that F was a sheaf, then the sheaf condition would imply that $F(\omega) = \{a, b\}^\omega$, that is, that in the limit any combination is also possible. However, this contradicts the definition of F as a fair scheduler. Thus, it is not a sheaf. On the other hand, if we require the indexing set to be finite, then this F does satisfy the finite sheaf and gluing conditions. This (non)example is interesting because it suggests that interesting phenomena can appear "at infinity" that do not show up in the finite approximations.

Returning to more general considerations, as the section on bundles already suggested, there is a standard procedure for completing a presheaf to make it a sheaf. Since there are two fundamental ways a presheaf can fail to be a sheaf, this process—dubbed "sheafification"—can be roughly thought of as doing one of two things: (1) *discarding* those extra sections that make the presheaf fail to satisfy the locality condition; (2) *adding* those missing sections which, had they been present, would allow the local sections to glue together into a unique global section, satisfying the gluability condition. In other words, with respect to the second of these two, we are adding functions to the global set that restrict to compatible functions on each of the opens, and then, recursively, we continue adding the restrictions of the newly generated global functions. We have already seen what this sheafification abstractly looks like in the setting of bundles. In chapter 10, we will look more closely at the process whereby an arbitrary presheaf can be turned into a sheaf of the same type.

For now, let us just sketch how the sheaf G from a moment ago would be sheafified. Basically, this involves expanding $G(\{p, q\})$ to $\mathbb{Z} \oplus \mathbb{Z}$, thus defining a new sheaf H, then letting the restriction maps be the appropriate projection maps $\pi_i : \mathbb{Z} \oplus \mathbb{Z} \to \mathbb{Z}$, thereby defining $H(\{p\}) = image(\pi_1) = \mathbb{Z}$ and $H(\{q\}) = image(\pi_2) = \mathbb{Z}$. Everything else can be defined just as it was for G. The resulting functor H now satisfies the gluability condition and so is a sheaf, usually called the *constant sheaf* on X valued in $\mathbb{Z}$.

8.3 Presheaves and Sheaves in Order Theory

As so often when working with categories, things are greatly simplified when dealing with orders. As we started to really appreciate in chapter 6, highly abstract results and concepts in category theory can have a particularly friendly showing when specialized to orders. Sometimes, posets are especially nice to us, in that important general distinctions (such as that between presheaves and sheaves) can be "collapsed" when dealing with posets. This in turn can sometimes make the more general distinction easier to grasp.

Before concluding this chapter, we will thus briefly cover a highly useful result that relates presheaves and sheaves on a poset (regarded as a category):

Proposition 219 Presheaves on a poset are *equivalent* to sheaves over that poset, once the latter has been equipped with a suitable topology (the "Alexandrov topology").

We will just sketch how this works. First recall from definition 147 (chapter 6) the notion of a *downset* (and its dual, an *upper set*). Recall also how we defined *principal downsets*, denoted $\mathcal{D}_p$ (or just $\downarrow p$): these were sets of the form

$$\downarrow p := \{q \in \mathcal{P} \mid q \leq p\},$$

for $p \in \mathcal{P}$ (and dually for the principal upper sets). It is fairly straightforward to show how the principal downsets can generate a topology, called the Alexandrov topology (or, sometimes, *lower Alexandrov topology*, to distinguish it from the topology generated by the principal upper sets).[129] While the poset just supplies us with "points" p from the underlying set, the topology generated by the principal down (upper) sets supplies us with a way of looking at points now in terms of "opens" (the language a topological sheaf will understand). Recall that any downset can be written as the union of principal downsets (taking unions of downsets is the same as taking colimits of representables in the poset category of downsets), so using the principal down (upper) sets as a basis, we can form the collection of all downsets of a poset $\mathcal{P}$, denoted by $\mathcal{D}(\mathcal{P})$, and this will define a topology on $\mathcal{P}$, where we take $\mathcal{D}(\mathcal{P})$ as our open sets $\mathcal{O}(\mathcal{P})$. As $\mathcal{D}(\mathcal{P})$ is stable or closed under arbitrary intersections, the "closed sets" (i.e., the upper sets of $\mathcal{P}$) will be stable under arbitrary unions. But this means that the upper sets also forms a topology, usually called the *upper Alexandrov topology*. Thus, dual to our downsets, we denote by $\mathcal{U}(\mathcal{P})$ the collection of all upper sets of $\mathcal{P}$, and this will yield another topology on $\mathcal{P}$.

With these notions in hand, it can be shown that for $\mathcal{P}$ a poset, it's essentially the same to look at presheaves on $\mathcal{P}$ as it is to look at sheaves on $\mathcal{D}(\mathcal{P})$, while (dually) copresheaves (variable sets) on $\mathcal{P}$ are essentially the same as sheaves on $\mathcal{U}(\mathcal{P})$.

Theorem 220

$$\mathbf{Set}^{\mathcal{P}^{op}} \simeq \mathbf{Sh}(\mathcal{D}(\mathcal{P})),$$

and

$$\mathbf{Set}^{\mathcal{P}} \simeq \mathbf{Sh}(\mathcal{U}(\mathcal{P})).$$

In other words, this tells us (in the first case) that, given a presheaf defined on a poset $\mathcal{P}$, it can already be regarded as a sheaf when $\mathcal{P}$ is equipped with the natural Alexandrov topology (induced by the downset construction). Instead of going through a full proof of this, we will sketch one way of looking at why this is true. Because of the example to follow, we will also focus on $\mathbf{Set}^{\mathcal{P}} \simeq \mathbf{Sh}(\mathcal{U}(\mathcal{P}))$.[130]

Recall the very close relationship between $\downarrow p$ and p established by the downset embedding (which specialized the Yoneda embedding); and we just saw that the principal downsets of $\mathcal{P}$ can be used to form a topology. There is a similar result for upper sets, where $p \leq p'$ iff $\uparrow p' \subseteq \uparrow p$ (note the reversal of order); likewise, we just mentioned that

129. This can be described as a functor from **Pre** to **Top**.

130. What follows is largely derived from Goldblatt (2006). On the whole, there are many hand-wavy "proofs" that functors from posets are equivalent to sheaves on the upper set topology—something that can fundamentally be derived from the folk theorem that sheaves are determined by their values on a basis. However, it can be difficult to track down a proper proof of this. Curry (2019) includes a proof that, once appropriately dualized, yields the desired result.

this construction gives another natural (upper Alexandrov) topology on $\mathcal{P}$, the basis of which is given by the principal upper sets $\uparrow p$. Suppose, now, that you have a (variable set or copresheaf) functor $F^* \in \mathbf{Set}^{\mathcal{P}}$. Then F^* can be seen as just a collection $\{F_p^*\}_{p \in P}$ of indexed sets—where F_p^* is the image of $F^*(p)$—equipped with the maps $F_{pp'}^* : F_p^* \to F_{p'}^*$ whenever $p \leq p'$. We can take this functor to a sheaf F on the topology supplied by the upper sets $\mathcal{U}(\mathcal{P})$, by defining F on a basis of the Alexandrov topology and setting

$$F(\uparrow p) = F_p^*,$$

where of course $\uparrow p \in \mathcal{U}(\mathcal{P})$. Another way of thinking of this situation is that, given a functor $F^* : \mathcal{P} \to \mathbf{Set}$, via the inclusion functor

$$\iota : \mathcal{P} \to \mathcal{U}(\mathcal{P})^{op},$$

we would like to produce a sheaf on $\mathcal{U}(\mathcal{P})$

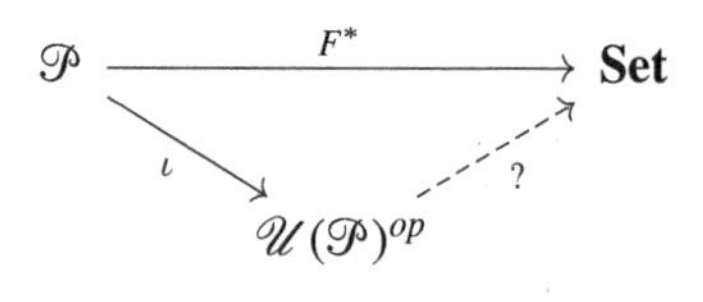

The most concise categorical way of accomplishing all this would be to use what is called *Kan extensions*, specifically the right Kan extension of F^* along ι, denoted $\mathrm{Ran}_\iota F^*$, to assign data to the opens in our poset "in a nice way" such that it is a sheaf, and declare that when this happens, then F^* itself can be seen as a sheaf. But instead, let's just describe things in more elementary terms. The idea is that, having set $F_p^* = F(\uparrow p)$, notice that whenever $p \leq p'$, we will have $\uparrow p' \subseteq \uparrow p$, so the map

$$F_{pp'}^* : F(\uparrow p) \to F(\uparrow p')$$

can be seen as the image of the inclusion $\uparrow p' \hookrightarrow \uparrow p$ under the action of the (contravariant) functor F. Note how the inclusion ι reverses the order, and then the underlying presheaf action of F reverses order once more. Now, for each $V \in \mathcal{U}(\mathcal{P})$, the set $\{\uparrow p \mid p \in V\}$ will be a cover of V, and so we can use

$$F(V) = \varprojlim_{p \in V} F(\uparrow p) = \varprojlim_{p \in V} F_p^*$$

to define $F(V)$ from the collection $\{F_p^*\}_{p \in V}$ (using the universal limiting cone for the relevant diagram formed from the various $F_p^*, F_{p'}^*, \ldots$). Going further, provided $U \subseteq V$, we will have $\{F_p^*\}_{p \in U} \subseteq \{F_p^*\}_{p \in V}$. The universal cone $F(V)$ for the diagram formed from the $\{F_p^*\}_{p \in V}$ will also be a cone for the diagram built out of the $\{F_p^*\}_{p \in U}$, thereby inducing a unique arrow $F(V) \to F(U)$, which we can use as our sheaf map F_U^V. This gives a rough sketch of how to think of (one direction of) this process.

Altogether, the main result informs us that we can pass freely from a sheaf over the topology $\mathcal{U}(\mathcal{P})$ induced on $\mathcal{P}$ and a "plain" copresheaf (variable set) on that original $\mathcal{P}$. By taking $\mathcal{P}^{op}$ instead—and recognizing that the upper sets of $\mathcal{P}^{op}$ are the same as the downsets of $\mathcal{P}$—we also have that we can move between a sheaf over the natural lower Alexandrov topology (given by $\mathcal{D}(\mathcal{P})$) and a plain presheaf on $\mathcal{P}$.

Of course, in the general case of categories that are not posets, things are not always so "nice" and presheaves are not automatically going to yield a sheaf in this straightforward way. In making a presheaf a sheaf in the general case, we are in a sense demanding that it "awaken to" the features of the underlying topology; specifically, in trying to find the sheaves among some presheaves, we want to restrict attention to those presheaves that are sensitive to the structure supplied by the cover. The above result that lets us blur the distinction between presheaves and sheaves, when $\mathcal{P}$ is a poset now equipped with its natural Alexandrov topology, in a sense tells us that presheaves in this setting are "automatically sensitive" to the structure of covers—and thus, a presheaf on a poset can already be regarded as a sheaf (with respect to its Alexandrov topology). The next chapter will make good use of this result. This chapter ends with a brief look at an example involving a different sort of poset where this result is of some utility.

Example 221 Let's start by defining a *(time)frame* as a structure $\mathcal{T} = (T, \leq)$ consisting of a nonempty set T of "times" (instants, events) on which the relation $\leq$ forms a reflexive and transitive order—that is, $\mathcal{T}$ is a preorder.[131] At the outset, we do not insist that a (time)frame be a partial order (i.e., that $\leq$ be antisymmetric as well), so the equivalence relation defined on T by $t \approx s$ iff $t \leq s$ and $s \leq t$ will be nontrivial in general. We could call the $\approx$-equivalence classes the *clusters* of $\mathcal{T}$, ordered by setting $\hat{t} \leq \hat{s}$ iff $t \leq s$, where $\hat{t}$ is the cluster containing t. The resulting order is an antisymmetric one, enabling us to regard the (time)frame as a poset of clusters. Finally, a frame is said to be *directed* provided any two elements have an upper bound, that is, for all $t, s \in T$, $\exists v \in T$ such that $t \leq v$ and $s \leq v$.

In section 7.2 and the appendix, we discuss some features of modalities and propositional modal logic. Propositional modal logic consists of sentences constructed from proposition letters $p, q, r, \ldots$, the usual Boolean connectives, and the modal operator $\Box$ (which is typically interpreted, in the context of tense logics which we are about to use, as "it will always be"). The Greek philosopher Diodorus of Megara is often thought to have held that the modalities "necessity" and "possibility" were best definable in terms of time. For instance, Diodorus held that the necessary should be understood as that which *is* (now) and *will always be* the case. One might then define $\Diamond$—interpreted as "it will (at some time) be"—as $\neg\Box\neg$. In terms of contemporary modal logics, while it is common for tense logics to regard time as an irreflexive ordering (so that "at all future times" does *not* include the present moment), Diodorus's approach suggests a temporal interpretation of $\Box$ that uses reflexive orderings instead, and we can apply this to nonlinear time structures. The reflexivity of $\leq$ in our time frame will give $\Box$ the Diodorean intepretation of "*is* (now) and *always will be*."

As Goldblatt (1980) first showed, one can develop a Diodorean logic of n-dimensional (for $n \geq 2$) Minkowskian special-relativistic spacetime and show that this is exactly the modal logic **S4.2**. In more detail, if $x = \langle x_1, \ldots, x_n \rangle$ is an n-tuple of real numbers, then let

$$\mu(x) = x_1^2 + \cdots x_{n-1}^2 - x_n^2.$$

Then, for $n \geq 2$, n-dimensional spacetime is the frame

$$\mathbb{T}^n = (\mathbb{R}^n, \leq),$$

131. The following example is derived from Goldblatt (1980, 1992).

with $\mathbb{R}^n$ the set of all real n-tuples, and the order $\leq$ defined, for all x and y in $\mathbb{R}^n$, as follows:

$$x \leq y \text{ iff } \mu(y-x) \leq 0 \text{ and } x_n \leq y_n$$

$$\text{iff} \sum_{i=1}^{n-1} (y_i - x_i)^2 \leq (y_n - x_n)^2 \text{ and } x_n \leq y_n.$$

The frame $\mathbb{T}^n$ is partially-ordered and directed. Then the usual Minkowski spacetime of special relativity is given by $\mathbb{T}^4$, for which a point will represent a spatial location $\langle x_1, x_2, x_3 \rangle$ at time x_4. In this case, the interpretation of $x \leq y$ is that a signal may be sent from event x to event y at a speed at most that of the speed of light, entailing that y comes after, or is in the causal future of, x. In other words, the (reflexive) relation is given by "can reach with a lightspeed-or-slower signal." For simplicity, we can look at the frame $\mathbb{T}^2$, and easily visualize the future cone $\{z \mid x \leq z\}$ for a point $x = \langle x_1, x_2 \rangle$, where the future cone contains all points on or above the directed rays of slopes ± 1 beginning from x (using a coordinate system for which speed of light is one unit of distance per time unit).

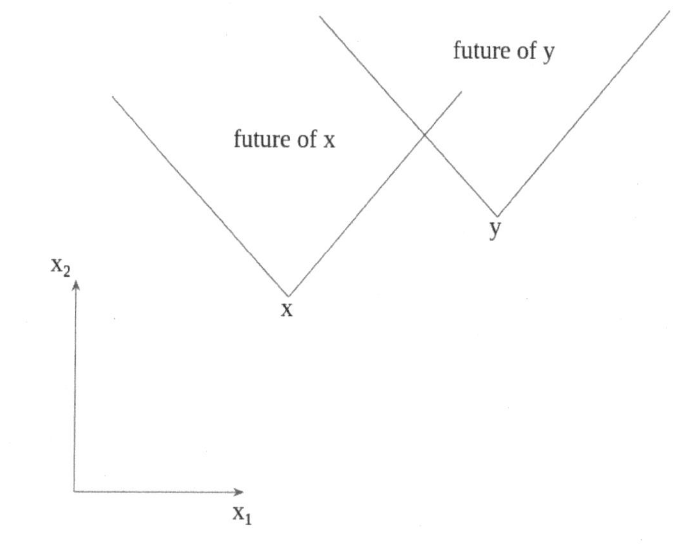

One can observe that the future cones of any two points will eventually intersect, which entails that the underlying order is *directed*, in the sense that for any two locations x, y, there is a third that is in the future of both x and y. This directedness forces the Diodorean interpretation of $\Box$ to validate the **S4.2** axiom schema $\Diamond\Box\phi \to \Box\Diamond\phi$, where **S4.2** just arises by adding that axiom schema to the usual axioms of **S4**.[132] Goldblatt (1980) shows that *each* of the frames $\mathbb{T}^n$ has the logic **S4.2** for its Diodorean modal logic, regardless of spatial dimensions.

Now, following Goldblatt, we may also call $T' \subseteq T$ *future-closed* under $\leq$ provided whenever $t \in T'$ and $t \leq s$, then also $s \in T'$. Note that this is just to say that T' is an upper set! In this case, $\mathcal{T}' = (T', \leq)$ will be a *subframe* of $\mathcal{T}$, and by the transitivity of $\leq$, for each t the set $\{s \mid t \leq s\}$ will supply the basis. We can use $\mathcal{T}^+$ to denote the ordered collection of all

132. See the appendix for more on these logics and their defining axioms.

future-closed subsets of $\mathcal{T}$.[133] But as $\mathcal{T}^+$ constitutes an (Alexandrov) topology—namely, it is just $\mathcal{U}(\mathcal{T})$—we can use the exact correspondence

$$\mathbf{Set}^{\mathcal{T}} \simeq \mathbf{Sh}(\mathcal{T}^+)$$

to move freely between sheaves on $\mathcal{T}^+$ and the usual, more direct "variable set" perspective on $\mathcal{T}$.

133. We could also define, as usual, a $\mathcal{T}$-*valuation* as a function $V : \Phi \to \mathcal{T}^+$, where Φ is the set of atomic formulae, with typical member p. A valuation sends each propositional variable p to a future-closed subset $V(p) \subseteq T$, interpreted as the set of times at which p is "true." We also have that $t \in V(\Box\phi)$ iff $t \leq s$ implies $s \in V(\phi)$, allowing the valuation to be extended to all propositions (using the usual connectives). A *model* based on $\mathcal{T}$ is then defined as a pair $\mathcal{M} = (\mathcal{T}, V)$, where V is a $\mathcal{T}$-valuation. Moreover, when $\mathcal{T}'$ is a subframe of $\mathcal{T}$, for any proposition ϕ, ϕ is valid on the frame $\mathcal{T}$ (i.e., ϕ is true in every model based on $\mathcal{T}$) only if ϕ is valid on $\mathcal{T}'$.

9 Cellular Sheaf Cohomology through Examples

In which we start to look at some more involved and computationally explicit examples, working up toward an extended introduction to cellular sheaf cohomology and giving a brief glimpse into the theory of cosheaves.

Roughly, if sheaves represent local data—or, more precisely, represent how to properly ensure that what is locally the case everywhere is in fact globally the case—sheaf cohomology can be thought of as a tool for systematically measuring and relating *obstructions* to such passages from the local to the global. In particular, sheaf cohomology in low degrees gives an explicit measure of the failure of local data to patch. Moreover, in sheaf theory more generally, one could argue (as does Grothendieck, for instance) that individual sheaves are only of secondary importance—the real power of sheaf theory emerges from the use of constructions involving various sheaves, linked together via sheaf morphisms. The exposition of cellular sheaf cohomology that follows will allow us to begin to appreciate such a perspective. There is a more general theory of sheaf cohomology in terms of homological algebra and derived categories, but the focus here will be on the simpler case of cellular sheaf cohomology, leaving the reader to explore the more general theory on their own.

As motivation for the general idea of cohomology, let us briefly consider *impossible objects*. Consider the "Penrose tribar," first devised by Roger Penrose:

Focusing one's eye on any sufficiently small pieces of this depicted figure—for instance, regarding the tribar as being assembled from three L-shaped pieces—suggests that the drawing is the projection of a visually viable spatial object. In other words, breaking it up into possible pieces, such a 2-D picture suggests an object that is *locally consistent*. However, while it may initially seem plausible that the pieces of the tribar may assemble together into the closed triangle depicted above, the local depth data cannot in fact be consistently merged, and there is no plausible interpretation of the entire perspectival drawing as a projection of a viable spatial object that respects standard visual expectations. The 3-D object that the drawing of the Penrose tribar would depict is not a possible object in ordinary Euclidean space—hence, the Penrose tribar is sometimes referred to as an "impossible object." In other words, *globally* such an object is inconsistent; but *locally* (i.e., as one considers any small enough region of the drawing), there is no such impossibility in what the depiction represents. To adopt another way of seeing the same issue: there is a property—the *impossibility* of such an object—that ceases to be valid locally, that is, is *nonlocal*.

Similarly, consider the "Penrose staircase," a 2-D depiction of a staircase where the stairs take four 90-degree turns as they descend in a continuous loop, something that is impossible in ordinary 3-D Euclidean space:

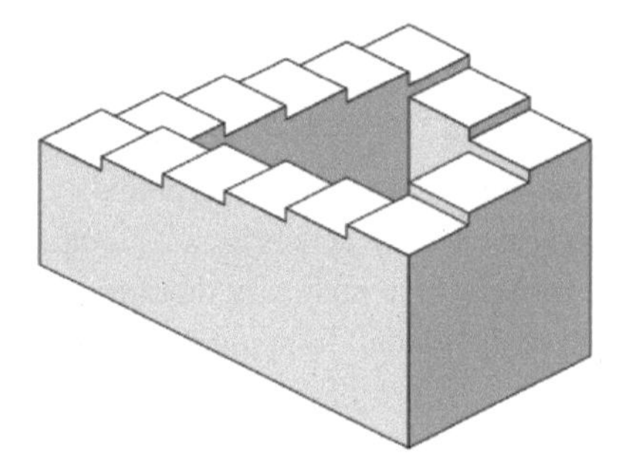

Visually, it is to be understood that we might have a number of local snapshots of four individual sides or corners of the staircase, where we would like to patch these together into a viable 3-D globally consistent staircase. Each part of the structure is viable as a representation of a directed flight of stairs. However, any attempt to merge all these partial representations must fail—as the steps are continually descending in a clockwise direction—and so the representation must be of an object that, globally, is inconsistent. Since the implied heights will not all match up, this inconsistency thus presents an obstruction to any such attempted patching. In short, (1) you have a number of local pictures, that is, snapshots of the individual sides of the staircase; which (2) you would like to patch together into a globally consistent object (the full staircase); yet (3) there exists an obstruction to step (2), as what one expects for the heights does not match.

With such impossible figures, we are fundamentally being presented with an impossibility of patching locally consistent data into a consistent whole. Locally, such impossible figures look entirely possible, but globally, they are not. Cohomology can help track such obstructions. A decisive feature of cohomology is that it is fundamentally nonlocal. Such impossible objects can help build some intuition for (sheaf) cohomology, for the "impossibility" of any correct 3-D representation can be captured by an element of cohomology. The obstruction described in the third step above shows up in the form of a nonzero element of some cohomology group.[134]

134. Penrose (1992) uses cohomology to make more precise the fact that such figures are impossible globally, while being viable when viewed locally. Example 231 below covers the main ideas of that paper.

One of the main tools in sheaf theory is sheaf cohomology, which can be seen as a generalization of some classical cohomology theories (de Rham cohomology and Čech cohomology). In a rough sense, sheaf cohomology is fundamentally an invariant that quantifies and tracks obstructions to extensions of local sections of a sheaf to global sections. The so-called first cohomology of a sheaf will accordingly capture the set of things that locally appear just like sections, yet globally may not come from a section.

To begin to tell this story—focusing ultimately on the case of cellular sheaf cohomology—let us first build up some necessary background.

9.1 Simplices and Their Sheaves

Of the many ways to represent a topological space, a particularly computationally friendly way is to perform a triangulation with entities called *simplices*, decomposing the space into simple pieces (thought of as being glued together) whose common intersections or boundaries are lower-dimensional pieces of the same kind. With simplices come certain simplicial maps that, moreover, approximate continuous maps. In this way, simplices play a role in bridging the gap between continuous figures and their discrete representation and approximation via decompositions of spaces into discrete parts. More than that, as we will see, they allow us to develop profound connections between algebra and geometry. Simplices are powerful and easy-to-use devices for understanding qualitative features of data collections, and in general they can be thought to represent n-ary relations between n vertices.

Basically, we use collections of simplices—for now, just think of points, line segments, generalized triangles, or tetrahedra generalized to arbitrary dimensions—to build up what are called *simplicial complexes*. Given a simplex σ, it is common to refer to a (nonempty) simplex τ whose vertices are a subset of the vertices of σ as a *face* of σ. A (geometrical) simplicial complex K is a collection of simplices such that (1) every face of a simplex of K is in K, and (2) the intersection of any two simplices of K is a face of each of them. Roughly, then, one can think of a simplicial complex K as comprising generalized triangles of various dimensions, glued together along common faces. This is really part of a more general story involving *cell complexes* (including cubical complexes, multigraphs, etc.), where one can roughly think of a cell complex as a collection of *closed disks* of various dimensions which are moreover glued together along their boundaries. But we will instead focus on the more computationally tractable combinatorial counterpart to the already simplified notion of a simplex: that of *abstract simplicial complexes* (ASC). Here, we reencode the information of a simplicial complex via the more computationally friendly notion of an ASC, where this is basically just a collection of finite subsets of vertices (elements), closed under the operation of taking subsets. This captures in a purely combinatorial way the geometrical notion of simplicial complex.[135]

Definition 222 An *abstract simplicial complex* $K = (A, K)$ is a set A equipped with a collection of ordered finite nonempty subsets $K \subseteq \mathbb{P}(A)$ that contains all singletons and that is

135. What has been lost is how the simplex is embedded in, say, Euclidean space; however, this specification retains all the data needed to reconstruct the complex up to homeomorphism.

closed under taking subsets (sublists), that is, every nonempty subset of a set in K is also in K. In other words, we must have

- for each $x \in A$, the singleton $\{x\} \in K$; and
- if $\sigma \in K$ and $\tau \subseteq \sigma$, then $\tau \in K$.

Terminologically, each member of K is called a *simplex* (or, looking ahead, *cell*), and given a $\tau \subseteq \sigma$ we say that τ is a *face* of σ. A simplex with $n+1$ elements is called an n-dimensional simplex of K. (But, as one would expect, a 0-face is usually just called a *vertex*, and a 1-face an *edge*.) If all the simplices of an ASC K are of dimension n or less, K is said to be an n-dimensional simplicial complex, that is, the dimension of an ASC is the maximal dimension of its constituent simplices. A *simplicial map* $f: K \to K'$ from an ASC K (defined on a set A) to an ASC K' (defined on the set B) is a function $f: A \to B$ with the property that for any σ of K, the image $f(\sigma)$ is an element of K'.[136]

Altogether, this data in fact lets us define the category **SCpx** that has (abstract) simplicial complexes as objects and simplicial maps as morphisms.

To check understanding, notice that a 1-D simplicial complex essentially recovers the notion of a simple graph.

ASCs are particularly easy to describe, but one might worry that certain topologically valuable information gets lost in encoding things in this simplified, set-theoretical fashion. We will ultimately be interested in certain topological information, so it makes sense to construct for a given ASC K a geometrical *realization* $|K|$ of K, allowing K to be realized (basically, "pictured") as some (generalized) triangles glued together in suitable ways, living in a subspace of $\mathbb{R}^n$. For every simplicial complex K, there in fact exists a unique (up to simplicial isomorphism) geometric realization $|K|$, so we will not bother to distinguish between ASCs and their geometric realizations.[137] Such a realization better explains why we think of objects (sets) in an ASC K as faces or simplexes, since via the realization, three-element sets correspond to filled-in triangles, two-element sets to edges, singletons to vertices. For instance, given a set $A = \{a, b, c\}$ for which we have an ASC $K = \{\{a, b\}, \{b, c\}, \{a, c\}, \{a\}, \{b\}, \{c\}\}$, then its realization $|K|$ will be the (hollow) triangle (with a natural orientation). Drawing such pictures yields the usual geometrical notion of a simplicial complex, obtained by gluing together the standard simplices along the boundaries; the usual approach then employs these n-simplices to probe a topological space via continuous maps into the space.[138]

In more detail, observe that an oriented 0-simplex thus corresponds to a signed $(+,-)$ point P, while an oriented 1-simplex is a directed line segment P_1P_2 connecting the points P_1 and P_2, where we assume that we are traveling in the direction from P_1 to P_2, that is, $P_1P_2 \neq P_2P_1$ (however, $P_1P_2 = -P_2P_1$). An oriented 2-simplex will be a triangular region $P_1P_2P_3$, with a prescribed order of movement around the triangle. An oriented 3-simplex is given by an ordered sequence $P_1P_2P_3P_4$ of four vertices of a solid tetrahedron. Similar

136. Simplicial maps between simplicial complexes are the natural equivalent of continuous maps between topological spaces.

137. Ensuring the uniqueness of this is precisely the reason for considering *ordered* sets (lists) in the definition of an ASC, instead of just unordered sets.

138. More details on these matters can be found in Ghrist (2014) or Hatcher (2002).

definitions hold for $n > 3$. By gluing together various simplices along their boundaries, we get simplicial complexes such as the following:

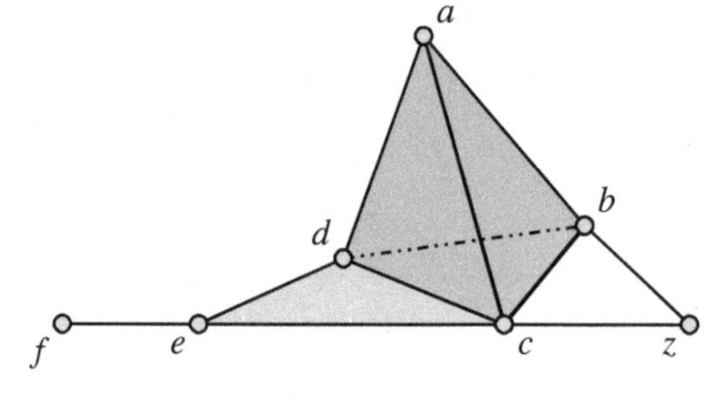

While it is perhaps useful to visualize things in this way, and while this perspective can be important for connections to other concepts, the alert reader might have observed that given the way ASCs were defined, they should already come with a natural topology, letting us bypass the geometric realization, and associate to each ASC a particular topological space. This will be useful to us in the construction of sheaves on such spaces toward which we are building.

To see this, first observe that ASCs come with a canonical partial order on faces, given by the face subset inclusion (or attachment) relation between vertices, edges, and higher-dimensional faces. We can use this fact to define the *face* (or *cell*) *category*, where this has for objects the elements of K, a cell complex, and (setwise) inclusions of one element/cell of X into another for its morphisms; if a and b are two faces in a complex X with $a \subseteq b$ and $|a| \leq |b|$, we will write $a \rightsquigarrow b$ and say that a is *attached* to b.[139] We will then identify a complex with its face poset, writing the incidence relation $a \rightsquigarrow b$. Building on the graphical construction of a complex, the attachments between the faces or cells of a complex can then be displayed in attachment diagrams, where the links represent attachments going from lower- to higher-dimensional cells (and where any additional attachments that arise as compositions of attachments are left implicit). Observe that the attachment diagram of a graphical complex is itself just a set partially ordered via the attachment relations, that is, it is a poset. The data of this face relation poset can be displayed with a Hasse diagram. Assume we have the following simplicial complex K, which we imagine has been realized thus:[140]

139. Note that technically to make the following construction work we need to use the more general notion of *cell complexes*, the full definition of which can be found in Curry (2014, chap.4); but since ultimately the realization $|K|$ of an ASC K on a finite set, which is the sort of thing we will be dealing with in our example, can be shown to be a cell complex, and simplicial maps $f : K \to K'$ induce a cellular map $|f| : |K| \to |K'|$, we will not worry too much about the distinction.

140. Note that the rightmost simplex $(abcd)$ is *not* meant to depict a hollow *tetrahedron*; each of the four component triangles are to be thought of as lying in the plane. We have simply spaced it this way to make the sheaf diagrams we build on top of this in a moment a little more readable.

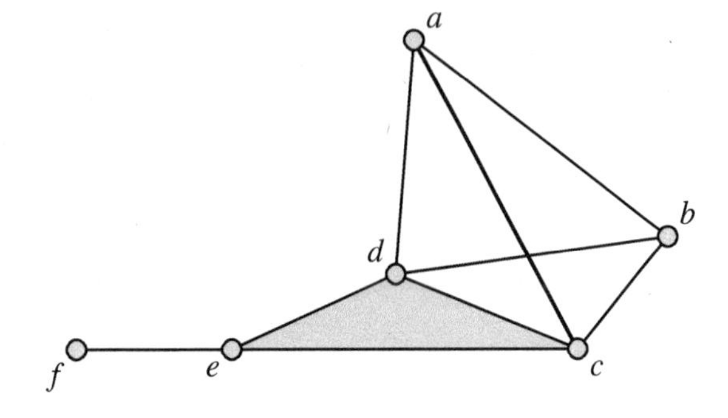

We can form the diagram of the face-subset relations, where the edges and higher faces are given the natural names (and are assumed to be ordered lexicographically):

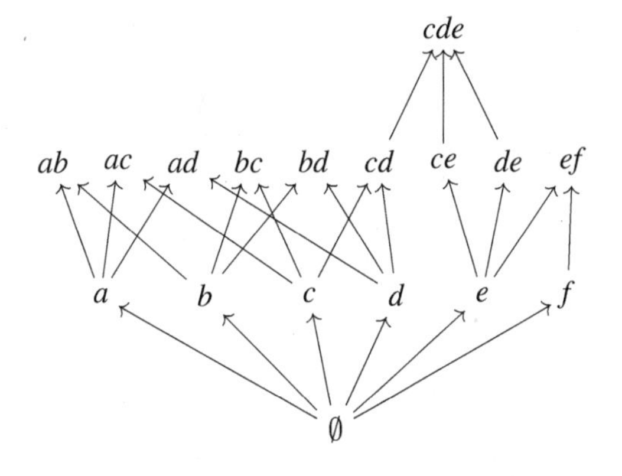

But it is more revealing to display this in the form of an attachment diagram, as follows:

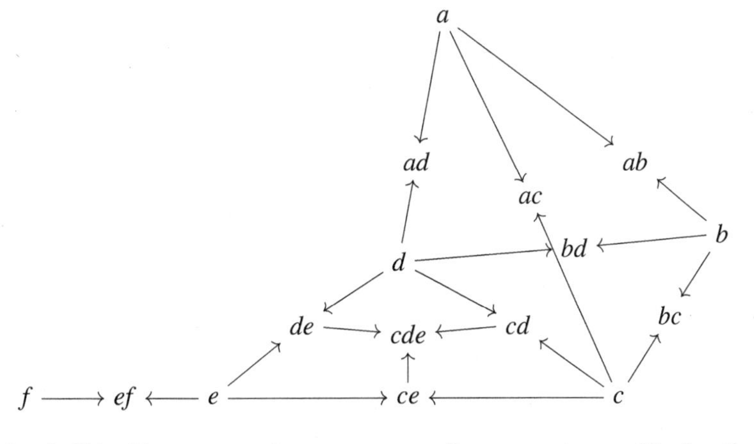

It is this face/cell incidence poset that we are regarding as a category, $\mathscr{F}_K$. In other words, to a simplicial complex we can associate a category, which is just the face-incidence poset viewed as a category. We can then put the Alexandrov topology on this poset of face-relations. Then, given this Alexandrov topology on a poset, the usual topological notions (such as interiors, boundaries, and closures) can be easily understood in terms of the poset itself.

The basic idea is this: pick a simplex; then look at all the other simplices that include that one as a face (i.e., higher-dimensional simplices adjacent to it); then regard such "upper sets" as the open sets. The sets of the form $\uparrow x = \{y \in \mathcal{P} \mid x \leq y\}$—that is, the principal upper sets—form a basis for the topology. We can also define the closure of x by $\overline{x} = \{y \in \mathcal{P} \mid y \leq x\}$; and, provided the poset $\mathcal{P}$ is finite, a basis of closed sets is given by these $\overline{x}$.[141]

In the cellular context, for σ a cell of a cell complex, the analogue of the principal upper set construction is called the *open star* of σ, where this is denoted st(σ) and consists of the set of cells τ such that $\sigma \rightsquigarrow \tau$—that is, it captures all the higher-dimensional cells containing that cell. Then, in terms of the topology, st(σ) will be the smallest open set of cells containing σ, or $\text{st}(\sigma) = \bigcup_{x \in \sigma} U_x$, with $U_x = \uparrow x$. Taking all the stars and the union of all the stars will give a topology for the complex/simplex—this is the Alexandrov topology. Every intersection of opens in the Alexandrov topology on a poset $\mathcal{P}$ is open. Thus, a star over $A \subseteq X$ is then defined to be the intersection of the collection of all open sets containing A. The resulting collection of stars will be a basis for the Alexandrov topology. While stars need not exist in general, in the Alexandrov topology on a poset, there will exist a star of every subset. There is accordingly a dictionary between cellular complexes and Alexandrov spaces, which can be seen by considering that for a cell complex, every cell Δ_σ has a star, where this is a set consisting of all those cells Δ_τ such that $\Delta_\sigma \leq \Delta_\tau$.

Note that with respect to the inclusions in the face relation poset $\mathcal{F}_K$, the containment relation for the open sets (stars) in the Alexandrov topology is order-reversing. In other words, the Alexandrov construction will yield an order-reversing inclusion functor $\mathcal{P} \to \mathcal{O}(\mathcal{P})^{op}$, just as we saw in chapter 8. More generally, we have effectively described a contravariant functor $Alxd$: **Pre** $\to$ **Top**—one that, upon applying $Alxd$ to X^{op}, yields a space that has for open sets unions of simplices.

It turns out that a sheaf (of sets) over an ASC K can be defined as a covariant functor from the face category $\mathcal{F}_K$ of its associated face-relation poset to **Set**. More generally, given a functor from a preorder or poset $\mathcal{P}$ to a category **D**, this functor can be used to produce a sheaf on the Alexandrov topology (via a right Kan extension).[142] With this construction, the sheaf gluing axiom for any cover is automatically satisfied! In other words, given a poset endowed with the Alexandrov topology, as anticipated in the last chapter, we do not even need to distinguish between presheaves and sheaves.

The following definition follows Shepard (1985), who defined sheaves for more general *cell complexes*, which are just a collection of closed disks of certain dimensionality that are glued together along the boundaries. Via its realization $|K|$, an ASC K is just a cell complex, so while the definition is more general, it can be applied to a simplicial complex (which is what we will work with in the coming example). Just as with ASCs, since cell complexes are built up from simple pieces (cells), the associated attachment diagram exhibiting the

141. A dual topology thus arises by considering, for a given simplex, all the other simplices that are attached to (included in) it—these are the downsets, which also can serve as the opens of this topology. In the Alexandrov topology, arbitrary intersections of opens are open and arbitrary unions of closed sets are closed; therefore, by exchanging opens with closed sets, we can pass from any Alexandrov space to its dual topology.

The Alexandrov topology construction is really appropriate when the elements of the underlying poset $\mathcal{P}$ represent finite pieces of information (i.e., are compact), something that is typical for many combinatorial and computer science applications; however, if $\mathcal{P}$ includes infinite elements, then the "Scott topology" is called for (see Vickers [1996, chap. 7] for details on this).

142. See Curry (2019) for details.

relations between cells contains the information of the cell complex itself. Attending to the face poset in particular, then, we will define a *cellular sheaf*, following Shepard, as a covariant functor from the face category of a complex K to some other category $\mathbf{D}$.[143] For concreteness, for the remainder we consider $\mathbf{D} = \mathbf{Vect}$.

Definition 223 A *cellular sheaf* (of vector spaces) F on a cell complex X is

- an assignment of a vector space $F(\sigma)$ to each cell σ of X,[144] together with
- a linear transformation

$$F_{\sigma \rightsquigarrow \tau} : F(\sigma) \to F(\tau)$$

  for each incident cell pair $\sigma \rightsquigarrow \tau$.

These linear maps have to further satisfy the identity relation $F_{\sigma \rightsquigarrow \sigma} = \text{id}$ and the usual composition condition, namely

$$\text{if } \rho \rightsquigarrow \sigma \rightsquigarrow \tau, \text{ then } F_{\rho \rightsquigarrow \tau} = F_{\sigma \rightsquigarrow \tau} \circ F_{\rho \rightsquigarrow \sigma}.$$

The reader may be wondering if there is a typo in the direction of the maps described in this definition of a cellular *sheaf*. A sheaf, after all, is a particular presheaf, so one would have expected (order-reversing) *restriction* maps. But this is not a typo, and in fact goes to the heart of the underlying construction. The reason for the seemingly "wrong" direction of the arrows—typically, sheaf restriction maps reverse the direction of the arrows, while cosheaves preserve them—is explained (as correct) by a result we have already encountered.

Recall that, in general, when dealing with a poset $\mathcal{P}$, we can regard a sheaf on $\mathcal{P}$—once this has been equipped with the upper Alexandrov topology—as a plain old pre-cosheaf (covariant functor) on $\mathcal{P}$ (which could, in turn, be regarded as a presheaf on $\mathcal{P}^{op}$).[145] But the face category $\mathcal{F}_X$ with which we identify a complex X is a poset. So, using our general result[146]

$$\mathbf{Sh}(\mathcal{U}(\mathcal{P})) \simeq \mathbf{D}^{\mathcal{P}}$$

letting us move freely between sheaves (valued in $\mathbf{D}$) on the upper Alexandrov topology placed on $\mathcal{P}$ and covariant $\mathbf{D}$-valued functors (pre-cosheaves) on $\mathcal{P}$, we know that we can freely regard a plain old *covariant functor* $\mathcal{P} \to \mathbf{D}$ as a *sheaf* on $\mathcal{U}(\mathcal{P})$, that is, on the upper Alexandrov topology on $\mathcal{P}$. The inclusion taking a poset into its upper sets is order-reversing, and the underlying functor of a sheaf (now on the upper sets) is itself order-reversing—their composition, equivalent to the original covariant functor, is accordingly covariant. This accounts for why the definition of a cellular *sheaf* seems to just contain the data of a definition of a covariant functor—that is indeed all it says! Theorem 220,

143. While this, and the subsequent cellular sheaf cohomology, is original to Shepard, his thesis was never published and fell into oblivion for some time. Part of the intent of Curry (2014) was to revive Shepard's contribution, while providing a more modern account, supplemented with missing details, demonstrating why cellular sheaves are actually sheaves. The account here thus owes a great debt to Curry (2014), which the reader is encouraged to consult for more extensive treatment of these matters.

144. This vector space $F(\sigma)$ is then the *stalk* of F at (or over) σ.

145. Also recall that, taking $\mathcal{P}^{op}$ instead, sheaves on $\mathcal{P}$—where this is equipped with the *lower* Alexandrov topology (using that $\mathcal{D}(\mathcal{P}) \simeq \mathcal{U}(\mathcal{P}^{op})$)—are just presheaves on $\mathcal{P}$.

146. Earlier, in chapter 8, we just described it in terms of set-valued functors; but in fact, the equivalences hold for (pre)sheaves valued in a category $\mathbf{D}$, provided that category is complete and cocomplete.

discussed at the end of the last chapter, lets us conflate these two perspectives. In short, we have the slogan:

> cellular sheaves are covariant functors from the face category into some other category **D**.

In particular, a covariant functor from $\mathcal{F}_X$ to **Vec** is already just a sheaf of vector spaces.

Dually, we could define:

Definition 224 A *cellular cosheaf* (of vector spaces) $\hat{F}$ on a cell complex X is

- an assignment of a vector space $\hat{F}(\sigma)$ to each of the cells σ of X—this vector space $\hat{F}(\sigma)$ is called the *costalk* of $\hat{F}$ at (or over) σ—together with
- a linear transformation $\hat{F}_{\sigma \rightsquigarrow \tau} : \hat{F}(\tau) \to \hat{F}(\sigma)$ for each incident cell pair $\sigma \rightsquigarrow \tau$.

These maps—called the *corestriction maps*—have to further satisfy the identity relation $\hat{F}_{\sigma \rightsquigarrow \sigma} = \text{id}$ and the usual composition condition, namely

$$\text{if } \rho \rightsquigarrow \sigma \rightsquigarrow \tau, \text{ then } \hat{F}_{\rho \rightsquigarrow \tau} = \hat{F}_{\rho \rightsquigarrow \sigma} \circ \hat{F}_{\sigma \rightsquigarrow \tau}.$$

In the extended example to follow, we focus on cellular sheaves; at the end of the chapter, an example with cellular cosheaves is presented. Before turning to the example, let us also highlight a few more explicit definitions of the corresponding cellular notions that are more or less as you would expect for a sheaf.

Definition 225 For F a cellular sheaf on X, we define a *global section* x of F to be a choice $x_\sigma \in F(\sigma)$ for each cell σ of X, where this satisfies

$$x_\tau = F_{\sigma \rightsquigarrow \tau} x_\sigma$$

for all $\sigma \rightsquigarrow \tau$.

Fundamentally, the data of a cellular sheaf on a complex X amounts to a specification of spaces of local sections on a cover of X (namely, the one given by open stars of cells). Ultimately, we will be able to form the category of *all sheaves* $\mathbf{Sh}(X)$ over a fixed complex X, adopting the only notion of morphisms that there could be, namely as the natural transformations between the corresponding functors.[147] Explicitly,

Definition 226 A *morphism* $f : F \to G$ *of sheaves* (or *sheaf morphism*) on a cell complex X is an assignment of a linear map

$$f_\sigma : F(\sigma) \to G(\sigma)$$

to each cell σ of X, where for each attachment $\sigma \rightsquigarrow \tau$, the usual (natural transformation) compatibility condition holds, making the diagram

$$\begin{array}{ccc} F(\sigma) & \xrightarrow{f_\sigma} & G(\sigma) \\ {\scriptstyle F(\sigma \rightsquigarrow \tau)} \downarrow & & \downarrow {\scriptstyle G(\sigma \rightsquigarrow \tau)} \\ F(\tau) & \xrightarrow[f_\tau]{} & G(\tau) \end{array}$$

commute.

147. In the next definition, we focus on the corresponding notion of morphism for sheaves of vector spaces; but if we were to work with sheaves valued in some other category **D**, then we would require that the maps defined below be the appropriate structure-preserving map (e.g., for sheaves of groups, just homomorphisms).

A *sheaf isomorphism*, then, is also defined in the inevitable way, as a morphism where each of the f_σ is an isomorphism.[148]

In brief: if X is a cell (or simplicial) complex with the associated face poset category $\mathscr{F}_X$, we identify the complex with its face category, and then a cellular sheaf is just a vector space-valued functor $F : \mathscr{F}_X \to \mathbf{Vect}$, while a cellular cosheaf is a vector space-valued functor $F : \mathscr{F}_X^{op} \to \mathbf{Vect}$. Thus, to avoid confusion, realize that in the above definitions of cellular (co)sheaves, X was really short for $\mathscr{F}_X$, to which we associate the cell complex, so that a cellular sheaf on X (per the definition) is really a sheaf on $\mathscr{F}_X$ where this has been equipped with the Alexandrov topology, which in turn is uniquely determined by a functor $\mathscr{F}_X \to \mathbf{Vect}$ (and *this* last functor is what the definition is describing).

In the example that follows, we illustrate these ideas in a concrete fashion by constructing a cellular sheaf on a particular simplicial complex.

Example 227 Recall our simplicial complex X from before, which we imagine has been realized thus:

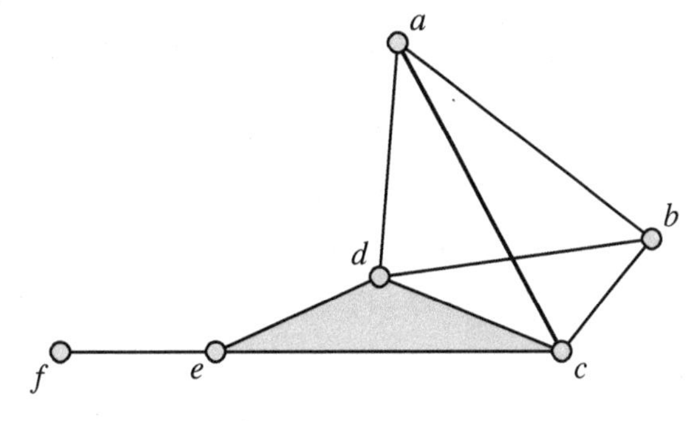

together with its associated attachment diagram displaying the poset of face relations (i.e., its face category $\mathscr{F}_X$):

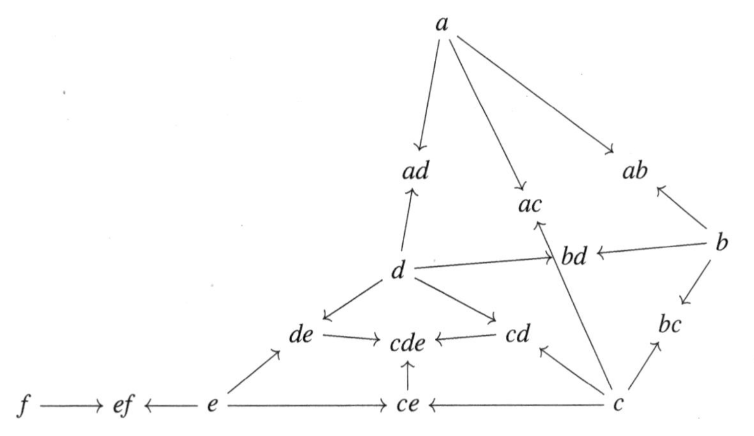

148. It is also easy to show that a morphism between sheaves of vector spaces, such as that given above, induces a linear map between the spaces of global sections of the sheaves; moreover, isomorphic sheaves will have isomorphic spaces of global sections.

Let us see what a sheaf (of vector spaces) on X—or rather, on the corresponding face attachment diagram $\mathcal{F}_X$—looks like. Well, we need to spell out all the data of the topology, covers, and the sheaf conditions, right? No! Using the main theorem, it will suffice to just describe a vector-valued covariant functor on this diagram—and this will already contain all the data of a sheaf! This is one of the many instances where very abstract category-theoretic results, which may be difficult to understand at first, can make our lives a lot easier in practice.

Thus, following the definition of a cellular sheaf, for the values of the sheaf over cells, this will just amount to the specification or assignment of values (spaces) to each of the cells of the simplices, data that comes in the form of vectors, for example,

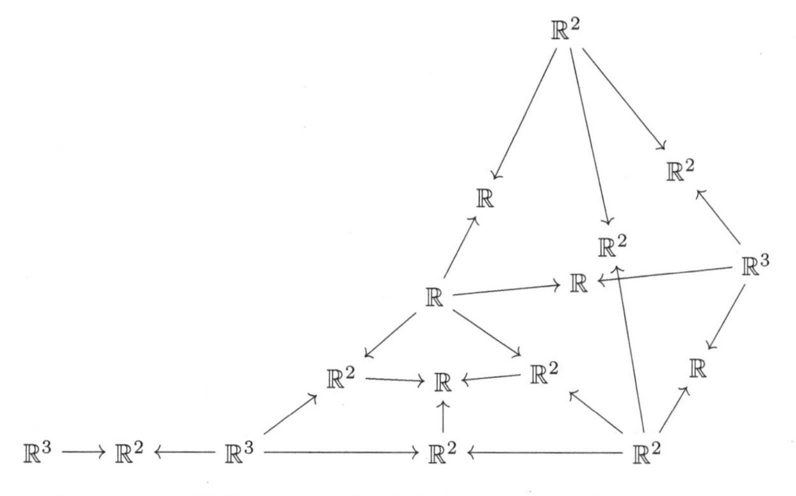

But we need maps as well. The maps, for their part, may be thought of as representing some sort of local constraints or as enforcing certain relations between the data. In general, when the stalks $F(\sigma)$ have structure—for instance, here they are vector spaces—then a sheaf of that type (i.e., valued in the relevant category) is obtained when the restriction maps preserve this structure. In other words, we should have a function (in our particular case of vector space assignments, these will be given by a linear map) assigned to each inclusion of faces in such a way that the diagram commutes, that is, the composition of functions throughout the diagram is path independent. The following sheaf diagram nicely displays all these ideas:

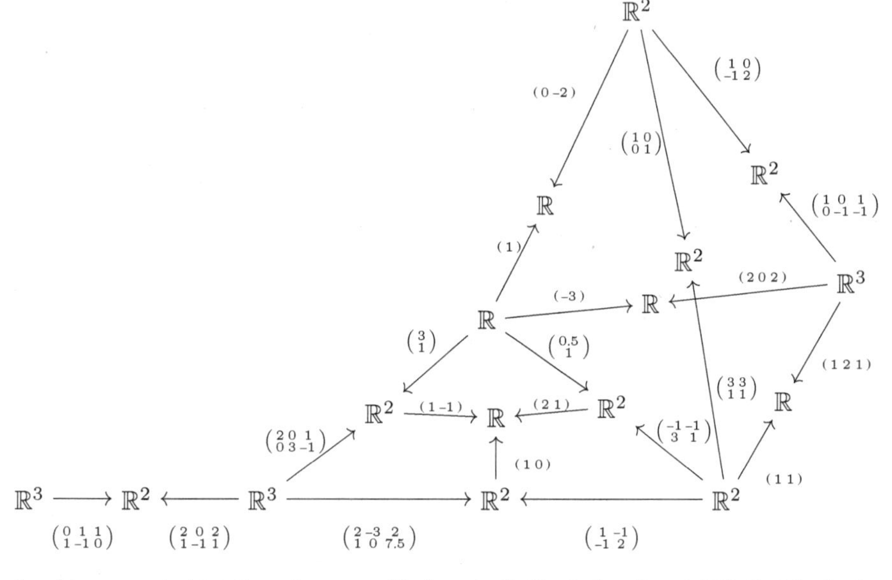

A sheaf is generated by its values specified on individual simplices of X, that is, by local sections specified on the vertices. But a sheaf is not just this data. The restriction maps of a sheaf are an essential part of the construction, as they encode how any local sections can be extended into sections over a larger part of the diagram (ultimately throughout the entire diagram), and so it is precisely via the restriction maps that it is made explicit how the local sections can be glued together. The sheaf assignment given over *all* of X will be specified by a collection of local sections that can be extended along all the restriction maps to higher-dimensional faces. There may be some flexibility or freedom in the actual data assignments over a vertex, but they are not entirely arbitrary, for the restriction maps encode how local assignments—values specified on certain parts of the diagram—can be extended to other parts of the diagram, and so the maps will constrain the assignments in various ways.

If the reader would like to get a good working understanding of the important distinction between a local section and a global section, it would be useful to closely consider what happens in this concrete case when we assign, for instance, the value $\begin{pmatrix} 1 \\ 0 \\ -1 \end{pmatrix}$ to the stalk over the vertex e versus what happens when we assign, for instance, the value $\begin{pmatrix} -1 \\ 2 \\ 2 \end{pmatrix}$ to the same vertex. In the first case, one finds that we can "extend" or propagate this particular selection along *some* of the arrows to those stalks highlighted in gray, but then there is a problem, an obstruction to our continuing this process any further upward along the edges of the diagram:

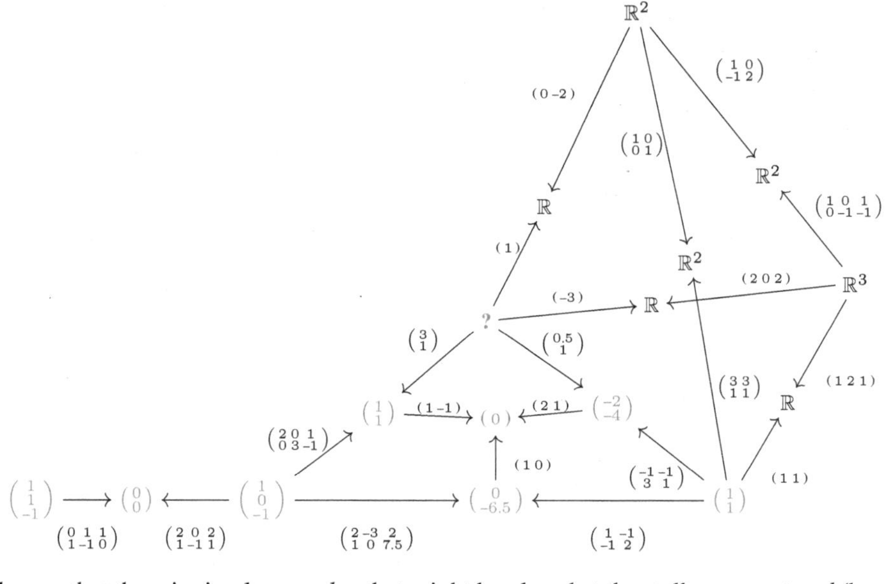

Observe that there is simply *no value* that might be placed at the stalk over vertex d (hence the "?") that would allow us to continue with this extension process. If we assigned (–4) to the stalk over d, this would indeed be consistent with the map $\begin{pmatrix} 0.5 \\ 1 \end{pmatrix}$ proceeding down and to the right and landing in the stalk over cd, which would in turn land us, perfectly consistently, in the stalk over cde with the value (0), as required by the other restriction maps. However, by following what happens to that same assignment –4 under the action of the map down and to the left via $\begin{pmatrix} 3 \\ 1 \end{pmatrix}$, we see that we would get $\begin{pmatrix} -12 \\ -4 \end{pmatrix}$, which, when further mapped under $\begin{pmatrix} 1 & -1 \end{pmatrix}$ would yield (–8). We thus cannot assign –4—or *anything* for that matter—to the stalk over d, given our original assignment over e.

The original assignment at the stalk over e, then, is said to describe a strictly *local section*, one that importantly cannot be extended globally, that is, over the entire complex. By beginning with other values at the same (or, if one desires, another) vertex, the reader can explore various other solutions that are merely local versus those that manage to be global. In this way, the reader will not only discover that some local solutions or sections are "more local" than others, but will discover other types of obstructions to the extension of local sections. For instance, whereas with our test assignment above it turned out that there was simply *no value at all* that could be assigned to the stalk over vertex d, while maintaining consistency with the other stalk assignments required by the restriction maps, another (less serious) issue one frequently encounters is that a specific assignment on one stalk ends up requiring two *different* assignments at some stalk.[149]

149. We might here take the opportunity to mention that in a variety of applications of sheaves, including beyond sheaves of vector spaces on complexes, it is possible that *exact equality* of assignments will be unattainable or the least valuable thing to consider. There are a few ways to develop machinery to accommodate this, but it is beyond the scope of this book to cover them in detail. Instead, for one particularly friendly approach, the reader is referred to the work of Michael Robinson, who has proposed to deal with this situation via formalizing a "consistency structure" with a corresponding notion of *distance* between assignments (say, with the structure of a pseudo-metric space). Moreover, the notion of *pseudosections* can be developed, and Robinson has shown that in fact pseudosections are already sections, just with respect to a different sheaf; for instance, pseudosections of a sheaf over an ASC X are veritable sections of another sheaf over the barycentric subdivision of X. See Robinson

In contrast to the above failures, one observes that by seeding the stalk over vertex e with the value $\begin{pmatrix} -1 \\ 2 \\ 2 \end{pmatrix}$, we encounter no such obstruction to the extension of this assignment to a consistent assignment of values over the *entire* diagram, thus yielding what is appropriately called a *global section*. A global section is just a selection of value assignments from each of the stalks over all the cells that is consistent with *all* the restriction maps of the diagram:

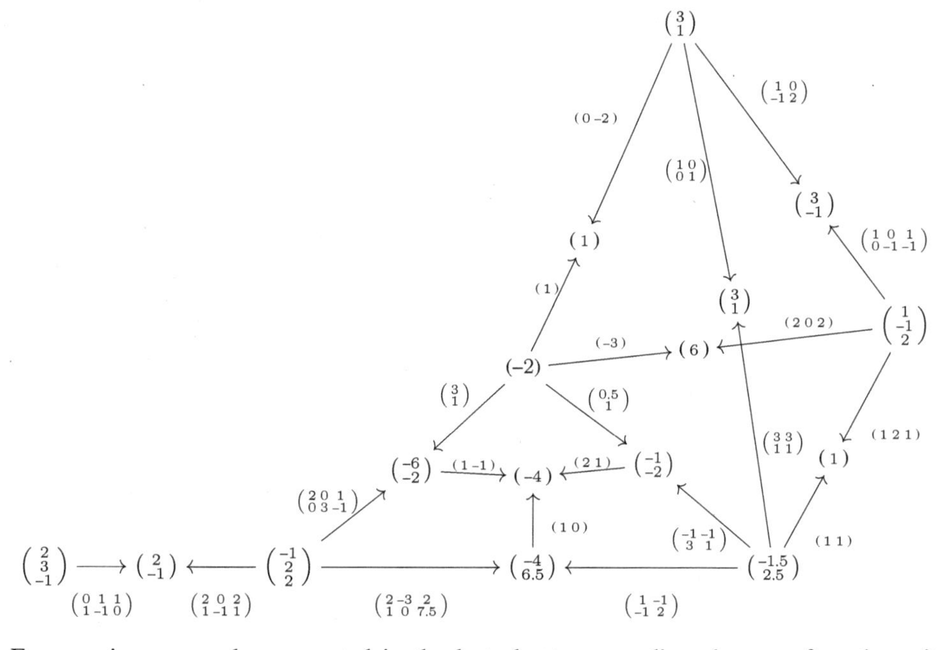

From various examples presented in the last chapter, regarding sheaves of sections, it should already be clear that some local sections of a sheaf (of *sets*) will not extend to global sections. As we just saw, in the cellular sheaf of vector spaces, the same sort of thing occurs—that is, sections can remain strictly local, when they cannot be defined across all the faces of the simplex or when they conflict with the constraints of the restriction maps. There might be interesting local solutions among the variable sets of solutions to a local problem, but only those solutions that can be consistently propagated along the entire diagram, respecting the sheaf restriction maps, will provide us with a global section or solution.

9.1.1 Sheaf Morphisms and Some Operations on Sheaves

Earlier, in discussing the definition of cellular sheaves, we mentioned the category **Sh**(X) of all cellular sheaves on a fixed cellular space X, where we observed that the morphisms of this category are, inevitably, just natural transformations between the functors defining the cellular sheaves. But suppose we no longer fix the cellular space, so that we are considering

---

(2015, 2016b) and Praggastis (2016) for more details. Pushing this a little further, we could analyze data using the *consistency radius* of the sheaf, that is, the maximum distance between the value in a stalk and the values propagated along the restriction maps. By imposing such a consistency structure on the sheaf, this could tell you "how far" a particular data instance was from conforming to the consistency requirements stipulated by the structure encoded by the entire sheaf. In other words, given a particular data assignment, it could be used to inform how to find the *closest global section* (where "closest," of course, would be given by, say, the pseudometric placed on the assignments). See Robinson (2018).

cellular sheaves on (possibly) different spaces. We would like to extend the notion of a morphism of cellular sheaves to provide maps between sheaves on different spaces. Of course, we could consider a morphism between sheaves on a fixed space X as a special case of this approach, but we have really been heading towards the more general case of a sheaf morphism involving different spaces.

Definition 228 A *sheaf morphism* $s: F \to G$ from a sheaf F over a space Y to a sheaf G over the space X consists of the following data:

- a cellular map $f: X \to Y$,
- a collection of (linear) maps $l_\sigma : F(f(\sigma)) \to G(\sigma)$ such that for each attachment map $\sigma \rightsquigarrow \tau$ in X, the following diagram commutes:

$$\begin{array}{ccc} F(f(\sigma)) & \xrightarrow{l_\sigma} & G(x) \\ {\scriptstyle F(f(\sigma)\rightsquigarrow f(\tau))}\downarrow & & \downarrow{\scriptstyle G(\sigma\rightsquigarrow\tau)} \\ F(f(\tau)) & \xrightarrow[l_\tau]{} & G(\tau). \end{array}$$

In other words, a sheaf morphism takes data in the stalks over two sheaves and relates them through linear maps in such a way that the resulting diagram commutes.

The reader may observe how this in fact makes use of the *pullback sheaf* notion, first mentioned in chapter 8, where for a map $f: X \to Y$, and a sheaf F on Y, the pullback f^*F will be a sheaf on X defined by $(f^*F)(\gamma) = F(f(\gamma))$ and $(f^*F)(\sigma \rightsquigarrow \tau) = F(f(\sigma) \rightsquigarrow f(\tau))$.

Such notions (and others, such as the pushforward sheaf) are clearly useful for switching base spaces. Sheaf morphisms can also be composed, under certain conditions, leading to *sequences of sheaves*, linked together by sheaf morphisms. Certain sequences will even exhibit special algebraic properties, like *exactness*, which will have significance in a number of applications. We take up these matters in the next section, after a brief discussion of a few further operations on sheaves.

We have seen a few constructions—such as that of subsheaves, pullback sheaves, and pushforward sheaves—where old sheaves are used to generate new ones. Here is a very brief look at just a few other important things one can do *to* or *with* sheaves, to generate new sheaves; specifically, we focus on indicating a few of the *algebraic* operations one can perform on sheaves. These notions can be defined in greater generality, but we will stick to the cellular context.

Definition 229 For F and G, two sheaves of vector spaces on a cell complex X, we can define $F \oplus G$, their *direct sum*, in the natural way:

$$(F \oplus G)(\gamma) = F(\gamma) \oplus G(\gamma)$$

and

$$(F \oplus G)(\sigma \rightsquigarrow \tau)(v, w) = (F(\sigma \rightsquigarrow \tau)v, G(\sigma \rightsquigarrow \tau)w)$$

for $v \in F(\gamma)$ and $w \in G(\gamma)$.

In a similar fashion, we could define $F \otimes G$, the *tensor product* of two sheaves F, G, in the expected way; but there is a subtlety here when we try to think of this in terms of sheaves, and we will not be making use of this, so instead we will just indicate an example of a direct sum of sheaves.

Example 230 Consider a network, that is, a 1-D cell complex with oriented edges. Ghrist (2014) provides some nice applications of cellular sheaves over such networks, called *flow sheaves*, where these represent the flow of a commodity (as in supply chains or various information or transportation of goods moving through networks). The underlying graph supports certain viable flow values, and one of the purposes of the sheaf is to encode these feasibility conditions or constraints. Basically a sheaf on such a network will supply some algebraic structure encoding a particular collection of logical, numerical, stochastic, or other constraints on the "flows" or transport of commodities through a network. One of the advantages of using sheaves, in this setting, is that we can easily generalize beyond numerical constraints on a network to other (perhaps noisy or logical) constraints.

Here is a very rough sketch of how this works.[150] A *flow* (or *flow sheaf*) on a network X is an assignment of coefficients (e.g., in $\mathbb{Z}$ or $\mathbb{N}$) to each edge of X, in such a way that a particular "conservation" condition is met (namely the sum of the incoming edge flow values equals the sum of the outgoing edge values, except at the special "external" vertex, where they are not conserved). Each such value can be imagined as representing an amount of a commodity or resource in transit at a location of the underlying graph. Restrictions $F(v \rightsquigarrow e)$ are then projections onto components. The *direct sum* of two flow sheaves $F \oplus G$ could then be used to represent the transportation of two *different* resources being transported along the same given network. In other words, the sum $(F \oplus G)$ would represent the number of both sorts of resources, so $(F \oplus G)(e) = \mathbb{N}^2$ would represent the number of F-items and G-items being carried along the edge e.

9.2 Sheaf Cohomology

The impatient reader might well be wondering at this point: "Okay, I understand what a sheaf is already! But what good is all this?" One glib answer, following Hubbard, might be that "without cohomology theory, they aren't good for much"![151] While this seems too pessimistic—after all, even if all sheaves on their own did was organize a wealth of particular and highly disparate constructions involving local data into a powerfully general framework, this would be immensely valuable—it is a perspective that gets at something important. If sheaves represent local data—or, more precisely, represent how to properly ensure that what is locally the case everywhere is in fact, more than that, globally the case—sheaf cohomology is a device that lets us extract global information from local data and systematically explore, represent, and relate obstructions to the extension of the local to the global. In this way, sheaf cohomology can cope with situations where the local-to-global passage breaks down; and this is of immense value, since we would like to be able to handle and talk about structures that somehow fall short of assembling into sheaves. Moreover, as we mentioned at the outset of the chapter, we would like to appreciate a fact that Grothendieck insisted on, namely that individual sheaves are only of secondary importance—the real power of sheaf theory emerging from the use of constructions involving various sheaves, linked together via sheaf morphisms. Sheaf cohomology will allow us to glimpse this.

150. The reader who desires a less rough sketch of these notions is invited to look at Ghrist (2014).
151. See Hubbard (2006, 383).

In line with our categorical approach thus far, in our presentation sheaf cohomology will emerge as a *functor*, specifically as one with domain the category of sheaves (together with their sheaf morphisms) and codomain the category of vector spaces. The cellular sheaves that we will continue to work with, together with their cohomology, have the nice property that it is just as easy to compute with them as to compute the usual cellular cohomology of a cell complex, something one can learn about in more elementary contexts. Sheaf cohomology is particularly important to understand, though, so we do not assume that the reader already knows or recalls all they need to know about the basic notions of (co)homology. Over the next few pages, we accordingly build up to sheaf cohomology by first reviewing the basic notions of (co)homology with respect to ordinary simplices.[152]

9.2.1 Primer on (Co)Homology

Given oriented simplices (or cell complexes), as described above, a very natural thing is to look at the *boundary* of a given n-simplex. As one might expect, the boundary of a 1-simplex P_1P_2 is simply the vertices of the edge; however, we must now carefully attend to the issue of orientation. Taking the boundary, an operation denoted by ∂, is more precisely defined as taking the formal "difference" between the endpoint and the initial point, that is, $\partial_1(P_1P_2)=P_2-P_1$. Similarly, the boundary of a 2-simplex $P_1P_2P_3$ is given by

$$\partial_2(P_1P_2P_3)=P_2P_3-P_1P_3+P_1P_2.$$

This in fact corresponds to traveling around what we intuitively think of as the boundary of a triangle in the direction indicated by the orientation arrow. The boundary of a 3-simplex is then defined as

$$\partial_3(P_1P_2P_3P_4)=P_2P_3P_4-P_1P_2P_3+P_1P_2P_4-P_1P_2P_3.$$

The pattern should be clear, allowing us to define the boundary operator ∂_n more generally for $n>3$:

$$\partial_k(\sigma)=\sum_{i=0}^{k}(-1)^i(v_0,\ldots,\widehat{v_i},\ldots,v_k),$$

where the oriented simplex $(v_0,\ldots,\widehat{v_i},\ldots,v_k)$ is the i-th face of σ obtained by deleting its i-th vertex. Notice that each individual summand (i.e., the positive terms) of the boundary of a simplex is just a *face* of the simplex.

We can associate some *groups* to a given complex X. The group $C_n(X)$ of oriented n-chains of X is defined to be the free abelian group generated by the oriented n-simplices of X. Every element of $C_n(X)$ is a finite sum $\Sigma_i m_i\sigma_i$, where the σ_i are n-simplices of X and $m_i\in\mathbb{Z}$. Then the addition of chains is carried out by algebraically combining the coefficients of each occurrence in the chains of a given simplex. For instance, considering the surface of a tetrahedron S (oriented in an obvious way), the elements of $C_2(S)$ will look like $m_1P_2P_3P_4+m_2P_1P_3P_4+m_3P_1P_2P_4+m_4P_1P_2P_3$, while an element of $C_1(S)$ will look like $m_1P_1P_2+m_2P_1P_3+m_3P_1P_4+m_4P_2P_3+m_5P_2P_4+m_6P_3P_4$.

152. Though, again, technically we ought to be working with the more general *cell complexes* and their cellular maps.

Now observe that if σ is an n-simplex, then applying the boundary operator to σ will land us inside the group of $(n-1)$-chains, that is, $\partial_n(\sigma) \in C_{n-1}(X)$.[153] Moreover, since $C_n(X)$ is a free abelian group—thus enabling us to describe a *homomorphism* of such a group by specifying its values on generators—it is clear that ∂_n describes a boundary homomorphism mapping $C_n(X)$ into $C_{n-1}(X)$. In other words,

$$\partial_n\Big(\sum_i m_i\sigma_i\Big) = \sum_i m_i\partial_n(\sigma_i).$$

For instance,

$$\begin{aligned}\partial_1(3P_1P_2 - 4P_1P_3 + 5P_2P_4) &= 3\partial_1(P_1P_2) - 4\partial_1(P_1P_3) + 5\partial_1(P_2P_4)\\ &= 3(P_2 - P_1) - 4(P_3 - P_1) + 5(P_4 - P_2)\\ &= P_1 - 2P_2 - 4P_3 + 5P_4.\end{aligned}$$

But since we have a homomorphism, we are naturally drawn to look at two things: the *kernel* and the *image*. The kernel of ∂_n will consist of those n-chains with boundary zero, and so the elements of the kernel are just n-cycles. We sometimes denote the kernel of ∂_n, the group of n-cycles, by $Z_n(X)$. So for instance, if $q = P_1P_2 + P_2P_3 + P_3P_1$, then $\partial_1(q) = (P_2 - P_1) + (P_3 - P_2) + (P_1 - P_3) = 0$. Note that this q corresponds to a cycle around the triangle with vertices P_1, P_2, P_3 (oriented in the obvious way). Furthermore, we can consider the image under ∂_n, namely the group of $(n-1)$-boundaries, which consists of those $(n-1)$-chains that are boundaries of n-chains. We sometimes denote this group $B_{n-1}(X)$.

Homomorphisms compose and it is a well-known fact that the composite homomorphism $\partial_{n-1}\partial_n$ taking $C_n(X)$ into $C_{n-2}(X)$ takes everything into zero, that is, for each $c \in C_n(X)$, we have that $\partial_{n-1}(\partial_n(c)) = 0$, or $\partial^2 = 0$. A corollary of this is that $B_n(X)$ is a *subgroup* of $Z_n(X)$, allowing us to form the quotient or factor group $Z_n(X)/B_n(X)$ which we denote $H_n(X)$ and call the n-dimensional *homology group* of X. While perhaps obvious, it is important to realize that this quotient simply puts an equivalence relation on Z_n with respect to B_n, that is, $\omega \sim \sigma \iff \omega - \sigma \in B_n$, and so is technically represented as some coset.

The important thing to realize now is that we can form the *sequence of chain groups* linked together by such boundary homomorphisms, a sequence we call the *chain complex*:

$$\cdots \xrightarrow{\partial_{n+2}} C_{n+1} \xrightarrow{\partial_{n+1}} C_n \xrightarrow{\partial_n} C_{n-1} \xrightarrow{\partial_{n-1}} \cdots$$

Moreover, if $C = \langle C, \partial\rangle$ is the previous (in principle doubly infinite) sequence of abelian groups together with the collection of homomorphisms satisfying the condition that each map descends by one dimension and that $\partial^2 = 0$, then we can extend all the above reasoning to the sequences themselves and immediately see that under these conditions the image under ∂_k will be a subgroup of the kernel of ∂_{k-1}. In brief, we can define the kernel $Z_k(C)$ of ∂_k as the group of k-cycles, the image $B_k(C) = \partial_{k+1}[C_{k+1}]$ as the group of k-boundaries, and then the factor group $H_k(C) = Z_k(C)/B_k(C) = \ker \partial_k/\text{image } \partial_{k+1}$ as the k-th homology group of C. In other words, H_k gives all the vectors that are annihilated in stage k that were not already present in stage $k+1$. If for all k in a sequence we have that the image under ∂_k

153. For the record, if we define $C_{-1}(X) = \{0\}$, the trivial group of one element, then $\partial_0(\sigma) \in C_{-1}(X)$.

is equal to the kernel of ∂_{k-1}, then we have what is called an *exact sequence*. While exact sequences are chain complexes, the converse is not true, since a chain complex need only satisfy that the image (of the prior map) is contained in the kernel (of the subsequent map). The important thing to realize here is that *homology just measures the difference* between the image and the kernel maps.

For simplicial complexes X and Y, a map f from X to Y induces a mapping (i.e., homomorphism) of homology groups $H_k(X)$ into $H_k(Y)$. This arises if we consider that for certain triangulations of X and Y, the map f will give rise to a homomorphism f_k of $C_k(X)$ into $C_k(Y)$, which moreover commutes with ∂_k, that is, $\partial_k f_k = f_{k-1}\partial_k$.

We can dualize this entire account to get an account of *co*homology, and we do so briefly now since it will be important for what follows. Consider a simplicial complex X. For an oriented n-simplex σ of X, we can define the *coboundary* $\delta^n(\sigma)$ of σ as the $(n+1)$-chain summing up all the $(n+1)$-simplices τ that have σ as a face. In other words, we are summing those τ that have σ as a summand of $\partial_{n+1}(\tau)$. For instance, if we let X be the simplicial complex consisting of the solid tetrahedron, then $\delta^2(P_3P_2P_4) = P_1P_3P_2P_4$, while $\delta^1(P_3P_2) = P_1P_3P_2 + P_4P_3P_2$.

We can also define the group $C^n(X)$ of n-cochains as the same as the group $C_n(X)$. However, the coboundary maps δ^n go the other way from the boundary maps, that is, we have $\delta^n : C^n \to C^{n+1}$, defined by

$$\delta^n\left(\sum_i m_i\sigma_i\right) = \sum_i m_i\delta^n(\sigma_i).$$

We can then build up sequences of cochain groups into cochain complexes, just as one would expect. Cochain complexes in general can be thought of as looking at how objects are related to larger superstructures instead of to smaller substructures (as was the case for chain complexes). Just as before, we could show that $\delta\delta = 0$, that is, that $\delta^{n+1}(\delta^n(c)) = 0$ for each $c \in C^n(X)$. Moreover, we can define the group $Z^n(X)$ of n-cocycles of X as the kernel of the coboundary homomorphism δ^n, the group B^n of n-coboundaries of $C^n(X)$ as the image of δ^{n-1}, that is, $\delta^{n-1}[C^{n-1}(X)]$; and since we have that $\delta\delta = 0$, again $B^n(X)$ will be a subgroup of $Z^n(X)$. This last fact allows us to define the n-dimensional *cohomology group* $H^n(X)$ of X as $Z^n(X)/B^n(X)$, that is, the kernel of the map "going out" mod the image of the map "coming in." A cocycle that is not a coboundary is topologically informative—and this difference gives rise to a "nontrivial" cohomology group (or nontrivial cocycle).

Example 231 Let us look a little closer at these ideas through the lens of the Penrose tribar, introduced at the beginning of the chapter. Let Q be the region of the plane on which the tribar is drawn. We can consider Q as being pasted together from three pieces, as in:[154]

154. This image, and the next, are modified from Phillips (2021). The exposition that follows is derived from Penrose (1992) and Phillips (2021).

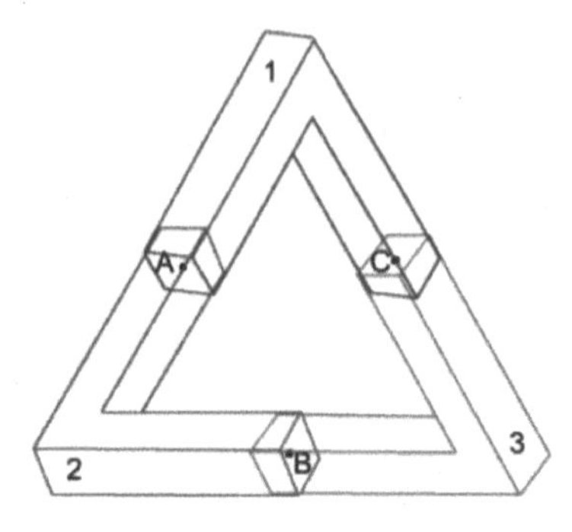

The three pieces overlap and we can imagine having chosen points in each of the three overlapping regions—for example, A in the overlap between piece 1 and piece 2; B in the overlap between piece 2 and piece 3; and C in the overlap between piece 1 and piece 3. Then imagine "pulling apart" the tribar into these components, and relabeling the points appropriately, so that A_1 in piece 1 and A_2 in piece 2 both correspond to A, points B_2, B_3 correspond to B, and C_1, C_3 correspond to C, as in:

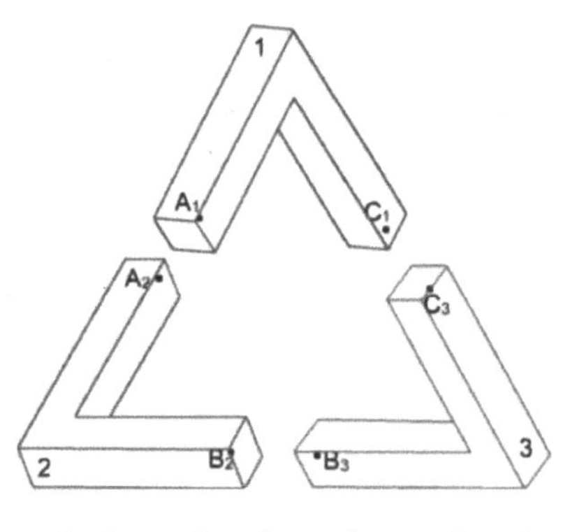

We take the tribar to be patched together from the overlapping smaller drawings, where each of these three components is taken to be a perspective drawing of a corresponding object in space. Observe how such a perspective drawing—representing 3-D space in the plane—implies that certain of the points are meant to be further from or closer to the eye E of a viewer than other points. For instance, letting $d(E, A_1)$ denote the implied distance from a viewer E to point A_1 in space, this may differ from $d(E, A_2)$—in accordance with conventions of perspective where, say, the corner of piece 2 is implied as closer than the corner of piece 1, forcing A_1 to be further from the eye than A_2. Thus, we can consider the ratio

$$d_{12} = \frac{d(E, A_1)}{d(E, A_2)},$$

which describes piece 1's implied distance in relation to that of piece 2. Similarly, we can let $d_{13} = \frac{d(E,C_1)}{d(E,C_3)}$ and $d_{23} = \frac{d(E,B_2)}{d(E,B_3)}$. Of course, the other ratios d_{21}, d_{31}, and d_{32} would then be defined by just taking the reciprocal, for example, $d_{21} = \frac{1}{d_{12}}$. Observe that these values d_{ij} do not depend on the chosen points—since any two points in the relevant pieces would yield the same value for the ratio.

While the drawing of each piece is a consistent rendering of a 3-D structure, there is some "ambiguity" in any interpretation of the perspective drawing of the spatial objects:

the distance of the object depicted, in relation to the viewer's eye, is not known. The implied object could be three times as big and three times further away and appear the same as the original. This perceived distance can accordingly be tracked with positive real numbers in $\mathbb{R}^+$, and we can use the factors λ to track how the perceived distances change, for example, if the perceived distance of piece 1 is changed by a factor of λ then d_{12} and d_{13} will have to be multiplied by λ, and d_{21} and d_{31} divided by λ. Clearly, if the hypothetical tribar (or any other drawing of a figure) could indeed be consistently realized in 3-D space—that is, if the pieces 1, 2, and 3 could indeed be fused into a viable 3-D object—then we would expect that we could effectively rescale each of the three pieces by factors $\lambda_1, \lambda_2, \lambda_3$ until they came together into one consistent structure. And their "coming together into one consistent structure" would be captured by rescalings that make each of the $d_{ij} = 1$. In other words, we would be able to find three positive real numbers such that $(\lambda_i/\lambda_j)d_{ij} = 1$ for each different i, j, or (what is the same)

$$d_{12} = \frac{\lambda_2}{\lambda_1}, d_{13} = \frac{\lambda_3}{\lambda_1}, d_{23} = \frac{\lambda_3}{\lambda_2}.$$

On the other hand, if no such factors $\lambda_1, \lambda_2, \lambda_3$ exist, then the hypothetical object would be impossible to realize in 3-D space.

This situation, and the decisive impossibility, can be codified in the language of cohomology, specifically attending to the first cohomology group. First, let us consider some terminology. Here, a collection $\{d_{ij}\}$ will be a *cocycle*. If it respects the equations shown above, then the cocycle is a *coboundary*. Two cocycles are regarded as equivalent if they can be converted to one another using a rescaling factor λ, that is, when we shift the distance from which object Q_i is being viewed by replacing the pair (d_{ij}, d_{ik}) with $(\lambda d_{ij}, \lambda d_{ik})$ for some positive real number λ. Under such an equivalence, we are left with the elements of the *cohomology group* $H^1(Q, \mathbb{R}^+)$, the quotient group of 1-cocycles modulo 1-coboundaries. The trivial "unit" element of $H^1(Q, \mathbb{R}^+)$ is thus given by the coboundaries—so, checking whether the figure depicted in Q is an impossible figure amounts to checking whether the element of $H^1(Q, \mathbb{R}^+)$ is the unit.

More explicitly, our object Q is the union of $n = 3$ subsets—given by the three parts Q_1, Q_2, Q_3—where, technically, each of the Q_i is topologically a solid ball, as are the intersections $Q_{ij} = Q_i \cap Q_j$. With our setup, a 0-dimensional cochain (taking values in $\mathbb{R}^+$) assigns a number $\lambda_i \in \mathbb{R}^+$ to each of the Q_i. A 1-dimensional cochain assigns a number (say, a_{ij}) to each of the Q_{ij}, where $a_{ji} = 1/a_{ij}$, allowing us to take account of orientation and treat Q_{ij} and Q_{ji} as different (even though they correspond to the same set). A two-dimensional cochain would assign a number to each threefold intersection—yet, in this example, there are no such intersections. Supposing some collection of λ_i is our 0-cochain, its coboundary $\delta\lambda$ just assigns to each Q_{ij} the number λ_j/λ_i. If Q had threefold intersections, then the coboundary of the 1-cochain $\{a_{ij}\}$ would be the 2-cochain δa that assigns the number $\delta a_{ijk} = a_{ij}a_{jk}/a_{ik}$ to Q_{ijk}. Observe that we would have $\delta\delta = 1$, which 1 is the unit in the group $\mathbb{R}^+$ of coefficients.

As before, a cochain with coboundary equal to 1 (the unit) is called a *cocycle*. Of course, if there happen to be no cochains in the next higher dimension—as is the case with 1-cochains of the tribar—then every cochain is automatically a cocyle. A cocycle that is *not* a coboundary is said to be "nontrivial"—and it is these cocycles that are of interest.

In particular, if a drawing is of an impossible object, then decomposing it into possible pieces, selecting points in the intersections of those pieces, and defining the numbers d_{ij} will yield a nontrivial cocycle, that is, one that is not a coboundary. In looking at the first cohomology group of Q, $H^1(Q, \mathbb{R}^+)$, we would ordinarily really be working with the cohomology of the *cover* $\mathcal{U} = \{Q_1, Q_2, Q_3, Q_{12}, Q_{13}, Q_{23}\}$—however, since each of these elements of the cover is topologically equivalent to a solid ball, we are able to work directly with Q.

In this construction, there is an implied evaluation map ϵ that takes a 1-cochain d to the product $d_{12}d_{23}d_{31}$, which will be an element of $\mathbb{R}^+$. A 1-cochain d will then be a coboundary iff $\epsilon(d) = 1$. To see this, observe that if there were a 0-chain $\lambda = \{\lambda_1, \lambda_2, \lambda_3\}$ with $d = \delta\lambda$ (i.e., such that $d_{ij} = \frac{\lambda_j}{\lambda_i}$), then all the factors of λ in the product $d_{12}d_{23}d_{31}$ would cancel; conversely, if $d_{12}d_{23}d_{31} = 1$, then one could just define $\lambda_1 = 1, \lambda_2 = d_{12}, \lambda_3 = d_{13}$ and then get $d = \delta\lambda$. It would remain to check that $d_{23} = \frac{\lambda_3}{\lambda_2}$, but since $\epsilon(d) = 1$, this would be $d_{13}/d_{12} = 1/(d_{31}d_{12}) = d_{23}$.

Considering the tribar, then, we would just calculate

$$\epsilon(d) = \frac{d(E, A_1)}{d(E, A_2)} \frac{d(E, B_2)}{d(E, B_3)} \frac{d(E, C_3)}{d(E, C_1)}.$$

Considering pairs of the three pieces, and adopting a counterclockwise orientation (starting at A), observe that ordinary perspective implies A_1 is further from the eye than A_2—making the first factor $\frac{d(E,A_1)}{d(E,A_2)}$ greater than 1. B_2 and B_3, for their part, are viewed as being the same distance away, making the factor $\frac{d(E,B_2)}{d(E,B_3)} = 1$. Finally, C_3 is regarded as further from the eye than C_1, making the last factor $\frac{d(E,C_3)}{d(E,C_1)}$ greater than 1. Altogether, we have one factor equal to 1 and the other two factors greater than 1—thus, whatever their particular values, their product $\epsilon(d) \neq 1$. In other words, there is no way of rescaling the distances of the pieces to make them all fit together to assemble into the tribar. This impossibility is codified by the nontriviality of the above cocycle d, that is, by the nontriviality of the first cohomology group H^1.

(Co)chain complexes can be thought of as representing a cell complex within the context of linear algebra, expressing the action of taking the boundary of a cell in terms of a linear transformation. If we put this latter approach together with the development of ASCs from before, we get what is usually called simplicial (co)homology. We can thus start with some arbitrary ASC X or its realization and turn it into a chain complex (for instance). This then allows us to compute its homology. Note that what we are doing here is moving *via functors* from **Top** (after having placed the appropriate topology on X) to the category of chain complexes **Chn**, finally landing in the category **Vect**. The idea here is that we can use the algebraic properties exhibited by the composite of functors $H_\bullet$ to view the topological properties of the original ASC. Initially, the C_k and their maps may be vector spaces over some field like $\mathbb{F}_2$ (the field of two elements, 0 and 1), and so the simplicial homology of X, denoted $H_\bullet(X; \mathbb{F}_2)$ will take coefficients in $\mathbb{F}_2$ ("on" or "off"). But we could also let (co)homology take coefficients elsewhere—for instance, as we saw a moment ago, in $\mathbb{R}$ or $\mathbb{R}^+$ (thereby describing simplices' intensities, say, as opposed to the simple "on-off" of $\mathbb{F}_2$). Eventually, the idea here is to abstract further and let (co)homology take *sheaves as coefficients*.

When we start computing simplicial homology, we must choose and fix an ordering on the list of vertices in each simplex. We use coefficients other than $\mathbb{F}_2$, like $\mathbb{R}$, and the boundary maps will be used to track orientation.[155] In brief, for an arbitrary simplicial complex X, $C_k(X)$ will be a vector space whose dimension is the number of k-simplices of X, that is, the vector space whose basis (each row as one of the k-simplices, with some coefficients) is the list (given some fixed ordering) of k-simplices of X. The boundary maps $\partial_k : C_k(X) \to C_{k-1}(X)$ go down in dimension, from the chain space built of the k-dimensional simplices to the chain space built of the $k-1$ simplices. In other words, the map acts on the ordered set $[v_0, \ldots, v_k]$ and should be a linear map. Indeed it takes some linear combination of k-simplices and returns some linear combination of $(k-1)$-simplices.

As a very simple example illustrating these ideas, consider the following ASC $X = \{[v_0], [v_1], [v_0 v_1]\}$, realized geometrically as

$$v_0 \circ\!\!\xrightarrow{[v_0 v_1]}\!\!\circ\, v_1$$

Then the chain complex associated to this is given by

$$C_2 \xrightarrow{\partial_2} C_1 \xrightarrow{\partial_1} C_0 \xrightarrow{\partial_0} C_{-1}$$

$$0 \xrightarrow{\partial_2} \mathbb{R} \xrightarrow{\partial_1 = \begin{pmatrix} -1 \\ 1 \end{pmatrix}} \mathbb{R}^2 \xrightarrow{\partial_0} 0.$$

By inspection, one can see that $H_2(X)$ (and all higher homology groups) must be zero (the trivial group). Moreover, the kernel of the ∂_1 map is trivial, and the image of ∂_2 (a 1×0 matrix map) is zero, so $H_1(X)$ must be zero. As for H_0, on the other hand, by inspection we observe that the kernel of ∂_0 is everything, that is, $\mathbb{R}^2$. The image of ∂_1 is spanned by the $\partial_1 = \begin{pmatrix} -1 \\ 1 \end{pmatrix}$ matrix. Taking the quotient, then, we see that the dimension of H_0 must be 1. Strictly speaking, H_0 is a coset, and has a vector space that is isomorphic to a single copy of $\mathbb{R}$. Thinking of it in terms of its coset representation, parameterized by one free parameter, one can think of this as *identifying* the two points on account of the fact that they happen to be connected via an edge. As H_0 effectively represents the connected components, this should make good sense. Since the dimension of $H_1(X)$ can be thought of as picking out the number of *cycles* in the graph (i.e, among the vertices and edges) that are not filled in by two-dimensional simplices, it should also be intuitively clear that H_1 ought to be trivial in this example. If we had found an H_1 not equal to zero, say for another simplex, this might be indicating that the simplices all fit together in some fashion but that they cannot be glued together into one big construct, on account of some kind of obstruction or "hole." The dimension of $H_k(X)$ is usually referred to as k-cycles, and for nontrivial values this can be thought of as picking out $(k+1)$-dimensional "voids" or holes.[156]

155. Really, though, we would like to generalize beyond *field* coefficients, say to $\mathbb{Z}$ coefficients, making each C_k a $\mathbb{Z}$-module. In this case, the chains will record finite collections of simplices with some orientation and multiplicity. We then move to a chain complex over an R-module, where R is a ring, the boundary maps being module homomorphisms. Then we have a sufficiently general definition: a chain complex $\mathscr{C} = (C_\bullet, \partial)$ is any sequence of R-modules C_k with homomorphisms $\partial_k : C_k \to C_{k-1}$ satisfying $\partial_k \circ \partial_{k+1} = 0$.

156. We could further develop this story in a number of directions, for instance going on to define the Betti numbers, which indicate various levels of obstructions, stringing them together as k varies; but we leave the curious reader to pursue these matters on their own.

We now return to simplicial or cellular maps.[157] Just as X can be unfolded into a chain complex $C_\bullet(X)$, cellular maps $f: X \to Y$ between cell complexes X and Y can be unfolded to yield a sequence $f_\bullet$ of homomorphisms $C_k(X) \to C_k(Y)$. Since f is continuous, it induces a *chain map* $f_\bullet$ that "plays nicely" with the boundary maps of $C_\bullet(X)$ and $C_\bullet(Y)$, meaning that the following diagram is made to commute:

$$\begin{array}{ccccccccc} \cdots \longrightarrow & C_{n+1}(X) & \xrightarrow{\partial} & C_n(X) & \xrightarrow{\partial} & C_{n-1}(X) & \xrightarrow{\partial} & \cdots \\ & \downarrow f_\bullet & & \downarrow f_\bullet & & \downarrow f_\bullet & & \\ \cdots \longrightarrow & C_{n+1}(Y) & \xrightarrow{\partial'} & C_n(Y) & \xrightarrow{\partial'} & C_{n-1}(Y) & \xrightarrow{\partial'} & \cdots. \end{array}$$

Since the squares in this diagram commute, meaning the chain map respects the boundary operator, we have that f will act not just on chains but on cycles and boundaries as well, which entails that it induces the homomorphism $H(f): H_\bullet(X) \to H_\bullet(Y)$ on homology. We are dealing with functors! In particular, the functoriality of homology means that the induced homomorphisms will reflect, algebraically, the underlying properties of continuous maps between spaces. Chain maps, in respecting the boundary operators, send neighbors to neighbors, and thus capture an essential feature of the underlying continuous maps. Via such functors, we can build up big "quiver" diagrams with composable maps between chain complexes, that is, between one complex representation and the next.

Overall, the idea is that homology should algebraically capture changes that happen to the underlying complex structure. Homology proceeds by first replacing topological spaces with complexes of algebraic objects. It then takes a hierarchically ordered sequence of these parts "chained together," that is, a chain complex, as input and returns the global features. Then other topological concepts—like continuous functions and homeomorphisms—have analogues at the level of chain complexes. *Cohomology*, for its part, is just the homology of the cochain complex.

In terms of the big picture, then, homology can be seen as a way of translating topological problems into algebraic ones. Specifically, it will (in the general approach) translate a topological problem into a problem about modules over commutative rings; but when we can take coefficients in a *field*, this actually amounts to a translation of the topological problem into one of linear algebra. And this is one of the principal motivations! Linear algebra is generally much simpler than topology, in part because of how dimension can classify finite-dimensional vector spaces (thus the centrality of "dimension formula" that relates the kernel of a linear transformation to its image). The "long exact sequences" we started to look at are basically fancy versions of the dimension formula. In the context of sheaf theory, we can develop powerful ways of building long exact sequences of cohomology spaces from short exact sequences of sheaves, where such sequences can already tell us a lot on their own.

9.2.2 Cohomology with Sheaves

We now proceed, at last, to the construction of the cochain complex for a sheaf F, where each C^k will comprise the stalks, stacked together, over the k-simplices, and the coboundary maps (denoted with δ^k) are built by gluing together a bunch of restriction maps. Once we

157. We ignore more sophisticated issues here, like maps between different dimensions.

have a sheaf, the tactic for computing all sections at once is to build a chain complex (where we go up in dimension, so that really we have a *cochain* complex) and examine its zero-th homology. The difference between the (co)homology of spaces and sheaf (co)homology mostly just has to do with the fact that with sheaves we are again looking at functions on a space, but the range of these functions is allowed to vary, that is, we will have a collection of possible outputs, where the output space of the functions can change as we move around the domain space.

The basic idea here can be nicely explained as follows.[158] Suppose we have a simplicial complex, say, for simplicity

$$v_1 \circ\!\!-\!\!-\!\!-\!\!\circ\, v_2$$

with the attachment diagram (displayed on top) and the corresponding data of the sheaf (below):

$$v_1 \longrightarrow e \longleftarrow v_2$$

$$F(v_1) \xrightarrow{F(v_1 \rightsquigarrow e)} F(e) \xleftarrow{F(v_2 \rightsquigarrow e)} F(v_2)\,.$$

Now suppose that s is a global section of this sheaf. Then obviously we must have

$$F(v_1 \rightsquigarrow e)s(v_1) = s(e) = F(v_2 \rightsquigarrow e)s(v_2),$$

where $F(v_i \rightsquigarrow e)$ is the restriction map, $s(v_1)$ is a section belonging to $F(v_1)$, and $s(v_2)$ is a section belonging to $F(v_2)$. But now we need only observe that this equation in fact holds in a vector space, which means that we can rewrite it $F(v_1 \rightsquigarrow e)s(v_1) - F(v_2 \rightsquigarrow e)s(v_2) = 0$, or in matrix form:

$$\left(+F(v_1 \rightsquigarrow e) \middle| -F(v_2 \rightsquigarrow e) \right) \begin{pmatrix} s(v_1) \\ s(v_2) \end{pmatrix} = 0.$$

Note that these $F(v_i \rightsquigarrow e)$ entries are just the restriction maps, so in the context of sheaves of vector spaces, they will in general be (potentially large) matrices the entries of which will be given by the linear (restriction) maps.[159]

Extending this reasoning to arbitrary simplicial complexes (which we assume comes with a listing of vertices in a particular order, say lexicographic for concreteness), we can observe that computing the space of global sections of a sheaf is equivalent to computing the kernel of a particular matrix. Moreover, the matrix

$$\left(+F(v_1 \rightsquigarrow e) \middle| -F(v_2 \rightsquigarrow e) \right)$$

generalizes into the coboundary map

$$\delta^k : C^k(X; F) \to C^{k+1}(X; F),$$

158. The next two paragraphs closely follow Robinson (2014).

159. Note that, in the context of the (co)homology groups defined earlier in terms of equivalence relations and cosets, saying that the difference of the two restriction maps is equal to their value along the edge (i.e., $s(e)$), is effectively the same as saying that their difference can be regarded as *zero*.

which takes an assignment s on the k-faces to another assignment $\delta^k(s)$ whose value at a $(k+1)$-face b is

$$(\delta^k(s))(b) = \sum_{\text{all k-faces } a \text{ of } X} [b:a]F(a \rightsquigarrow b)s(a),$$

where $[b:a]$ is defined to be 0 if a (a k-simplex) is not a face of b (a $(k+1)$-simplex) and $(-1)^n$ if you have to delete the n-th vertex of b to get back a.[160] This makes sense since a and b must differ by one dimension, so either a is not a face of b, or a is a face of b, in which case they will differ by exactly one vertex (i.e., one need only delete one of the vertices of b to get a). The sign mechanism tied to the deleted vertex allows us to track and respect the chosen orientation given to the complex.

The matrix kernel construction enables us to extend this approach to higher dimensions and over simplices that are arbitrarily larger. Then we can show that we have a cellular cochain complex for a sheaf F on some simplicial complex X. To form the cochain spaces $C^k(X;F)$, we just collect the stalks over vertices (the domain) and edges (the codomain) together via direct sum, so that an element of $C^k(X;F)$ comes from the stalk at each k-simplex:

$$C^k(X;F) = \bigoplus F(a),$$

where a is a k-simplex. Moreover, the coboundary map $\delta^k : C^k(X;F) \to C^{k+1}(X;F)$ is defined as the block matrix where for row i and column j, the (i,j)-th entry is given by $[b_i : a_j]F(a_j \rightsquigarrow b_i)$, where the $[b_i : a_j]$ term is either $0, +1$, or -1, depending on the relative orientation of a_j and b_i, assuming one is a face of the other. Carrying on in this way, we have defined the cellular cochain complex:

$$\cdots \xrightarrow{\delta^{k-2}} C^{k-1}(X;F) \xrightarrow{\delta^{k-1}} C^k(X;F) \xrightarrow{\delta^k} C^{k+1}(X;F) \xrightarrow{\delta^{k+1}} \cdots.$$

Using the cellular sheaf cohomology group definition,

$$H^k(X;F) = \ker \delta^k / \text{image } \delta^{k-1},$$

we will recover all the cochains that are *consistent* in dimension k (kernel part) but that did not yet show up in dimension $k-1$ (image part). Notice that $H^0(X;F) = \ker \delta^0$, that is, the zero-th cohomology classes, will be those assignments s that are global sections of the sheaf F. For $k > 0$, the nontrivial elements of H^k will represent some calculable obstruction to globally consistent fusions. For instance, the appearance of nontrivial cohomology in H^1 supplies an explicit measure of the failure of local data to patch.[161]

All of this is perhaps better illustrated with our running example.

Example 232 We reproduce our running example below for convenience:

160. We start counting at 0.

161. In this vein, we saw how Penrose (1992) displayed the global impossibility of the locally realistic figure of his tribar in terms of a nontrivial element of first cohomology—that is, the first cohomology group $H^1(Q, \mathbb{R}^+)$ of some region of the plane on which the drawing of the impossible figure is made, with coefficients in $\mathbb{R}^+$. But that example could have also been presented, in fundamentally the same way, by taking coefficient values in the sheaf of positive functions (under pointwise multiplication). For a number of interesting examples and discussion of further aspects of sheaf cohomology, see Curry (2014) and Robinson (2014).

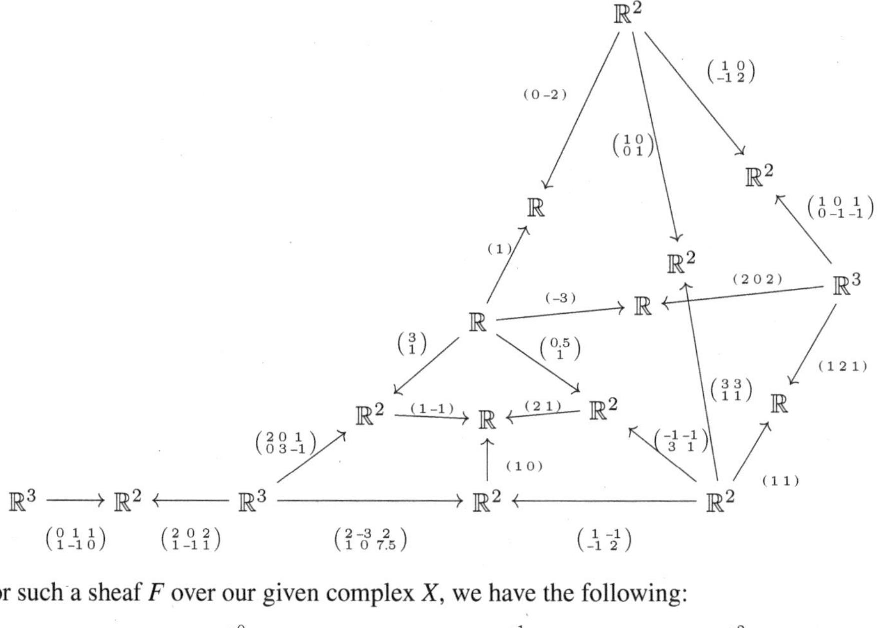

For such a sheaf F over our given complex X, we have the following:

$$H^0(X;F) \xrightarrow{\quad\delta^0\quad} H^1(X;F) \xrightarrow{\quad\delta^1\quad} H^2(X;F) \xrightarrow{\delta^2} H^3(X;F)$$

$$\mathbb{R}^2 \oplus \mathbb{R}^3 \oplus \mathbb{R}^2 \oplus \qquad\qquad \mathbb{R}^2 \oplus \mathbb{R}^2 \oplus \mathbb{R} \oplus \mathbb{R} \oplus$$

$$\mathbb{R} \oplus \mathbb{R}^3 \oplus \mathbb{R}^3 \xrightarrow{\delta^0} \mathbb{R} \oplus \mathbb{R}^2 \oplus \mathbb{R}^2 \oplus \mathbb{R}^2 \oplus \mathbb{R}^2 \xrightarrow{\quad\delta^1\quad} \mathbb{R} \xrightarrow{\quad\delta^2\quad} 0$$

$$\begin{pmatrix} s(a) \\ s(b) \\ s(c) \\ s(d) \\ s(e) \\ s(f) \end{pmatrix} \xrightarrow{\qquad\delta^0\qquad} \begin{pmatrix} s(ab) \\ s(ac) \\ s(ad) \\ s(bc) \\ s(bd) \\ s(cd) \\ s(ce) \\ s(de) \\ s(ef) \end{pmatrix} \xrightarrow{\qquad\delta^1\qquad} \left(s(cde) \right) \xrightarrow{\quad\delta^2\quad} 0$$

and where δ^0 is given by

	a ↓	b ↓	c ↓	d ↓	e ↓	f ↓
[ab] →	$-F(a \rightsquigarrow ab)$	$F(b \rightsquigarrow ab)$	0	0	0	0
[ac] →	$-F(a \rightsquigarrow ac)$	0	$F(c \rightsquigarrow ac)$	0	0	0
[ad] →	$-F(a \rightsquigarrow ad)$	0	0	$F(d \rightsquigarrow ad)$	0	0
[bc] →	0	$-F(b \rightsquigarrow bc)$	$F(c \rightsquigarrow bc)$	0	0	0
[bd] →	0	$-F(b \rightsquigarrow bd)$	0	$F(d \rightsquigarrow bc)$	0	0
[cd] →	0	0	$-F(c \rightsquigarrow cd)$	$F(d \rightsquigarrow cd)$	0	0
[ce] →	0	0	$-F(c \rightsquigarrow ce)$	0	$F(e \rightsquigarrow ce)$	0
[de] →	0	0	0	$-F(d \rightsquigarrow de)$	$F(e \rightsquigarrow de)$	0
[ef] →	0	0	0	0	$-F(e \rightsquigarrow ef)$	$F(f \rightsquigarrow ef)$

and δ^1 by

	[ab] ↓	[ac] ↓	[ad] ↓	[bc] ↓	[bd] ↓	[cd] ↓	[ce] ↓	[de] ↓	[ef] ↓
[cde] →	0	0	0	0	0	$F(cd \rightsquigarrow cde)$	$-F(ce \rightsquigarrow cde)$	$F(de \rightsquigarrow cde)$	0

and δ^2 is trivial. Computing the cohomologies here will require doing some linear algebra and row reduction, which is left to the reader. Observe that in general by adding in more higher-dimensional consistency checks—that is, filling in data corresponding to the "missing" higher-dimensional simplices of the underlying simplex—we would be able to reduce the kernel of δ^1, and thus get closer to reducing the nontrivial H^1.

It is worth emphasizing a more general truth, namely that the space of global sections of a sheaf F on a cell complex X will be isomorphic to $H^0(X; F)$. Moreover, this H^k (i.e., cohomology with sheaves as coefficients) is a functor, which means that when we have sheaf morphisms between sheaves, they will induce linear maps between their cohomology spaces, allowing us to extend things still further. Ultimately, it is also worth noting that while we have been focused on cellular sheaves, *general sheaf cohomology* can be shown to be isomorphic to cellular sheaf cohomology, so this sort of example is really part of a much more general story. And the vector space version of homological algebra we have used in order to provide a concrete example—while useful for many applications and for building intuition—does not display the full power of these notions. Ultimately one would like to (and could) retell the story using rings, modules, and other categories.

Via cohomology, global (in)compatibilities between pieces of local data can display global qualitative features of the data structure. At a very high level, the general idea of all this is suggested by the following figure, illustrating how we get a comparatively simple algebraic representation of features of spaces:

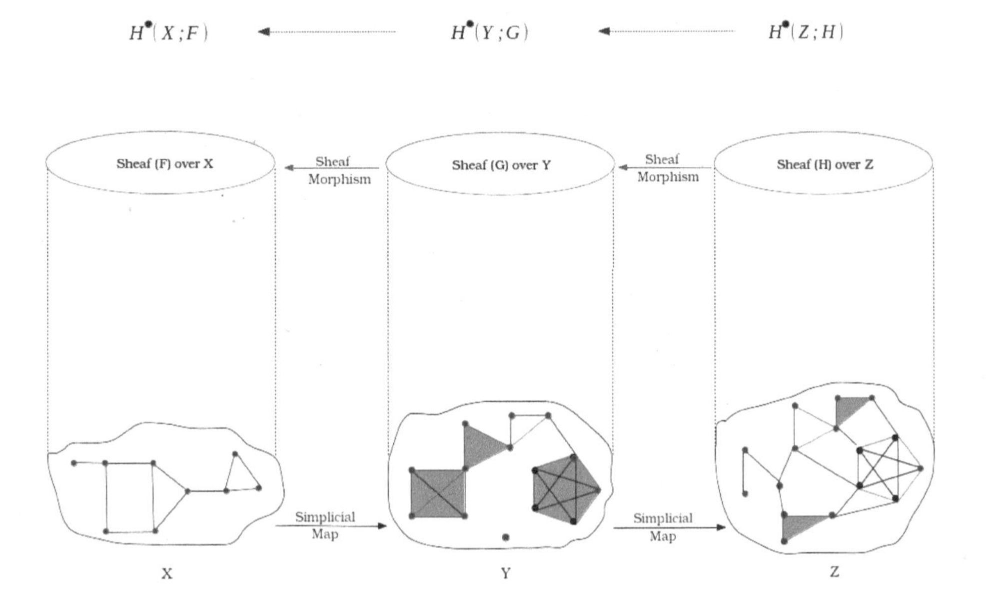

9.3 Philosophical Pass: Sheaf Cohomology

Box 9.1
Philosophical Pass: Sheaf Cohomology

If the sheaf compatibility conditions require controlled transitions from one local description to another, enabling progressive patching of information over overlapping regions until a unique value assignment emerges over the entire region—something that is captured by the vanishing of the group H^0 (yielding our global sections)—higher (nonvanishing) cohomology groups basically detect and measure (in an algebraic fashion) obstructions to such local patching and consistency relations among various dimensional subsystems. In other words, it can be thought of as measuring (for some cover) how many incompatible (purely local) systems we would have to "discard" in order to be left with only the compatible systems. In this way, sheaf cohomology moreover allows us to examine the relationship between information valid globally and the underlying topology of the space.

A proper discussion of the possible invariants that emerge in cohomology, especially as we ascend in dimension, would require a much longer and more detailed discussion. Instead, here are some very general reflections on the idea of sheaf cohomology. Referencing the high-level figure on the previous page, in forming sheaves (middle level) over the discrete approximations of spaces via their triangulations (bottom level), a more "continuous" perspective is recaptured. However, the nonvanishing cohomology groups (top level) give an algebraic (more "discrete") representation of something like the *resistance* of certain information (assigned to a part of a space) to integration into a more global system. In short, if the collation condition in the sheaf construction aligns them with continuity in the sense that it ensures smooth passage from the local to the global, cohomology with sheaves is something like its discrete counterpart providing us an algebraic measure of when such local-global passages might be blocked. In this respect, sheaf cohomology could intuitively be thought of as capturing—in a dialectic between the continuous and discrete—the nonglobalizability or nonextendibility of a given information structure in relation to other overlapping structures. Both in its algebraic representation and in this general interpretation, then, the nonvanishing cohomology groups, like H^1, might be thought of as giving us a picture of just how nonintegrated a system of information over a space may be. On the other hand, vanishing cohomology groups indicate the mutual compatibility or globalizability of local information systems (since they tell us about the global sections). In this way, sheaf cohomology emerges as a tool for representing (algebraically) what might be thought of as the degree of *generality* (or lack thereof) of a given system of locally specified data or interlocking ways of assigning information to a space.

Some years before the invention of sheaf theory, Charles Peirce argued that "continuity is shown by the logic of relations to be nothing but a higher type of that which we know as generality. It is relational generality" (Peirce 1997, 6.190). Such a suggestive, if somewhat cryptic, remark provokes us to take a closer look at the connections between generality and continuity that emerge in the context of sheaves. We know that a sheaf enables a collection of local sections to be patched together uniquely given that they agree (or that there exists a translation system for making them agree) on the intersections. Consider the satellite image mosaic sheaf introduced in example 127 (chapter 5). Recall the way in which the sheaf (collation/gluing) condition ensures a systematic passage from local sections (images of parts of the glacier) to a unique global section (the image of the entire glacier). Where the localizing step of the sheaf construction might be thought of as analytic, decomposing an object into a multitude of individual parts (local), the gluing steps are synthetic in restoring systematic relations between those parts and thereby securing a unique assignment over the entire space (global). The global sections of such a sheaf should not be thought of as a single (topmost) image but rather as the entire network of component parts welded together via certain compatibility relations or constraints. In this connection, generality can be understood in terms of the systematic passage from the local to the global, a passage that is strictly *relational* in that the action of the component restriction maps is precisely an enforcing of certain relations or mutual constraints between the local sections (that are then built up, along the lines of these relations, into a global section), and these are an ineliminable part of the construction.

9.4 A Glimpse into Cosheaves

In the cellular case, we can easily talk about *sheaf homology* by just reversing the direction of the arrows, technically producing a *cosheaf* (with its "corestriction" maps). Simply by observing that the vector space dual of every corestriction map in a cosheaf produces a sheaf over the poset, we arrive at homology for a cosheaf. More explicitly, this reversal gives us a *cosheaf* $\hat{F}$ of vector spaces on a complex, as in definition 224. As with a sheaf, a cosheaf can come to serve as a system of coefficients for homology that varies as the space varies. However, the globality of the cosheaf sections will be found in the top dimension (unlike how the global sections were built up from the *vertices* in the case of cellular sheaves).

More generally, following Curry (2014), we can define a pre-cosheaf as one might expect:

Definition 233 A *pre-cosheaf* is a functor $\hat{F} : \mathcal{O}(X) \to \mathbf{D}$.[162] Whenever $V \subseteq U$, the co-restriction (or extension) map for the pre-cosheaf is written as $r^V_U : \hat{F}(V) \to \hat{F}(U)$.

To construct a cosheaf, we again need the notion of open cover, and a cosheaf will merge the notion of covers with that of data (given by the pre-cosheaf). We can think of an open cover here in terms of a function from the "nerve" construction, which basically acts on covers to produce the ASC consisting only of those finite subsets of the cover whose intersection is nonempty. More formally, if we suppose $\mathcal{U} = \{U_i\}$ is an open cover of U, then we can take the *nerve* of the cover to yield an ASC $N(\mathcal{U})$, which will have for elements the subsets $I = \{i_0, \ldots, i_n\}$ for which it holds that $U_I = U_{i_0} \cap \cdots \cap U_{i_n} \neq \emptyset$. $N(\mathcal{U})$ is then the category with objects the finite subsets I where $U_I \neq \emptyset$, and unique arrows from I to J whenever $J \subseteq I$. Finite intersections of opens are open, so we get the functors $\iota_{\mathcal{U}} : N(\mathcal{U}) \to \mathcal{O}(X)$ and $\iota^{op}_{\mathcal{U}} : N(\mathcal{U})^{op} \to \mathcal{O}(X)^{op}$. In general, as the colimit of a cover $N(\mathcal{U}) \to \mathcal{O}(X)$ is the union $U = \bigcup_i U_i$, the data we associate to U ought to be expressible as the colimit of data assigned to the nerve. With this expectation in mind, we arrive at the following definition:

Definition 234 $\hat{F}$ is a *cosheaf on* $\mathcal{U}$ if the unique map from the colimit of $\hat{F} \circ \iota_{\mathcal{U}}$ to $\hat{F}(U)$, supplied by

$$\hat{F}[\mathcal{U}] := \varinjlim_{I \in N(\mathcal{U})} \hat{F}(U_I) \to \hat{F}(U),$$

is in fact an isomorphism. Then $\hat{F}$ is a *cosheaf*, period, if for every open set U and every open cover $\mathcal{U}$ of U, the map $\hat{F}[\mathcal{U}] \to \hat{F}(U)$ is an isomorphism.

162. Looking ahead to the cosheaf conditions, technically we ought to insist that **D** be not just any category, but rather a category with enough (co)limits.

For simplicity, suppose $\mathbf{D} = \mathbf{Set}$ and take a cover $\mathscr{U} = \{U_1, U_2\}$ of U by just two open sets. The sheaf condition would of course stipulate that for two functions or sections $s_1 \in F(U_1), s_2 \in F(U_2)$ to give an element in $U = U_1 \cup U_2$, the sections must agree on the overlap $U_1 \cap U_2$. This constraint serves to pick out the consistent choices of elements over the local sections that can then be glued together into a section over the larger set. With a similar setup—that is, $\mathbf{D} = \mathbf{Set}$ and $U = U_1 \cup U_2$—the cosheaf condition requires not that we find consistent choices, but rather that we use *quotient objects*. We do indeed still form the union of the two sections, but in the process we identify those elements that would be double-counted on account of coming from the intersection. Formally,

$$\hat{F}(U) \cong \Big(\coprod_{i=1,2} \hat{F}(U_i) \Big) / \sim,$$

where $s_1 \sim s_2$ iff there exists an s_{12} (a section over the intersection) such that $s_1 = r^{U_{12}}_{U_1}(s_{12})$ and $s_2 = r^{U_{12}}_{U_2}(s_{12})$. This makes sense, since in accordance with duality, we would expect that the equalizer definition of the sheaf condition would be converted, in passing to cosheaves, into an underlying coequalizer diagram.

A sheaf is constructed in such a way that the values of its sections on larger sets in the Alexandrov topology will determine values on smaller sets. A cosheaf basically reverses this dependence. While so far we have seen many examples of sheaves in contexts where it makes sense to perform *restrictions* of assignments of data from larger spaces to data over smaller spaces, building up global assignments from the "bottom up" via local pieces, cosheaves may roughly be thought of as instead proceeding "top down," involving *extensions* of data given over smaller spaces to larger spaces.[163] While in some sense the paradigmatic example of a sheaf was given by the restriction of continuous functions, the paradigmatic example of a cosheaf might be given by the cosheaf of compactly supported continuous functions where, instead of restricting along inclusions, we extend by zero (in the other direction). Ghrist (2014, chap. 9) gives one nice way of appreciating the duality between sheaves and cosheaves, in terms of an application to sensing problems: in this setting, a sheaf fundamentally amounts to *sensing* (the global sections of the sheaf yielding the sensorium), while a cosheaf arises in terms of what the sensors allow one to *infer* (the global sections of the cosheaf being supplied by constraint satisfactions that are consistent with sensing).

Importantly, while in the cellular context the difference between sheaf and cosheaf is somewhat immaterial, simply a matter of which direction makes the most sense for the framing of the problem, things can be far more subtle in the context of sheaves and cosheaves over opens sets for a continuous domain. In insisting on the more general functorial perspective, allowing us to make use of duality, one might suspect that the differences between sheaves and cosheaves are merely formal and not worth discussing. For instance, a sheaf is a particular functor that commutes with limits in open covers. As one might expect, a cosheaf is a functor that preserves colimits in open covers. However, in more

163. In the context of simplices, because of the reversal of direction involved in the topology given on the face relation construction, such extensions will go from higher-dimensional simplices to lower.

general contexts than the cellular one, especially with open sets coming from a continuous domain, the differences can reflect much more than a preference for direction of arrows.[164]

The next example explores a particularly fascinating connection between sheaves and cosheaves in the context of probabilities and Bayes nets.

Example 235 Imagine we are given a set of random variables $X_0, X_1, \ldots, X_n$.[165] We can consider the set $P(X_0, X_1, \ldots, X_n)$ of all joint probability distributions over these random variables, that is, the nonnegative measures or generalized functions with unit integral. Now, there is a very natural map, one that will be familiar to anyone with some exposure to probability theory:

$$P(X_0, X_1, \ldots, X_n) \to P(X_0, X_1, \ldots, X_{n-1}).$$

This map is accomplished via *marginalization*, that is, we have

$$f(X_0, X_1, \ldots, X_{n-1}) = \int f(X_0, X_1, \ldots, X_n) dX_n,$$

and where there are similar maps for marginalizing out the other random variables. The important point is that this in fact yields a cosheaf on the complete n-simplex. We elaborate on this with an example.

Assume given the random variables $X_0 = W$, $X_1 = S$, $X_2 = R$ (the reason for renaming these random variables thus will be clear in a moment). Then the space of probability measures $P(W, S, R)$ is a function from the Cartesian product of the random variables to the (nonnegative) reals such that the integral is zero. Now, we can perform the marginalization operation, for instance, marginalizing out the R by integrating along R, yielding a map $P(W, S, R) \to P(W, S)$. We can do this for each of the random variables, and continue down in dimension until we reach the measurable functions over a single random variable. In other words, we have:

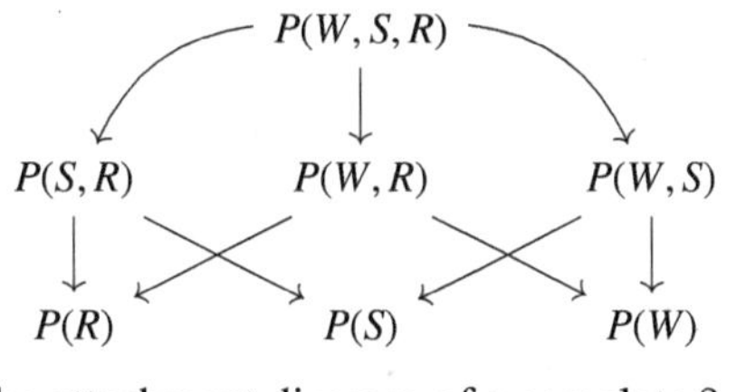

which in fact represents the attachment diagram of a complete 2-simplex,

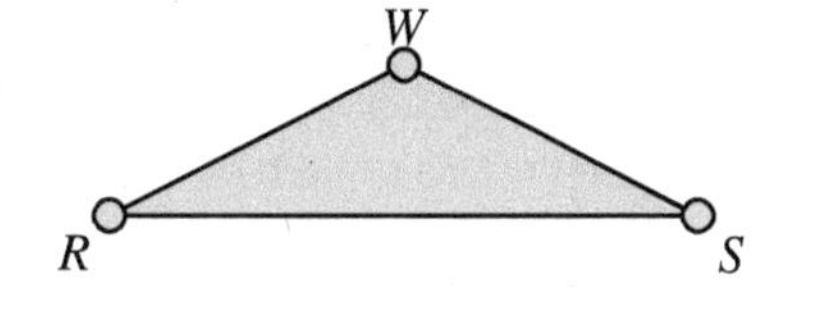

164. For instance, when working with the Alexandrov topology on a poset (or when working with locally finite topological spaces), we can ignore the distinction between pre(co)sheaves and (co)sheaves; however, while for general topological spaces, there is a sheafification functor that allows us to pass from a presheaf to the unique smallest sheaf consistent with the given presheaf, there is no analogous cosheafification functor for general topological spaces. For much more on cosheaves, and a number of interesting connections and differences with sheaves, we again refer the reader to Curry (2014). Together with Robinson (2016a), Curry contains more details on some of the dualities in the sheaf-cosheaf perspective, as well as instances of asymmetry (when certain constructions are natural for sheaves but not for cosheaves).

165. The idea for this example was inspired by Robinson (2016a).

The commutative diagram, together with the appropriate marginalization maps, gives a cosheaf on this ASC.

Now, the reader may have already wondered about maps going the other way, namely:

$$P(X_0, X_1, \ldots, X_{n-1}) \to P(X_0, X_1, \ldots, X_n),$$

an operation that is parameterized by functions C

$$P(X_0, X_1, \ldots, X_n) = C(X_0, X_1, \ldots, X_n) f(X_0, X_1, \ldots, X_{n-1}),$$

where usually one writes the arguments to C as follows

$$C(X_n | X_0, X_1, \ldots, X_{n-1}).$$

The reader may recognize that we are just describing *conditional probabilities* and Bayes's rule. The key observation is that such conditional probability maps yield a sheaf over a portion of the n-simplex.

The reader may be familiar with the construction of a *Bayes net*, given by a directed acyclic graph (with an induced topology) together with local conditional probabilities. A Bayes net encodes joint distributions and does so as a product of local conditional distributions, that is,

$$P(X_1, X_2, \ldots, X_n) = \prod_{i=1}^{n} P(X_i | \text{ parents}(X_i)).$$

For the sake of concreteness, we consider the following very simple example of a Bayes net, the standard one given in most introductions to the device, involving probabilities of "grass being wet" given that it rained, for instance, or that the sprinkler was running, illustrated with some sample probability assignments:

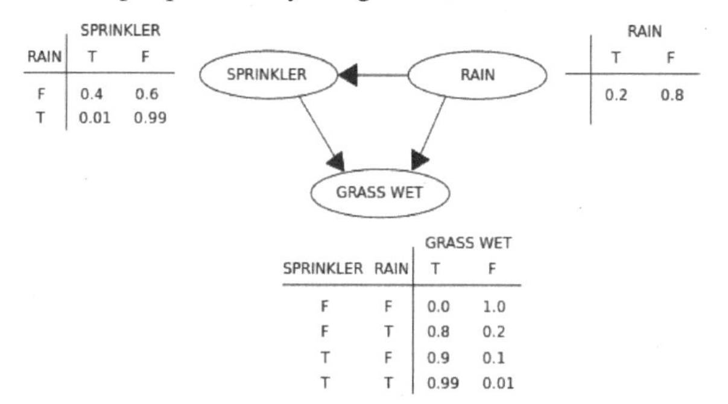

RAIN	SPRINKLER T	SPRINKLER F
F	0.4	0.6
T	0.01	0.99

RAIN T	RAIN F
0.2	0.8

SPRINKLER	RAIN	GRASS WET T	GRASS WET F
F	F	0.0	1.0
F	T	0.8	0.2
T	F	0.9	0.1
T	T	0.99	0.01

The perhaps surprising result is that the content of this Bayes net is in fact entirely captured by the paired sheaf-cosheaf construction given below, where the marginalization cosheaf is given by the entire diagram (arrows going down the page), while the paired conditional probability sheaf is given in bold (going up the page) over a part of the underlying attachment diagram.

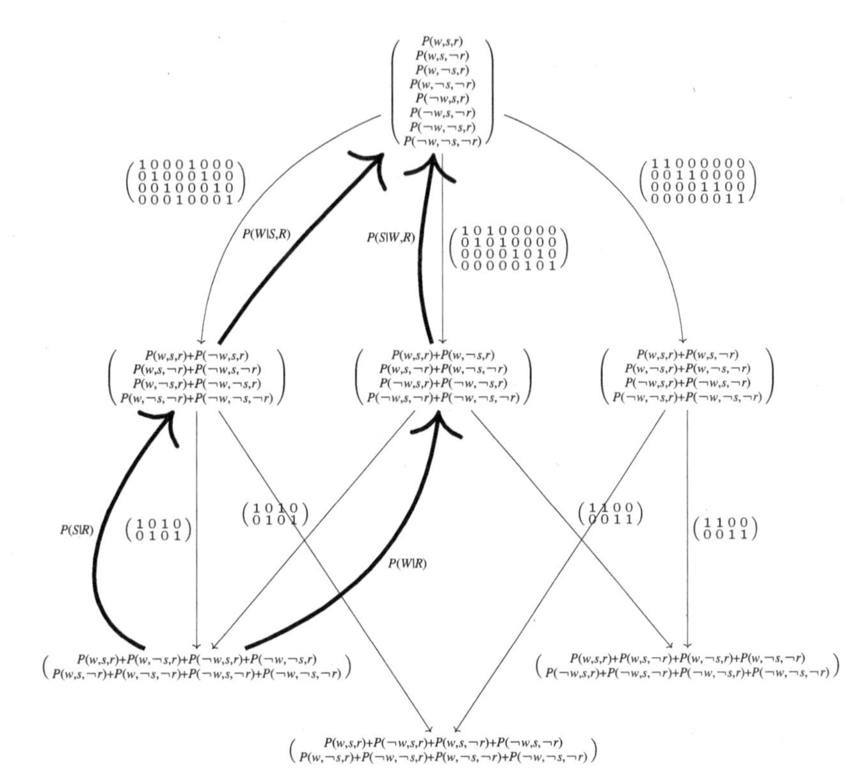

The paired sheaf-cosheaf construction contains all the data of a Bayes net; and, in fact, a solution to the Bayes net is just a global section that is a section for *both the sheaf and the cosheaf*!

Before ending this chapter and moving into discussion of toposes, we take the opportunity to make an important but frequently overlooked observation. Consider the (co)sheaf construction above. Now, consider that relatively small Bayes nets, say, one with only eight or nine nodes, are already rather simple compared to those that will be of use in practice. One might thus be suspicious of just how complicated the corresponding (co)sheaf might look for even only slightly more involved examples, not to mention the issue of storing the relevant sections for such sheaves. This indeed seems to be a real issue. One might also suspect that computing the sheaf cohomology (and global sections) on "monster" (extremely large) sheaves would be extremely difficult. As the discussion of this Bayes net sheaf-cosheaf construction suggests, most real-life (co)sheaves one would meet in practice may very well turn out to be monsters in the sense of being so large as to cause difficulties in storage, representation, or computation—difficulties we have simply avoided by confining our attention to more modest constructions. This is an issue that deserves to be recognized and pondered.[166]

166. For some ways to reduce the difficulty, in one setting, see Smith (2014, 527–533); see Curry (2014, 64–66) for some ways to think about preprocessing the input data so as to deal with the "too many sections" problem. The reader may also find Curry, Ghrist, and Nanda (2015) highly relevant, as this shows how you can "collapse" the data structure if your restriction maps are nice enough. Thanks to Michael Robinson (personal correspondence) for pointing me in the direction of this paper and observing the connection.

10 Sheaves on a Site

In which we turn to various powerful generalizations of the sheaf notion, moving beyond the topological case to consider sheaves on a site (Grothendieck toposes).

The chapter begins by exploring *Grothendieck topologies*, moving beyond the usual topological notion of a cover to a more general categorical setting. Using the resulting notion of a *site*—a category together with a choice of Grothendieck topology—the chapter then introduces the notion of a *Grothendieck topos* in terms of sheaves on a site. The rest of the chapter is devoted to developing a particular angle on such sheaves when specialized to posets, and to considering further examples. The chapter concludes by dwelling with the *idea* of such Grothendieck toposes and looking ahead to the even more general notion of *elementary toposes*, via a *Lawvere-Tierney topology*.

10.1 Revisiting Covers: Toward General Sheaves

Sheaf theory emerged as the investigation of the global consequences of properties that are defined locally; and in the classical definition of a sheaf, this notion of *local* ultimately derives from the underlying topology. In the classical definition of a sheaf on a topological space X, sheaves were basically an association of information to the open sets of X, satisfying a gluing axiom specified in terms of a pointwise covering—where, of course, by "pointwise covering" we mean that for $U \subseteq X$, we have that open subsets $\{U_i\}_{i \in I}$ *cover* U iff $\bigcup_i U_i = U$, where every point that is in U comes from some U_i.

In the appendix, we consider some matters addressing why, in the usual topological definition of sheaves, *open* sets are always emphasized; but, as mentioned there, sheaves were historically defined over closed sets before they were defined over open sets. Considerations of this sort—regarding the objects used to define topologies—is one way to prompt a more careful consideration of questions surrounding *covers*. The objects of study in general (point-set) topology are collections of open sets that are stable (closed) under arbitrary union and finite intersection, where such collections are used above all to study continuity, especially those properties of spaces which are invariant under continuous deformation. In short, as we have seen, a topology is basically a structure that enables us to define objects locally and then look at how these objects are continuously transformed into one another. However, while it has a good deal of power for its simplicity, general topology is not useful for all purposes. For one thing, many interesting structures simply do not fit into this framework or satisfy the relevant axioms. For instance, attempting to treat all the substrings of a string as open sets would not work, since such a collection will not be stable under union. But it seems sensible to think of the substrings {"*Groth*", "*thend*", "*ndieck*"} as being some sort of covering of the string "*Grothendieck*," since we can join together the substrings along the overlapping parts ("*th*" and "*nd*") to yield the original string.

While general topology is, despite the name, perhaps insufficiently *general*, it remains one of the cornerstones of modern mathematics because of the immense utility and wide applicability of many of its notions and the prevalence of situations where topological intuitions come in handy. There are contexts in mathematics where some form of topological intuition appears relevant or useful, yet where no description of the situation in terms of a space (or any other common ways, e.g., in terms of "locales") can be given.

Moreover, from the perspective of the *association of information* to the open sets of a topological space, we know that the definition of a presheaf, for its part, has a straightforward generalization beyond the topological case to the use of an arbitrary (small) category $\mathbf{C}$. These sorts of considerations naturally lead one to wonder whether the concept of a sheaf—presented thus far in terms of topological covers—can also be extended beyond the usual topological settings, admitting a definition on more general "topologies." As you can imagine, given this preamble, the answer is yes!

Grothendieck was responsible for a more general notion of a "topology" on a category, one that can now be defined for any category and which we will be able to use to extend the sheaf notion. This notion largely arose from a desire to study entities—in particular, from algebraic geometry, arising in the study of cohomology theories—that appeared to exhibit properties that *looked like* those of a sheaf, yet where the underlying categories involved were not the lattice of open sets of a topological space. The cohomology theory is built on a structure that *seems* like a space, insofar as it comes with some cover-like structure, yet which has features that cannot be found in topology. Grothendieck gave the right description of such contexts, using categories equipped with a general notion of coverings, allowing for a more unified account.

To begin to appreciate how this works, first notice that the open neighborhoods of a topological space are really just topological maps $U \rightarrowtail X$ that are monic (cancellable on the left).[167] Grothendieck was led to think that perhaps, in place of the neighborhoods U in a topological space X, we could use more general maps into X, where these are not necessarily monic. In this manner, the old-style pointwise covers by open sets, mentioned in the beginning of this section, would come to be replaced with a "covering" by a family of maps that simply satisfies certain conditions or axioms. Using this to consider an arbitrary category, the points of the usual pointwise covering approach will then vanish, leaving only some abstract "open sets," related no longer by inclusion arrows but by arbitrary arrows. The "topology" making the objects behave like the usual open sets is then entirely captured by the specification of the generalized covering in terms of abstract conditions on such families of maps.

A Grothendieck topology can initially be thought of—where, really, we are currently describing what is called a pretopology—as a rule for specifying when certain objects of a category "ought to" cover another object of the category, but purifying (by axiomatizing)

167. Recall that monic maps, or monomorphisms, are the category-theoretic generalization of the notion of injective maps from sets; see definition 65 (chapter 3).

the usual topological notion of an open cover. As such, we are basically putting a structure on a category that allows its components to be decomposed, demanding that the objects "behave like" the open sets of a topological space, except that rather than looking at intersections, we can look at pullbacks, and unions don't even have to come into play. While this is already a powerful generalization, we can actually go further, dropping the assumption that our category has pullbacks. The Grothendieck topology that emerges will effectively tell us about when abstract morphisms act as one might hope any cover would, formalizing the notion of being *locally the case*, without confining us to the notions or assumptions from the usual topological setting. A major difference between this new "topology" and the old notion of open covers from general topology is that our "open sets" (now objects in a category) need not be stable under union (i.e., coproducts need not exist in the category), nor even under intersection (i.e., pullbacks need not exist).[168]

Via this wider notion of a Grothendieck topology—where, instead of the usual inclusion relation between open sets, we can consider arbitrary arrows in a category—we can make use of many topological intuitions and devices in situations where there does not appear to be any topology (in the traditional sense) at play. *That*—and not just generality for its own sake!—is part of what is behind the real power of all this. The fundamental takeaway here is that there is no need to restrict ourselves to topological spaces to do sheaf theory: as long as the category gives information *similar to* open covers (which similarity is formalized with the notion of a Grothendieck topology), we can generalize the constructions involved in creating a sheaf beyond the category of open sets of a topological space. Using Grothendieck topologies, we develop a more powerful notion of a sheaf defined on a (small) category, disposing of the condition that the base be formed of open sets and the lattice formed by inclusions arrows between these sets.[169]

Before diving into the relevant definitions and constructions, here's a last way of thinking of all this. Suppose you have an (arbitrary) presheaf. You may want to say that if you know the value of that presheaf on some family of objects X_i—let's not assume anything else of those objects—then that "ought to" be enough to tell you the value of the presheaf on some X (not necessarily a space), which we would thus like to think of as being "covered" by the family X_i. In other words, you want to know which families of arrows into X are such that the presheaf data valued locally on that family *extends* to presheaf data on X along those arrows. But your domain category might not be topological at all, and so the objects might not present as "open sets" in the strict sense. By developing the notions discussed above, we will be able to address these needs. The important notion of a *site* will emerge as a category together with a choice of Grothendieck topology—using this, we will be able to define the all-important notion of a *Grothendieck topos* in terms of sheaves on a site.

168. While Grothendieck's use of the term "topology" for his own notion is entirely sensible, we should emphasize that it is not literally a topology in the usual sense. A Grothendieck topology, strictly speaking, has nothing to do with—or, rather, does not rely on—open sets or closed sets in the usual topological sense (i.e., as treated in chapter 4).

169. Sheaf theory cast in terms of this notion of Grothendieck topologies was first described in Artin, Grothendieck, and Verdier (1972, SGA 4 Exposé I–IV).

10.2 Grothendieck Toposes

10.2.1 Grothendieck Topologies: First Pass

The first way of commonly approaching this new sort of "covering" is by considering how the construction can be accomplished in any category $\mathbf{C}$ assumed to have pullbacks. We will use the ingredients about to be introduced to define what is technically a *pretopology*. As the name suggests, one can think of it as "on the way toward" the still more general notion toward which we are building.

For each object c of $\mathbf{C}$, let's start by considering a set S of indexed families of morphisms to c, that is,

$$S=\{f_i : c_i \to c \mid i \in I\}.$$

If you think of an arrow into c as a "perspective" on c, taking such an S is like bunching together a number of perspectives on, or ways of seeing, c. The poet Wallace Stevens has a poem "Thirteen Ways of Looking at a Blackbird," which consists of thirteen short parts, each of which present a distinct "perspective" on a blackbird. For intuition, S is the poem which comprises thirteen (so index set $I=13$) ways of looking at $c=$blackbird. Suppose next that for each object c of $\mathbf{C}$ we assign a collection

$$K(c)=\{S, S', S'', \dots\}$$

consisting of a selection of certain of the families of morphisms (each of the same form as S). On the "perspective" way of seeing things, this is like selecting certain "poems" about our blackbird—as if to say, "I am interested in *these* particular families of ways of seeing c." But suppose we make you abide by a rule: you can only make such a selection provided it meets certain (three) conditions. Provided it does meet those conditions, then your collection $K(c)$ is allowed, and we agree to call it a *covering* and the families in it *covering families* (or *covers* of c). Such a $K(c)$ then makes up our "(pre)topology."

Definition 236 Let $\mathbf{C}$ be a category that has pullbacks. Then a *Grothendieck pretopology* (or *basis* (for a Grothendieck topology)) on $\mathbf{C}$ is a function K that assigns to each $c \in \mathrm{Ob}(\mathbf{C})$ a collection $K(c)$ of families of morphisms into c, as above, where this satisfies three conditions:

1. Whenever an arrow $f : c' \xrightarrow{\cong} c$ in $\mathbf{C}$ is an isomorphism, then the family $\{f : c' \xrightarrow{\cong} c\}$ consisting of just that arrow is in the collection $K(c)$.[170]
2. If $\{f_i : c_i \to c\}$ is a covering family (i.e., is in $K(c)$) and $g : b \to c$ is any morphism in $\mathbf{C}$, then the family of pullbacks $\{\pi_2 : c_i \times_c b \to b\}_{i \in I}$ exists and is a covering family (i.e., is in $K(b)$).[171]
3. If $\{f_i : c_i \to c\}$ is a covering family (i.e., is in $K(c)$) and for each i we also have a family $\{g_{ij} : b_{ij} \to c_i\}_{j \in J_i}$ that is a covering family (i.e. is in $K(c_i)$), then the family of composites $\{f_i \circ g_{ij} : b_{ij} \to c_i \to c\}_{i \in I, j \in J_i}$ is also a covering family (i.e., is in $K(c)$).[172]

170. Think of this as the "solipsist" stipulation: if you're going to collect different "poems" about c, you can only do so as long as you agree to "let c tell her own story from her own perspective (or from the perspective of anyone who essentially shares her perspective)."

171. Attempt to finish the analogy for yourself (this is an important one!): Given a "poem" with a bunch of perspectives on c, and then given another isolated perspective on c (from the vantage of some b), ...

172. Try formulating for yourself the right analogy here as well.

To formulate what it means to have a sheaf for the coverings comprising such a pretopology, we can actually rehash the classical (topological) definition of a sheaf. We just have to make one change. Recall from the definition of a sheaf on a topological space the stipulation that for each open cover $\{U_i\}_{i\in I}$ of some U, every family of elements $\{x_i \in P(U_i)\}_{i\in I}$ matching on the intersections $U_i \cap U_j$ for all i and j can be glued together into a unique element $x \in P(U)$. The same sort of thing works for a definition using covering families of an object c; all we need to do is replace intersection $c_i \cap c_j$ by the pullback $c_i \times_c c_j$, that is,

$$\begin{array}{ccc} c_i \times_c c_j & \xrightarrow{h_{ij}} & c_j \\ {\scriptstyle v_{ij}}\downarrow & & \downarrow{\scriptstyle f_j} \\ c_i & \xrightarrow[f_i]{} & c. \end{array}$$

Now, to define a sheaf, we need a *presheaf* to work with. But we can of course just use a functor $P : \mathbf{C}^{op} \to \mathbf{Set}$, where we only assume of $\mathbf{C}$ that it has pullbacks. Then, applying P to the diagrams in $\mathbf{C}$ of the above sort, we would get associated diagrams living in $\mathbf{Set}$:

$$\begin{array}{ccc} P(c_i \times_c c_j) & \xleftarrow{P(h_{ij})} & P(c_j) \\ {\scriptstyle P(v_{ij})}\uparrow & & \uparrow{\scriptstyle P(f_j)} \\ P(c_i) & \xleftarrow[P(f_i)]{} & P(c). \end{array}$$

We then just follow the lead given by the topological definition. The matching or gluing condition for a sheaf will thus read: if $\{x_i \in P(c_i)\}_{i\in I}$ is a family of elements that match, in the sense that $P(v_{ij})(x_i) = P(h_{ij})(x_j)$ for all i, j, then that family determines a unique element $x \in P(c)$ such that

$$P(f_i)(x) = x_i$$

for all $i \in I$. And this is equivalent to requiring that the arrow e is an equalizer

$$P(c) \xrightarrow{e} \prod_i P(c_i) \rightrightarrows \prod_{i,j} P(c_i \times_c c_j),$$

where $e(x) = (P(f_i)(x))_{i\in I}$, for every covering family $\{c_i \xrightarrow{f_i} c\}_{i\in I}$.

Assuming a category $\mathbf{C}$ has pullbacks, we will thus have a way of talking about sheaves on it! No topology (in the usual sense) needed!

The definition of pretopologies via covering families just sketched requires that pullbacks exist in the underlying category. We can just as well provide a more general definition that does not rely on their existence, and this is done not just for the sake of giving the most general definition but also because there are certain categories of interest that do not have pullbacks. This generalization is typically accomplished by replacing the indexed families S of morphisms into an object with the *sieves* they generate, a construction the reader is in fact already familiar with (under another name). By using sieves, we will see that we can dispense (in principle) with the assumption that $\mathbf{C}$ has pullbacks.

But in fact, without yet getting into sieves, we can point out that the three conditions given above are not created equal: it is really the second condition in the definition that is most decisive, as far as sheaves are concerned. Following Johnstone (2002), we might thus isolate the second condition and define the following notion.

Definition 237 Let **C** be a category. Then a *coverage* on **C** is a function T that assigns to each $c \in \mathrm{Ob}(\mathbf{C})$ a collection $T(c)$ of families of morphisms $\{f_i : c_i \to c\}_{i \in I}$ into c (these are called *T-covering families*, or just *covering families*, when T is understood), where this satisfies:

- If $\{f_i : c_i \to c\}_{i \in I}$ is a covering family and $g : b \to c$ is any morphism with codomain c, then there exists a covering family $\{h_j : b_j \to b\}_{j \in J}$ such that each composite $g \circ h_j$ factors through some f_i, in the sense that

$$\begin{array}{ccc} b_j & \xrightarrow{h_j} & b \\ {\scriptstyle k}\downarrow & & \downarrow{\scriptstyle g} \\ c_i & \xrightarrow[f_i]{} & c. \end{array}$$

Remark 238 The traditional way of defining a *Grothendieck topology*, as we shall see, is to use three conditions (similar to those found in the definition of a Grothendieck pretopology—see definition 236). But, really, the second condition in 236 is decisive for the introduction of the sheaf notion, while the other conditions (usually called "saturation conditions" or "closure properties") are conditions that are often met in practice (and so, typically assumed), yet fundamentally do not affect the sheaf notion. In particular, the collection of families of morphisms for which a given functor satisfies the associated sheaf axiom will typically have a number of closure properties (essentially encoded by the notion of a sieve), which are thus commonly added in the definition itself.

However, following Johnstone (2002), one can take the approach of isolating this essential condition and giving the definition of a coverage, as in 237, and then reserving the name *Grothendieck coverage* for a coverage that also satisfies the two other saturation conditions (where a Grothendieck coverage is the same thing as what is traditionally called, and that we will call, a Grothendieck topology).[173] Grothendieck himself originally considered coverages that also obey the additional saturation conditions.

Finally, observe that a Grothendieck pretopology differs from the above in that it assumes the category **C** has pullbacks and replaces the condition in the definition 237 by the stronger condition:

> If $\{f_i : c_i \to c\}$ is a covering family and $g : b \to c$ is any morphism with codomain c, then the family of pullbacks $\{\pi_2 : c_i \times_c b \to b\}_{i \in I}$ exists and is a covering family.

This condition is satisfied in many examples; however, strictly speaking, it is not needed for defining sheaves—the more general condition given in 237 will be enough.

Using this notion of coverage, we can already define the notion of a site, and then define sheaves for this, just as one would expect.

Definition 239 A *site* will mean a pair $(\mathbf{C}, T)$ consisting of a category **C** equipped with a coverage T. (A *small site* is a site for which the underlying category is small.)

Definition 240 For **C** a category, a functor $F : \mathbf{C}^{op} \to \mathbf{Set}$ is said to satisfy the *sheaf axiom* for a family of morphisms $\{f_i : c_i \to c\}_{i \in I}$ if, whenever we have a family of elements $s_i \in$

173. Johnstone 2002 renamed things to avoid the "emotional baggage" associated with the language of a "topology."

$F(c_i)$ that are *compatible*—in the sense that whenever $g : b \to c_i$ and $h : b \to c_j$ satisfy $f_i \circ g = f_j \circ h$ (for any i, j, not necessarily distinct), we have $F(g)(s_i) = F(h)(s_j)$—then there exists a unique $s \in F(c)$ such that $F(f_i)(s) = s_i$ for each $i \in I$.[174]

Then, given T a coverage on **C**, F is said to be a T-*sheaf* provided it satisfies the sheaf axiom for every T-covering family.

Again, the collection of all families of morphisms for which a functor satisfies the sheaf axiom, as above, will generally have some further "closure properties." On account of this, it is typical to add such conditions to the single condition defining a coverage, obtaining an expanded definition. We will now expand on the covering families approach via the notion of a *sieve*, which will lead to the original (and common) definition of a Grothendieck topology. Moreover, as we will see—and as the name "pretopology (or basis for a Grothendieck topology)" suggests—a pretopology on a category can ultimately be used to generate a Grothendieck topology (though different pretopologies may generate the same topology).

10.2.2 Sieves

Definition 241 A *sieve* S on an object c of **C** is a family of morphisms in **C**, all of which have codomain c, that satisfies

$$f \in S \text{ implies } f \circ g \in S,$$

whenever such a composition makes sense.

In other words, a sieve S on c is a collection of arrows with codomain c, where this is "closed under composition on the right" (or "closed under precomposition"). A generic picture of a sieve might look something like:

$$\ldots \qquad \ldots$$

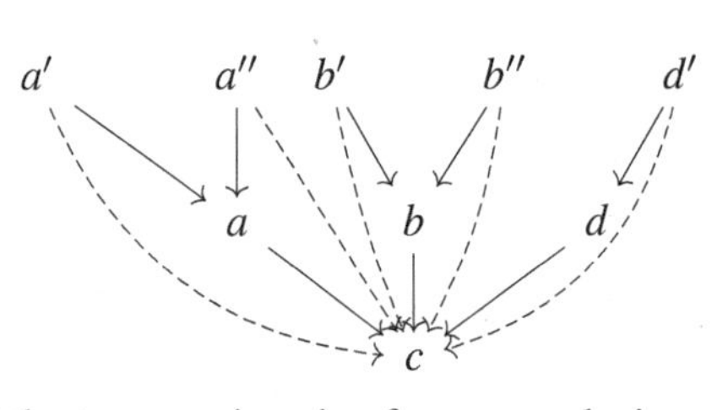

In the ordinary sense of the term, a sieve is of course a device used to strain, sift or sort some material, filtering out undesired pieces from a larger input. In a similar fashion, if we start with all arrows into c, a sieve (in our sense) on c is thus a collection of arrows that sifts or filters certain arrows from that set of all arrows into c: if there is an arrow $b \to c$ going through to c, then whenever there is an arrow $b' \to b$, the arrow $b' \to c$ "finds its way" through to c as well. In slogan form (explaining the term "sieve"):

If b goes through the sieve, then so too does anything "smaller than" b.

174. By the way, notice how the definition simplifies in the case where **C** has pullbacks: specifically, one need only check the compatibility of s_i and s_j on the pullback of f_i along f_j, instead of on arbitrary pairs of functions (g, h), as stipulated by this more general definition.

This defining condition of a sieve being closed under pre-composition is sometimes called a "saturation condition." Moreover, given a collection of morphisms landing in c, it can always be saturated in this way, leaving us with a sieve on c.

Here is another important saturation feature of the notion of a sieve. The definition implies that a sieve S is such that for any $b \to c$ in S, any path to b followed by the path from b to c will itself be a path in S. In particular, then, if S is a sieve on c and $h : b \to c$ is any arrow to c, we will have that

$$h^*(S) = \{g \mid \text{codom}(g) = b \text{ and } h \circ g \in S\}$$

is a sieve on b. This is another closure or saturation condition, in that it says that sieves are "closed under pulling back." It is straightforward to show that $h^*(S)$ is itself a sieve.

This is a powerful notion that makes sense in any category. But to get a better handle on this notion, it may again help to specialize to posets. Recall that a *downset* is a subset $D \subseteq P$ for which, for all $x \in D$ and for all $y \in P$, if $y \leq x$, then $y \in D$. Moreover, recall that for each element p of the poset, the downset generated by p—its *principal downset*—is

$$\downarrow p := \{q \in \mathcal{P} \mid q \leq p\}.$$

Applying the definition of a sieve in the context of posets, we can thus say that a subset $S \subseteq P$ will be a sieve on p if $q \leq p$ for each $q \in S$ and $r \in S$ if $r \leq q$ for some q. In other words, a subset $S \subseteq \mathcal{P}$ is a sieve on p provided S belongs to $\mathcal{D}(\downarrow p)$, the collection of all downsets of the principal downset $\downarrow p$. Moreover, in a poset, regarded as a category, recall that there is an arrow $q \to p$ precisely when $q \leq p$, and there is *at most* one arrow from q to p. Accordingly, in such a setting, we can identify the underlying families of morphisms $S = \{f_i : c_i \to c\}_{i \in I}$, consisting of arrows into c closed under precomposition, with a family of *elements* $\{c_i \mid c_i \leq c \text{ for all } i \in I\}$ that is downward closed.

In a moment, we will use this formulation for posets to present the main definitions in terms of posets. But let us first make a few other, more general observations. On account of the intimate connection between principal downsets and representable functors—as explored in chapter 6—it should come as no surprise to the reader that, in the more general case, each sieve S on an object c can be identified with a subpresheaf $S \subseteq \text{Hom}_{\mathbf{C}}(-, c) := Y_c$ of the representable functor (at least in settings where $\mathbf{C}$ is locally small).

Proposition 242 A sieve S on an object c of a category $\mathbf{C}$ can be exhibited as a subfunctor of the representable hom-functor $Y_c := \text{Hom}_{\mathbf{C}}(-, c)$, via the following specification:

$$b \longmapsto \{f \in \text{Hom}_{\mathbf{C}}(b, c) \mid f \in S\}$$

$$(a \xrightarrow{g} b) \longmapsto g^*, \text{ such that } g^* : f \mapsto f \circ g.$$

In other words, given a sieve S on c, defining $Q(b) = \{f \mid f : b \to c \text{ and } f \in S\} \subseteq \text{Hom}_{\mathbf{C}}(b, c)$ will produce a functor $Q : \mathbf{C}^{op} \to \mathbf{Sets}$ that is a subfunctor of Y_c. Conversely, via the Yoneda embedding $\mathbf{y} : \mathbf{C} \to \mathbf{Set}^{\mathbf{C}^{op}}$ taking an object $c \in \mathbf{C}$ to Y_c, we can further consider that if we have a subfunctor $Q \subseteq Y_c$, then the set

$$S = \{f \mid \text{ for some object } b, f : b \to c \text{ and } f \in Q(b)\}$$

will form a sieve on c.

Altogether, since we can pass from S to Q and from Q to S in reciprocal fashion, this informs us of the fact that:

$$\text{sieve on } c = \text{subfunctor of } Y_c.$$

This already gives us a more category-theoretic way of formulating the notion of a sieve, as compared to the definition that takes a sieve to be some *set* of arrows satisfying certain constraints. A third characterization of sieves is also in this more category-theoretic spirit.

Proposition 243 A sieve $S = \{c_i \xrightarrow{f_i} c \mid i \in I\}$ on an object c of a category $\mathbf{C}$ can be seen as a subcategory of the slice (comma) category $\mathbf{C}/c$, where the objects are just the arrows $c_i \xrightarrow{f_i} c$ of S, and the morphisms are given by arrows u making triangles of the following form commute:

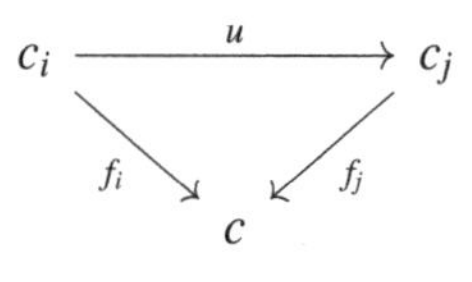

In short, a sieve has three (equivalent) characterizations:

- as a set of arrows into some object (where these arrows satisfy a condition of closure under composition on the right);
- as a subfunctor of the representable hom-functor;
- as a subcategory of the slice category.

Before putting this notion of a sieve to work, let us make a simple observation that will allow us to define one other notion that will be of use to us. Observe that a sieve on some object clearly does not need to include *all* the arrows into that object. However, taking all arrows into an object will indeed amount to a sieve, and we reserve a special name for such a sieve.

Definition 244 For an object c of $\mathbf{C}$, the set

$$M_c = \{f \mid \text{codomain}(f) = c\}$$

of *all* arrows into c will be a sieve, one that (for reasons that should be clear) is called the *maximal sieve on c*.

Observe that, specialized to posets, the maximal sieve on $p \in P$ is just given by taking $\downarrow p$, the principal downset itself.

Finally, relating this back to the earlier discussions using families of morphisms into c: observe that any family $\{f_i : c_i \to c\}_{i \in I}$ generates a sieve on c, by taking all those morphisms with codomain c which factor through at least one of the f_i. And it is straightforward to demonstrate that given a family $\{f_i : c_i \to c\}_{i \in I}$ of morphisms of $\mathbf{C}$ and F a presheaf on $\mathbf{C}$, F satisfies the sheaf axiom for $\{f_i : c_i \to c\}_{i \in I}$ iff it satisfies the sheaf axiom for the sieve generated by that family.

10.2.3 Grothendieck Topologies and Sites Defined

Equipped with the notion of a sieve, we are now in a position to define the following concept of a Grothendieck topology.

Definition 245 A *Grothendieck topology* (or *Grothendieck coverage*) on a category **C** is a function J that assigns to each object c of **C** a collection $J(c)$ of sieves on c, in such a way that

1. *identity cover* (or *maximality axiom*): the maximal sieve M_c is in $J(c)$;
2. *stability under change of base* (or *stability axiom*): if $S \in J(c)$ and $h: b \to c$ is any morphism with codomain c, then $h^*(S)$ is in $J(b)$;[175]
3. *local character condition* (or *transitivity axiom*): if $S \in J(c)$ and R is another sieve on c such that, for each f in S (e.g., $f: b \to c$), the sieve $f^*(R)$ is in $J(b)$, then R belongs to $J(c)$.

At first glance, this might seem unmotivated or difficult to parse. But observe that the conditions on a Grothendieck topology are rather intuitive: (1) is just an inclusion of representable functors as covers; (2) is really a disguised way of insisting that J itself be functorial (a presheaf); and (3) requires that things are transitive.

Terminologically, if $S \in J(c)$, we say that S is a *covering sieve*, or that S *covers* c (or, sometimes, for explicitness, that S is a J-*cover* of c). We will also say that a sieve S on c *covers an arrow* $f: b \to c$ if $f^*(S)$ covers b. In other words, S covers c iff S covers the identity arrow on c. In these terms, the three axioms for a Grothendieck topology given above can equivalently be formulated in terms of a relation between the sieves and the arrows of the category (this is called the *arrow form* of the definition in Mac Lane and Moerdijk 1994):

- (1a) *identity*: if S is a sieve on c and f is in S, then S covers f; informally, this says that any "open set" covers itself, or (in terms of sets) that any set is covered by all its possible subsets.
- (2a) *stability*: if S covers an arrow $f: b \to c$, it also covers the composition $f \circ g$, for any arrow $g: a \to b$; informally, this says that coverings "pull back."
- (3a) *transitivity*: if S covers an arrow $f: b \to c$ and R is a sieve on c which covers all arrows of S, then R covers f; informally, this ensures that a cover of a cover will be a cover.

A further feature that one might wish to impose on such covering sieves in fact already follows from the three main axioms (in either form): namely, that any two covers have a common *refinement*, that is, $J(c)$ is closed under finite intersections. Explicitly, we mean

- (4) if $R, S \in J(c)$, then $R \cap S \in J(c)$,

or in arrow form:

- (4a) if R and S both cover $g: b \to c$, then $R \cap S$ covers g.

Equipped with the notion of a Grothendieck topology, we can define the following:

Definition 246 A *site* will mean a pair $(\mathbf{C}, J)$ consisting of a category **C** equipped with a Grothendieck topology J.[176]

175. Observe that, in the presence of the other two conditions, this essentially amounts to a simplification of the coverage condition in definition 237.

176. In such contexts, C is typically assumed to be small.

Both because it offers a more immediately accessible version of the above definition and because we will make particular use of this specialized version in what follows, let us also give a version of the definition specialized to posets.

Definition 247 (*Grothendieck topology specialized to posets*) A *Grothendieck topology* J on a poset $\mathcal{P}$ is a map

$$p \mapsto J(p)$$

assigning to each element $p \in \mathcal{P}$ a collection $J(p)$ of sieves on p—where, as indicated above, $S \subseteq \mathcal{P}$ is a sieve on p provided $q \leq p$ for each $q \in S$ and if $r \leq q$ for some $q \in S$, then also $r \in S$—that satisfies the following conditions:

1. the maximal sieve $\downarrow p$ is an element of $J(p)$;
2. if $S \in J(p)$ and $q \leq p$, then $S \cap \downarrow q \in J(q)$;
3. if $S \in J(p)$ and R is a sieve on p such that $R \cap \downarrow q \in J(q)$ for each $q \in S$, then also $R \in J(p)$.[177]

Earlier it was mentioned that the new definition of a Grothendieck topology in terms of sieves, as in definition 245, has some advantages, including the advantage of allowing us to dispense with the assumption that the underlying category **C** has pullbacks. Even if the individual morphisms in **C** cannot be pulled back, a sieve can *always* be pulled back along a morphism of **C**—definition 245(2) tells us how this should go. This is ultimately the case because each sieve on an object c of the category can be identified with a subpresheaf of the representable hom-functor Y_c, and the presheaf category $\mathbf{Set}^{\mathbf{C}^{op}}$ in which this lives itself has pullbacks. Being able to identify sieves with subpresheaves of the representable hom-functor Y_c will be especially useful to us in defining sheaves for such topologies built from covering sieves. Another reason for the new definition in terms of sieves is that—referring back to the notion of a Grothendieck pretopology—two different pretopologies may yield the exact same sheaves, so there was a lingering imprecision in the definition of a Grothendieck pretopology. It is partly in order to overcome this imprecision that we move from focusing on covering families to covering sieves.

On the other hand, in practice, it is often the case that covering families are easier to specify and work with than the sieves that they generate; especially when the underlying category **C** has pullbacks, it also appears to be simpler to work with the generating covering families, at least when assessing whether or not a particular functor is a sheaf.

Before coming back to some of these niceties, let us look at some examples of Grothendieck topologies. There will generally be a number of Grothendieck topologies that could be attached to a category. The next two examples represent two important extremes.

177. With this, it becomes especially evident that J in fact must be a functor $\mathcal{P}^{op} \to \mathbf{Set}$, once we define $J(q \xrightarrow{\leq} p)(S) = S \cap \downarrow q$ for each $S \in J(p)$. If we have $r \leq q \leq p$ in $\mathcal{P}$ and $S \in J(p)$, then

$$J(r \xrightarrow{\leq} p)(S) = S \cap \downarrow r = (S \cap \downarrow q) \cap \downarrow r = J(r \xrightarrow{\leq} q) J(q \xrightarrow{\leq} p)(S).$$

Defining things in this way, the stability axiom ((2) above) then just expresses that $J \in \mathbf{PreSh}(\mathcal{P})$. However, observe that, in general, as a functor, J itself is not a sheaf—for, given a cover $\{c_i \xrightarrow{f_i} c\}_{i \in I}$ of c and a compatible family of covers $\{x_i \in J(c_i)\}_{i \in I}$, there may in fact be a number of covers of c that extend this latter family.

Example 248 The *minimal topology*—also going under the names *trivial*, *indiscrete*, or *coarse* topology—J_{ind} on a category $\mathbf{C}$ is the one in which the *only* sieve covering an object c is the maximal sieve M_c. In other words, we are declaring that only sieves of the form 245(1)—that is, $\mathrm{Hom}_{\mathbf{C}}(-, c)$—are covering sieves. In particular, when dealing with $\mathcal{P}$ a poset, since we know that representable functors are the principal downsets, the indiscrete Grothendieck topology on $\mathcal{P}$ is just given by taking the principal downsets, that is,

$$J_{ind}(p) = \{\downarrow p\}.$$

This topology is the *coarsest* (smallest) of all topologies that can be put on $\mathbf{C}$.

Example 249 The *maximal topology*—also *discrete* topology—on $\mathbf{C}$ is such that *every sieve* is declared a covering sieve. Specialized to $\mathcal{P}$ a poset, the discrete Grothendieck topology on $\mathcal{P}$ will be given by

$$J_{dis}(p) = \mathcal{D}(\downarrow p).$$

This is clearly the *finest* (biggest) topology on $\mathbf{C}$. In practice, this topology can sometimes lead to rather nonintuitive covers. As the names suggest, the maximal and the minimal topologies provide the two extremes on the spectrum of possible topologies.

Example 250 As is commonly known in differential geometry, the category **Man** of smooth manifolds and their smooth maps does not have all pullbacks—in particular, pullbacks of manifolds are not generally manifolds. (It has a number of other deficiencies, from a categorical perspective.) In general, one needs to impose some structure on the maps in order to ensure the existence of pullbacks.

But if we take for covering families the families of "open embeddings" $\{f_\alpha : U_\alpha \hookrightarrow M\}_{\alpha \in A}$ such that the family of open sets $\{f_\alpha(U_\alpha)\}_{\alpha \in A}$ is an open cover of M, this gives us a Grothendieck topology on **Man**—and using such a site, one can observe that the representable functors are sheaves.

The notion of a Grothendieck topology and its covering sieves is sometimes said to be a vast generalization of the usual notions of a topological space and its covers.[178] For this to be a valid perspective, we should at the very least be able to check that the latter notion is indeed captured, as a particular case, by the former.

Example 251 By viewing the lattice (poset) $\mathcal{O}(X)$ of open subsets of a topological space X as a category, a sieve on U is just a family $S = \{U_i\}_{i \in I}$ of open subsets of U satisfying the condition that $W \subseteq V \in S$ implies that $W \in S$. One then says that S covers U provided U is contained in the union of the open sets of S. The usual topological notion of a cover is thus recovered by taking as our Grothendieck topology that specified by

$$J_{\mathcal{O}(X)}(U) = \left\{S \in \mathcal{D}(\downarrow U) \mid \bigcup S = U\right\},$$

where, as before, $\mathcal{D}(\downarrow U)$ is the collection of all the downsets of $\downarrow U$. You can verify that the standard open cover definition in fact satisfies the axioms for a Grothendieck topology,

178. This is not exactly the right way to think about it, since it is not at all a straightforward abstraction from the topological notions. But, as the next example shows, it does indeed embrace the usual topological notions.

and that the above Grothendieck topology corresponds to the usual topological notion of an open cover.[179]

In greater generality, given a *complete Heyting algebra H* (that is, a Heyting algebra with arbitrary joins $\bigvee$ and meets)—or a *frame*, introduced below—we can define the Grothendieck topology J_H by taking

$$\{c_i\}_{i \in I} \in J_H(c) \text{ iff } \bigvee_{i \in I} c_i = c.$$

This is sometimes called the *canonical topology on H*.

Example 252 (*The dense topology*) The dense topology J_{dense}—also called the double-negation topology, for reasons we will see shortly—can be defined for an arbitrary category **C** in terms of collections of sieves $J_{dense}(c)$ of the form:

$$S \in J_{dense}(c) \text{ iff for any } f : b \to c \text{ there exists a } g : a \to b \text{ such that } f \circ g \in S.$$

It is useful to look at this in the special case when we are dealing with a poset $\mathcal{P}$. In general, for a poset $\mathcal{P}$, a subset $D \subseteq \mathcal{P}$ is said to be *dense* if, for every $p \in P$, there is a $q \leq p$ with $q \in D$. The subset $D \subseteq \{q \in P \mid q \leq p\}$ is then said to be *dense below p* if for every $p' \leq p$, there exists a $q \leq p'$ with $q \in D$. Observe that since $D \subseteq \{q \in P \mid q \leq p\}$, D is already a sieve. If D is dense below p, we can call it a dense sieve (on p). The collection of dense sieves will supply us with a topology J_{dense} on a poset by taking

$$J_{dense}(p) = \{D \mid q \leq p \text{ for all } q \in D, \text{ and } D \text{ is a sieve dense below } p\}.$$

Observe that another way of saying this is

$$J_{dense}(p) = \{D \in \mathcal{D}(\downarrow p) \mid \downarrow p \subseteq \uparrow D\},$$

that is, $J_{dense}(p)$ will consist of a selection of sets D from the collection of all downsets of the principal downset $\downarrow p$, in such a way that the principal downset itself is contained by $\uparrow D = \{x \in \mathcal{P} \mid x \geq d \text{ for some } d \in D\} = \bigcup_{d \in D} \uparrow d$, the upper set generated by such D.

It may be helpful to prove that, defined thus, such objects indeed supply us with a Grothendieck topology.

Exercise 22 Prove that such dense sieves form a Grothendieck topology.

Solution

Proof. We need to check the three requirements:

1. the maximal sieve (principal downset) $\downarrow p$ is an element of $J(p)$;
2. if $S \in J(p)$ and $q \leq p$, then $S \cap \downarrow q \in J(q)$ (stability);
3. if $S \in J(p)$ and R is any sieve on p such that $R \cap \downarrow q \in J(q)$ for each $q \in S$, then also $R \in J(p)$ (transitivity).

179. However, there is a subtlety here. As we know, in the usual context of a topological space, an open cover of U is typically described as a family $\{U_i \mid i \in I\}$ of open subsets of U whose union $\bigcup_i U_i$ is equal to U. Note that such a family is not necessarily a sieve. However, it does *generate* a sieve, via taking the collection of all those open $V \subseteq U$ where $V \subseteq U_i$ for some U_i. This motivates a more careful consideration of the notion of *generating a covering sieve*, something that can be described in the more general context of an arbitrary category (with or without pullbacks), via a *basis* for a Grothendieck topology. See below and Mac Lane and Moerdijk (1994, III.2) for details.

Requirement 1 is basically immediate: certainly $\downarrow p$ itself is (trivially) in the collection of downsets of $\downarrow p$, and $\downarrow p \subseteq \uparrow (\downarrow p)$. Thus, $\downarrow p \in J_{dense}(p)$.
Requirement 2: let $D \in J_{dense}(p)$ and consider $q \leq p$. But $\downarrow p \subseteq \uparrow D$ (since $D \in J_{dense}(p)$), so if $r \in \downarrow q$, we will have that $r \in \downarrow p$; hence, there will have to be an $s \in D$ with $r \leq s$. But then $s \in D \cap \downarrow q$, as $q \geq r$, from which we must have $\downarrow q \subseteq \uparrow (D \cap \downarrow q)$, which means $D \cap \downarrow q \in J_{dense}(q)$.
Requirement 3: let $D \in J_{dense}(p)$ and $E \in \mathfrak{D}(\downarrow p)$ so that $E \cap \downarrow q \in J_{dense}(q)$ for $q \in D$. Thus, $\downarrow p \subseteq \uparrow D$ and $\downarrow q \subseteq \uparrow (E \cap \downarrow q)$ for $q \in D$. If we now consider $r \in \downarrow p$, we can show that $r \in \uparrow E$ as follows: $\downarrow p \subseteq \uparrow D$, so there exists a $q \in D$ with $q \leq r$; we also have that $\downarrow q \subseteq \uparrow (E \cap \downarrow q)$; thus, there exists an $s \in (E \cap \downarrow q)$ with $s \leq q$, which means that $r \in \uparrow E$, with $s \leq q \leq r$. This shows that $E \in J_{dense}(p)$. □

For concreteness, consider the following poset

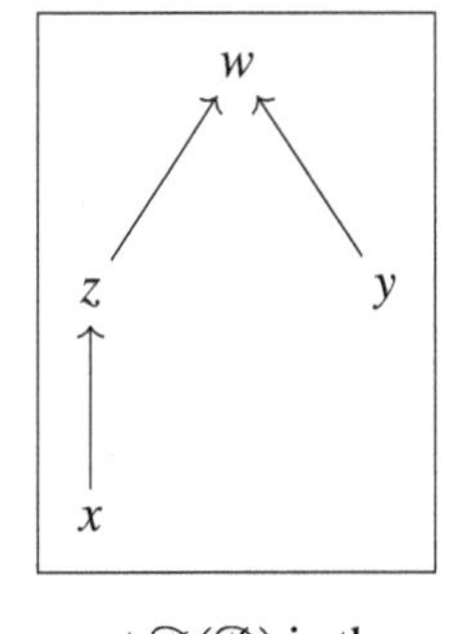

Its associated downset completion poset $\mathfrak{D}(\mathcal{P})$ is then

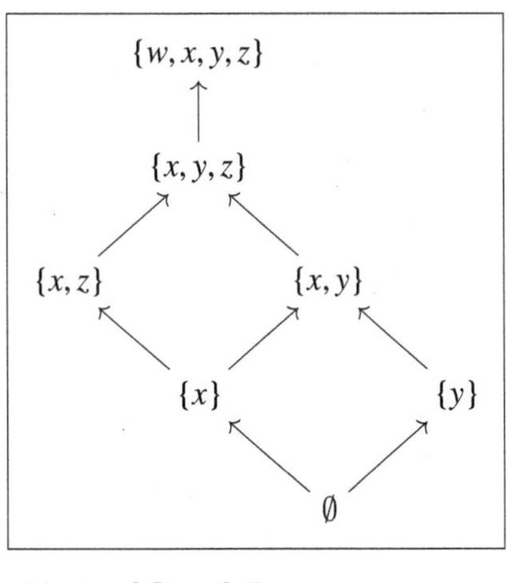

J_{dense} is then given on the objects of $\mathcal{P}$ as follows:

$$J_{dense}(x) = \{\{x\}\}$$

$$J_{dense}(y) = \{\{y\}\}$$

$$J_{dense}(z) = \{\{x\}, \{x, z\}\}$$

$$J_{dense}(w) = \{\{x, y\}, \{x, y, z\}, \{w, x, y, z\}\}.$$

Consider, for instance, why $\{x, z\}$ does not belong to $J_{dense}(w)$. While both x and z are below w, $D = \{x, z\}$ is not a sieve *dense below* w. In particular, given that y is an element in the poset below w, we know that if $\{x, z\}$ *were* dense below w, then there must exist some element $q \in \{x, z\}$ that gets below y in the poset. But there is no such element, except for y itself, which is not in $\{x, z\}$.

Intuitively, one can picture how these four J_{dense} assignments of dense below downsets altogether supply the pieces of something that acts to "cover" the poset,

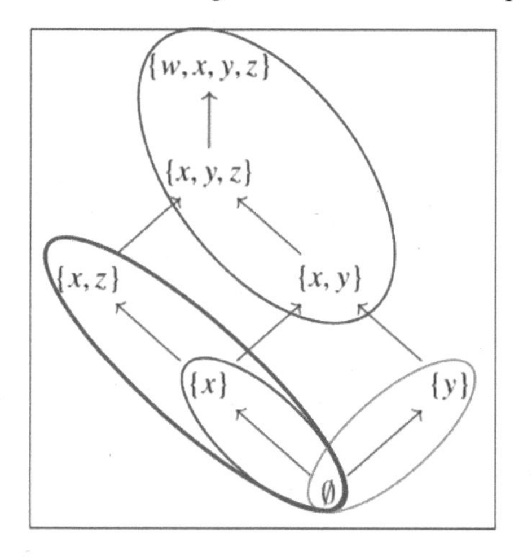

An important special case of the dense topology is the following.

Example 253 (*The atomic topology*) The atomic topology J_{atom} is specified by saying that $S \in J_{atom}(c)$ whenever the sieve S is nonempty. Then axiom (2) in the definition of a Grothendieck topology will be satisfied if we make the assumption (sometimes called the *right Ore condition*)[180] that any two morphisms $f : d \to c$ and $g : e \to c$, both with codomain c, can be completed to a commutative square

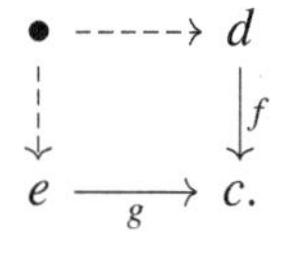

Specialized to posets, again, the atomic topology on $\mathcal{P}$ can be defined only when $\mathcal{P}$ is downward directed, where in general a (nonempty) $N \subseteq \mathcal{P}$ is *downward directed* if for every $x, y \in N$ there exists a lower bound $z \in N$.[181] In such a case, the atomic topology is defined by taking

$$J_{atom}(p) = \mathcal{D}(\downarrow p) \setminus \{\emptyset\}.$$

To get a slightly better handle on this, consider the following poset that is *not* downwards directed, for which the stability axiom for the candidate J_{atom} would then fail:[182]

180. Actually, it turns out that we do not even need this assumption to define the atomic topology—we can just define it as the smallest Grothendieck topology that contains all the nonempty sieves, that is, as the intersection of all Grothendieck topologies with the property that all nonempty sieves cover. See Caramello (2012) for details.

181. In other words, the existence of (finite) meets will suffice to guarantee that a poset is downward directed.

182. This example is derived from Lindenhovius (2014), a nice resource for Grothendieck topologies applied to posets and examples of sheaves on the resulting sites.

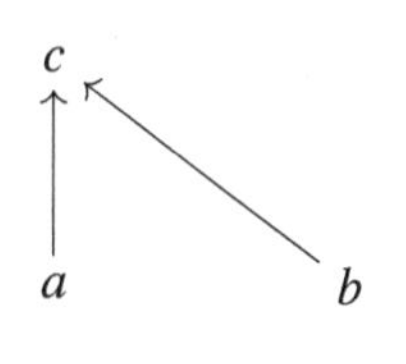

Here, $\downarrow a \in J_{atom}(c)$, and as $b \leq c$, we ought to have that $\downarrow a \cap \downarrow b \in J_{atom}(c)$, that is, that $\emptyset \in J_{atom}(c)$. But if it is really the atomic topology, this cannot be.

Conversely, if a poset *is* downward directed, then one can verify that the stability axiom will necessarily hold.

If a poset $\mathcal{P}$ is downward directed, the dense topology is nothing other than the atomic topology (which is why we referred to it as a "special case" of the dense topology). To see this, assume $\mathcal{P}$ is downward directed, and let S be a nonempty sieve on an element p. Then, as $S \in \mathcal{D}(\downarrow p)$, there is an $x \leq p$. Now let $q \in \downarrow p$. By downward directedness, there exists an $r \leq q$, $r \leq x$, from which we have that $r \in S$, and so $q \in \uparrow S$—altogether implying that $S \in J_{dense}(p)$. Going the other way: supposing $S \in J_{dense}(p)$, we must have $p \in \downarrow p \subseteq \uparrow S$; but this can be the case only if $S \neq \emptyset$. Thus, $S \in J_{atom}(p)$.

A few other examples of Grothendieck topologies will be given throughout this chapter. But with the notion of Grothendieck topology, and thus a site, we are already in a position to define a sheaf in greater generality.

10.2.4 Sheaves on a Site

Sheaves on a site can be defined in a very similar way to sheaves on a topological space, according to the "compatible families of sections can be uniquely patched together" model. Suppose we have a site $(\mathbf{C}, J)$, and of course, a presheaf is simply a functor $P : \mathbf{C}^{op} \to \mathbf{Set}$. Let S be a (J-covering) sieve of $c \in \mathbf{C}$. We first define the following notions:

Definition 254 A *matching family* (or *compatible family*) for S of elements of P is a function that assigns to each arrow f of S an element $x_f \in P(\mathrm{dom}(f))$ in such a way that given any arrow $g \in \mathbf{C}$ with $\mathrm{cod}(g) = \mathrm{dom}(f)$, we have

$$P(g)(x_f) = x_{f \circ g}.$$

Observe that this makes sense, as $f \circ g$ will automatically be an element of S, by virtue of the fact that S is a sieve.

Recall the alternate formulation of a sieve $S = \{c_i \xrightarrow{f_i} c \mid i \in I\}$ on an object c of a category $\mathbf{C}$ from proposition 243, in terms of a subcategory of the slice category $\mathbf{C}/c$, where the objects are just the arrows f_i of S, and the morphisms are given by arrows u making triangles of the following form commute:

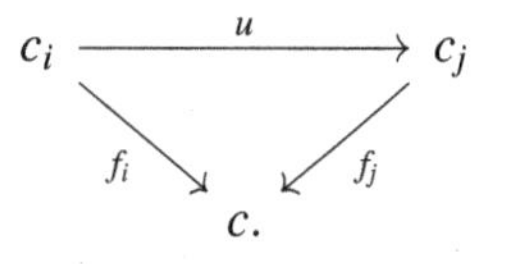

In these terms, a *matching family* of elements of the presheaf P with respect to the sieve S would then be defined as a collection of elements $\{s_i \in P(c_i) \mid i \in I\}$, one for each arrow of the sieve S, such that for any arrow $u : c_i \to c_j$ in S for which the above triangle commutes, the function $F(u)$ sends s_j onto s_i.

Finally, note that since a covering sieve S of c can be construed as a subfunctor of the representable functor Y_c, when we insist on seeing the sieve S in this way, then a matching family $\{x_f\}_{f \in S}$ is just a natural transformation

$$x : S \to P$$
$$f \mapsto x_f,$$

where the assignment of f to x_f is the component of the natural transformation at $\mathrm{dom}(f) \in \mathbf{C}$. If we want to emphasize this perspective, where a matching family of a presheaf P with respect to a sieve S is regarded as a natural transformation $S \to P$, we will sometimes denote the collection of matching families of P with respect to S by $\mathrm{Nat}(S, F)$ or $\mathrm{Match}(S, F)$.

Definition 255 An *amalgamation* for a matching family $\{x_f\}_{f \in S}$ for S, with S a sieve on c, is a unique element $x \in P(c)$ such that

$$P(f)(x) = x_f \text{ for all } f \in S.$$

Making use of both of these notions—matching families and amalgamations—lets us define when a presheaf is a sheaf with respect to a site.

Definition 256 P is a *sheaf* (with respect to J)—or J-*sheaf*—iff each matching family $\{x_f\}_{f \in S}$ with respect to any J-covering sieve $S \in J(c)$ of any object c in $\mathbf{C}$ has a *unique amalgamation.*

In terms of diagrams, a presheaf P will be a sheaf if the diagram

$$P(c) \xrightarrow{\ e\ } \prod_{f \in S} P(\mathrm{dom}(f)) \overset{p}{\underset{q}{\rightrightarrows}} \prod_{f,g \in S, \mathrm{cod}(g) = \mathrm{dom}(f)} P(\mathrm{dom}(g))$$

is an equalizer for each object $c \in \mathbf{C}$ and each cover $S \in J(c)$. Here, e is the map that takes $x \in P(c)$ to $\{P(f)(x)\}_{f \in S}$, while the products range over all composable pairs f, g with $f \in S$ (thus also $f \circ g \in S$), so that $p(\{x_f\}_{f \in S})_{f,g} = x_{f \circ g}$ and $q(\{x_f\}_{f \in S})_{f,g} = P(g)(x_f)$.

Using the fact that a sieve S on c is the same thing as a subfunctor of Y_c, we get another way of presenting the definition. We just saw how a matching family taking $f \mapsto x_f$ for $f \in S$ is the same thing as specifying a natural transformation $S \to P$. Thus, that the matching family $\{x_f\}_{f \in S}$ has a unique amalgamation is equivalent to requiring that $S \to P$ can be uniquely extended as in

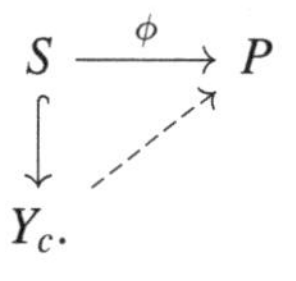

And thus, a presheaf P is a sheaf iff, for every covering sieve S of c, any natural transformation $\phi : S \to P$ has a unique extension to a morphism $Y_c \to P$, in the sense that given the diagram

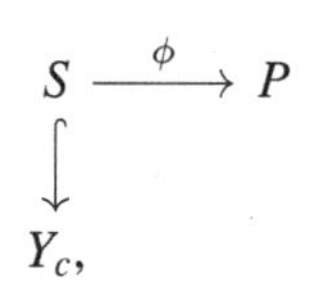

there is always exactly one extension of this to the following commutative diagram:

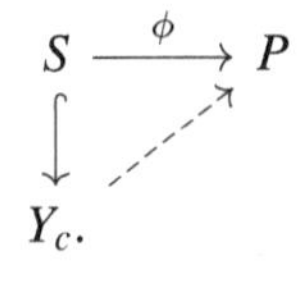

But what does this mean? This is just to say that P is a sheaf precisely when for every covering sieve S on c (where this S is regarded as a functor), the inclusion $i_S : S \hookrightarrow Y_c$ induces an isomorphism

$$\mathrm{Nat}(Y_c, P) \overset{\cong}{\longleftrightarrow} \mathrm{Nat}(S, P).$$

But since the natural transformations are morphisms in the presheaf category, by the Yoneda lemma this $\mathrm{Nat}(Y_c, P) := \mathrm{Hom}_{\mathbf{PreSh(C)}}(Y_c, P)$ is the same as $P(c)$. Moreover, $\mathrm{Nat}(S, P)$ are just the matching families of P with respect to S. Thus, the above just says that a presheaf P on $\mathbf{C}$ is a J-sheaf iff the map

$$\kappa_S : P(c) \to \mathrm{Match}(S, P)$$

$$x \mapsto x \circ i_S$$

is bijective for any object $c \in \mathbf{C}$ and any covering sieve $S \in J(c)$.

Definition 257 (*Alternate definition of sheaf on a site*) A sheaf on a site $(\mathbf{C}, J)$ is a presheaf $P : \mathbf{C}^{op} \to \mathbf{Set}$ such that for every object c of $\mathbf{C}$ and every covering sieve $S \in J(c)$, each morphism $S \to P$ in $\mathbf{Set}^{\mathbf{C}^{op}}$ has exactly one extension to a morphism $Y_c \to P$.

Covering sieves can be generated by covering families, and we saw already, in definition 236, how such covering families are used to define the notion of a basis (pretopology) of a Grothendieck topology.[183] Moreover, in terms of a coverage, the Grothendieck topology generated by a coverage will be the smallest collection of sieves containing it which also happens to be closed under the maximal and transitivity conditions (1 and 3 in the definitions). In section 10.2.1, we then saw how to provide a definition of a sheaf in terms of a basis K. We used the notions of a matching family and an amalgamation, now with respect to a basis K; let's gather these definitions together again.

Definition 258 Given a presheaf P, a basis (pretopology) K, and a family of morphisms $R = \{f_i : c_i \to c \mid i \in I\} \in K(c)$, we call a family of elements $\{x_i \in P(c_i)\}_{i \in I}$ a *matching family* for R iff

$$P(\pi_1)(x_i) = P(\pi_2)(x_j) \ \forall i, j \in I,$$

where $\pi_1 : c_i \times_c c_j \to c_i$ and $\pi_2 : c_i \times_c c_j \to c_j$ are the projections from the pullback, as in

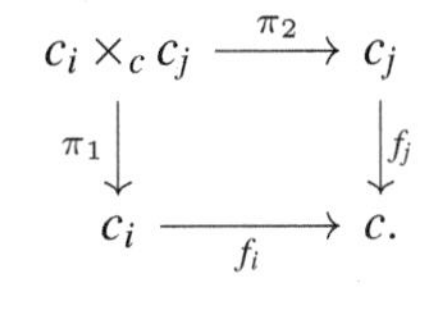

183. We can also define a basis (pretopology) even when $\mathbf{C}$ does not have pullbacks. To do so, replace the second condition in 236 with the condition from the definition of coverage (definition 237).

Definition 259 Given a matching family $\{s_i\}_{i\in I}$ for $R=\{f_i : c_i \to c\}_{i\in I} \in K(c)$, we define an *amalgamation* for that matching family as an element $s \in P(c)$ such that

$$P(f_i)(s) = s_i \;\; \forall i \in I.$$

A sheaf is then defined as it was in section 10.2.1. But now note that given a basis (pretopology) K, this will *generate* a Grothendieck topology J by taking

$$S \in J(c) \text{ iff } \exists R \in K(c) \text{ s.t. } R \subseteq S.$$

Often, in practice, it is more convenient to verify the sheaf condition for the sieve generated by a covering family in terms of the covering family itself (and the associated sheaf definition).

Proposition 260 (*Sheaf with respect to a basis*) Given a site $(\mathbf{C}, J)$, a presheaf P on $\mathbf{C}$ will be a sheaf for J iff for any family of morphisms $R=\{f_i : c_i \to c\}_{i\in I} \in K(c)$ in the basis (pretopology) K, any matching family $\{x_i\}_{i\in I}$ has a unique amalgamation.

In the particular cases where $\mathbf{C}$ has pullbacks, things are simplified and the above proposition can be expressed by saying that a presheaf P on a category $\mathbf{C}$ is a sheaf for J iff for any family of morphisms $R=\{f_i : c_i \to c\}_{i\in I} \in K(c)$ in the basis, the diagram

$$P(c) \xrightarrow{\;e\;} \prod_i P(c_i) \rightrightarrows \prod_{i,j} P(c_i \times_c c_j),$$

is an equalizer.

Moreover, using a coverage and appealing to the definition 240, if a presheaf P satisfies the sheaf condition with respect to a coverage, then it satisfies the sheaf condition with respect to the Grothendieck topology generated by it.[184]

If this is your first time seeing these definitions (and even if it isn't!), they probably remain very abstract. We will try to get a better handle on things by first specializing things to posets, and then considering how these definitions look in the case of the corresponding definition of **2**-enriched J-sheaves on a poset. But, before doing this, it is worth observing that when working with a site consisting of the atomic topology (defined in example 253), sheaves can be given an especially simple description:

Proposition 261 A presheaf P is a sheaf for the atomic topology on $\mathbf{C}$ iff for any morphism $f : b \to c$ and any $y \in P(b)$, if $P(a \xrightarrow{g} b)(y) = P(a \xrightarrow{h} b)(y)$ for all diagrams

$$a \underset{h}{\overset{g}{\rightrightarrows}} b \xrightarrow{\;f\;} c$$

with $f \circ g = f \circ h$, then $y = P(b \xrightarrow{f} c)(x)$ for a unique $x \in P(c)$.[185]

10.2.5 Simpler Version with Posets

As we have said many times now, by regarding a poset $\mathcal{P}$ as a category, there will be an arrow (and at most one arrow) $q \to p$ precisely when $q \leq p$, and we can identify indexed families S of arrows into p with a collection of *elements* "below" p in the poset. Then,

184. For further discussion of some subtleties here, having to do with bases (pretopologies) and the like, we refer the reader to Johnstone (2002, C.2.1).

185. A proof of this fact can be found in Mac Lane and Moerdijk (1994, 127).

taking a poset $\mathcal{P}$ with J a Grothendieck topology, and $F: \mathcal{P}^{op} \to \mathbf{Set}$ a presheaf, we can define the following:

Definition 262 Let $p \in \mathcal{P}$ and $S \in J(p)$. Then a *matching family* for the cover S is a family $\{x_q \mid q \in S\} \in \prod_{q \in S} F(q)$ such that

$$F(r \leq q)(x_q) = x_r$$

for each $r, q \in S$ with $r \leq q$.

Definition 263 An element $x \in F(p)$ for which we have

$$F(q \leq p)(x) = x_q$$

for each $q \in S$ will be called an *amalgamation*.

Using the previous definitions, we can then define a J-sheaf as follows.

Definition 264 (*J-sheaf for posets*) F will be a *J-sheaf* provided for each $p \in \mathcal{P}$, each $S \in J(p)$, and each matching family $\{x_q\}_{q \in S}$ there is a *unique* amalgamation $x \in F(p)$.

It will help to see some examples, working with particular Grothendieck topologies.

Example 265 Take for J the indiscrete (minimal) topology on $\mathcal{P}$:

$$J_{ind}(p) = \{\downarrow p\}.$$

Suppose we have a presheaf $F \in \mathbf{Set}^{\mathcal{P}^{op}}$ and let $p \in \mathcal{P}$. But with J_{ind} as our topology, we know that, for each p, there is only *one* cover of p, namely that given by the principal downset $\downarrow p$ itself. Now, suppose that we have a matching family $\{x_q\}_{q \in \downarrow p}$ for $\downarrow p$. Then, since in particular p itself is in $\downarrow p$, we must have

$$F(q \leq p)(x_p) = x_q.$$

But this just means that x_p is an amalgamation of the matching family. In fact, it must be unique as well. For, suppose $x \in F(p)$ were another amalgamation. Then, by definition, we would need in particular to have

$$x = F(p \leq p)(x) = x_p.$$

Thus, our amalgamation x_p is unique.

In the previous example of a sheaf, notice how we have basically just described what it is to be a presheaf. The more or less trivial nature of the previous example—where it seems that being a sheaf with respect to the given topology is the same as being a presheaf on the original poset—is not an accident. In fact, it reflects a general result, one that also calls back to earlier special results about being able to treat presheaves on a poset as the same thing as sheaves on the poset (once it has been equipped with an appropriate topology). Namely, with the site $(\mathcal{P}, J_{ind})$, where J_{ind} is the indiscrete topology, sheaves are the same thing as presheaves—that is, we have that $\mathbf{Sh}(\mathcal{P}, J_{ind})$ is precisely $\mathbf{Set}^{\mathcal{P}^{op}}$.

Example 266 Suppose $\mathcal{P}$ is downward directed, and take the atomic topology J_{atom} on it. The reader should convince themselves that the sheaves on such a site are constant (up to isomorphism), essentially recovering what it is to be a *set*—that is, that $\mathbf{Sh}(\mathcal{P}, J_{atom}) \simeq \mathbf{Set}$.

The next, more extended, example will give us an opportunity to present a new angle on how to think about sheaves. In building to this construction, which will be developed over the course of the next few pages, we will need to appeal to the following general facts, and some new definitions.

Proposition 267 Let $\{J_i \mid i \in I\}$ be a set of Grothendieck topologies on $\mathbf{C}$. Then

1. $\bigcap_{i \in I} J_i$ is a topology and is the infimum of the J_i;
2. there exists a topology $\bigvee_{i \in I} J_i$ which is the supremum of the J_i.[186]

The Grothendieck topologies on $\mathbf{C}$ are partially ordered by inclusion. Focusing on a poset category in particular, the set of Grothendieck topologies on a poset $\mathcal{P}$ can be partially ordered (pointwise) as follows: for any two Grothendieck topologies J, K on $\mathcal{P}$, we define $J \leq K$ provided $J(p) \subseteq K(p)$ for all $p \in \mathcal{P}$. Then, the poset of Grothendieck topologies on a poset in fact forms a complete lattice, with the infimum $\bigwedge_{i \in I} J_i$ of each collection $\{J_i\}_{i \in I}$ of Grothendieck topologies on $\mathcal{P}$ just given by pointwise intersection $(\bigwedge_{i \in I} J_i)(p) = \bigcap_{i \in I} J_i(p)$.

Returning to the more general account,

Proposition 268 For any presheaf F, there will exist a unique *largest* topology J_M for which F is a sheaf.[187]

Putting this last proposition together with proposition 267 (1), we can deduce that *there is a unique largest topology for which each of a given class of presheaves is a sheaf.* One class of presheaves that are naturally of special interest is the representable functors. This result allows to ask about the largest topology for which, in particular, all the representable functors are sheaves. Such a topology is called the *canonical topology.*

Definition 269 For a category $\mathbf{C}$, we know that the Yoneda embedding provides us, for each object c of $\mathbf{C}$, with the functor $Y_c = \mathrm{Hom}_{\mathbf{C}}(-, c)$. We define the *canonical topology* to be the finest (largest) Grothendieck topology such that every presheaf of the form Y_c is a sheaf—that is, where all the representable functors are sheaves.[188]

Definition 270 We call a Grothendieck topology J *subcanonical* whenever every representable presheaf is itself a sheaf (with respect to J), that is, if Y_c is a J-sheaf for each $c \in \mathbf{C}$. In particular, then, subcanonical topologies are in general smaller (coarser) than the canonical topology (hence the name), which is the finest (largest) of the subcanonical topologies.

In addition to supplying us with further examples of Grothendieck topologies, we are discussing such matters in order to build toward a particular construction and result, the

186. This is lemma 0.34 in Johnstone (2014). This proof of the first item is immediate by definition; and for the second item, one just applies the first result to the set of upper bounds for the J_i.

187. This is lemma 0.35 in Johnstone (2014); a proof of this can also be found in Johnstone (2014).

188. It is important to realize that the canonical topology is not the indiscrete topology, even though representable functors are involved in the determination of both. Recall that the indiscrete topology was the one for which the *only* sieves covering an object $c \in \mathbf{C}$ are things of the form $\mathrm{Hom}_{\mathbf{C}}(-, c)$; moreover, for this topology, sheaves were the same as presheaves. The canonical topology, by contrast, is the largest of the topologies requiring that *all* representable functors be sheaves. This means, in particular, that this may involve other functors as well—just that all the representable ones are sheaves.

full force of which may not become fully evident for a few pages. Recall from chapter 6, section 6.5, how the **2**-enriched presheaves (**2**-presheaves, for short) on a poset $\mathcal{P}$ emerged the same thing as the poset $(\mathcal{D}(\mathcal{P}), \subseteq) := \mathbf{Down}(\mathcal{P})$ formed from the collection $\mathcal{D}(\mathcal{P})$ of all downsets of $\mathcal{P}$, ordered by inclusion, that is,

$$\mathbf{Down}(\mathcal{P}) \cong \mathbf{2\text{-}PreSh}(\mathcal{P}),$$

a fact discussed in the context of the **2**-enriched version of the Yoneda-embedding

$$\mathcal{P} \xhookrightarrow{y=\downarrow} \mathbf{2\text{-}PreSh}(\mathcal{P}) \cong \mathbf{Down}(\mathcal{P}).$$

This so-called downset completion of $\mathcal{P}$ is not just a poset, but a *lattice* (where the meet is just set intersection and the join is union). As it turns out, $\mathbf{Down}(\mathcal{P})$ is not just a lattice, but it is a special one, where finite meets distribute over arbitrary joins (it is what is called a "frame").

Definition 271 A *frame* is a complete lattice in which finite meets distribute over arbitrary joins, that is, in which the following distributivity law

$$a \wedge \left(\bigvee_{i \in I} b_i\right) = \bigvee_{i \in I} (a \wedge b_i)$$

holds, where I is an arbitrary indexing set, and a, b_i are elements of the lattice.

Maps between frames are given by *frame homomorphisms*, where these are functions f that both preserves all joins (including the empty join 0 or "bottom"), that is,

$$f\left(\bigvee_{i \in I} b_i\right) = \bigvee_{i \in I} f(b_i)$$

for any $\{b_i\}$ in the lattice, and preserves binary (finite) meets (including the empty meet 1 or "top"), that is,

$$f(a \wedge b) = f(a) \wedge f(b), \quad f(\top) = \top \tag{10.1}$$

for any a, b in the lattice.

Frames together with their frame homomorphisms assemble into a category, **Frm**.

While frames are rather general entities, observe that every topology X (in the usual sense of topology discussed in chapter 4) forms a frame, in the sense that the lattice of open sets $\mathcal{O}(X)$ is a complete lattice which moreover satisfies the distributivity law, with $\wedge$ becoming set-theoretic intersection and $\bigvee$ the union.[189] Moreover, it can be shown that every frame is in fact a (complete) Heyting algebra—that is, that a complete lattice satisfies the distributivity law above iff it is a Heyting algebra—and object-wise they are identical; the distinction between them has to do only with the relevant homomorphisms, as a frame homomorphism is not necessarily a Heyting algebra homomorphism (one that preserves the Heyting arrow). Thus, while "frame" and "complete Heyting algebra" mean the same thing when considering *objects* (i.e., frames, cHas), they become distinct, and must be treated as such, when we involve the morphisms.

Now recall that a meet-semilattice is a poset in which all finite (hence also empty) meets (greatest lower bounds) exist. A homomorphism f between meet-semilattices S, T

189. More explicitly, one can show that the law holds in $\mathcal{O}(X)$ by using the fact that it holds in the Boolean algebra $\mathbb{P}(X)$, and then considering that the inclusion $\mathcal{O}(X) \to \mathbb{P}(X)$ preserves finite meets and all joins.

is just a map that preserves all finite (including empty) meets (see equation 10.1). This data assembles into a category, **Meet**, the category of meet-semilattices. One can observe that the homomorphisms serving as the morphisms of this category will automatically be monotone; yet observe also that in general the converse need not hold—a monotone map between meet-semilattices need not be a homomorphism in this category (i.e., need not preserve all finite meets). Definitionally, a meet-semilattice that has *all* joins, and where these are distributive, is nothing other than a frame.

Before relating **Meet** and **Frm**, and discussing the point of all this, let us record another important definition that will be of some use to us.

Definition 272 Let $\mathcal{P}$ be a meet-semilattice. A family $\{b_i\}_{i\in I}$ of elements in $\mathcal{P}$ is said to be *join distributive* (or have a *distributive join*) if.

- its join $\bigvee_{i\in I} b_i$ exists in $\mathcal{P}$, and
- for every $a \in \mathcal{P}$,

$$a \wedge \left(\bigvee_{i\in I} b_i \right) = \bigvee_{i\in I} (a \wedge b_i)$$

  for I an arbitrary indexing set.[190]

Then, a map $f : P \to Q$ between two meet-semilattices is said to *preserve distributive joins* provided whenever $\{b_i\}_{i\in I}$ is join distributive in P, then $\{f(b_i)\}_{i\in I}$ is join distributive in Q, and

$$f\left(\bigvee_{i\in I} b_i \right) = \bigvee_{i\in I} f(b_i).$$

Again, one can observe that morphisms that preserve distributive joins in the above sense are automatically monotone. If we take as our morphisms those that preserve distributive joins (and also finite meets), we get the category $\mathbf{Meet}_{\vee dist}$, which of course forms a subcategory of **Meet**.

Join distributivity might seem like a natural condition, but distributive lattices are in fact rather special—and, historically, it is curious to note that it was originally believed, quite wrongly, that *every* lattice was distributive. Here is a simple example of a lattice that does *not* have this property, so that you have something more concrete to hang on to as we press forward with abstractions.

Example 273 The finite lattice depicted below is *not* distributive:

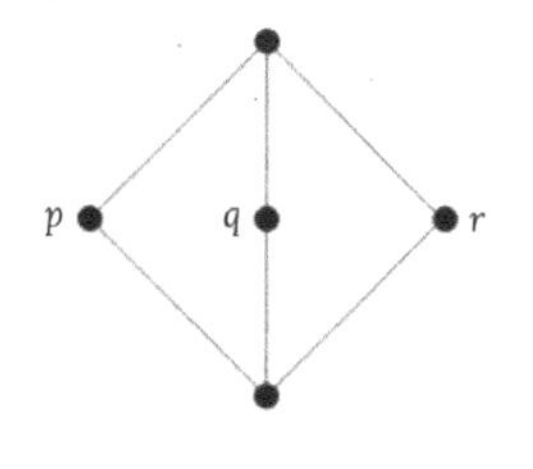

190. In case it helps, for finite joins, note that this just reduces to the usual

$$a \wedge (b \vee d) = (a \wedge b) \vee (a \wedge d).$$

Proof. Observe that $r \wedge (p \vee q) = r$. However, $(r \wedge p) \vee (r \wedge q) = \bot$ (the bottom of the lattice). Thus,

$$r \wedge (p \vee q) \neq (r \wedge p) \vee (r \wedge q),$$

making the lattice fail to distributive over joins. □

Now, as a frame is a lattice with a certain structure—so that, in particular, it is a meet-semilattice (and a join-semilattice, for that matter)—by forgetting about the join-structure of a frame, we should just recover the data of a meet-semilattice. Doing so just describes a forgetful functor

$$U : \mathbf{Frm} \to \mathbf{Meet}$$

that lets us view each frame as a meet-semilattice and each frame morphism as a meet-semilattice morphism. **Frm** is of course also a subcategory—but not a *full* one—of the category **Meet** of meet-semilattices. But **Frm** in fact forms a *full* subcategory of $\mathbf{Meet}_{\vee dist}$ (by just taking those of the meet-semilattices that are frames). As we go forward, unpacking what is going on here, we will need the following general category-theoretic notion:

Definition 274 A (full) subcategory **C** of a category **D** is said to be a *reflective subcategory* when its inclusion $\mathbf{C} \to \mathbf{D}$ has a left adjoint $L : \mathbf{D} \to \mathbf{C}$, called the *reflector*. The unit of the adjunction has components $r_D : D \to L(D)$ for $D \in \mathbf{D}$, where these are called the *reflections*.

The reflector left adjoint to the inclusion typically acts as a sort of "completion" operator. In the frame theory literature, one can find the result that **Frm** is a reflective subcategory of **Meet**.[191] Effectively, this informs us that for every meet-semilattice S there exists a frame F_S together with an embedding $i_S : S \hookrightarrow F_S$ in **Meet**—meaning that i_S is an injection that, being a map in **Meet**, moreover preserves finite meets—where this has the universal property that for any other frame H and any other morphism $g : S \to H$ in **Meet** there will exist a unique morphism $\overline{g} : F_S \to H$ in **Frm** that extends g along i_S, in the sense that $\overline{g} \circ i_S = g$.

$$\begin{array}{ccc} S & \xrightarrow{g} & H \\ & \searrow^{i_S} & \uparrow \overline{g} \\ & & F_S \end{array}$$

Now, given U the forgetful functor, we would expect there to be an associated "free functor" going in the other direction, left adjoint to U. This functor is in fact given by the embedding that takes each element s of a meet-semilattice to its principal downset $\downarrow s$, altogether taking meet-semilattices to the downset frame. As was already seen in chapter 6, the embedding $\downarrow(-) : \mathcal{P} \to \mathbf{Down}(\mathcal{P})$ is monotone, and the construction can be carried out for any poset $\mathcal{P}$. With this embedding, arbitrary infima are preserved (just as the analogous Yoneda embedding will in general *preserve all limits* that exist in **C**) and finite suprema (joins) are preserved. While this obtains for any poset, when $\mathcal{P} = S$ happens to be a meet-semilattice, this same embedding will yield a meet-semilattice morphism. To appreciate this, first observe that, trivially, $\downarrow(\top) = S$, the top of $\mathcal{D}(S)$. Moreover, taking any $a, b \in S$, for each $x \in S$ we will have $x \in \downarrow(a \wedge b)$ iff $x \leq a \wedge b$ iff $x \leq a$ and $x \leq b$ iff $x \in \downarrow(a)$

191. See, for instance, Johnstone (1986).

and $\downarrow(b)$ iff $x \in \downarrow(a) \cap \downarrow(b)$—which shows that $\downarrow(a \wedge b) = \downarrow(a) \cap \downarrow(b)$. In fact, the embedding $\downarrow(-)$ gives us precisely the universal "frame completion," as described in the previous paragraph.

While the embedding $\downarrow(-): \mathcal{P} \to \mathbf{Down}(\mathcal{P})$ must preserve all meets that exist in $\mathcal{P}$, it need not preserve arbitrary suprema. To appreciate this latter failure, it suffices to consider the poset of natural numbers (together with $\{\infty\}$) under the natural ordering. Moreover, when $\mathcal{P}$ again happens to be a meet-semilattice, the embedding will *not* yield a morphism of $\mathbf{Meet}_{\vee dist}$—for, if $\mathcal{P}$ has distributive joins, these are not necessarily going to be preserved by $\downarrow(-)$. In the usual setting of frame theory, this problem is addressed by construing **Frm** as a *full* reflective subcategory of $\mathbf{Meet}_{\vee dis}$, which effectively informs us how for every meet-semilattice (i.e., poset for which all finite meets exist) S, there will exist a universal frame completion in $\mathbf{Meet}_{\vee dist}$ involving $i_S : S \hookrightarrow F_S$, where F_S is a frame and the map i_S is an injection that both preserves finite meets and also preserves any distributive joins that exist in S.

The idea here is to proceed by using an equivalence relation on the frame $\mathbf{Down}(\mathcal{P})$ to generate a quotient, denoted $\mathbf{Down}_{\vee dist}(\mathcal{P})$, whose elements are all those downsets that are in fact *closed under the taking of distributive joins*; then the same construction of the universal frame completion in an appropriate category can be exhibited as a situation where the embedding $\mathcal{P} \hookrightarrow \mathbf{Down}(\mathcal{P})$ factors through $\mathbf{Down}_{\vee dist}(\mathcal{P})$, as in

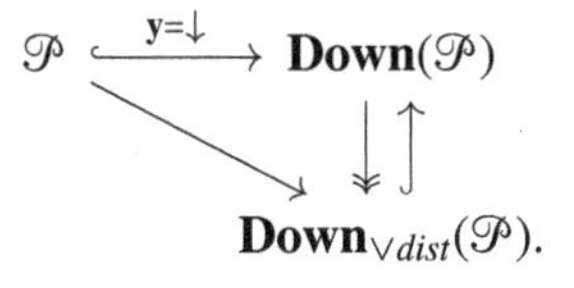

More explicitly, given a meet-semilattice $\mathcal{P}$, we entertain the following binary relation R on $\mathbf{Down}(\mathcal{P})$: if $\{b_j\}_{j \in J}$ is a family with distributive join in $\mathcal{P}$, then $\bigcup_j \downarrow b_j$ will be regarded as R-related to $\downarrow(\bigvee_j b_j)$. Then the *congruence* (equivalence relation on the frame) generated by R can be used to get a quotient of $\mathbf{Down}(\mathcal{P})$, denoted $\mathbf{Down}_{\vee dist}(\mathcal{P})$, the elements of which are those downsets *closed under distributive joins*. In other words, whenever $\{b_i\}_{i \in I}$ is a family with distributive join in $\mathcal{P}$, the elements b_i of which are in $D \in \mathbf{Down}_{\vee dist}(\mathcal{P})$, then $\bigvee_i b_i$ will also be in D. This congruence allows us to give the universal frame completion in the category of meet-semilattices (with join-distributive-preserving maps) as a factorization of the embedding $P \hookrightarrow \mathbf{Down}(\mathcal{P})$ through $\mathbf{Down}_{\vee dist}(\mathcal{P})$.

So we consider just those downsets that are closed under distributive joins by forming the congruence on $\mathbf{Down}(\mathcal{P})$; and as a quotient, this situation in fact assembles into a Galois connection (or adjunction)

$$\mathbf{Down}_{\vee dist}(\mathcal{P}) \underset{\hookrightarrow}{\overset{\twoheadleftarrow}{\bot}} \mathbf{Down}(\mathcal{P}).$$

The surjection here is left adjoint to the embedding, so it the preserves suprema/joins (as well as finite meets, in fact), by the LAPC result of proposition 175; while the embedding, as a right adjoint, preserves infima/meets, by RAPL.

Now we come to the purpose of all this. Just as $\mathbf{2}$-presheaves on a poset $\mathcal{P}$ are essentially the same as the downsets in $\mathbf{Down}(\mathcal{P})$, we can show that the $\mathbf{2}$-*sheaves* on the so-called canonical topology (the largest topology for which all representable presheaves are sheaves) recover precisely those of the downsets that are closed under distributive joins.

Thus, the quotient frame $\mathbf{Down}_{\vee dist}(\mathcal{P}) \hookrightarrow \mathbf{Down}(\mathcal{P})$ exhibited above emerges as a special case of the more general $\mathbf{Sh}(\mathcal{P}, J) \hookrightarrow \mathbf{PreSh}(\mathcal{P})$, and the factorization sketched above will be seen as a special case of a more general process of sheafification of presheaves on a meet-semilattice.

That the downset embedding does not necessarily preserve distributive joins that exist in $\mathcal{P}$ can accordingly be seen as a special case of a more general matter. In general, we know that the Yoneda embedding $\mathbf{y}$ will preserve all *limits* that exist in $\mathbf{C}$. However—and this is the relevant point—$\mathbf{y}$ need not behave as well with respect to *colimits*. To address this, the idea is that one could attempt to restrict the codomain of the embedding, the presheaf category $\mathbf{PreSh}(\mathbf{C})$, by requiring that (at least certain) colimits that happen to exist in $\mathbf{C}$ get preserved in moving to $\mathbf{PreSh}(\mathbf{C})$. The category $\mathbf{PreSh}(\mathbf{C})$ in fact has many nice properties—so the aim of such a restriction of the codomain, then, would ultimately be to describe something that, while treating colimits in the desired fashion, did not restrict so much as to give up any of those nice properties of $\mathbf{PreSh}(\mathbf{C})$.

By making use of the notions of Grothendieck topologies and sheaves on a site, we can make better sense of the underlying situation. In very broad strokes, the idea is this: for each Grothendieck topology J on $\mathbf{C}$, we can consider the sheaves $\mathbf{Sh}(\mathbf{C}, J)$ on the site, and the resulting inclusion functor

$$\mathbf{Sh}(\mathbf{C}, J) \hookrightarrow \mathbf{PreSh}(\mathbf{C})$$

in fact admits a left adjoint which moreover preserves finite limits. As a left adjoint, it will automatically preserve all colimits. By bringing the Yoneda embedding $\mathbf{C} \to \mathbf{PreSh}(\mathbf{C})$ into this setup, we can consider the subcanonical topologies (where all representable presheaves are sheaves), and form the "corestriction" of the Yoneda embedding to an embedding $\mathbf{C} \to \mathbf{Sh}(\mathbf{C}, J)$, exhibiting the category $\mathbf{C}$ as a subcategory of the category of sheaves on the site $(\mathbf{C}, J)$. By taking the *finest* topology for which all the representable presheaves are sheaves—that is, using the canonical topology for our J—we will be left with a universal sort of solution to the underlying problem: we will have the smallest category $\mathbf{Sh}(\mathbf{C}, J)$ into which the category $\mathbf{C}$ may be embedded using a corestriction of the Yoneda embedding. Altogether, this can be thought of as a powerful generalization of the story just sketched of frames and "frame completions"—and such a generalization not only gives us a nice and concrete way of thinking about sheaves in terms of properties of lattices (like distributivity), but it will also allow us to put to work some earlier definitions from the context of $\mathbf{2}$-enriched category theory and to provide an explicit description of an additional Grothendieck topology. It will be worth the reader's while to grasp the particular connection at the heart of this generalization. But we need a few more definitions first.[192]

In what follows, we will take $\mathcal{P}$ to be a poset with finite and empty meets, i.e., a meet-semilattice. We view this as a category in the usual way: regard the underlying poset as a category $\mathcal{P}$, and then stipulate that it has finite limits (and a terminal object). We now

192. In his insightful paper, Isar Stubbe covers the main ideas behind the observations of the last few paragraphs and those that follow on the next few pages (see Stubbe 2005). The main observation concerning how the "frame completion" situation discussed above can be seen as a special case of the more general sheafification of presheaves on a poset equipped with the canonical Grothendieck topology (involving some sort of covers that dealt with distributivity) was something I had stumbled on independently, but Stubbe's paper is a very nice and thorough presentation of the core relevant facts, and includes an explicit description of the canonical topology in question, so the account that follows leans on that paper and refers the reader there for further details.

describe the particular Grothendieck topology J on $\mathscr{P}$ that will be the most relevant for our purposes, that is, the topology according to which the resulting **2**-sheaves on $\mathscr{P}$ paired with the topology J will emerge as nothing other than those downsets of $\mathscr{P}$ that are closed under distributive joins.

Definition 275 We define the Grothendieck topology $J_{\vee dist}$ by regarding this as a function assigning to each $p \in \mathscr{P}$ the collection of *distributive covers* of p in $\mathscr{P}$, where this means what you might expect, namely

$$J_{\vee dist}(p) = \{\{p_i\}_{i \in I} \mid \{p_i\} \text{ is a family in } \mathscr{P} \text{ that has } p \text{ as its distributive join}\}.$$

Note that a family $\{p_i\}_{i\in I}$ in $\mathscr{P}$ has distributive join provided (1) its join $\bigvee_i p_i$ *exists*, and (2) for every $q \leq \bigvee_i p_i$ in $\mathscr{P}$,

$$q = \bigvee_i (q \wedge p_i).$$

In other words, for such a topology, $\bigvee_i p_i = p \in \mathscr{P}$ and $\forall q \in \mathscr{P}$, the (join-)distributivity law

$$q \wedge \left(\bigvee_i p_i\right) = \bigvee_i (q \wedge p_i)$$

holds.[193]

Since we have been specializing matters to posets (lattices), where the relevant maps are ultimately maps between posets, the idea is now to reconstrue the sheaf definition in its **2**-enriched version, in which setting it becomes rather straightforward to discern what a sheaf is, namely the presheaves that are "J-continuous."

Definition 276 (**2**-*enriched sheaf*) A **2**-*sheaf* on a site $(\mathscr{P}, J)$ is a presheaf $\phi : \mathscr{P}^{op} \to \mathbf{2}$ for which for every $p \in \mathscr{P}$ and every $\{p_i\}_{i\in I} \in J(p)$, if $\phi(p_i) = 1$ for all $i \in I$, then $\phi(p) = 1$.

The poset of **2**-sheaves is then denoted $\mathbf{2}\text{-}\mathbf{Sh}(\mathscr{P}, J)$; this is a subposet of $\mathbf{2}\text{-}\mathbf{PreSh}(\mathscr{P})$.

In the isomorphism $\mathbf{2}\text{-}\mathbf{PreSh}(\mathscr{P}) \cong \mathbf{Down}(\mathscr{P})$ first introduced in chapter 6, the isomorphism takes any $\phi \in \mathbf{2}\text{-}\mathbf{PreSh}(\mathscr{P})$ to $\phi^{-1}(1) \in \mathbf{Down}(\mathscr{P})$; and, given a downset D of $\mathscr{P}$, one defines $\phi_D : \mathscr{P}^{op} \to \mathbf{2}$ by setting $\phi_D(x) = 1$ precisely when $x \in D$. For any element $p \in \mathscr{P}$, there will be a representable **2**-presheaf

$$\phi_p : \mathscr{P}^{op} \to \mathbf{2}$$

that makes $q \mapsto 1$ iff $p \leq q$. In this way, the representable presheaves act as the "characteristic maps" of the principal downsets of $\mathscr{P}$, and the Yoneda embedding taking each $p \mapsto \phi_p$ is essentially the same as the inclusion of the elements of the poset into its downsets. The above **2**-enriched sheaf definition just tells us that ϕ (a **2**-presheaf) is a sheaf if whenever ϕ sends every element in a cover to "true" (i.e., 1), then it sends the element covered by those elements to "true" as well. This gives us a particularly easy way of thinking about a sheaf as involving a gluing together of its local assignments.

In general, as indicated earlier, for a category $\mathbf{C}$ and for every Grothendieck topology J, the inclusion functor $\mathbf{Sh}(\mathbf{C}, J) \hookrightarrow \mathbf{PreSh}(\mathbf{C})$ from the sheaves on the site $(\mathbf{C}, J)$ to the

193. Once we allow that joins exist, notice that this is effectively the stability condition (condition 2) in the general definition of a Grothendieck topology. The other conditions making $J_{\vee dist}$ count as a Grothendieck topology are evident.

presheaves on $\mathbf{C}$ has a left adjoint that preserves finite limits. Since a topology is subcanonical when all representable presheaves are sheaves, that a given topology J is subcanonical amounts to stipulating that the Yoneda embedding $\mathbf{y}:\mathbf{C}\to\mathbf{PreSh}(\mathbf{C})$ can be "co-restricted" to an embedding $\mathbf{C}\to\mathbf{Sh}(\mathbf{C},J)$ revealing $\mathbf{C}$ as a subcategory of $\mathbf{Sh}(\mathbf{C},J)$. Now, with a subcanonical topology J, a meet-semilattice (or poset) $\mathcal{P}$ can be embedded in the poset $\mathbf{2}\text{-}\mathbf{Sh}(\mathcal{P},J)$ by simply sending $p\in\mathcal{P}$ to $\phi_p\in\mathbf{2}\text{-}\mathbf{Sh}(\mathcal{P},J)$, making this embedding a "corestriction" of the Yoneda embedding. It is straightforward to see that the topology $J_{\vee dist}$ will be subcanonical, and that the following result obtains:

Proposition 277 $\mathbf{2}\text{-}\mathbf{Sh}(\mathcal{P},J_{\vee dist})\cong\mathbf{Down}_{\vee dist}(\mathcal{P})$.

Thus, in taking

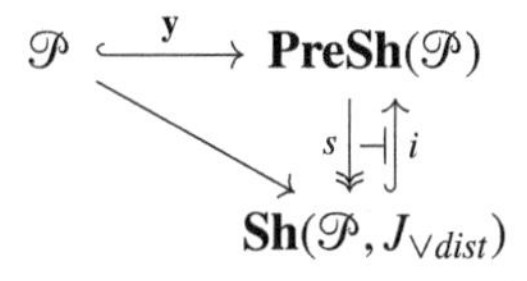

to $\mathbf{2}$-enriched category theory, we have

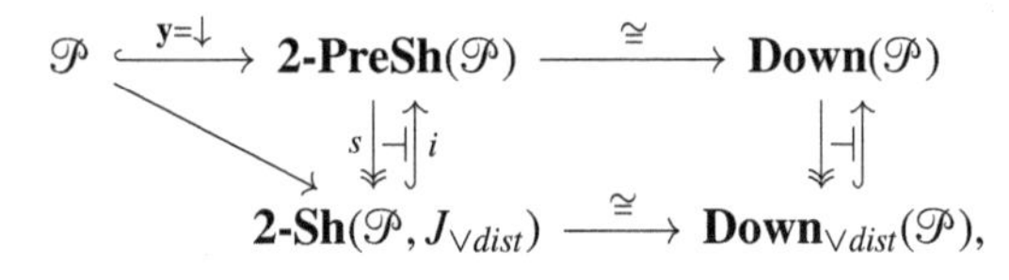

and the desired frame completion is just given by this corestriction $\mathcal{P}\hookrightarrow\mathbf{2}\text{-}\mathbf{Sh}(\mathcal{P},J_{\vee dist})$, which acts to sheafify the $\mathbf{2}$-presheaves along the topology of distributive covers on $\mathcal{P}$. This map will preserve not just all meets but all distributive joins that exist in $\mathcal{P}$.

The category $\mathbf{Sh}(\mathcal{P},J_{\vee dist})$ is in fact the *smallest* corestriction of the Yoneda embedding for which $\mathcal{P}$ can still be considered a subcategory. This can be seen by showing that the topology $J_{\vee dist}$ is the *finest* (largest) of the subcanonical topologies on $\mathcal{P}$, in turn demonstrating universality in the frame completion. The overall idea here is that the larger the topology, in general, the fewer sheaves there will be on the associated site—after all, the larger the topology, the more covers there are for which one must verify the sheaf condition is met, and so the more chances for there to be a failure to meet the condition! Since we know that, in general, the canonical topology gives us the largest of the subcanonical topologies, if our topology $J_{\vee dist}$ can be shown to be the canonical topology on $\mathcal{P}$, in considering the sheaves on the resulting site, we will be left with the universal solution—in the sense of the *least* (smallest) category $\mathbf{Sh}(\mathcal{P},J)$ in which $\mathcal{P}$ can be embedded via "corestricting" the Yoneda embedding.

Proposition 278 $J_{\vee dist}$ is the *finest* subcanonical topology on $\mathcal{P}$, making it the *canonical* topology.

Proof. $J_{\vee dist}$ is subcanonical. Let J' be another subcanonical topology on $\mathcal{P}$. But if $\{p_i\}_{i\in I}\in J'(p)$, then we must have that the join exists and $p=\bigvee_i p_i$, since if $\{p_i\}_{i\in I}\in J'(x)$, then $p_i\leq p$ for all p_i, making p an upper bound for its covering family; and if p were not the *least* upper bound, then there would have to exist a $q<p\in\mathcal{P}$ such that $p_i\leq q$ for all p_i, but since J' is subcanonical, the representable ϕ_q is a sheaf, and this entails that $\phi_q(p)=1$ (as $\phi_q(p_i)=1$ for all p_i and ϕ_q is a sheaf), yet this contradicts the assumption that $q<p$.

The join of such a covering family is, moreover, necessarily distributive. Thus, $J'(p) \subseteq J_{\vee dist}(p)$ for every p. By definition, this makes $J_{\vee dist}$ finer than any other subcanonical topology.[194] □

This topology of distributive covers $J_{\vee dist}$ on a meet-semilattice $\mathscr{P}$ (qua category), as the canonical Grothendieck topology, is in fact the canonical Grothendieck topology even for more general sheaves, landing in **Set** (as opposed to **2**-enriched sheaves).[195]

We leave it to the reader to continue to ponder the impacts of thinking about sheaves and sheafification in terms of the topology of distributive covers and maps that preserve all distributive joins. Working with particularly simple lattices should help to uncover some of the significance of this special way of thinking about the more general result, and the process of obtaining a sheaf as one that engages distributivity—or, rather, as one of adding in all the joins, while keeping the meets, and then demanding that distributive joins are preserved.

10.2.6 Grothendieck Toposes

Returning to a more general approach, the sheaves on a site $(\mathbf{C}, J)$ form a category, where the maps are the natural transformations (between presheaves). As such, this category of J-sheaves, which we denote as $\mathbf{Sh}(\mathbf{C}, J)$, forms a full subcategory of the presheaf category

$$\mathbf{Sh}(\mathbf{C}, J) \rightarrowtail \mathbf{Sets}^{\mathbf{C}^{op}}.$$

Definition 279 A *Grothendieck topos* is a category which is equivalent to the category $\mathbf{Sh}(\mathbf{C}, J)$ of sheaves on some site $(\mathbf{C}, J)$—that is, a category $\mathbf{E}$ is a Grothendieck topos if there exists a site $(\mathbf{C}, J)$ such that $\mathbf{E}$ is equivalent to $\mathbf{Sh}(\mathbf{C}, J)$.

On account of this, Grothendieck toposes also sometimes go under the name *sheaf toposes*. Let us first look at some (mostly) trivial examples.

Example 280 In the previous section, in example 265, we already saw how if we take the site consisting of a category $\mathbf{C}$ with the indiscrete (trivial) topology J_{ind}, then J_{ind}-sheaves are the same thing as presheaves on $\mathbf{C}$. Thus, $\mathbf{Sh}(\mathbf{C}, J_{ind})$ reduces to the presheaf category $\mathbf{Set}^{\mathbf{C}^{op}}$. In particular, then, for any (small) category $\mathbf{C}$, by taking J as the indiscrete topology, any category of the form $\mathbf{Set}^{\mathbf{C}^{op}}$—that is, any presheaf category—will be a Grothendieck topos. This gives us a large stock of examples.

Example 281 Specializing the previous example, one can verify that $\mathbf{Set}^{\mathbf{1}^{op}} = \mathbf{Set}$ is a Grothendieck topos, equivalent to the category of sheaves on the terminal category $\mathbf{1}$.

To see this more explicitly, we can specialize this to the one-point topological space $\mathbf{1} = \{*\}$ with its unique topology, the topology in which all subsets are open. Observe that for the space $\mathbf{1}$—or just the category with one object and one morphism, the identity morphism—there are just two opens, $\emptyset$ and $\{*\}$ itself. For any sheaf $F \in \mathbf{Sh}(\mathbf{1})$, being a sheaf requires that for the empty set, which is covered by the empty collection of subsets, the only matching family for the empty cover of $\emptyset$ is the trivial empty tuple, that is, $F(\emptyset) = \{()\}$; but then the only information supplied by the sheaf F is the *set* $F(\{*\})$. As

194. Our proof of this mimics Stubbe (2005).
195. This is the substance of theorem 1 in Stubbe (2005); a proof can be found there.

such, a set may be regarded as a sheaf on a point. $\mathbf{Set} = \mathbf{Sh}(\{*\})$ accordingly plays the role of a point in topos theory, and is thus referred to as the *punctual topos*.

Example 282 For a topological space X, the Grothendieck topos $\mathbf{Sh}(\mathcal{O}(X), J_{\mathcal{O}(X)})$ of sheaves on the site $(\mathcal{O}(X), J_{\mathcal{O}(X)})$ (as defined in example 251) is the same as the usual category $\mathbf{Sh}(X)$ of sheaves on X.

Example 283 Recall the Sierpiński space $\mathbb{S}$, first introduced in exercise 4 (chapter 4). This was formed from a two-point set $\{0,1\}$, with the collection of open sets given by taking just one open point, that is, $\{\emptyset, \{1\}, \{0,1\}\}$. This open set category can accordingly be displayed by

$$\emptyset \to \{1\} \to \{0,1\}.$$

A presheaf P on such an open set category just amounts to the specification of

$$P(\{0,1\}) \to P(\{1\}) \to P(\emptyset),$$

that is, three sets together with two functions between them. But recall from theorem 129 (chapter 5) that when P is a sheaf, we must have that $P(\emptyset) \cong \{1\}$, that is, that $P(\emptyset)$ is a set with just one element. This means that the category of sheaves on the Sierpiński space reduces to the data of two sets $F(\{0,1\}), F(\{1\})$ and one function $F(\{0,1\}) \to F(\{1\})$ between them, where $\{0,1\}$ is the whole space and $\{1\}$ is the sole open singleton. In other words, the objects (sheaves) of this category are just functions between sets, and the morphisms are necessarily natural transformations between such objects (sheaves). Altogether, this informs us that the category of sheaves on the Sierpiński space $\mathbb{S}$ is nothing other than the arrow category $\mathbf{Set}^{\to}$ of $\mathbf{Set}$.[196] $\mathbf{Sh}(\mathbb{S})$ is accordingly a Grothendieck topos, called the *Sierpiński topos*.

A Grothendieck topos has certain desirable properties—in particular, it is complete and cocomplete. To appreciate this, it suffices to realize that given a site $(\mathbf{C}, J)$, the presheaf category $\mathbf{Set}^{\mathbf{C}^{op}}$ is complete and cocomplete; one computes limits and colimits pointwise. One can show that if a category is complete and cocomplete, then so too is the reflective subcategory. $\mathbf{Sh}(\mathbf{C}, J)$ is a (full) reflective subcategory of the presheaf category $\mathbf{Set}^{\mathbf{C}^{op}}$, and hence inherits all limits and colimits. Such properties can help us rule out certain categories as being Grothendieck toposes.

Example 284 **FinSet** is *not* a Grothendieck topos. In fact, various toposes arising from finite sets—like $\mathbf{FinSet}^{\mathbf{C}^{op}}$ (with $\mathbf{C}$ a fixed finite category)—will not count as one either, since they lack infinite colimits.[197]

Before moving on to some more explicit and exciting examples, we will take the opportunity to look more closely at the important adjoint "sheaf functor" associated to the inclusion of presheaves into sheaves on a site—a functor that sends each presheaf P to the sheaf that most closely "approximates" it, and exhibits the sheaf topos as a reflective subcategory.

196. See definition 20 (chapter 1) for the definition of the arrow category.

197. There are also prominent non-Grothendieck toposes that are treated in topos-theoretical approaches to nonstandard analysis. How can something be a topos but not a Grothendieck topos? We will look at that shortly.

10.2.7 The Plus Construction

In the previous two sections, we have repeatedly appealed to the fact that the inclusion functor

$$\iota : \mathbf{Sh}(\mathbf{C}, J) \rightarrowtail \mathbf{Sets}^{\mathbf{C}^{op}},$$

taking J-sheaves to their underlying presheaves, admits a left adjoint

$$\mathbf{a} : \mathbf{Set}^{\mathbf{C}^{op}} \to \mathbf{Sh}(\mathbf{C}, J),$$

which sends an arbitrary presheaf to a sheaf. This **a** is called the *associated sheaf functor*. In other words, we will have

$$\mathrm{Hom}_{\mathbf{Sh}(\mathbf{C},J)}(\mathbf{a}(P), F) \cong \mathrm{Hom}_{\mathbf{PreSh}(\mathbf{C})}(P, \iota(F)).$$

To show such an adjunction, what do we need? Well, first of all we would of course need to show that such a functor indeed exists, and then show that it is indeed left adjoint to the stated inclusion. While this adjunction is of some importance in its own right—and we return to it in the next chapter—this section will be mainly devoted to constructing the functor in question. Such a functor is valuable in that it effectively generalizes, to sites, the functor $\Gamma\Lambda$ introduced in equation 8.1, and gives us a canonical way of upgrading a presheaf to a sheaf.

The process involved in constructing the functor **a** in question is carried out in two main steps, the first of which involves converting a presheaf into a *separated presheaf*, using something called the *plus construction*, denoted $(-)^+$. The plus construction turns out to be a functor $(-)^+ : \mathbf{Set}^{\mathbf{C}^{op}} \to \mathbf{Set}^{\mathbf{C}^{op}}$ in its own right, but we will also show that for any presheaf P, applying this functor to get P^+ will leave us with a presheaf that is moreover separated, that is,

$$\mathbf{PreSh}(\mathbf{C}) \xrightarrow{(-)^+} \mathbf{SepPreSh}(\mathbf{C}).$$

While a separated presheaf is not yet necessarily a sheaf, it gets us "one step closer" to a sheaf. As it turns out, it can be shown that P^+ is in fact a sheaf provided P is already separated. Thus, for any presheaf (including those that are not already separated), by applying the plus construction to the presheaf twice—this is the second of the two main steps—we will be sure to end up with a sheaf, since regardless of whether or not P was itself separated, P^+ will be separated and applying the plus construction again to this P^+ will leave us with a sheaf. This composite $((-)^+)^+$ indeed gives us the desired sheafification left adjoint functor **a**

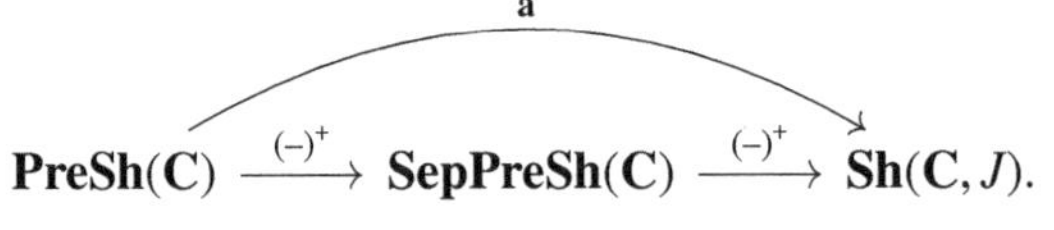

$$\mathbf{PreSh}(\mathbf{C}) \xrightarrow{(-)^+} \mathbf{SepPreSh}(\mathbf{C}) \xrightarrow{(-)^+} \mathbf{Sh}(\mathbf{C}, J).$$

But what does an explicit description of this plus construction look like? The first thing to realize is this process takes an arbitrary presheaf and ultimately tells us something about how it behaves with respect to covering sieves, so we will need to bring covering sieves into the mix. The relevant functor can be defined in a number of (equivalent) ways, but fundamentally one can think of this construction as taking a presheaf and replacing its elements by covering sieves that are themselves equipped with matching families, adjoining unique amalgamations (gluings) to every matching family.

Given P a presheaf on a category $\mathbf{C}$, for each object $c \in \mathbf{C}$, we will define

$$P^+(c) = \varinjlim_{S \in \mathbf{Cov}(c)^{op}} \text{Match}(S, P),$$

where the colimit is taken over all covering sieves of c, ordered by reverse inclusion, and where Match(S, P) is the set of matching families for the cover S of c. Let us unpack this.

First of all, observe that the covering sieves of c form a poset, with the natural order inherited from the inclusion order on sieves: $S \leq S'$ iff $S(d) \subseteq S'(d)$ for all $d \to c$. But we know that we can also regard covering sieves as subfunctors of the representable functor $Y_c = \text{Hom}(-, c)$, and so it is also natural to consider the poset of covering sieves ordered by reverse inclusion (refinement). This category of all covering sieves of c ordered by reverse inclusion $(J(c), \subseteq^{op})$ will be denoted $\mathbf{Cov}(c)^{op}$, where an inclusion $S \subseteq S'$ of covering sieves for c is a morphism $S' \to S$ in the category $\mathbf{Cov}(c)^{op}$. Moreover, we know from the induced condition (4) of definition 245 that the intersection of any two covering sieves on an object will be a covering sieve on that object—that is, any two covering sieves have a common refinement. In terms of $\mathbf{Cov}(c)^{op}$, the way of highlighting this refinement property is to say that $\mathbf{Cov}(c)^{op}$ is *filtered.*

Definition 285 A category $\mathbf{J}$ is *filtered* (or *directed*) if

- it is not empty;
- for every pair of objects $j, j' \in \mathbf{J}$, there exists an object k and two morphisms $f : j \to k$ and $f' : j' \to k$ in $\mathbf{J}$; and
- for every two parallel arrows $u, v : i \to j$ in $\mathbf{J}$, there exists an object k and an arrow $w : j \to k$ such that $w \circ u = w \circ v$.

This is a generalization of the notion of a (upward) directed poset, from order theory, where that means a poset that is inhabited (nonempty) and for which every finite subset has an upper bound.

A *filtered colimit* is a colimit of a functor $F : \mathbf{J} \to \mathbf{C}$, where $\mathbf{J}$ is a filtered category. And in particular, a colimit over a filtered poset $\mathscr{P}$ is the same as the colimit over a *cofinal* subset Q of that poset, where this means that for every element $p \in \mathscr{P}$, there exists an element $q \in Q$ with $p \leq q$.

Proposition 286 $\mathbf{Cov}(c)^{op}$ is *filtered.*

Proof. To see this, observe:

- Y_c is always a covering sieve, so $J(c)$ is nonempty.
- We are using the reverse order, so the requirement amounts to saying that any two covering sieves must have a common refinement, which we know is the case.
- Again reversing the arrows in the requirement, and using pullbacks, one can easily see that this is also true.

□

Thus, $P^+(c)$ will be a filtered colimit, which matters as we are taking a colimit over (finer and finer) covering sieves.

Now suppose $R = \{c_i \xrightarrow{f_i} c \mid i \in I\}$ is a covering sieve of c, for an object $c \in \mathbf{C}$. We know that a sieve can be regarded as a functor, specifically a subfunctor $i_R : R \hookrightarrow \text{Hom}_{\mathbf{C}}(-, c)$ of the representable functor on c, which is in turn an object of **PreSh(C)**. In other words, we have a functor

$$i_c : \mathbf{Cov}(c)^{op} \rightarrowtail \mathbf{PreSh}(\mathbf{C})$$

taking our covering sieve R into the presheaf category. But this inclusion in fact induces a map of natural transformations, for any presheaf F,

$$\mathrm{Nat}(\mathrm{Hom}_{\mathbf{C}}(-, c), F) \to \mathrm{Nat}(R, F). \tag{10.2}$$

By the Yoneda lemma, each morphism in $\mathrm{Nat}(\mathrm{Hom}_{\mathbf{C}}(-, c), F)$ just corresponds to an element $x \in F(c)$. How about the other side? What is a natural transformation $\sigma : R \Rightarrow F$? Well, it will just amount to a matching family of elements $\{x_i \in F(c_i) \mid i \in I\}$! We can appreciate this by simply unpacking the definition of a natural transformation. As such, $\sigma : R \Rightarrow F$ will take each arrow $f_i : c_i \to c$ of the sieve R to an element $\sigma_{c_i}(f_i) \in F(c_i)$. Moreover, for any commutative triangle in the sieve,

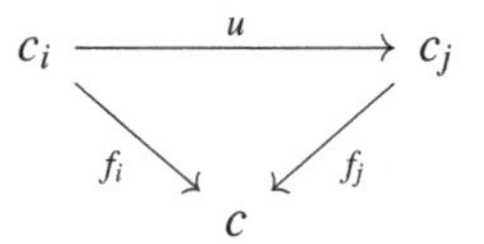

we will have that σ_{c_i} takes $f_i \mapsto x_i \in F(c_i)$, σ_{c_j} takes $f_j \mapsto x_j \in F(c_j)$, and $F(u)$ takes $x_j \mapsto x_i$. And this just informs us—viewing the sieve in terms of its slice category description—that each morphism from R to F amounts to a matching family of elements on R. Moreover, we know that F will be a sheaf precisely when the map in equation 10.2 is an isomorphism.

Altogether, we have a composite functor

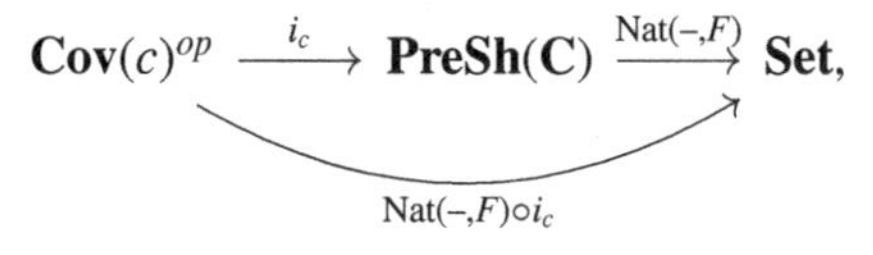

where this takes a covering sieve R to the set of natural transformations from R to F, where each such morphism is the same as a matching family of elements for R. For $f \in R$, this functor will then map $f \mapsto x_f$, where x_f is a component of the natural transformation $R \to F$ at $\mathrm{dom}(f) \in \mathbf{C}$.

Now, given a presheaf P, $P^+(c)$ will be exactly the colimit of the composite functor just described. Given an $R \in \mathbf{Cov}(c)$, we will designate the associated morphism of the colimit by

$$x_R^P : \mathrm{Nat}(R, P) \to P^+(c).$$

Recall that one way of reading the sheaf condition on a presheaf P is as saying: corresponding to each matching family $\{x_f \in P(\mathrm{dom}(f))\}_{f \in R}$ for the sieve R on c, there is a unique element $x \in P(c)$ such that $x_f = P(f)(x)$ for all $f \in R$. In this way, we can start to appreciate that this composite functor gives us a presheaf that sends a covering sieve $R \hookrightarrow \mathrm{Hom}(-, c)$ to the set $\mathrm{Nat}(R, P)$—or, alternatively, to the set $\mathrm{Match}(R, P)$ of matching families for P on R. The difference between this presheaf and the original $P(c)$ can be seen as supplying us with a measure of the degree to which P fails to obey the sheaf conditions at c.

Now, any morphism $f : c' \to c$ will induce (by pulling back) a functor $\mathbf{Cov}(c) \to \mathbf{Cov}(c')$. How does this work? Suppose given a morphism $f : c' \to c$ in $\mathbf{C}$. Then each sieve R on c determines, by pulling back,

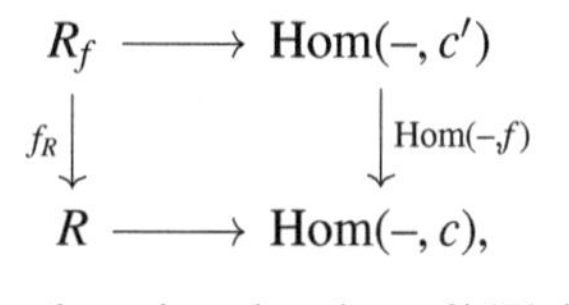

a sieve on c', where we are here denoting the sieve $f^*(R)$ by $R_f := \{g \mid f \circ g \in R\}$. And we can take

$$\begin{array}{ccc} \mathrm{Nat}(R_f, P) & \longleftarrow & \mathrm{Nat}(\mathrm{Hom}(-, c'), P) \\ {\scriptstyle \mathrm{Nat}(f_R, P)}\uparrow & & \uparrow \\ \mathrm{Nat}(R, P) & \longleftarrow & \mathrm{Nat}(\mathrm{Hom}(-, c), P), \end{array}$$

which is the same (using the Yoneda lemma) as

$$\begin{array}{ccc} \mathrm{Nat}(R_f, P) & \longleftarrow & P(c') \\ {\scriptstyle \mathrm{Nat}(f_R, P)}\uparrow & & \uparrow \\ \mathrm{Nat}(R, P) & \longleftarrow & P(c). \end{array}$$

Altogether, we have a functor

$$\mathbf{Cov}(c) \to \mathbf{Cov}(c')$$
$$R \mapsto R_f,$$

and this induces a unique morphism $P^+(c) \to P^+(c')$, from which we can deduce the functoriality of P^+.

Let us see how this works in more detail. Given a morphism $f : c' \to c$ in $\mathbf{C}$ as above, as we range through the R in $\mathbf{Cov}(c)$, the induced morphisms

$$\mathrm{Nat}(R, P) \xrightarrow{\mathrm{Nat}(f_R, P)} \mathrm{Nat}(R_f, P) \xrightarrow{x^P_{R_f}} P^+(c') \qquad \text{(composite: } x^P_{R_f} \circ \mathrm{Nat}(f_R, P)\text{)}$$

actually form a co-cone on the diagram defining $P^+(c)$, that is, on

$$\mathrm{Nat}(R, P) \xrightarrow{\mathrm{Nat}(f_R, P)} \mathrm{Nat}(R_f, P) \xrightarrow{x^P_{R_f}} P^+(c). \qquad \text{(composite: } x^P_R\text{)}$$

But by definition, $P^+(c)$ is the colimit, and so it is the universal co-cone:

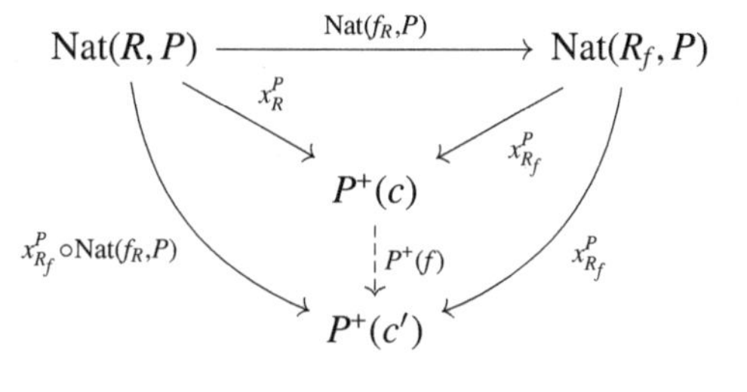

And this supplies us with a unique morphism $P^+(f): P^+(c) \to P^+(c')$ making the diagram commute, in the sense that it satisfies

$$P^+(f) \circ x_R^P = x_{R_f}^P \circ \mathrm{Nat}(f_R, P).$$

But then the functoriality of P^+ is basically immediate from this uniqueness condition in the definition of $P^+(f)$. In particular, given another morphism $g: c'' \to c'$ of $\mathbf{C}$, we can see that

$$\begin{aligned} P^+(f \circ g) \circ x_R^P &= x_{R_{f \circ g}}^P \circ \mathrm{Nat}((f \circ g)_R, P) \\ &= x_{(R_f)_g}^P \circ \mathrm{Nat}(g_R, P) \circ \mathrm{Nat}(f_R, P) \\ &= P^+(g) \circ x_{R_f}^P \circ \mathrm{Nat}(f_R, P) \\ &= P^+(g) \circ P^+(f) \circ x_R^P. \end{aligned}$$

But, using the uniqueness condition in the definition of $P^+(f \circ g)$ inherited from the universality of the colimit construction, this just tells us that

$$P^+(f \circ g) = P^+(g) \circ P^+(f)$$

whenever $\mathrm{cod}(g) = \mathrm{dom}(f)$. It is also evident that $P^+(\mathrm{id}_c) = \mathrm{id}_{P^+(c)}$. Altogether, this tells us that the P^+ we defined is in fact itself a presheaf (functor).

Before going forward, let us reconsider the description of $P^+(c)$ once again. Suppose we have two covers $R, S \in \mathbf{Cov}(c)$. Since finite intersections of elements of $\mathbf{Cov}(c)$ are in $\mathbf{Cov}(c)$, and since for any two covers we have a common refinement $Q \subseteq R \cap S$, with $Q \in \mathbf{Cov}(c)$, the induced maps $S \to Q$ and $R \to Q$ in our relevant category $\mathbf{Cov}(c)^{op}$

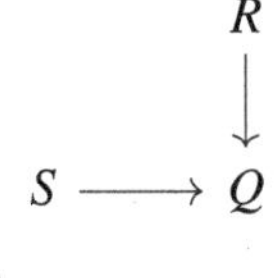

allow us to form the pullback diagram

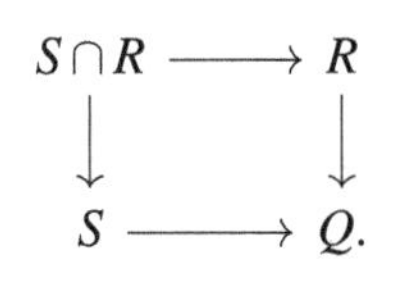

In terms of presheaves, the diagram of covering sieves then becomes

the colimit of which is then the pushout (the categorical dual of the pullback)

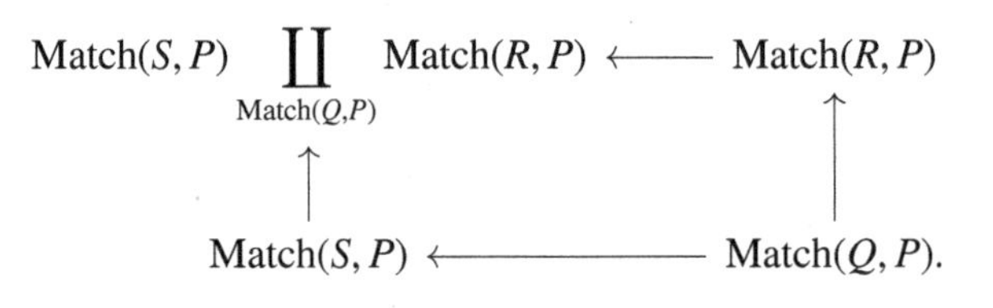

And this $\text{Match}(S, P) \coprod_{\text{Match}(Q,P)} \text{Match}(R, P) \cong \text{Match}(S \cap R, P)$ just amounts to equivalence classes of matching families. Specifically, two matching families will belong to the same equivalence class if their restriction along a common refinement agrees. In other words, $P^+(c)$ consists of equivalence classes $[\{x_f\}_{f \in S}]$—where, as you would expect, the $\{x_f\}_{f \in S}$ are such that $x_f \in P(c')$ and for any $g : c'' \to c'$, we have $P(g)(x_f) = x_{f \circ g}$. On this setup, two families, $(S, \{x_f\}_{f \in S})$ and $(S', \{y_h\}_{h \in S'})$, are then taken to be equivalent—$(S, \{x_f\}_{f \in S}) \sim (S', \{y_h\}_{h \in S'})$—precisely when there is a common refinement $Q \subseteq S \cap S'$ on which the restrictions of $\{x_f\}_{f \in S}$ and $\{y_h\}_{h \in S'}$ agree, that is, where $x_i = y_i$ for every $i \in Q$.

Now, given a morphism $l : c' \to c$ in $\mathbf{C}$, there will then be an induced restriction map $P^+(c) \to P^+(c')$ between the presheaves, given by

$$P^+(c) \to P^+(c')$$

$$[\{x_f\}_{f \in S}] \mapsto P^+(l)([\{x_f\}_{f \in S}]) = [\{x_{l \circ \hat{f}}\}_{\hat{f} \in l^*(S)}].$$

One can verify that this map preserves equivalence classes and is well defined. It is also straightforward to check the functor conditions, ensuring that this P^+ is itself a presheaf (functor).

Altogether, we have seen (in more than one way) how the construction of the presheaf P^+ works, and that it is a presheaf.

10.2.7.1 $(-)^+$ is a functor as well We can actually show that $(-)^+ : \mathbf{PreSh}(\mathbf{C}) \to \mathbf{PreSh}(\mathbf{C})$, the process we used to construct the presheaf P^+ from a presheaf P, is itself functorial. To see this, let us first define the functor. Again observe that a natural transformation $\alpha : F \Rightarrow G$ between two presheaves F, G will induce a natural transformation

$$\text{Nat}(-, F) \circ i_c \Rightarrow \text{Nat}(-, G) \circ i_c,$$

which will supply us with a unique factorization

$$\alpha_c^+ : F^+(c) \to G^+(c)$$

between the colimits, making is so that for each covering sieve R, we have

$$\alpha_c^+ \circ x_R^F = x_R^G \circ \text{Nat}(R, \alpha).$$

We can show that α^+ is natural as follows. If we have $f : c' \to c$ in $\mathbf{C}$, then using the definition of a colimit, we need only establish that

$$G^+(f) \circ \alpha_c^+ \circ x_R^F = \alpha_{c'}^+ \circ F^+(f) \circ x_R^F$$

for each $R \in \mathbf{Cov}(c)$. But the left-hand side of this equation

$$\begin{aligned}
&= G^+(f) \circ x_R^G \circ \text{Nat}(R, \alpha)\\
&= x_{R_f}^G \circ \text{Nat}(f_R, G) \circ \text{Nat}(R, \alpha)\\
&= x_{R_f}^G \circ \text{Nat}(R_f, \alpha) \circ \text{Nat}(f_R, F)\\
&= \alpha_{c'}^+ \circ x_{R_f}^F \circ \text{Nat}(f_R, F)\\
&= \alpha_{c'}^+ \circ F^+(f) \circ x_R^F,
\end{aligned}$$

as desired.

Having defined $(-)^+$, we can show that it is functorial. Given another natural transformation $\beta : G \Rightarrow H$, one can show—in effectively the same way as we showed P^+ to be functorial, using the uniqueness condition in the definition of $(-)^+$ applied to the composite $\beta \circ \alpha$—that $(\beta \circ \alpha)^+_c = \beta^+_c \circ \alpha^+_c$. We can also construct the natural transformation

$$\eta : \mathrm{id} \Rightarrow (-)^+,$$

by constructing, for each presheaf F, a natural transformation $\eta_F : F \Rightarrow F^+$. On objects $c \in$ **C**, set

$$\eta^c_F : F(c) \to F^+(c)$$
$$s \mapsto x^F_{Y_c}(\hat{s}),$$

where $\hat{s} : Y_c \Rightarrow F$ is just the natural transformation associated to F (using the Yoneda lemma). In particular, given $c' \in \mathbf{C}$,

$$\hat{s}_{c'} : \mathrm{Hom}(c', c) \to F(c')$$
$$f \mapsto F(f)(s),$$

which completes the definition of η^c_F. Finally, one can show that η_F and η are natural (which we leave to the reader),[198] altogether leaving us with a functor $(-)^+$.

10.2.7.2 When is P^+ a sheaf? The point of all this, we suggested, was to generate a sheaf! Running the plus construction on an arbitrary presheaf P, will P^+ be a *sheaf*? Well, not always! But it does get us "closer" to having a sheaf, in a precise sense to be unpacked. Observe how if P^+ were a sheaf, we could show that for any matching family there exists an amalgamation, where this amalgamation is moreover unique. Thus, if something fails to give us a sheaf, what may have gone wrong is that there may not be an amalgamation for a matching family, or there may be "too many."

The important sense in which a presheaf may be "closer" to being a sheaf is that it is *separated*, where fundamentally a separated presheaf is one that satisfies the uniqueness part of the sheaf amalgamation condition but not necessarily the existence part. Referring back to the equalizer diagram defining a sheaf in definition 256, for a general presheaf P, if P fails to be a sheaf, it will thus fail to be an equalizer for the diagram

$$P(c) \xrightarrow{\ e\ } \prod_{f \in S} P(\mathrm{dom}(f)) \overset{p}{\underset{q}{\rightrightarrows}} \prod_{f,g \in S, \mathrm{cod}(g) = \mathrm{dom}(f)} P(\mathrm{dom}(g)).$$

Since P^+ may fail to be a sheaf, it can thus fail to be an equalizer. However—and this is the point!—even if it fails to be a sheaf, we will see that P^+ will always be separated, where this can be captured more formally by saying that the map e above must still be injective.

Definition 287 A presheaf F is *separated for* $Q \in \mathbf{Cov}(c)$ if

$$F(c) \to \prod_{f \in Q} F(\mathrm{dom}(f)) \quad .$$

is injective, that is, if a section of F is entirely determined by its restrictions along the elements of a cover. F is *separated* (period) if it is separated for every covering sieve Q.

198. See, for instance, Borceux (1994) for details.

Observe how, in terms of the map in equation 10.2, we could have said that a presheaf P is separated if $\mathrm{Hom}_{\mathbf{Set}^{\mathcal{C}^{op}}}(Y_c, P) \to \mathrm{Hom}_{\mathbf{Set}^{\mathcal{C}^{op}}}(S, P)$ is a monomorphism for each covering sieve.

Thus, a presheaf is separated if a matching family can have *at most one* amalgamation. Before proving the main result, let us first appreciate, with an example, how the plus construction does not always leave us with a sheaf.

Example 288 Recall, from chapter 8, section 8.2, the constant presheaf P on X a topological space consisting of two elements, endowed with the discrete topology (i.e., every set is open). In other words, we have a presheaf P on X that assigns any set (or abelian group) with more than one element to each of the four open sets and the identity map to each of the nine restriction maps (five plus the four trivial self-maps), for example, $P(\{0,1\}) = P(\{0\}) = P(\{1\}) = P(\emptyset) = \mathbb{Z}$. Observe that this presheaf is not only not a sheaf but in particular it is not separated (precisely for the empty cover of the empty set).

What is P^+? I claim that it is

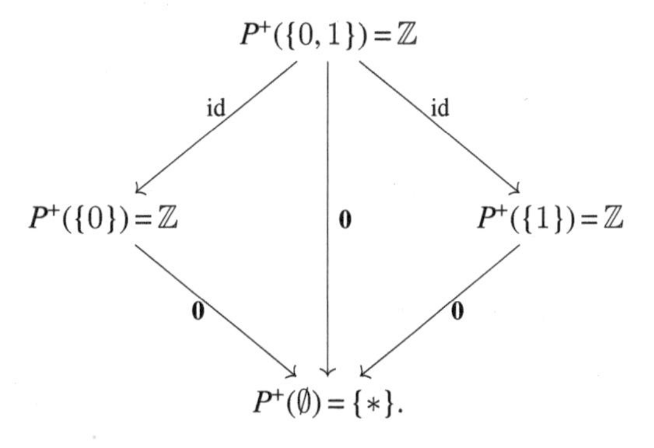

Observe what has changed. While we have $P^+(U) = \mathbb{Z}$ whenever $U \neq \emptyset$, this construction forces $P^+(\emptyset) = \{*\}$, the terminal object. This is because, for the empty cover, there is exactly one matching family.

At this point, we should make two observations: First, observe that P^+ is still not yet a sheaf. For, let $U_1 = \{0\}$ and $U_2 = \{1\}$ be the disjoint opens of X. Together, $\{U_1, U_2\}$ covers X. But then $s_1 \in P^+(U_1)$ and $s_2 \in P^+(U_2)$ will both have, for their restriction to $P^+(\emptyset)$, the unique element of $P^+(\emptyset) = \{*\}$. But this means that $\{s_1, s_2\}$ is actually now a matching family for the cover $\{U_1, U_2\}$ of $U_1 \cup U_2 = \{0, 1\}$. Yet, as we saw in chapter 8, section 8.2, if s_1 and s_2 are taken to be different elements of $\mathbb{Z}$, they cannot have an amalgamation in P^+. This brings us to the second observation: Notice how, in running this plus construction, we may introduce matching families—and thus, families that ought to have amalgamations—that were not there in the original presheaf P (and so we need not have amalgamations for them).

Again, the plus construction does not necessarily leave us with a sheaf, but it does take us in the right direction, as running this $(-)^+$ once gets us "halfway" toward a sheaf—in the precise sense that it leaves us with a separated presheaf. Here is the main result:

Proposition 289 Given P an arbitrary presheaf, P^+ is always a separated presheaf.

Proof. To show that P^+ is separated, it will suffice to show that the map e in

$$F(c) \xrightarrow{e} \prod_{h \in Q} F(\mathrm{dom}(h))$$

is injective. In particular, consider two elements $\{x_f\}_{f \in S}$ and $\{y_g\}_{g \in R}$ in $P^+(c)$ for which $P^+(h)(\{x_f\}_{f \in S}) = P^+(h)(\{y_g\}_{g \in R})$ for all collections of morphisms $h : c' \to c$ in Q, with Q a cover of c. One must then show that the matching families $\{x_f\}_{f \in S}$ and $\{y_g\}_{g \in R}$ are in fact the same. The proof is left to the reader. □

Thus, the plus construction takes an arbitrary presheaf and will make it a separated presheaf. But what happens if the presheaf on which we run the plus construction is *already* separated? In that case, running the plus construction will actually leave us with a sheaf!

Proposition 290 If a presheaf P is already separated, then P^+ will be a sheaf.

Proof. Again left to the reader. □

But by putting these last two results together, we have that for *any* presheaf F, $(F^+)^+$ will be a sheaf! In other words, we can functorially upgrade any presheaf on a site to a sheaf by just two applications of the plus construction! While sometimes one can get away with just running it once, in order to take an arbitrary presheaf (separated or not) to a sheaf, it will always suffice to apply it just twice.

Using the double application of the plus functor as our definition of the sheaf functor **a**, we can now prove the following:

Theorem 291 **a** is left adjoint to the inclusion functor,

$$\mathbf{Sh}(\mathbf{C}, J) \underset{\iota}{\overset{\mathbf{a}}{\rightleftarrows}} \mathbf{PreSh}(\mathbf{C}).$$

Proof. The adjunction amounts to saying that to any map $P \to \iota(F)$ there will correspond a unique map $\mathbf{a}(P) \to F$. But consider the natural transformation $\eta : P \to P^+$. On components $c \in \mathbf{C}$, this is defined as

$$\eta_c : P_c \to P^+(c)$$
$$x \mapsto \eta_c(x) = \{P(f)(x) \mid f \in M_c\},$$

using M_c the maximal sieve. As we know, applying η twice gives us a map from P to $(P^+)^+ = \mathbf{a}(P)$. We would like to show that any map from P to a sheaf F will factor uniquely through the map $P \xrightarrow{\eta \circ \eta} \mathbf{a}(P)$, in the sense that

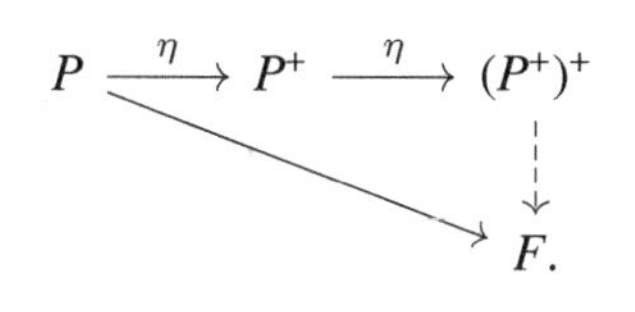

But the same map η is being applied twice, so it would be enough to show that for

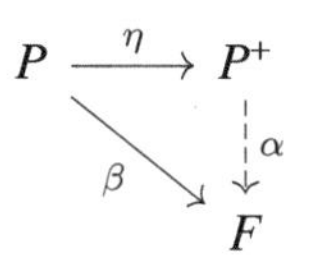

there is a unique map α making the diagram commute.

Take an element $\{x_f\}_{f\in R} \in P(c)$ for a cover R of c. Then, for a map $g: c' \to c$ in R, we will have that $\eta_{c'}(x_g) = \{P(h)(x_g) \mid h \in M_{c'}\}$. But since $\{x_f\}_{f\in R}$ is a matching family, we will have $P(g)(\{x_f\}_{f\in R}) = \{x_{g\circ f'} \mid f' \in g^*(R)\}$. As $g \in R$, $g^*(R)$ will be equal to the maximal sieve $M_{c'}$, giving $\eta_{c'}(x_g) = P(g)(\{x_f\}_{f\in R})$, which will hold for all $g \in R$.

Now, if there indeed existed the map α, as desired, it would preserve the above equality, in the sense that there would exist a unique $\alpha(\{x_f\}_{f\in R}) \in F(c)$ satisfying

$$F(g)(\alpha(\{x_f\}_{f\in R})) = \alpha([P(g)(\{x_f\}_{f\in R})]) = \alpha(\eta_{c'}(x_g)) = \beta(x_g)$$

for all $g \in R$. Since F is a sheaf and $\{\beta(x_g) \mid g \in R\}$ is a matching family, there will exist such a unique element $\alpha(\{x_f\}_{f\in R}) \in F(c)$ satisfying the above equation.

Thus, given a map $P \to \iota(F)$, this will uniquely determine a map $h: \mathbf{a}(P) \to F$ making

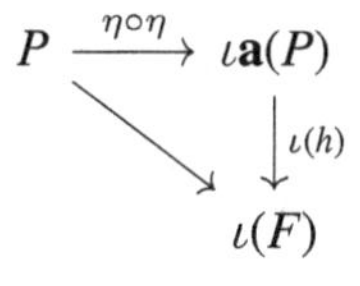

commute, and telling us that $\eta \circ \eta$ is in fact the unit of the adjunction. □

As **a** is a left adjoint, by LAPC, **a** preserves colimits—thus, all small colimits will exist in $\mathbf{Sh}(\mathbf{C}, J)$, since they exist in $\mathbf{PreSh}(\mathbf{C})$. In effect, we thus ensure that the "nice" features of the presheaf category (topos) carry over to our sheaf topos. In fact, we could show that the functor $(-)^+$ also preserves finite limits of presheaves; using this, and running $(-)^+$ twice, it further follows that $\mathbf{a}: \mathbf{PreSh}(\mathbf{C}) \to \mathbf{Sh}(\mathbf{C}, J)$ preserves finite limits as well. These facts will become important in the next chapter.

Example 292 Let us return to example 288. What if we run $(-)^+$ once more, on P^+ itself? Then we should have

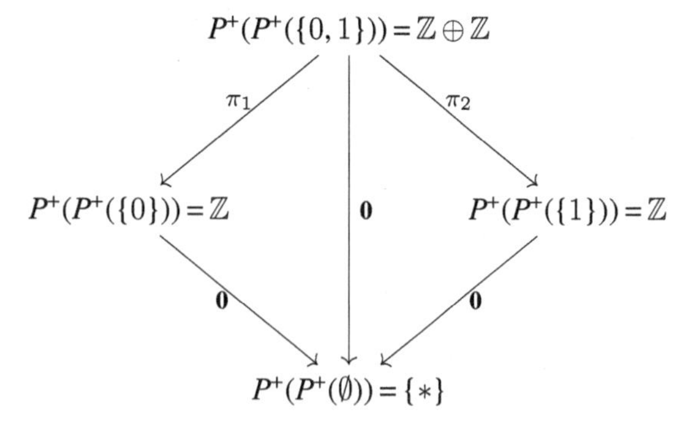

which is indeed a sheaf.

For a related example, again taking X the discrete topological space on a two-element set, readers should convince themselves that for *any* set-valued presheaf P on X, $P^+(X)$ will be equal to the pullback of

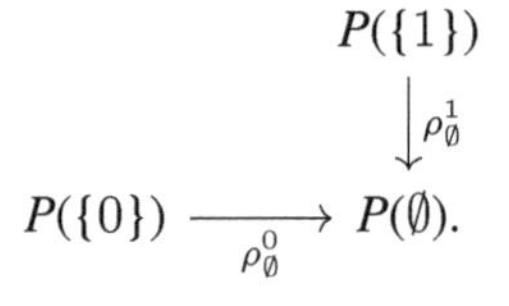

For instance, suppose the presheaf P has been given by $P(X) = \emptyset$ and $P(\{1\}) = P(\{0\}) = P(\emptyset) = A$, where A is some set with (at least) two elements and where the restriction maps $\rho_{\emptyset}^{\{1\}}$ and $\rho_{\emptyset}^{\{0\}}$ are both the identity map. In other words, we have

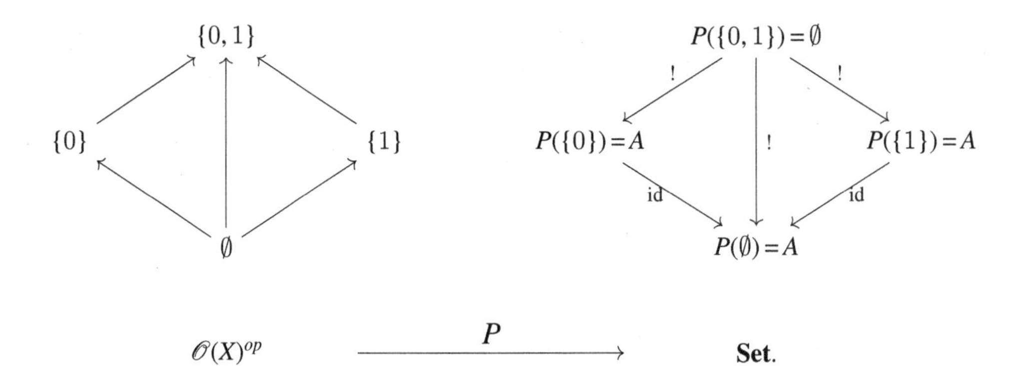

This is, of course, not a sheaf. To figure out how P^+ is determined, let us first consider what the plus construction does over the whole space X. As mentioned above, $P^+(X)$ should be the pullback of

$$\begin{array}{ccc} & & P(\{1\}) = A \\ & & \big\downarrow \rho_{\emptyset}^{1} = \mathrm{id} \\ P(\{0\}) = A & \xrightarrow[\rho_{\emptyset}^{0} = \mathrm{id}]{} & P(\emptyset) = A \end{array}$$

which set is canonically isomorphic to A itself. Thus, $P^+(X) = A$. What about for the rest of the opens? P^+ will not change anything for the two nontrivial open subsets of X. However, as in the earlier example with abelian groups, $P^+(\emptyset)$ will be forced to be $\{*\}$, the terminal object in **Set**, that is, a singleton set. This is again on account of the fact that $\emptyset$ is covered by the empty cover; and for the empty cover (a covering by no sets at all), there is exactly one matching family, and so any two sections of $F(\emptyset)$ will agree on this cover. In more detail, recall how products over the empty index set reduce to $\{*\}$, so the relevant equalizer diagram for this cover forces $F^+(\emptyset) = \{*\}$.

Altogether, then, this leaves us with

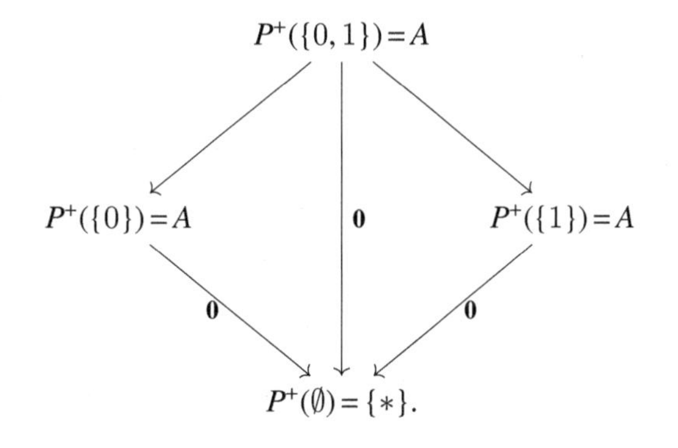

Again, P^+ is not yet a sheaf. But if we run $(-)^+$ once more, on P^+ itself, then $P^{++}(X)$ is again the same as the pullback, now of

$$\begin{array}{ccc} & & P^+(\{1\})=A \\ & & \downarrow \mathbf{0} \\ P^+(\{0\})=A & \xrightarrow[\mathbf{0}]{} & P(\emptyset)=\{*\}. \end{array}$$

With this P^{++}, we will indeed have a sheaf.

Altogether, observe how, in general, P, P^+, and $P^+(P^+)$ may all be different—and given a presheaf we may truly need to run the plus construction twice to end up with a sheaf. On the other hand, for certain presheaves, it will suffice to run the plus construction once, as the following example demonstrates.

Example 293 Recall the presheaf B of all bounded continuous functions on a topological space, for instance,

$$B(U) := \{\text{bounded continuous functions } U \to \mathbb{R}\},$$

defined on $X=\mathbb{R}$. This is clearly a presheaf, with the usual restriction maps. Moreover, it is separated. However, as you were asked to contemplate in chapter 8, section 8.2, this is not a sheaf. For, consider the open cover of $\mathbb{R}$ given by $\{U_n=(-n,n)\}_{n\in\mathbb{N}}$. And take the identity function $f_n(x)=x$. This will give us a sequence of functions, bounded on each U_n, that moreover agree on intersections—that is, a matching family. However, the only possible amalgamation of $\{f_n\}$ to all of $X=\mathbb{R}$ is $f(x)=x$, which is unbounded. Hence, we have a matching family with no amalgamation. The moral here was that even if we have a collection of bounded functions whose domain of definition cover all of $\mathbb{R}$ and which are well behaved on intersections, we cannot be sure that we will have a function that is *bounded* and defined on all of $\mathbb{R}$, when these are glued together. There are many functions that are locally bounded yet not globally bounded—hence, you cannot expect to be able to glue together locally bounded pieces, since such a gluing might leave you with an unbounded function.

If there is some matching family that does not have an amalgamation, the plus construction effectively adds it. One can turn this presheaf B of all bounded continuous functions into a sheaf—specifically, the sheaf of all continuous functions—with just one application of the plus construction to B. One way to appreciate this is to first realize that there is a map $B^+(U) \to C(U)$, since each matching family of bounded functions $f_i : U_i \to \mathbb{R}$ on a cover U_i

of U can be joined to supply a unique continuous function $f: U \to \mathbb{R}$, and equivalent matching families will yield the same function f. Going the other way, $C(U) \to B^{+}(U)$, every continuous function $f: U \to \mathbb{R}$ can be got from a matching family of bounded functions $f_n: U_n \to \mathbb{R}$, where $U_n = \{x \in U : |f(x)| < n\}$, and f_n is the restriction of f to such U_n.

10.3 A Few More Examples

Let us now take a break from the more abstract results and look at a few more examples of sheaves, some of which have the added benefit of exhibiting the utility of the more general notion of Grothendieck topologies.

Example 294 Recall from the n-coloring graph discussions how, in the case of the category of undirected, connected graphs, we can define a subgraph G of a graph H as a graph such that $\text{Edges}(G) \subseteq \text{Edges}(H)$, and in that context, to define a cover of a graph G it sufficed to specify a family of subgraphs $\{G_i \hookrightarrow G \mid i \in I\}$ satisfying the condition that

$$\bigcup_{i \in I} \text{Edges}(G_i) = \text{Edges}(G).$$

The present example will also work with such notions and the implied site on the category of undirected, connected graphs.[199]

The well-known chess "n-queens problem" presents the task of finding configurations of n queens on an $n \times n$ chessboard that satisfy the condition that none of them can attack one another, where two queens can *attack* each other, or are *in conflict*, in the usual chess sense (i.e., if they are placed on the same row, column or diagonal). We can encode this data by working over the subgraphs of the complete graph K_n with n nodes. Each node i, $1 \leq i \leq n$, is assigned a variable q_i, corresponding to a placement of a queen in the i-th column. In the concrete case of $n = 4$, we then have the set

$$F = \{F(q_1), F(q_2), F(q_3), F(q_4)\},$$

where the $F(q_i)$ are valued in the set $\{1, 2, 3, 4\}$, each of the values $u_i \in F(q_i)$ corresponding to the *row* where the queen in the i-th column has been placed. Initially, we can think of value assignments such as $(1, 2, 4, 3)$ in terms of subgraphs ordered by inclusion. For instance, for this last assignment, we take $(1, 0, 0, 0)$ to correspond to a chosen node (labeled 1); $(1, 2, 0, 0)$ for the chosen node together with the edge from 1 to the node labeled 2; $(1, 2, 4, 0)$ for the chosen node 1 and its edge to 2 together with additional edges from 2 to the node labeled 4 (and then from 4 to 1); and then $(1, 2, 4, 3)$ for the complete graph K_4.

What a given ordered set of numbers represents then is a given *configuration* of queens, where such assignments correspond to a particular subgraph of the complete graph. For instance, the selection of 1 in the "stalk" over q_1 would represent the placement of a queen in the first row of the first column—or, in terms of the underlying subgraphs, the assignment of data to the specified node labeled 1 in a subgraph. Then, for instance, if 2 were selected in the stalk over q_3, yielding $(1, 0, 2, 0)$, this would correspond to a configuration with a queen in the first row of the first column and another queen in the second row of the third column. But since ultimately exactly one queen will be placed in each column (up to 4 in our current case), we can restrict our attention to the subgraphs corresponding to

199. The idea for this example, and many of the details of its presentation, comes from Srinivas (1993).

sequences that proceed stepwise from q_1 through q_4. At the i-th step, we add a queen to the i-th column, ensuring that it does not attack any of the $i-1$ queens that have already been placed. In other words, in constructing the sheaf over such subgraphs ordered by inclusion, we first consider the restricted set

$$\{c_{ij} \mid (q_i \neq q_j) \wedge (|q_i - q_j| \neq |i-j|) \ \forall i,j \in \{1,2,3,4\}, i \neq j\}$$

which exhibits the set of constraints describing the problem (ruling out configurations such as $(1,2,4,3)$). In the case of selecting 1 in the stalk at q_1, then, the constraint demands that for such a selection, only 3 or 4 could be selected in the stalk at q_2. In turn, in the case of selecting 3, this could be extended no further; while in the case of selecting 4, we could extend one more step, assigning the value 2 to the stalk at q_3—however, such an assignment could not then be extended over the complete graph K_4, for the constraint prohibits any further assignment to q_4.[200] On the other hand, one of the (two possible) global sections is given by selecting 2 at the stalk at q_1, then 4 at q_2, 1 at q_3, and 3 at q_4. Since the selection of an integer $1, \ldots, 4$ over the nodes of the subgraphs of K_4 stands for a queen placement that respects the constraint, we might represent the data of such an assignment thus:

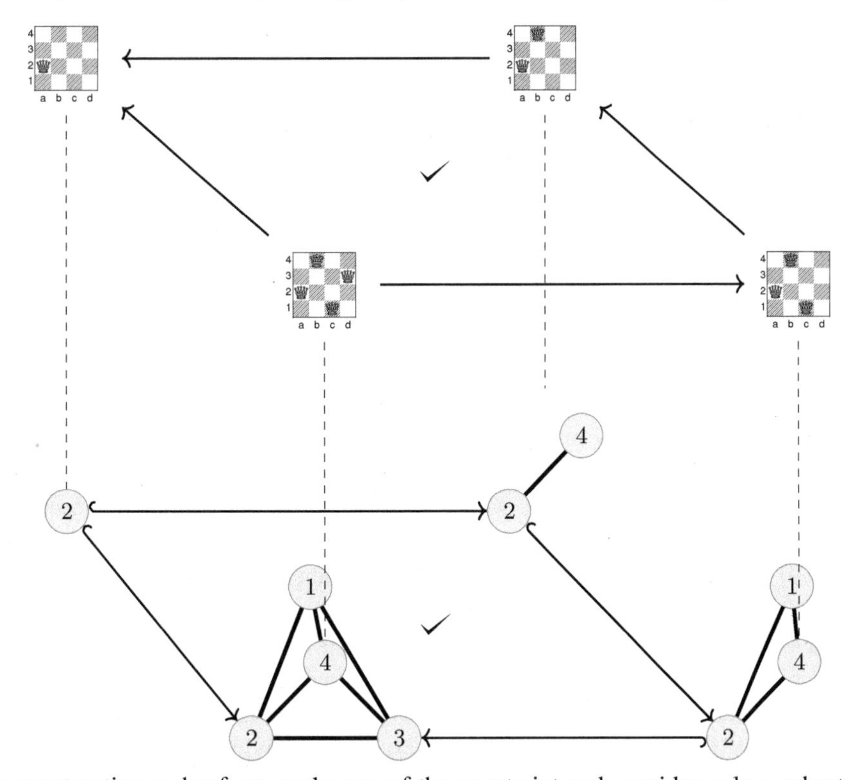

In constructing a sheaf, we make use of the constraint and consider only a subset of the possible sequences of q_i, so that, for instance, $(1,1,0,0), (1,2,0,0)$ are not allowed. More explicitly, a site for the sheaf is built on the poset of subgraphs of the complete graph K_n on n nodes, using the covering introduced at the outset. A configuration of queens

200. Recalling the discussion from chapter 9, such assignments could thus be seen to represent (strictly) local sections.

corresponding to a given graph G is then just a function that assigns to each node in G a position on an $n \times n$ board. In other words, we let Sub(K_n) be the poset of subgraphs of K_n, $N(G)$ and $E(G)$ the nodes and edges of a given (sub)graph G, and *Chessboard* the set of all pairs $\langle x, y \rangle$, where x and y range between 1 and n. But since we are not interested in all possible (arbitrary) configurations of queens, we can use **valid**(c, e) a predicate describing the special situations in which the positions in a given configuration (specified by a map c) of a pair of queens connected by the edge e do not present a conflict. In this way, we can ultimately define a contravariant functor

$$\mathbf{Config} : Sub(K_n)^{op} \to \mathbf{Set}$$

that assigns the set of valid configurations of queens to each subgraph, where a valid configuration is one in which none of the queens conflict along any of the edges of the given subgraph. In a little more detail, the functor acts to assign to each subgraph of the complete graph the set of all pairwise compatible (along each component edge) arrangements of queens on the chessboard. It is easy to see how the functor acts contravariantly with respect to inclusions of subgraphs, and the sheaf conditions are met, for the covering components will in each case intersect in at least one node, and at such nodes the queen represented there will automatically prevent conflicts with subconfigurations.

Continuing to confine our attention to $n = 4$, we display the configuration data of the n-queen sheaf, where the thick arrows trace out the two "global sections" (which, as we should expect, give the two solutions to the n-queen problem when $n = 4$), and the two grayed-out paths highlight two of the local sections which cannot be extended to global sections:

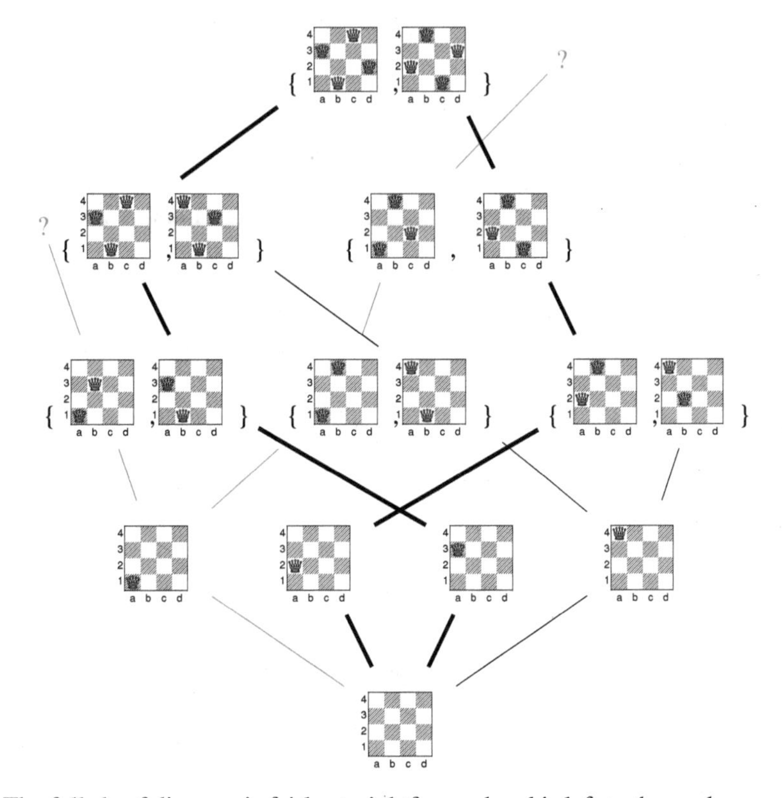

The full sheaf diagram is fairly straightforward and is left to the reader.

Example 295 *Self-similar groups* show up in a number of areas and are of particular use in modeling various phenomena involving some self-similarity, for example, certain "self-similar" melodies in musical composition. For self-similar groups (such as the well-known Basilica group, presented below) acting on a set (or infinite rooted binary tree, etc.), and given some subset of that set, we can form an interesting general construction called the *Schreier graph*, which is basically a generalization of a Cayley graph, but whose geometry can be much more interesting than that found in a Cayley graph. Cayley graphs are like "pictures" of a group, the geometric counterpart to an otherwise deeply algebraic entity (a group).[201] Similarly, Schreier graphs display, in a picture, how the cosets of a subgroup of G act in relation to the overall group G, specifically how the cosets "tile" the overall group. For this reason, Schreier would first refer to his graphs as *Nebengruppenbilder*, basically "coset pictures."

In abstract algebra, one often strives to obtain, for a given group G, a characterization of its subgroups (together with their properties). In terms of the associated geometrical approach, such tasks amount to finding characterizations of a given group G's Schreier graphs (together with their properties). In constructing the Schreier graphs with various actual self-similar groups in particular, one quickly observes that, in passing from the graph on one fixed level to the graph on the next level, some copying, rotating, and gluing are

201. More formally, a Cayley graph is a group G that has been geometrically realized as a G-torsor, where the latter is a space that G acts on *transitively* and *freely*.

involved; one moreover observes that yet another layer of self-similarity and restriction emerges in this passages of levels. These sorts of things lead one to wonder whether there is a sheaf lurking here.

One can indeed construct a sheaf here, however there are some difficulties and the construction requires some cleverness, specifically in the construction of the covering sieves for the objects of the category associated to the graphs to which Schreier graphs are to be assigned. First, we will better motivate and describe the basic objects, Schreier graphs.

Let us begin by considering a very familiar sort of object, an infinite rooted binary tree T (part of which is displayed):

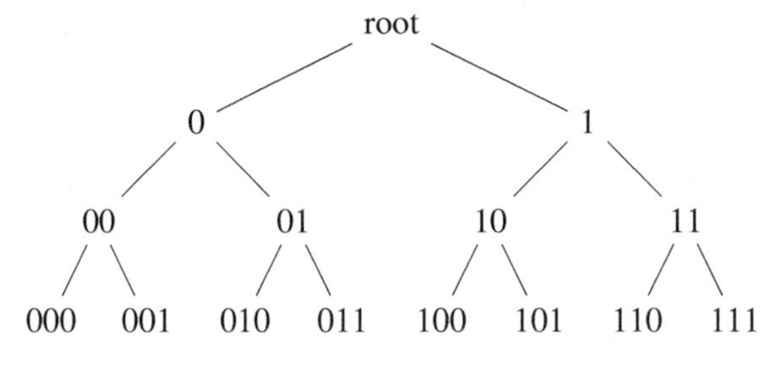

An *automorphism* of T is a bijective map from the nodes to the nodes which preserves adjacencies, that is, if the nodes are joined by an edge, then the nodes they are mapped to under the map are again joined by an edge.[202] If a and b are automorphisms of T, the notation ab means first apply the a transformation, then b. Now, since automorphisms are by definition bijective maps, they have inverses. The *inverse* of an automorphism x is an automorphism y such that $xy = e(= yx)$, where e is the identity/unit map. Of course, if we let G be a set of automorphisms of T together with their inverses, such that for each $x, y \in G$, the products xy and yx are also in G, then we obtain a *group*. We can guarantee this closure under multiplication by taking a set of automorphisms, say a and b, together with their inverses (which for the moment, we assume, for simplicity, are just a and b themselves), and then let G be the set of all finite products of these automorphisms. We then say that G is *generated* by the set $\{a, b\}$, and a and b are called the *generators*. Whenever a group is generated by a finite set of elements, in this case automorphisms, we call it a *finitely generated group*.

We could also define automorphisms via *rules*, for instance, for w an arbirary binary string, we might have:

$$a(0w) = 1.e(w), \qquad a(1w) = 0.e(w);$$

$$b(0w) = 0.e(w), \qquad b(1w) = 1.a(w),$$

where $e(w)$ means apply the identity map ("do nothing") to the suffixed string w. Products of a and b can also be expressed with rules, for example, $ba(0w) = a(0.e(w)) = 1.ee(w)$. Observe that in the above rules, the rules for a uses only e while the rules for b uses a and e. In other words, the set $\{a, b, e\}$ of automorphisms is described by a set of self-referential or *self-similar* rules.

202. The definition is far more general and works for any graph Γ, but we confine our attention to binary trees like T.

Definition 296 Let G be a group of automorphisms of T.[203] Then G is called a *self-similar group* if for each $g \in G$, each $x \in \{0,1\}$, and each binary string w, there is a $y \in \{0,1\}$ and an $h \in G$ such that

$$g(xw) = y.h(w)$$

Another particularly suggestive way of describing automorphisms of T is via *automata*. For instance, the automaton for the $\{a, b, e\}$ group from above looks like:

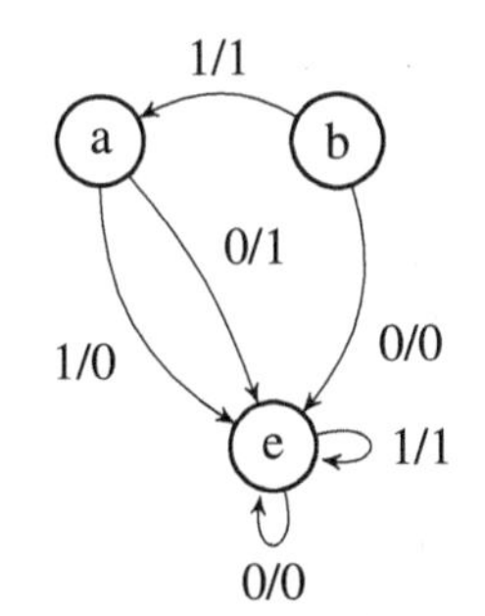

The idea then is that if we feed, for instance, the string 10110 in to the automaton, starting at b, then b will send 10110 along to a, keeping the 1 in the first position the same; then a will switch the second letter 0 to 1 and send the new string along to e; e in turn will preserve the subsequent three letters as they were, ultimately spitting out the string 11110.

Now, in general, whenever we have a group G generated by a finite set $\mathcal{G}$, acting on a set X, and given M some subset of X, there is a Schreier graph. For concreteness, in illustrating this notion, we confine our attention to X being the nodes of a tree T, M the set of nodes at some fixed level k of T, and G as a self-similar group acting on T. Roughly, the idea is that for each element of the subset M of T, we first draw a node labeled by this element. We then connect the nodes m_i, m_j with a directed edge assigned some label $s \in \mathcal{G}$ whenever $m_j = sm_i$. This graph is obviously *connected* if for any two nodes in M there is some group element (which need not be a single generator, but might be some combination or finite product of generators from $\mathcal{G}$) which carries us from m_i to m_j. In this case, we say that G acts transitively on M.

In more generality, given a group G with generating set $\mathcal{G}$, and any subgroup H, the set of left cosets G/H is a set on which G will act transitively, which entails that we can construct Schreier graphs for G acting on G/H. This will result in a graph Γ whose vertex set is G/H and in which two cosets Hg and Hg' are connected with a directed edge labelled with a generator $a \in \mathcal{G}$ iff $Hga = Hg'$. In the special case where H is the trivial subgroup, the Schreier graph coincides with the usual Cayley graph. If G is a self-similar group acting transitively on each level of T, let H be the subgroup of G containing all those elements which fix a node at level k. Then the Schreier graph for G acting on this G/H will be the graph for G acting on T when M is held to be the set of nodes at level k. More formally, a *Schreier graph* is a (connected and rooted) $2n$-regular (each of whose vertices has degree $2n$) graph Γ the edges of which are colored (using n different colors), and where for each vertex there is exactly one incoming and one outgoing edge of a given color attached to it.

203. This definition could be extended to groups of more general automorphisms, in particular to rooted n-ary trees or graphs.

We give a concrete illustration of this with the so-called Basilica group B. Let B be the group finitely generated by the set of automorphisms consisting of a, b, and e, acting on the rooted binary tree T. These generators are described by the following self-similar rules:

$$a(0w) = 1.b(w), \qquad a(1w) = 0.e(w);$$

$$b(0w) = 0.a(w), \qquad b(1w) = 1.e(w).$$

The automaton encoding these rules is given below:

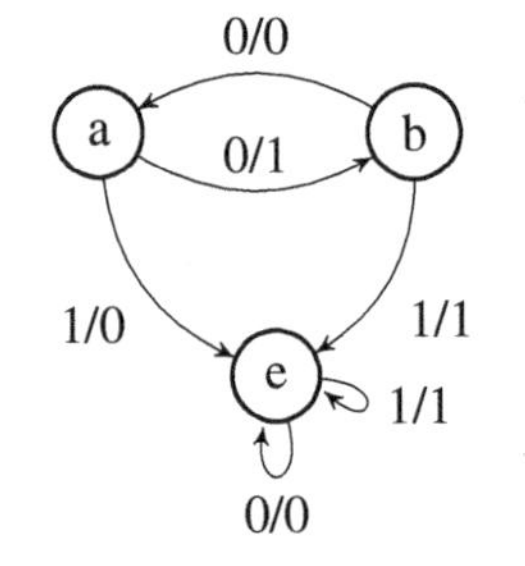

We know then that given the group B acting on the set of nodes of T an infinite rooted binary tree, we can look at the subset $M \subseteq T$ where M is the set of nodes at some fixed level k. For concreteness, let us consider $k \leq 5$. Then, for each element of M, we will draw a node. If $m_i = sm_j$ for some $s \in \mathscr{B}$ (where $\mathscr{B}$ is the generating set of the group acting on M), that is, if from any node we can get to any other node via some (combination of) action in the group, we say that B acts transitively on M or that the induced Schreier graph is *connected* at each level k. Concretely, for our Basilica group B acting on T, and M as just described, we have:

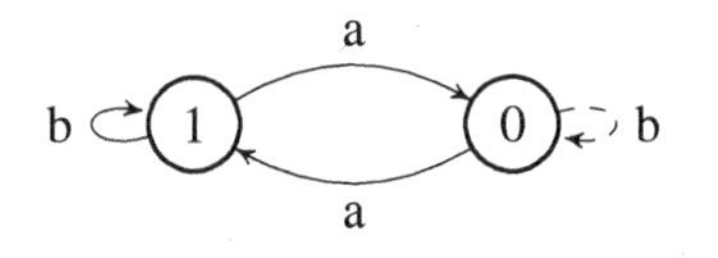

$k = 1$. To get the next level, duplicate, flip, and glue along dashed arrow

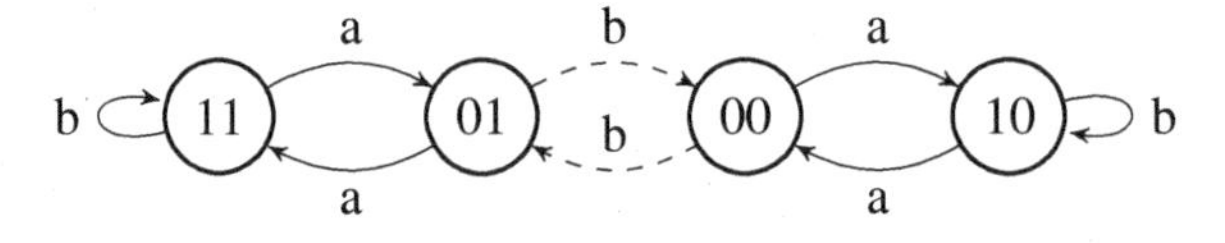

$k = 2$. To get back to level 1, reconnect the duplicates along dashed arrows

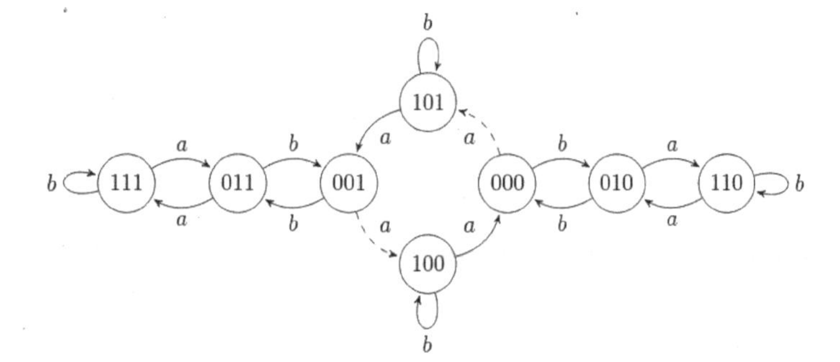

$k = 3$. To get back to level 2, reconnect the duplicates along dashed arrows

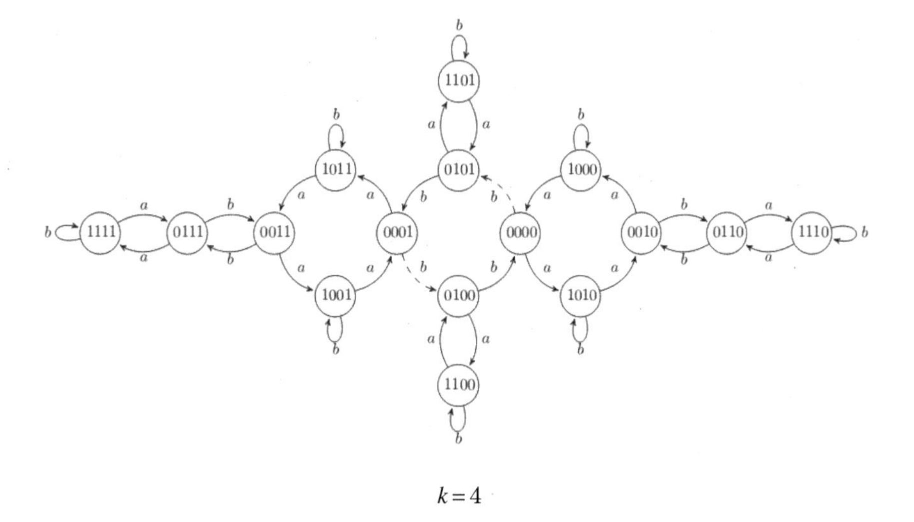

$k = 4$

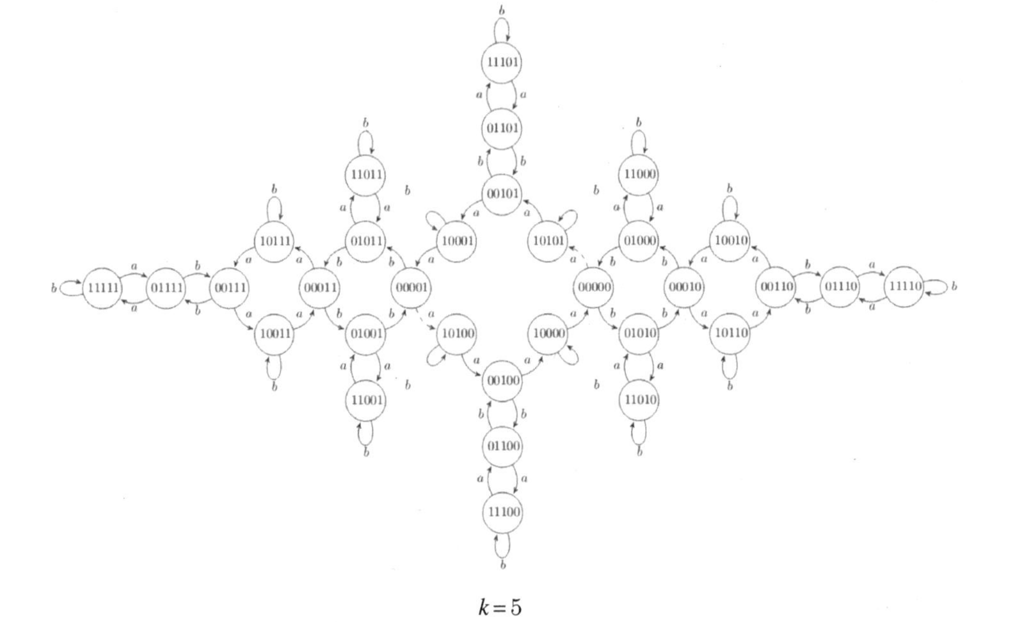

$k = 5$

The sorts of manipulations involved here prompt the notion that such Schreier structures over a graph might form a sheaf. However, while assigning Schreier graphs to the subgraphs of a given graph (or 1-simplicial complex), endowed with discrete topology (every subset open), describes a *presheaf*, it does not give us a sheaf. Yet, by using the more general notion of a Grothendieck topology, we can in fact describe things so as to produce a sheaf on an appropriate site. We will sketch how this works.[204]

Before developing this, we mention a few important features of Schreier graphs. Schreier graphs are graphs provisioned with a distinguished base point—the *root*. Moreover, Schreier graphs are *regular*, that is, each vertex has the same degree. For finite generating sets, the degree of vertices in a Schreier graph is always even. Since the action of G on G/H is transitive, Schreier graphs are connected. Furthermore, the quotient map construction $p: G \to G/H$ that takes the cosets of H and collapses them will induce a covering of Schreier graphs $p: G \to \Gamma$. Finally, if we denote by $L(G)$ the lattice of subgroups of G, and by $\Lambda(G)$ the corresponding space of Schreier graphs of G, we can observe that the space $L(G)$ is a lattice ordered by inclusion, while $\Lambda(G)$ is a lattice ordered by coverings. This latter claim means that $\Gamma \leq \Delta$ if there exists a covering map $p: \Gamma \to \Delta$ that takes the root of Γ to the root of Δ. Then, the map $f: L(G) \to \Lambda(G)$ that sends a subgroup $H \leq G$ to its Schreier graph is a lattice isomorphism sending an ordered pair $H \leq H'$ to the covering map $p: \Gamma \to \Gamma'$, where Γ and Γ' are the Schreier graphs of H and H'. The inverse map $f^{-1}: \Lambda(G) \to L(G)$ just takes Γ a Schreier graph of G to the subgroup H, which is the stabilizer of Γ under the action of G (i.e., the set of elements of G which fix the root of Γ).

204. I discovered a construction of a Schreier graph sheaf in an earlier draft of this book, but later discovered Cannizzo (2014) had done something similar. Moreover, I subsequently found an error in my own earlier approach, so the following account, including many details, largely derives from Cannizzo (2014).

In short, a Schreier graph is a connected and rooted $2n$-regular graph Γ whose edges are labeled by the generators of a group—we can think of the edges as coming in n different "colors," after assigning a color to each of the generators in the generating set $\mathscr{G}$ and painting the edges of Γ accordingly—and that is such that, for each vertex of Γ, there is exactly one incoming and one outgoing edge of a given color attached to it.[205] A *Schreier structure* Σ on Γ, a $2n$-regular rooted graph, is just a labeling of its edges by the generators of the free group $\mathbb{F}_n = \langle a_1, \ldots, a_n \rangle$, whereby Γ is made into a Schreier graph. In other words, a Schreier structure is just a map

$$\Sigma : E_0(\Gamma) \to A,$$

where $E_0(\Gamma)$ specifies a choice of orientation attached to each edge $(x, y) \in \Gamma$, and where for each $x \in \Gamma$ and each $1 \leq i \leq n$, there is, attached to x, exactly one incoming edge labeled with a_i (from A) and one outgoing edge labeled with a_i. Finally, we can define the *space of Schreier graphs* $\Lambda(G)$ of a given (finitely generated) group G by defining $\Lambda_r(G)$ as the set of isomorphism classes of r-neighborhoods (defined below) centered at the roots of the Schreier graphs in $\Lambda(G)$, and then setting $\Lambda(G) = \varprojlim \Lambda_r(G)$. We can, moreover, identify the space of Schreier graphs $\Lambda(G)$ and the lattice of subgroups $\mathscr{L}(G)$ of G.

It has been known for some time that every $2n$-regular graph admits a Schreier structure. Suppose we have Γ some $2n$-regular graph. We then want to describe the space of Schreier structures $Schr(\Gamma)$ over Γ. We can show that the association of Schreier structures to subgraphs of a given graph Γ supplies the data of a presheaf. However, by assigning the Schreier graphs to the subgraphs of a graph endowed with the discrete topology (in which every subset is open), or realized as a simplicial complex, one cannot make this a sheaf.

Suppose you place the discrete topology on a graph Γ, that is, the natural metric topology. Consider the graph Γ

v_0v_1 v_1v_2
v_0 v_1 v_2

The two edges (v_0v_1) and (v_1v_2) together form an open cover of the entire graph G. Their intersection is the sole vertex v_1. We can then assign a Schreier structure to the edge (v_0v_1) by first directing the edge from v_0 to v_1 and then labeling this with a generator a; similarly, we can assign a Schreier structure to the edge (v_1v_2) by first directing the edge from v_2 to v_1 and then also labeling this with the generator a, so that the assignments are to

v_0v_1 v_1v_2
v_0 $\rightarrow$ v_1 $\leftarrow$ v_2

On the overlap $v_1 = (v_0v_1) \cap (v_1v_2)$, the assigned Schreier structures trivially agree. However, the sheaf gluing axiom then requires that there exists a gluing $s \in \Lambda(\Gamma)$ such that its restriction to each edge equals a. But both edges are directed towards v_1 and have the same label, so the gluing condition cannot be satisfied and we do not have a sheaf.

But by reworking what it means for a family of subsets of a graph to serve as an open cover, specifically using a Grothendieck topology, we can arrive at a viable description of a sheaf of Schreier structures on such graphs. We begin by considering only sets that are

205. Technically, this graph will then be a Schreier graph of the free group generated by the group's generating set, or its Cayley graph.

r-neighborhoods, where this means sets $U \subseteq \Gamma$ of the form

$$U = \{x \in \Gamma \mid \rho(K, x) \leq r\},$$

where K is an arbitrary subset of Γ and ρ the standard metric on graphs. We say then that a subset $K \subseteq U$ of an r-neighborhood $U \subseteq \Gamma$ in a graph Γ is an *r-basis* of U provided the r-neighborhood of K (inside of Γ) is equal to U. This lets us define a cover as follows:

Definition 297 Let Γ be a graph with the usual graph metric. A system of r-neighborhoods $\{U_i\}_{i \in I}$ is said to be a *cover* of Γ provided there exist bases $K_i \subseteq U_i$ covering Γ is the ordinary sense, that is,

$$\bigcup_{i \in I} K_i = \Gamma.$$

While the open sets of covers thus defined will overlap in a nice way, rectifying our previous problem, allowing the assignment of Schreier structures to sets to be a sheaf, there is an issue in that the defined collection of r-neighborhoods is not actually a *topology*, since it is not closed under intersections. This is where the fully abstract notion of a Grothendieck topology comes to the rescue. This will, in turn, enable us to define a proper sheaf of Schreier structures on a graph.

For Γ a graph, we proceed by defining a category $\mathbf{C}_\Gamma$ associated to Γ, that has

- objects: the 1-neighborhoods in Γ; and
- morphisms: inclusions.

Let us now define a Grothendieck topology J_1 on $\mathbf{C}_\Gamma$ thus: for U an object of $\mathbf{C}_\Gamma$, we let $J_1(U)$, the collection of covering sieves of U, be the collection of all sets of inclusions $\{U_i \hookrightarrow U\}_{i \in I}$ subject to the requirement that each U_i is a 2-neighborhood and that the system $\{U_i\}_{i \in I}$ covers U in the sense of the definition above (taking $r = 2$). With the resulting Grothendieck topology, the functor assigning to a given 1-neighborhood U the set of Schreier structures over U of Γ is in fact a sheaf.

Theorem 298 Let Γ be a graph and let $(\mathbf{C}_\Gamma, J)$ be its associated site. Then the contravariant functor $Schr : \mathbf{C}_\Gamma \to \mathbf{Set}$ that assigns to a 1-neighborhood U in $\mathbf{C}_\Gamma$ the set of Schreier structures $Schr(U)$ over U is in fact a sheaf.[206]

Proof. The assignment of Schreier structures to a given graph is a contravariant assignment, so the functor *Schr* is a presheaf. Take U an object, that is, a 1-neighborhood, in $\mathbf{C}(\Gamma)$, and $S \in J(U)$ a covering sieve that consists of inclusions $\{U_i \hookrightarrow U\}_{i \in I}$. We can check that it is a sheaf by verifying the two usual sheaf conditions.

First, uniqueness: suppose $\Sigma, \Sigma' \in Schr(U)$ are two Schreier structures on U such that their restrictions to U_i agree for all $i \in I$. Suppose, moreover, that $\Sigma \neq \Sigma'$. This inequality means that there will have to exist an edge $(xy) \subseteq U$ that has different labels in Σ and Σ'. The definition of covering sieves stipulates that the point x belongs to the basis of some U_i, and since U_i is a neighborhood of this basis, it must also contain the point y and thus also the edge (x, y) connecting them. But this is a contradiction, so we get that, provided there are such gluings, there can only be one.

206. This theorem, and its proof, follows Cannizzo (2014) very closely.

Second, gluing: suppose that $\{\Sigma_i \in Schr(U_i)\}_{i \in I}$ is a system of Schreier structures with the property that, for all $i, j \in I$, the restrictions of Σ_i and Σ_j to the fibered product $U_i \times_U U_j$ agree. Take an arbitrary $x \in U$. If the set of all edges incident with x (called *star*(x)) is contained in the 1-basis of one of the U_i, then the Schreier structure of this *star*(x) that comes from the Σ_i can be extended to the entire neighborhood U. On the other hand, if x lies on the boundary of the 1-basis K_i of some U_i, making *star*(x) contain a point y *not* in K_i, then this point y will have to be contained in the 1-basis K_j of some U_j. Such an edge (x, y) will then belong to the fibered product $U_i \times_U U_j$, and this fibered product is itself a 1-neighborhood.[207] Thus, a Schreier structure can be extended to all of U. □

Fundamentally, the theorem says that the functor assigning Schreier structures to subgraphs of a given graph can be regarded as a sheaf, provided the subgraph is furnished with an r-neighborhood structure, and provided one then defines an appropriate Grothendieck topology.

The next two constructions are left to the reader.

Example 299 In general, for a topological group G equipped with the prodiscrete topology, the category of continuous G-sets or **Cont**(G)—having for objects the continuous actions of G on discrete sets and for morphisms the equivariant maps—forms a Grothendieck topos; this category is in fact equivalent to the topos of sheaves $\mathbf{Sh}(\mathbf{C}, J_{At})$, where $\mathbf{C}$ consists of the (nonempty) transitive actions of **Cont**(G) and J_{At} is the atomic topology on this subcategory.

A cellular automaton (CA) is a sort of machine that takes in some input and produces some output.[208] They are excellent devices for modeling complex behaviors with minimal building blocks. A CA consists, roughly, of the following data:

- An *alphabet* A: the set of symbols we allow, for example, $\{0, 1\}$, {blue, orange, white}, and so on; individual elements of this alphabet set are called *states* or colors.
- A *group* G, called the *universe*, typically displayed in the form of an ordered grid of cells (i.e., tessellation with some basic shape, e.g., squares); elements of the universe are called *cells*.
- A notion of *neighborhood*, for example, the eight cells immediately surrounding a given cell in a 2-D tessellation with squares;
- A *local transition function* or *rule* that provides instructions for moving from one assignment of elements of the alphabet to the lattice of cells to another; these instructions for producing the output state of a given cell are given in terms of the input states of a finite number of a cell's neighboring cells, which might include the cell itself.

We consider A^G, the set consisting of all maps from the universe G to the alphabet set A:

$$A^G = \prod_{g \in G} A = \{x : G \to A\}.$$

207. For $U \hookrightarrow W$ and $V \hookrightarrow W$ two morphisms in $\mathbf{C}_\Gamma$, the fibered product $U \times_W V$ is specifically the 1-neighborhood of the intersection of the maximal bases of U and of V, that is, $U_1(U_{max} \cap V_{max})$. See Cannizzo (2014) for details.

208. Much of the basic background theory concerning CA's can be found in Ceccherini-Silberstein and Coornaert (2010), a book that would serve as an excellent reference for anyone desiring to learn more about cellular automata.

Elements of this set A^G are called *configurations*. In other words, a configuration is a way of attaching an element of the alphabet to each member of the underlying group (i.e., to each cell in the universe), where each cell is assigned one state at a time. Informally, a configuration is a specific coloring of all the cells of the universe. There is a natural action of the group on the set of configurations, called the *shift action*. In brief, a CA is a self-mapping of the set of configurations given by a system of local rules which commutes with this shift action. Since G acts continuously on the set A^G (equipped with the prodiscrete topology)[209] it would appear that we are dealing with the category of continuous G-sets, $\mathbf{Cont}(G)$, whose objects are the continuous actions of G on the discrete sets and whose arrows are the equivariant maps between them; one could then construct a Grothendieck topos here by constructing a site out of the open subgroups and using the atomic topology on the full subcategory of $\mathbf{Cont}(G)$ that consists of the nonempty transitive actions. We leave readers to explore this on their own.

Example 300 For those who know about combinatorial game theory, try to construct a sheaf out of data of the *thermographs* of games.[210]

10.4 Philosophical Pass: The Idea of Toposes

Box 10.1
The Idea of Toposes

This chapter offered a first look at Grothendieck toposes, where these are categories equivalent to a category of sheaves on some site; moreover, every Grothendieck topos can be shown to arise in this way. When Grothendieck first introduced the essential concepts, it was in order to be able to treat cohomology within algebraic geometry—in this manner, toposes were able to play an important role in problems with a more arithmetic bent. But ultimately, this powerful concept of a Grothendieck topos allows us to deal with situations or theories wherein seemingly topological notions would appear to be useful or relevant, but where the ordinary topological spaces are not present. Grothendieck toposes accordingly allow us to study a variety of "space-like" situations, to regard space as something ubiquitous throughout mathematics, and even to have a unified framework we can use to study arithmetic and geometry. As Grothendieck stressed (see promenade 13 of Grothendieck 1986 in particular), the concept of a Grothendieck topos in a sense joins together the continuous and structures that would appear to be thoroughly algebraic and discrete.

From one perspective, the story of Grothendieck toposes can be seen as a decisive step in the story of the metamorphosis of the concept of space. The associated notion of a Grothendieck topology in terms of systems of covers is an important purification of what is arguably fundamental to the spatiality of a space—and the more general notion allows us to capture this even in situations quite distant from literal topological spaces. The essence of what is "spatial"

209. The prodiscrete topology is where A^G has the product topology, each factor A of A^G given the discrete topology (all subsets of A open).

210. See Siegel (2013), Berlekamp (1982), Berlekamp and Wolfe (2012), or Guy, Conway, and Berlekamp (1982) for details on combinatorial game theory and the notion of "temperature" and thermographs; the reader can also visualize some of these ideas, and perform some computations that would be too difficult to do by hand, with the open-source program *CGSuite*.

about space is arguably to be found wherever there is a structure where localization makes sense. Grothendieck topologies, and the sites built out of them, allow us to speak about this. When we can define something on covers and check agreement on intersections of such covers, it makes sense to speak of localization. And, fundamentally, the idea of covering sieves is one that enables us to examine the relations between the local and global features of some system of objects. When we can uniquely glue data assigned to covers, we can talk about sheaves on such a system. By considering all the sheaves on a site—and so a Grothendieck topos—we are recapturing structures that behave like sheaves on a topological space, even in situations where there may not be a literal topological space (and so, where there was not already a notion of what sorts of assignments of data to that space are "sheaf-like").

In this framework, the characteristic relations between local and global features of some system of objects emerge as more fundamental than any sort of *points*; the concept of a site accordingly starts to tease out, in a particularly powerful fashion, this idea that one can examine something spatial, and things defined on such a space, without looking at its *points*. The focus shifts away from points and behavior on points to how a localization or covering system can be made to define a "variable structure" that varies over that system of covers.

The usual (point-set) notion of a topology was ultimately designed to capture and understand two things: locally defined phenomena and continuous transformations. If Grothendieck's notion of a site, built out of a transformed notion of a topology, supplies us with a more general notion of a local system, consideration of the sheaves on a site provides some sort of continuous or continuously variable perspective on a category that may in principle be quite "nontopological." The notion of *geometric morphisms* between toposes, which we cover in the next chapter, further extends the generalized notion of continuous transformations. With the notion of a Grothendieck topos, we ultimately have a framework in which the (discrete) combinatorial diagrammatic approach (via categories) is paired harmoniously with (continuous) concepts native to topological spaces.

Furthermore, the introduction of Grothendieck toposes can also be seen as motivated from observations that a number of important properties of topological spaces themselves can be reformulated as certain invariant properties of their associated category of sheaves. In certain cases, as we have already seen, a topological space X can even be recovered from its associated category of sheaves $\mathbf{Sh}(X)$; and regarding a topological space in terms of its associated category of sheaves has various merits (since the latter has a rather rich categorical structure, as we will better appreciate in the next chapter).

It is often observed how attempting to obtain a topos appropriate for a particular sort of math by constructing the domain of variation (via a site), and then considering the category of sheaves over that site, gives rise to the perspective of *variable set theory*, wherein the standard constant universe of sets is replaced by something like a universe of continuously variable sets. As Johnstone (2014, xvii) claims, this is in fact the very essence of the topos-theoretic view of things:

> it consists in the rejection of the idea that there is a fixed universe of "constant" sets within which mathematics can and should be developed, and the recognition that the notion of "variable structure" may be more conveniently handled within a universe of *continuously variable* sets than by the method, traditional since the rise of abstract set theory, of considering separately a domain of variation (i.e., a topological space) and a succession of constant structures attached to the points of this domain. In the words of Lawvere, "Every notion of constancy is relative, being derived perceptually or conceptually as a limiting

> case of variation, and the undisputed value of such notions in clarifying variation is always limited by that origin."

When dealing with an ordinary topological space X, the variable set (indexed by the opens of X) just yields a sheaf of sets on X. Instead of using a topological space X, we can insist on the perspective of X as a parameter space, yielding the category of variable sets with parameters in X. In this way, the space withdraws into the background, providing the form of variability or cohesion for the objects in the foreground. (This perspective ultimately leads to variable topological algebra, which has deep connections with arithmetic.) By enforcing different axioms over and above those specified by topos theory in general (on which more below), one can obtain different "universes" of variable sets within which to do certain kinds of math.

Grothendieck toposes can also be seen as *unifying* in their ability to house disparate mathematical constructions, useful not only for studying relationships between different mathematical theories, but also for transferring information and ideas across different situations. (Readers are encouraged to consult the work of Olivia Caramello, who has developed the idea of Grothendieck toposes as "bridges" for unifying different theories and transferring information between them.) There can also be a number of different relevant sites for a given Grothendieck topos, which sometimes appears to amount to letting us look at the same theory or situation in different ways.

The definition of a Grothendieck topos can be applied to any category of sheaves associated to a site, and in particular we saw that this notion of topos encompasses any presheaf (or variable set) category $\mathbf{Set}^{\mathbf{C}^{op}}$ (just set J as the minimal topology). In particular, we looked at $\mathbf{Set}^{\mathbf{1}^{op}} = \mathbf{Set}$. But the notion of a Grothendieck topos in fact depends on the assumed model of set theory, and this suggests the need for even greater generality. Lawvere and Tierney first introduced the notion of an *elementary topos* in part to provide a characterization of categories that resemble Grothendieck toposes from one perspective, but that can be defined strictly by elementary axioms that are independent of set theory. In the next chapter, we develop the story, pursuing some of these further generalizations and giving some examples that put the underlying ideas to work.

11 Elementary Toposes

In which we explore the even more general notion of elementary toposes and consider what sheaves look like in this setting, after which we take up the topic of morphisms of toposes, including an extended section, full of concrete examples, that gives a glimpse into the theory of cohesive toposes.

Lawvere and Tierney introduced the definition of an elementary topos largely in order to characterize structures that behave like sets—and, in particular, those categories that apparently behave as Grothendieck toposes do—but that can be described by elementary axioms independent of set theory. The following definition is the one typically presented.

Definition 301 An *elementary topos* is a category $\mathscr{E}$ that

1. has all finite limits—equivalently, has pullbacks and a terminal object;
2. is *Cartesian closed*—that is, for each object X there is a functor, called the *exponential* $(-)^X : \mathscr{E} \to \mathscr{E}$, that is right adjoint to the functor $(-) \times X$; and
3. has a *subobject classifier* satisfying certain conditions.

A category that satisfies the first two conditions is a Cartesian closed category—that is, it has finite products and exponentials for each pair of objects—and the significance of such categories had been appreciated before the definition of an elementary topos.[211] However,

211. To give some small glimpse into the importance of Cartesian closed categories, consider that **Man** (which lacked finite inverse limits, as we already saw, since pullbacks of manifolds were not always manifolds) is not Cartesian closed either—since in particular it lacks exponentials, as the space of smooth (C^∞) maps between two smooth manifolds is not in general a manifold. In short, this means that in general for manifolds A and B and C, there is no equivalence between $A \times B \to C$ and $B \to C^A$. The significance of this is very nicely explained by Lawvere (2011), who notes that in addition to presenting some serious foundational problems for calculus of variations, this failure can be interpreted, physically, as saying that if we treat A as all possible states of time, B as some physical body, and C as some space, then the motion described by $A \times B \to C$ will not in general be equivalent to the assignment of a body to its path through the space, given by $B \to C^A$. However, even if it were cartesian closed, **Man** would still lack the resources for dealing with infinitesimal objects or structures. Incidentally, observations of these sorts of deficiencies have led to attempts to rectify such deficiencies (due to Lawvere) by constructing a category of spaces **Space**—sometimes called *smooth worlds* or *smooth toposes*, with objects *smooth spaces*—which extends or enlarges the usual category of manifolds in various ways so as to overcome the above-mentioned deficiencies. In brief, the construction usually proceeds by first embedding **Man** in the category of "formal varieties" **L**, or "loci," a category that does have finite inverse limits and also contains infinitesimal spaces; then, since in general one still cannot construct function spaces in the new category **L**, one next endows **L** with a specific Grothendieck topology J, after which the resulting smooth topos **Sh**(**L**, J) of sheaves on **L** allows for function spaces with certain desirable properties to be constructed. One can then use the smooth category **Space**, which is in fact a topos, where each object has a differentiable structure and each morphism a derivative (and where calculus reduces to exact algebraic calculations with infinitesimals), to do

the introduction of the concept of a subobject classifier was the real conceptual advance that made possible the development of topos theory along elementary lines, for a topos may also be defined as a Cartesian closed category with a subobject classifier. In what immediately follows, we focus predominantly on constructing and familiarizing ourselves with the third object in this definition—the subobject classifier.[212]

11.1 The Subobject Classifier

We begin by recalling that since in category theory everything can be specified in terms of arrows, we do not need to consider traditional "elements" of objects. However, there is a corresponding category-theoretic notion of an element.

Definition 302 For A an object in **C**, we call an arrow in **C** with codomain A an *element* (or *generalized element* or *variable element*) of A.

The domain of such an arrow is accordingly sometimes called the *stage of definition* (or *domain of variation*) of that element. For example, for an arrow $f: B \rightarrow A$, the object B is regarded as a stage or as the domain of variation in order to indicate that A is being defined over B or that it is where A is being viewed from. We can then vary the domain of variation, considering different stages of A. The identity arrow, $\mathrm{id}_A : A \rightarrow A$, for its part, is called the *generic element* of A. Elements defined over a category's terminal object—that is, generalized elements $1_{\mathbf{C}} \rightarrow A$—are special, and in the simple case of **Set** recover the usual (set theoretic) notion of elements of the set A.

Definition 303 If the category has a terminal object, which we will denote with $1_{\mathbf{C}}$ (or just 1), we can use this to define the notion of a *global element* or *point* as an arrow $1 \rightarrow A$ of A—that is, one that does not depend on any particular domain of reference or stage of variation.

By contrast, given an arrow $B \rightarrow A$, if B is not isomorphic to the terminal object, such an arrow is sometimes called the *local element* of A (at stage B).[213]

Even if a category has a terminal object, while some objects may have global elements, others may not. And in general, not all elements (in the above sense) of a given object have to be global: while the global elements of an object can be thought of as corresponding to the set-theoretic notion of points, in general there will exist elements of an object that are *not points*, which in part explains why one insists on the "domain of variation" perspective.

We will make some use of this perspective in what follows, in the development of the subobject notion, where a subobject of an object A will basically emerge as an equivalence class of certain morphisms (monomorphisms) with codomain A. First, recall that a monomorphism is the category-theoretic generalization of the notion of injective maps from sets, where it requires that whenever we have

---

differential geometry. In this category, analysis is called *smooth infinitesimal analysis* (SIA). See Bell (2008) for a brief and very accessible introduction to SIA; Moerdijk and Reyes (1991) and Kock (2006) are the standard references in this field (if the reader wishes to pursue these matters in more depth).

212. Detailed treatments of the three defining features of an elementary topos can be found in any standard text on topos theory, such as Johnstone (2014) or Mac Lane and Moerdijk (1994).

213. This language of "global (local) element" ultimately derives from sheaf theory, where the global elements of a sheaf on a space X are the global sections, defined on the entire space.

$$C \overset{f_1}{\underset{f_2}{\rightrightarrows}} B \xrightarrow{i} A$$

so that $i \circ f_1 = i \circ f_2$, then we have $f_1 = f_2$. In this case, we call such a morphism $i : B \to A$ a monomorphism (or monic arrow). In logical notation:

$$\forall C \forall f_1, f_2 : C \to B \qquad (i \circ f_1 = i \circ f_2) \Rightarrow f_1 = f_2.$$

In the case of sets, this "for all" reduces to C being the terminal object 1 (a singleton set $\{*\}$) in **Set** and the above recovers the notion of an injective function. Moreover, in **Set** an arrow $x : 1 \to B$ into B just corresponds to an element (in the usual sense) of the set B, and we can use such elements to examine whether two parallel arrows f, g are equal, by just checking that $f(x) = g(x)$ for every global element x of the domain. But in a number of other toposes, it need not be the case that such diagrams (of parallel arrows) commute iff they commute for every global element of the domain. The need to construe things in a more general way is part of what motivates the definition of generalized elements in the first place, but speaking of *elements* suggests that we still expect that operating with a topos will be something like operating with sets.

The notion of a monic arrow is really a way of saying that the arrow preserves distinctness in the sense that it keeps items that are seen as distinct in the domain, distinct in the codomain. It moreover allows us to define the notion of a *part* in a rather general way. The idea is that we can think of a map i with codomain A, where i is moreover a monomorphism, as supplying us with a *part of A*.

We make use of such morphisms in the following:

Definition 304 Two monomorphisms f and g satisfying

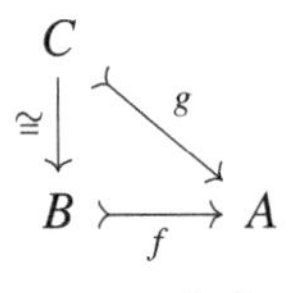

are *equivalent*, denoted $f \sim g$. Then we can define the *equivalence class* of f, denoted $[f] = \{g \mid f \sim g\}$.

Definition 305 A *subobject* of an object A is defined to be an equivalence class of monomorphisms with codomain A modulo the relation that identifies monomorphisms into A whenever one factors through the other (in the sense of definition 304). Then the *class of subobjects* of an object A will be

$$\mathrm{Sub}(A) := \{[f] \mid \mathrm{codom}(f) = A \text{ and } f \text{ is monic}\}.$$

There is a partial order on $\mathrm{Sub}(A)$, using the inclusion ordering $[f] \subseteq [g]$ (also just written $f \subseteq g$), giving us the poset $(\mathrm{Sub}(A), \subseteq)$ of subobjects of the object A. In fact, for each object A in a topos, the poset $\mathrm{Sub}(A)$ of subobjects of A forms a lattice.

As you might imagine, relying on the notion of monic arrows as supplying us with *parts*, in the category **Set** subobjects of a set X correspond precisely to the *subsets* of X—in this way, the notion of a subobject may be seen as a categorical generalization of the notion of a subset from set theory. But let's sit with this connection for a moment. Observe how, in the category **Set**, we generally think of arrows $X \to \{0, 1\}$ into the truth-value set

$\{0,1\}$ as *predicates* for X, or as *properties* of generalized elements of X. Moreover, we know that we can identify the subsets S of a given set X with their characteristic functions $\chi_S : X \to \{0,1\}$ into the truth-value set $\{0,1\}$, and picking out 1 from the truth-value set amounts to a function $\top : \{*\} \to \{0,1\}$. With such data, we in fact have a diagram

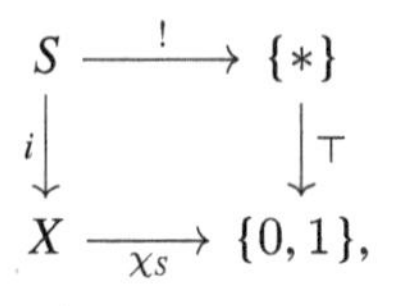

where $i : S \hookrightarrow X$ is the inclusion arrow and $! : S \to \{*\}$ the unique arrow to the terminal object $\{*\}$ of **Set**. This diagram is in fact a pullback square, since pulling $\top$ back along χ_S just yields the inverse image set $\{x \mid \chi_S(x) = 1\}$, which is the same as S. This situation serves as motivation for the following more general notion.

Definition 306 Suppose **C** has finite limits (so, in particular, a terminal object $1_{\mathbf{C}}$). A *subobject classifier* (or a *generalized truth-value object*) is an object Ω in **C** together with a monomorphism (called the *truth arrow*) $\top : 1_{\mathbf{C}} \to \Omega$, such that for every object $A \in \mathbf{C}$ and every monomorphism $m : B \rightarrowtail A$, there exists a unique morphism $\chi_B = \chi_m = \chi(m) : A \to \Omega$—called the *characteristic (or classifying) arrow* of the monic m (or of the subobject B of A)—making the diagram

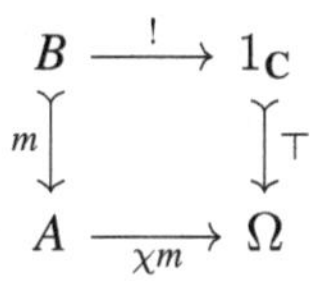

form a pullback square.

We often denote the composite arrow $(\top \circ !)$ by $\top_B : B \to \Omega$.

In **Set**, we already mentioned how $\Omega = \{0,1\}$, $\top(*) = 1$, and the morphism χ is the characteristic map of S in X, that is, $\chi(y) = 1$ if $y \in S$ and $\chi(y) = 0$ if $y \notin S$. The object Ω^X of arrows $X \to \Omega$ can be identified with the set of subsets of X, and the global elements $1 \to X$ just correspond to the actual elements of X.

It is straightforward to show that, in a general category with finite limits, a subobject classifier, when it exists, is unique up to isomorphism. Moreover, given $\mathscr{E}$ a topos, and two objects A, B of $\mathscr{E}$, by Cartesian closedness the global elements of B^A (i.e., the arrows $1 \to B^A$) correspond bijectively with $1 \times A \to B$ (i.e., with the arrows $A \to B$). Thus, in particular, for $\mathscr{E}$ a topos, and given an object $A \in \mathscr{E}$, the global elements of Ω^A will correspond bijectively with the subobjects of A.

Proposition 307 In a topos $\mathscr{E}$, given an object $A \in \mathscr{E}$, we have

$$\mathrm{Sub}(A) \cong \mathrm{Hom}_{\mathscr{E}}(A, \Omega).$$

Proof. By the fact that, in a topos, global elements of B^A correspond bijectively with the arrows $A \to B$, the global elements of Ω^A correspond bijectively with the arrows $\chi : A \to \Omega$. The pullback of χ with the subobject classifier $\top : 1 \to \Omega$ will yield a subobject $m : S \rightarrowtail A$. By the definition of a subobject classifier, this correspondence is bijective. □

We are going to want to see how these notions apply in the setting of presheaves. In short, since the category $\mathbf{Set}^{\mathbf{C}^{op}}$ has functors as objects, we might expect the subobjects will be defined in terms of *subfunctors*. A subfunctor F of a functor $G : \mathbf{C}^{op} \to \mathbf{Set}$ amounts to an inclusion $F \to G$ that is in fact a monic map in the presheaf category $\mathbf{Set}^{\mathbf{C}^{op}}$, meaning that each subfunctor is indeed a subobject—moreover, all subobjects are supplied by subfunctors. In such a setting, we would thus be looking to isolate a *functor* Ω that can be treated as a subfunctor classifier. Since a subobject was just defined as a class of equivalent monic arrows sharing the same codomain, a subobject classifier in the presheaf category, should there be such a thing, ought to classify, in particular, the subobjects (subpresheaves) of each representable presheaf. The proposition above, together with the Yoneda lemma, implies that if the presheaf category has a subobject classifier $\Omega : \mathbf{C}^{op} \to \mathbf{Set}$, it should satisfy

$$\mathrm{Sub}(\mathrm{Hom}_{\mathbf{C}}(-, a)) \cong \mathrm{Hom}_{\mathbf{Set}^{\mathbf{C}^{op}}}(\mathrm{Hom}_{\mathbf{C}}(-, a), \Omega) \equiv \mathrm{Nat}(\mathrm{Hom}_{\mathbf{C}}(-, a), \Omega) \cong \Omega(a)$$

for every $a \in \mathrm{Ob}(\mathbf{C})$. In this way, we can see that the subobject classifier Ω is defined on objects $\Omega(a)$ by taking subfunctors of the representable functor $\mathrm{Hom}_{\mathbf{C}}(-, a)$.

With such a description, the attentive reader may be reminded of the notion of a *sieve*. We already saw that there is a natural bijection between sieves over a given stage and the subfunctors of the Yoneda functor of that stage, suggesting that $\Omega(a) = \{S \mid S \text{ is a subfunctor of } \mathrm{Hom}(-, a)\}$. Thus,

$$\text{set of all sieves on } a \cong \mathrm{Sub}(\mathrm{Hom}(-, a)) \cong \mathrm{Nat}(\mathrm{Hom}(-, a), \Omega) \cong \Omega(a).$$

If we let sieves(a) denote the collection of all sieves on an object a of a category $\mathbf{C}$, the presheaf $\Omega : \mathbf{C}^{op} \to \mathbf{Set}$ is thus defined by $\Omega(a) = \mathrm{sieves}(a)$. For any arrow $f : b \to a$, we know that the action of $\Omega(f)$ will be given by pulling back along $\mathrm{Hom}(-, a)$, that is, $\Omega(f) = f^*$, where $f^* : \Omega(a) \to \Omega(b)$ maps sieves S on a to sieves on b by $S \mapsto \{c \xrightarrow{g} b \mid f \circ g \in S\}$, as in

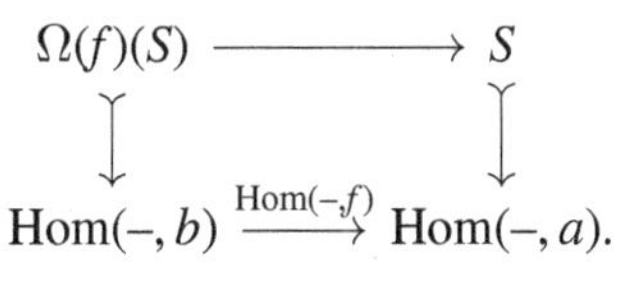

To finish the identification of the subobject classifier here, we need the truth arrow $\top : 1_{\mathbf{PreSh(C)}} \to \Omega$, and thus the terminal object. But the terminal presheaf 1 is defined by $1(a) = \{*\}$ for each $a \in \mathbf{C}$, and the truth arrow $\top : 1 \rightarrowtail \Omega$ will come from the maximal sieves, that is, by taking $\top_a(*) = \mathrm{Hom}(-, a) = M_a$ (i.e., the maximal sieve, or set of *all* arrows into a). Then, considering a subfunctor $R \rightarrowtail F$, it is fairly straightforward to construct a characteristic mapping $\chi : F \to \Omega$ for R, and then show this to be unique. The fully detailed construction of the corresponding characteristic arrows and the verification that the pair $(\Omega, \top)$ really does give a subobject classifier for the presheaf category is routine and can be found in many places in the literature. It can also be shown that the presheaf category $\mathbf{Set}^{\mathbf{C}^{op}}$ has finite limits and exponentials, which, together with the fact that this category has a subobject classifier, suffice to show that $\mathbf{Set}^{\mathbf{C}^{op}}$ meets the requirements for being an elementary topos. Instead of going through the details of this, let us now turn to make some observations regarding the subobject classifier for the presheaf category, and then consider these matters in more detail via an example.

Remark 308 Recall the result from proposition 132 (chapter 5), letting us recover the poset of open sets of a topological space X from the category $\mathbf{Sh}(X)$ of sheaves on X, specifically by attending to the subobjects of the terminal sheaf 1. We now have another way of appreciating why this is the case. As any object will have a unique arrow to the terminal object, an object S of a topos $\mathscr{E}$ will be a subobject of the terminal object 1 exactly when the unique arrow $S \to 1$ is in fact monic. Those U for which $U \to 1$ is monic can be called *open* in $\mathscr{E}$, giving us subobjects of 1, and if we let $\mathscr{E} = \mathbf{Sh}(X)$ for a topological space X, such "open sheaves" just correspond to the usual open subsets of X.

Let us make another observation. The generalized elements of an object A that belong to B at all stages may be regarded as those whose characteristic arrows have truth-value 1 throughout, or which is assigned the maximal sieve at each stage, while those elements of A that do not belong to B at any stage should have truth-value 0, where the value 0 is given by the empty sieve at each stage. Of course, in the category **Set**, that is the end of the story: we have $\Omega = \{0, 1\}$, and subobjects of a set X correspond bijectively to subsets U of that set, where the subsets can be identified with their characteristic functions $\chi_U : X \to \{0, 1\}$, indicating that every element of a set either belongs to another set or does not. But in considering sets varying on categories other than the terminal category—that is, considering presheaves over an arbitrary category—those elements of A that belong to B at some stage will in general be captured by partial or intermediate truth-values, and in general Ω will not reduce to two truth-values.

The correspondence between sieves and subfunctors of Ω effectively allows us to provide an assignment of generalized truth-values to every sieve. In this way, sieves enable us to probe *when* an element or subobject "becomes a part of" another object, rather than just say *whether or not* it belongs, connecting up with the movement beyond classical logics with two truth-values. Whenever there are more than just the maximal ("true") and the empty ("false") sieves on an object, we will have the ability to deal with nonglobal elements, determining at which stage an element or subobject "falls into" an object. In general, in considering presheaves that vary on a category, elements of a presheaf set can be regarded as belonging to a subset *to a certain degree*, or *at some stage*. Different sieves in the classifier set Ω will correspond to different degrees of belonging, a partial value representing the fact that perhaps an assertion holds at one stage but not at an earlier stage.

Let us look a little closer at this way of seeing the subobject classifier concept via a particular instance, examining the structure of the topos $\mathbf{Set}^{[3]^{op}}$.

Example 309 Consider $\mathbf{Set}^{[3]^{op}}$, with the linear order category **[3]** (we might also call it **4**):[214]

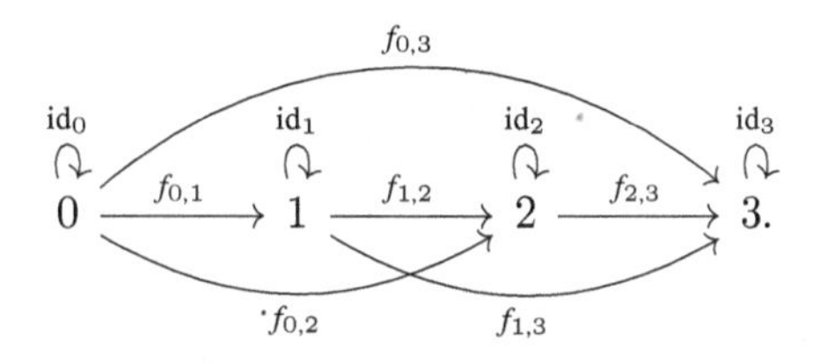

214. This example is essentially from Spivak (2014, 416).

In the spirit of the preceding discussion, we would like to understand each of the objects of this category by looking at the set of arrows coming into that object, or equivalently, by understanding the sieves. For instance, focusing on the object 2, we can consider the hom-functor $\text{Hom}(-, 2) = Y_2$. But this functor is defined, on objects, simply by plugging in the objects of the category and looking at the resulting set of arrows, that is, we have $\text{Hom}(0, 2) = \{f_{0,2}\}, \text{Hom}(1, 2) = \{f_{1,2}\}, \text{Hom}(2, 2) = \{\text{id}_2\}$, and $\text{Hom}(3, 2) = \emptyset$. Joining these together into the set $\{\{\text{id}_2\}, \{f_{1,2}\}, \{f_{0,2}\}\} \in \mathbf{Set}^{\mathbf{C}^{op}}$, we have an explicit construction of Y_2.

But observe that such a set is in fact none other than the maximal sieve on 2. Note, moreover, that the functor Y_2 supports the subfunctor construction. Indeed, there are four subfunctors of Y_2:

$$Y_2 \supseteq \emptyset = \text{"none"}$$

$$Y_2 \supseteq Y_2 \cong \{\{\text{id}_2\}, \{f_{1,2}\}, \{f_{0,2}\}\} = \text{"all"}$$

$$Y_2 \supseteq Y_2 \setminus (Y_2(2)) \cong \{\{f_{1,2}\}, \{f_{0,2}\}\} = \text{"back1"}$$

$$Y_2 \supseteq Y_2 \setminus (Y_2(1)) \cong \{f_{0,2}\} = \text{"back2"}$$

We can perform the same sort of analysis for the other three objects of **[3]**:

Data			
Object/Stage (n)	Sieves	Subfunctors $\Omega_{\mathbf{C}}(n)$	Truth-values
3	$\emptyset$	$\ulcorner\emptyset\urcorner$	0
	$\{f_{0,3}\}$	Back3	$\frac{1}{8}$
	$\{\{f_{1,3}\}, \{f_{0,3}\}\}$	Back2	$\frac{1}{4}$
	$\{\{f_{2,3}\}, \{f_{1,3}\}, \{f_{0,3}\}\}$	Back1	$\frac{1}{2}$
	$\{\{\text{id}_3\}, \{f_{2,3}\}, \{f_{1,3}\}, \{f_{0,3}\}\}$	All $\cong Y_3$	1
2	$\emptyset$	$\ulcorner\emptyset\urcorner$	0
	$\{f_{0,2}\}$	Back2	$\frac{1}{4}$
	$\{\{f_{1,2}\}, \{f_{0,2}\}\}$	Back1	$\frac{1}{2}$
	$\{\{\text{id}_2\}, \{f_{1,2}\}, \{f_{0,2}\}\}$	All $\cong Y_2$	1
1	$\emptyset$	$\ulcorner\emptyset\urcorner$	0
	$\{f_{0,1}\}$	Back1	$\frac{1}{2}$
	$\{\{\text{id}_1\}, \{f_{0,1}\}\}$	All $\cong Y_1$	1
0	$\emptyset$	$\ulcorner\emptyset\urcorner$	0
	$\{\text{id}_0\}$	All $\cong Y_0$	1

Because of how the terminal object of $\mathbf{Set}^{\mathbf{C}^{op}}$ is defined, the morphism $\mathbf{1} \to \Omega_{[3]}$ will pick out the functor corresponding to the maximal sieve from each row. In terms of truth-values and the contravariant Ω functor, the general idea could be pictured thus:

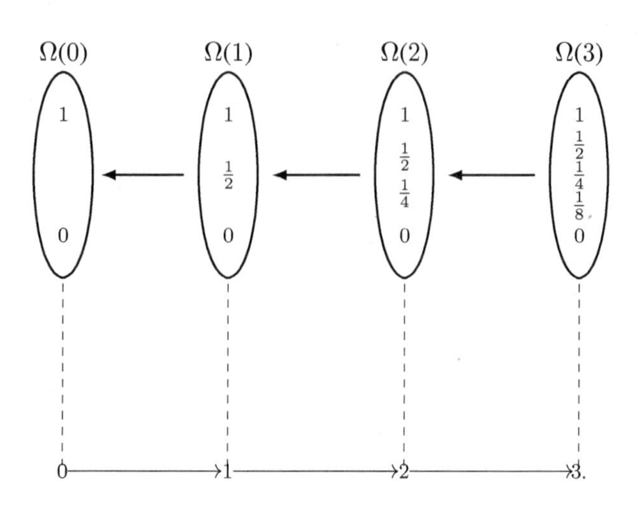

We see that under the action of Ω (regarded as a functor), the degrees of truth are reduced as we move backwards to earlier stages. On the other hand, one might interpret this to mean that as time goes on, the notion of truth obtains more "shades" or nuances.

Now that we have $\Omega_{\mathbf{C}} \in \mathrm{Ob}(\mathbf{Set}^{\mathbf{C}^{op}})$, given any other functor $X : \mathbf{C}^{op} \to \mathbf{Set}$, we know that the subfunctors of X will be in bijective correspondence with the morphisms $X \to \Omega_{\mathbf{C}}$. Suppose given the following instance $X : \mathbf{[3]}^{op} \to \mathbf{Set}$:

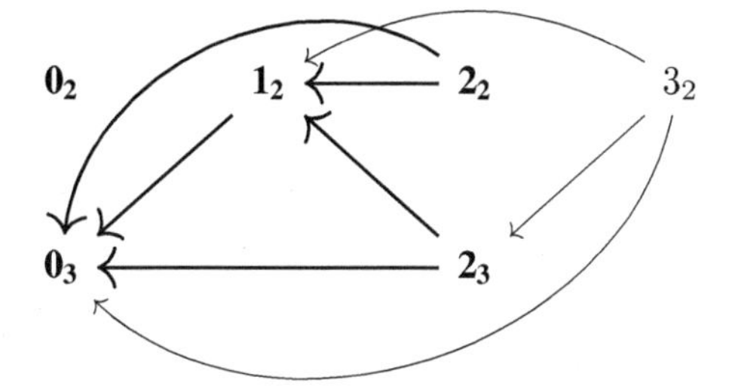

The portion of the graph in bold corresponds to a particular subfunctor $A \subseteq X$. This subfunctor in fact is just a natural transformation $\chi(A) : X \to \Omega_{[3]}$. For example, $\chi(A)(3)$ sends $3_2 \in X(3)$ to *back1* $\in \Omega(3)$, but sends 3_1 to *none*; while $\chi(A)(1)$ sends $1_1 \in X(1)$ to *none* $\in \Omega(1)$, and sends 1_2 to *all*. As for the sieve $\{f_{0,3}\}$, which only belongs to Ω at stage 3, this can be thought of as capturing the element of the set A which only finally falls into some other subset at the zero-th stage, meaning that it takes three stages for it to arrive in the given subset—thus it could be thought of as taking on a truth-value of $\frac{1}{2^3}$.

There is of course no reason to restrict ourselves to linear categories like the $\mathbf{C} = \mathbf{[3]}$ given above. We can consider poset categories or more complicated categories, and the concept of sieves and truth-values will apply just as one would expect.

11.2 Examples of Elementary Toposes

What follows are some notable examples of elementary toposes, where only a very abridged sketch is given in most cases. In subsequent sections, we will look at particular toposes in more detail.

Example 310 **Set** is an elementary topos (and an important one at that, since properties found there largely motivate the three requirements in the definition of an elementary topos). It has finite limits and given any two sets X and Y, we can always form the set Y^X, where this is just the set of functions from X to Y—and this set will enjoy the universal property that the bijection $\mathrm{Hom}_{\mathbf{Set}}(Z, Y^X) \cong \mathrm{Hom}_{\mathbf{Set}}(Z \times X, Y)$ will be natural (in both Y and Z), thus giving rise to an adjunction between (the left adjoint functor) $- \times X : \mathbf{Set} \to \mathbf{Set}$ and (right adjoint) $(-)^X : \mathbf{Set} \to \mathbf{Set}$. Moreover, as already discussed, $\Omega = \{0, 1\}$, and Ω^X can be shown to be isomorphic to the set $\mathbb{P}(X)$ of *subsets* of X, the subobject classifier being defined as we saw.

Example 311 The category $\mathbf{Set}^{\mathbf{C}^{op}}$ of presheaves on a small category $\mathbf{C}$ is an elementary topos.[215] If $\mathbf{C}$ is a small category, we know that the category of presheaves on $\mathbf{C}$ is complete. As for exponentials: supposing we have two presheaves F, G, the Yoneda lemma demands that the exponential G^F is defined as

$$G^F(c) \cong \mathrm{Nat}(\mathrm{Hom}(-, c), G^F) \cong \mathrm{Nat}(\mathrm{Hom}(-, c) \times F, G)$$

for every object $c \in \mathbf{C}$. By taking the right-hand side as a definition, we still have to verify that the adjunction

$$\mathrm{Hom}(H, G^F) \cong \mathrm{Hom}(H \times F, G)$$

holds for all presheaves H, not just the representable ones. However, by proposition 162 we know that we can express any presheaf as a colimit of representable functors, colim Y_c. Thus, given a third functor H, and writing it as a colimit of representable functors,

$$\begin{aligned}
\mathrm{Hom}(H, G^F) &\cong \mathrm{Hom}(\mathrm{colim}(\mathrm{Hom}(-, c_i)), G^F) \\
&\cong \lim \mathrm{Hom}(\mathrm{Hom}(-, c_i), G^F) \text{ by hom-functor transforms colimits into limits} \\
&\cong \lim G^F(c_i) \text{ by the Yoneda lemma} \\
&\cong \lim \mathrm{Hom}(\mathrm{Hom}(-, c_i) \times F, G) \text{ by definition of } G^F \\
&\cong \mathrm{Hom}(\mathrm{colim}\ (\mathrm{Hom}(-, c_i) \times F), G) \\
&\cong \mathrm{Hom}((\mathrm{colim}\ \mathrm{Hom}(-, c_i)) \times F, G) \text{ as } (-) \times F \text{ preserves colimits in } \mathbf{Set} \\
&\cong \mathrm{Hom}(H \times F, G).
\end{aligned}$$

In the previous section, we saw a sketch of how the subobject classifier Ω looks for a presheaf category, where for $c \in \mathbf{C}$, we will have $\Omega(c) = \{S \mid S \text{ is a subfunctor of } Y_c\}$.

The terminal presheaf 1 is defined as we have already seen, namely by $1(c) = \{*\}$ precisely when $c \in \mathbf{C}$, and the monic map $\top : 1 \rightarrowtail \Omega$ is given by $\top_c(*) = Y_c$.

215. Henceforth, I will stop adding "elementary"; topos will mean "elementary topos." Going forward, if Grothendieck topos is ever meant, this will be explicitly indicated.

Because of the powerful fact that so many categories of interest can be constructed as a presheaf category, this last example will supply us with a large store of examples. We will explore a few of them in more detail in coming sections, but for now note that all the presheaf categories introduced so far will thus fit into this mold. To take one more or less at random, the category of G-sets (for a fixed group G), as nothing other than the category of presheaves on (the categorified version of G) $\mathscr{G}$, is a topos.

Example 312 The category **FinSet** of finite sets *is* a topos.

If X, Y are finite sets, then Y^X will be finite; moreover, $\Omega = \{0, 1\}$ is finite, and a finite limit of finite sets is finite as well. Thus, the topos structure of the category of sets effectively carries over to the category of finite sets.

But notice that the category of finite sets lacks infinite products. There are precisely two morphisms from the terminal 1 to Ω, so if $\Omega \times \Omega \times \cdots$ were to exist, with the α factors infinite, then there would be precisely 2^α morphisms from 1 to $\Omega \times \Omega \times \cdots$. But the assumption was that $\Omega \times \Omega \times \cdots$ will be a finite set, so this cannot be. Incidentally, this helps account for why we said in example 284 (chapter 10) that the topos of finite sets is *not* a Grothendieck topos.

Similarly, *the category of finite presheaves on a finite category is a topos* (but, again, *not* a Grothendieck topos, due to a lack of completeness). By a finite presheaf on **C** finite, we of course just mean a functor $\mathbf{C}^{op} \to \mathbf{FinSet}$ that is valued in the category of *finite* sets.[216]

Example 313 Let us call the (large) category consisting of the ordinal numbers, partially ordered in the natural way, **On**.[217] Then $\mathbf{Set}^{\mathbf{On}}$ is not a topos, since in particular any representable functor will have a nonsmall class of subfunctors, so it cannot have a subobject classifier. $\mathbf{Set}^{\mathbf{On}^{op}}$, by contrast, *is* a topos. Note that in $\mathbf{Set}^{\mathbf{On}^{op}}$ there will still be a proper class of subfunctors of the terminal constant functor (with value 1)—thus, it is not locally small, as there will be a proper class of morphisms $1 \to \Omega$.

Incidentally, readers who know about surreal numbers (a construction that includes both the real numbers and the infinite ordinal numbers) might wish to consider how this all unfolds using (the appropriate category of) such numbers.[218]

Example 314 The category **Ring** of rings has rings (with identity) for objects and ring homomorphisms (preserving the identity) for morphisms.[219] This category is both complete and cocomplete, with the zero (trivial) ring **0** for terminal object. However, there are no nontrivial morphisms in **Ring** out of **0**, that is, from the terminal object to any nonzero

216. The constructions basically work as they do for general presheaf toposes, provided **C** itself is finite. In general, when **C** is an infinite category, one will run into difficulties in trying to use the usual presheaf topos constructions for *finite* presheaves. An interesting exception to this is with the category of finite G-sets (on an arbitrary group G), which *is* a topos, even without imposing a finiteness condition on G.

217. This differs from the category of finite ordinals, or the category of natural numbers, by generalizing to numbers of possibly infinite magnitudes.

218. For the proper class **No** of surreal numbers, built on top of a broader combinatorial theory of games, see Conway (2000). Joyal (1977) later gave a category theoretic description of the situation, defining a category of games where objects are sets of games, morphisms are given by winning strategies (for the "Left" player playing second in a particular game), the identity morphism is the "copycat" strategy, and composites are a little more complex to describe. See Joyal (1977) and Cockett, Cruttwell, and Saff (2010) for more details.

219. This category differs from **Rng**, the category of rings (without unit) for objects and ring homomorphisms for morphisms.

ring. This is enough to show that even though this category has a terminal object, it cannot have a subobject classifier. Thus, **Ring** is not a topos.

We have seen a few examples of toposes that are not Grothendieck toposes. However, it can be shown that a Grothendieck topos $\mathbf{Sh}(\mathbf{C}, J)$ is always a topos.

Example 315 Every Grothendieck topos (sheaf on a site) is an example of a topos.

In the last chapter, we saw how the inclusion functor $i : \mathbf{Sh}(\mathbf{C}, J) \to \mathbf{PreSh}(\mathbf{C})$ has a left adjoint $\mathbf{a}$, which preserves finite limits and (small) colimits. In fact, a Grothendieck topos is complete since $\mathbf{PreSh}(\mathbf{C})$ is complete, and limits in $\mathbf{Sh}(\mathbf{C}, J)$ are computed as in the presheaf category. As such, it actually has all (small) limits, which are preserved by the inclusion functor i; (small) limits are computed pointwise and the terminal object $1_{\mathbf{Sh}(\mathbf{C},J)}$ of $\mathbf{Sh}(\mathbf{C}, J)$ is the functor $t : \mathbf{C}^{op} \to \mathbf{Set}$ that sends each object $c \in \mathbf{C}$ to the singleton set $\{*\}$. Moreover, since $\mathbf{PreSh}(\mathbf{C})$ is cocomplete, so too is $\mathbf{Sh}(\mathbf{C}, J)$, where colimits in $\mathbf{Sh}(\mathbf{C}, J)$ are got by applying the associated sheaf functor $\mathbf{a}$ to the colimits in the presheaf category.

$\mathbf{Sh}(\mathbf{C}, J)$ also has exponentials, where these are formed just as in the presheaf topos $\mathbf{PreSh}(\mathbf{C})$. Moreover, it can be shown that if G is a sheaf for a topology J on $\mathbf{C}$, then for any presheaf F the presheaf G^F will itself be a sheaf.

Finally, the category $\mathbf{Sh}(\mathbf{C}, J)$ has a subobject classifier. One way of describing the subobjects is as follows: given a sieve S on c, we say that S is *J-closed* provided, for any arrow $f : c' \to c$, the pullback $f^*(S)$ being in $J(c')$ implies that $f \in S$. Then, we define $\Omega : \mathbf{C}^{op} \to \mathbf{Set}$ on objects $c \in \mathbf{C}$ as the set of all J-closed sieves on c, and on morphisms by the pullback $f^*(-)$ by a sieve. Then, the arrow $\top : 1_{\mathbf{Sh}(\mathbf{C},J)} \to \Omega$ is defined by $\top(*)(c) =$ the maximal sieve on c, and this is a subobject classifier. The classifying arrow $\chi_{F'} : F \to \Omega$ of a subobject $F' \subseteq F$ in $\mathbf{Sh}(\mathbf{C}, J)$ is then given by

$$\chi_{F'}(c)(x) = \{f : c' \to c \mid F(f)(x) \in F'(c')\}$$

for any $c \in \mathbf{C}$ and $x \in F(c)$.[220]

So while every Grothendieck topos is an elementary topos, the converse is not necessarily true: an elementary topos need not be a Grothendieck topos. Elementary toposes accordingly supply a proper generalization of Grothendieck toposes. Some of these non-Grothendieck toposes are of particular use or relevance in logic, particularly in the study of higher-order intuitionistic type theory. In practice, one important general difference between the two notions seems to be that the presence of sites in Grothendieck toposes provides a setting where one can more readily rely on or make use of geometric intuitions. In the next section, we explore a little more about elementary toposes, consider some translations of the relevant notions back to the realm of Grothendieck toposes, and then look closely at a particular example.

11.3 Lawvere-Tierney Topologies and Their Sheaves

Constructing a Grothendieck topos can generally be thought of as breaking down into two steps:

220. The reader can consult Mac Lane and Moerdijk (1994, III.6–7) for further details on how a Grothendieck topos satisfies the properties of an elementary topos.

1. given **Set** and some (small) category **C**, we pass to the category $\mathbf{Set}^{\mathbf{C}^{op}}$ of presheaves on **C**; then
2. we pass to the category $\mathbf{Sh}(\mathbf{C}, J)$ of sheaves for a particular Grothendieck topology J.

Via the notion of an elementary topos, we are essentially able to replace **Set** in step (1) by any topos $\mathscr{E}$. The purpose of the notions introduced in the present section, in particular that of a *Lawvere-Tierney topology*, is to similarly generalize the second step, ultimately enabling us to provide an even more general account of sheaves. We can moreover translate between earlier notions: in particular, a Grothendieck topology J on a category **C** can be shown to correspond to a Lawvere-Tierney topology (or local operator), defined as an invariant in the context of the topos $\mathbf{Set}^{\mathbf{C}^{op}}$.

To motivate this, suppose we are working in the context of a topological space. Suppose S is any collection of open sets U_i of the space X. Then, consider the collection of all the open sets *covered by* the U_i, and denote this for now by a generic symbol $\blacksquare(S)$. But then $\blacksquare$ of the entire space X should of course be everything, as taking the collection of all the open sets covered by the entire space should of course return the entire space itself. Moreover, we would expect that this operation of taking such collections will be idempotent, in the sense that $\blacksquare(\blacksquare(S)) = \blacksquare(S)$. What about intersections? It is also reasonable to expect that $\blacksquare(S \cap S') \subseteq \blacksquare(S) \cap \blacksquare(S')$—and, whenever both S and S' are sieves, this would become an equality.

But what if the collection S is in fact a *sieve* on an open set $U \subseteq X$? In that case, $\blacksquare(S)$ will also be a sieve and S will just be an element in $\Omega(U)$, where Ω is the subobject classifier for the topos $\mathscr{E} = \mathbf{Set}^{\mathcal{O}(X)^{op}}$ of presheaves on a topological space X. But altogether, this amounts to saying that the $\blacksquare$ just described is essentially an endomap $\blacksquare : \Omega \to \Omega$ satisfying certain equations, where the map itself effectively specifies what each sieve covers. This line of reasoning can serve as motivation for the following definition for toposes more generally (where we now denote the abstract operator $\blacksquare$ by j).

Definition 316 Let $\mathscr{E}$ be a topos, and let Ω together with the arrow $\top : \mathbf{1} \to \Omega$ be its subobject classifier. A *Lawvere-Tierney topology* (or sometimes *local operator*) on $\mathscr{E}$ is a map $j : \Omega \to \Omega$ in $\mathscr{E}$ such that the following three properties are satisfied:

1. $j \circ \top = \top$, that is,

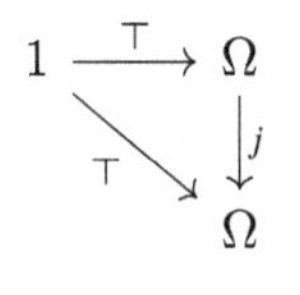

2. $j \circ j = j$, that is,

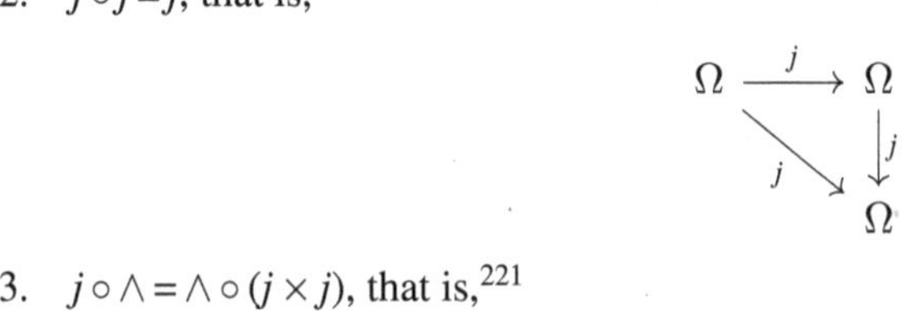

3. $j \circ \wedge = \wedge \circ (j \times j)$, that is,[221]

221. The $\wedge : \Omega \times \Omega \to \Omega$ here is just the meet operation (in the internal Heyting algebra formed by Ω). With $\leq$ the internal partial order on Ω, note that the first condition could also be expressed by saying $\mathrm{id}_\Omega \leq j$.

$$\begin{array}{ccc} \Omega \times \Omega & \xrightarrow{\wedge} & \Omega \\ {\scriptstyle j\times j}\downarrow & & \downarrow{\scriptstyle j} \\ \Omega \times \Omega & \xrightarrow[\wedge]{} & \Omega. \end{array}$$

One then sometimes refers to the pair $\mathscr{E}_j = (\mathscr{E}, j)$ as an *elementary site*.

It may help to think of j in a different way. We can effectively transfer the *algebraic* structure supplied by j and $\wedge$ on the special object Ω to a corresponding *functorial* structure on hom-sets, via the fact that in any topos, given an object E, we will have $\mathrm{Sub}(E) \cong \mathrm{Hom}(E, \Omega)$. Moreover, for each object E in a topos, it can be shown that the poset $\mathrm{Sub}(E)$ of subobjects of E forms a lattice (actually, it is a Heyting algebra!). In this way, j can equivalently be displayed as a particular *closure operator* on the Heyting algebra of subobjects.

Definition 317 A closure operator on a topos $\mathscr{E}$ will be an operator on the subobjects $A \rightarrowtail E$ of each object E of $\mathscr{E}$

$$\overline{(-)} : \mathrm{Sub}_{\mathscr{E}}(E) \to \mathrm{Sub}_{\mathscr{E}}(E)$$
$$A \mapsto \overline{A}$$

natural in E. It can be shown that j is a Lawvere-Tierney topology if and only if this operator has, for all $A, B \in \mathrm{Sub}(E)$, the following ("universal closure operator") properties

1. $A \subseteq \overline{A}$,
2. $\overline{\overline{A}} = \overline{A}$,
3. $\overline{A \cap B} = \overline{A} \cap \overline{B}$.[222]

The converse is also true: an operator defined on all of $\mathrm{Sub}(E)$, natural in E, with those three properties—that is, a closure operator on the subobjects of each object of $\mathscr{E}$—always arises from a unique Lawvere-Tierney topology j.

The last two equalities are basically direct translations of the last two conditions on a Lawvere-Tierney topology. For the first, the operator j determines a closure operator on the subobjects $A \rightarrowtail E$ of each object E by

$$\begin{array}{ccc} \mathrm{Hom}(E, \Omega) & \xrightarrow{\cong} & \mathrm{Sub}(E) \ni A \\ {\scriptstyle \mathrm{Hom}(\mathrm{id}, j)}\downarrow & & \downarrow \\ \mathrm{Hom}(E, \Omega) & \xrightarrow{\cong} & \mathrm{Sub}(E) \ni \overline{A}. \end{array}$$

This means that, given a subobject $s : A \rightarrowtail E$ with the classifier $\chi_s : E \to \Omega$, then, referring to the diagram (with both squares formed by pullbacks along the truth arrow $\top$),

222. Another condition would be that closure commutes with pullback along morphisms of $\mathscr{E}$, in the sense that given $f : Y \to X$ and a subobject $X' \rightarrowtail X$ (the closure of which is denoted $\overline{X'} \rightarrowtail X$), we will have $f^*(\overline{X'}) \cong \overline{f^*(X')}$ as subobjects of Y.

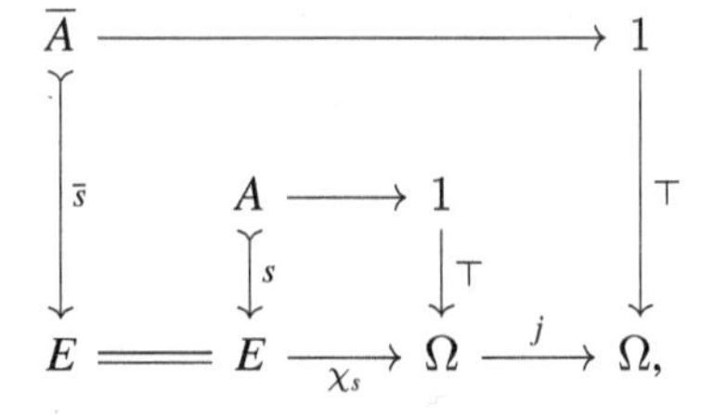

the composition $j \circ \chi_s$ defines another subobject $\overline{s}:\overline{A} \rightarrowtail E$, the one we call the j-closure $\overline{A}$, such that s is a subobject of $\overline{s}$. Conversely, a closure operator $\overline{(-)}$ on a topos will give a unique topology j via the following pullback diagram:

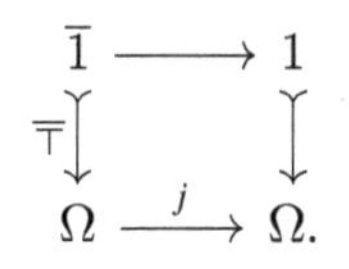

This description can be used to translate back and forth between the statement $j(\top) = \top$ and the property $A \subseteq \overline{A}$. Because of this equivalence between the two perspectives, we can freely refer to Lawvere-Tierney topologies in terms of *universal closure operators.*

Observe how, from property (3) of definition 317, and since $A \subseteq B$ iff $A \wedge B = A$, we can further obtain that $A \subseteq B$ implies $\overline{A} \subseteq \overline{B}$, which informs us that such an operator is also a closure operator in the earlier sense of definition 182—where this meant an operator that was monotone, extensive, and idempotent.[223] Recall also how we have seen that the modalities of modal logic could be described as closure operators on the poset of propositions (with their truth-valuations). But this, together with the fact that the subobject Ω can be interpreted as a collection of generalized "truth-values," suggests that we may regard j as a kind of modal operator! Let us roughly sketch the idea.

In the usual setting of topological spaces, when we say that a property holds *locally* at a point x of a space X, we mean that it holds at all points "nearby," or throughout some neighborhood V of x. Another way of saying this is that a property holds locally for an object U provided *it can be covered by open sets that all have the property as well.* A natural suggestion, due ultimately to Lawvere, is thus to regard j as a sort of modal operator that says "it is *locally* the case that."

For $E \in \mathscr{E}$, observe that we can view a morphism $p: E \to \Omega$ as a predicate, where in particular if $E = \mathbf{1}$, then the elements $p: \mathbf{1} \to \Omega$ may be seen as propositions. We will also have a function $p(U): E(U) \to \Omega(U)$ for any object U. For any object $E \in \mathscr{E}$, there will be a poset formed out of the predicates $p, q: E \to \Omega$ on E, where $p \leq q$ provided p implies q. In this way, we might then read the three conditions of definition 316 as saying

1. if p is true, then p is locally true;
2. p locally is locally true iff p is locally true;

223. However, a universal closure operator in the above sense should not be confused with the notion of a Kuratowski closure on the lattice of all subsets of a topological space. In particular, note that Kuratowskian closure always commutes with finite unions, but in general not with intersections. On the other hand, a universal closure operator commutes with intersections, but does not typically commute with unions. As such, while such a closure operator in a sense combines various of the properties of the interior and closure operators from standard topological spaces, the reader should be careful to note that the closure operators discussed here are distinct from "closed subsets" from topological spaces.

3. $p \wedge q$ is locally true iff p and q are each locally true.

In this way, the modality j can be seen as generalizing what it means to be *local* to the setting of toposes.[224]

Example 318 Recall the particular downset completion poset of the poset discussed in example 252:

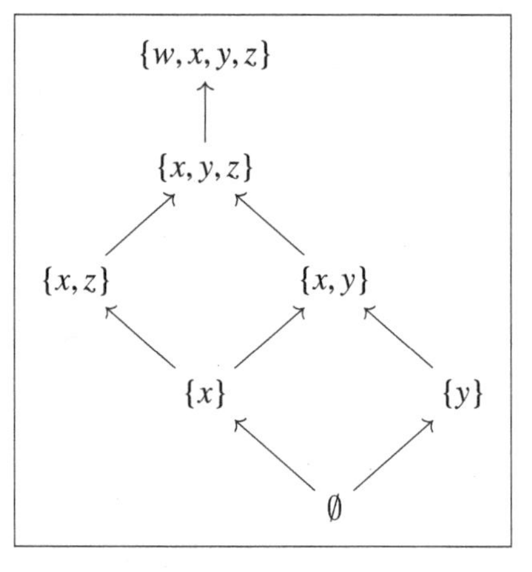

j will range over the elements of this downset poset $\mathbf{Down}(\mathscr{P})$. A particular Lawvere-Tierney topology may then be given by

$$j(\emptyset) = \emptyset$$
$$j(\{x\}) = \{x, z\}$$
$$j(\{y\}) = \{y\}$$
$$j(\{x, y\}) = \{w, x, y, z\}$$
$$j(\{x, z\}) = \{x, z\}$$
$$j(\{x, y, z\}) = \{w, x, y, z\}$$
$$j(\{w, x, y, z\}) = \{w, x, y, z\}.$$

You can verify that this constitutes a Lawvere-Tierney topology on the poset of downsets.

Stepping back, the reader should by now fully appreciate that to define sheaves, what is really important is what gets "covered" and how the data of the sheaf behaves with respect to (what formally acts as) covers. With Grothendieck toposes, when defining a sheaf, we did so in terms of coverages or coverings by sieves. Ultimately, our new "topology" j will similarly track such information, functioning to specify what each sieve covers, but now in the more general setting of elementary toposes. Accordingly, this notion of a Lawvere-Tierney topology will enable us to define (more general) sheaves for any topos. We can

224. An interesting question we leave to the reader: following this idea of j being regarded as a modal operator of sorts, since (following the ideas explored in chapter 7) we are used to modal operators coming in dual pairs, what (if anything) might the dual pair of "it is locally valid" be? For more details on j as a modal operator, see Goldblatt (2006, chap.14).

already appreciate the greater generality of this notion, though, by considering the following proposition concerning how Lawvere-Tierney topologies include the Grothendieck topologies.

Proposition 319 Each Grothendieck topology J on a small category $\mathbf{C}$ determines a Lawvere-Tierney topology j on the corresponding presheaf topos $\mathbf{Set}^{\mathbf{C}^{op}}$.

Without giving a proper proof of this fact, the idea here is this: remember that the subobject classifier Ω for $\mathbf{Set}^{\mathbf{C}^{op}}$ is the functor that takes an object c to the collection of sieves on c in $\mathbf{C}$, and the action of morphisms of $\mathbf{C}$ on Ω is by "pulling back," in the sense that for $f : c' \to c$, we take

$$f^* : \Omega(c) \to \Omega(c'), \quad f^*(S) = \{g \mid f \circ g \in S\}.$$

Recall also that a Grothendieck topology J is just an assignment, for each object c, of a collection of sieves that are "covering" and, as such, meet certain properties. A subobject $J \hookrightarrow \Omega$ is thus just a selection from the collection of sieves, which is moreover closed under pullback. Given a Grothendieck topology J, we accordingly define $j : \Omega \to \Omega$ by taking

$$\begin{aligned} j_c(S) &= \{g \mid S \text{ covers } g : d \to c\} \\ &= \{g \mid g^*(S) \in J(\text{dom} g)\}. \end{aligned}$$

Skimming over the details of how this works in the general case, let us look at our running example and see how to convert from a particular Grothendieck topology on $\mathcal{P}$ to a Lawvere-Tierney topology on the associated presheaf topos.

Example 320 Recall the dense (Grothendieck) topology on $\mathcal{P}$ from example 252:

$$\begin{aligned} J_{dense}(x) &= \{\{x\}\} \\ J_{dense}(y) &= \{\{y\}\} \\ J_{dense}(z) &= \{\{x\}, \{x, z\}\} \\ J_{dense}(w) &= \{\{x, y\}, \{x, y, z\}, \{w, x, y, z\}\}. \end{aligned}$$

This induces a particular Lawvere-Tierney topology on the associated downset completion $\mathbf{Down}(\mathcal{P})$. In particular, let us look at $j(\{x\})$. Consider that $\downarrow x \cap \{x\} = \{x\} \in J_{dense}(\{x\})$ and also $\downarrow z \cap \{x\} = \{x\} \in J_{dense}(\{z\})$; yet for no other p do we have $\downarrow p \cap \{x\} \in J_{dense}(p)$. Accordingly, we take $j(\{x\}) = \{x, z\}$. Similarly for the other elements. Altogether, the J_{dense} Grothendieck topology on $\mathcal{P}$ induces a corresponding Lawvere-Tierney topology j given by that already described in example 318.

The general result informs us, incidentally, that any Grothendieck topology J can be given a description in terms of a closure operator.[225]

In fact, for presheaf toposes, we can go the other way as well, recovering a Grothendieck topology on $\mathbf{C}$ from a Lawvere-Tierney topology on $\mathbf{Set}^{\mathbf{C}^{op}}$. In other words, it can be shown that every Lawvere-Tierney topology j on a presheaf topos arises (as above) from a Grothendieck topology.[226]

225. For further discussion of the explicit description of this, see Mac Lane and Moerdijk (1994).
226. Though, for more general toposes, there can be other Lawvere-Tierney topologies.

Proposition 321 If $\mathbf{C}$ is a small category, then the Grothendieck topologies J on $\mathbf{C}$ correspond precisely to Lawvere-Tierney topologies on the presheaf topos $\mathbf{Set}^{\mathbf{C}^{op}}$.

Again skimming over the details of a proof of this fact, any Lawvere-Tierney topology $j : \Omega \to \Omega$ on the presheaf topos classifies the subobject $J \hookrightarrow \Omega$ by taking $S \in J(c)$ iff $j_c(S) = M_c$, the maximal sieve. Such a J can be shown to be a Grothendieck topology.[227]

In particular, given a poset $\mathcal{P}$ and a Lawvere-Tierney topology j on $\mathcal{D}(\mathcal{P})$, the family

$$\{J(p) \mid J(p) = \{\downarrow p \cap X \mid p \in j(X) \text{ and } p \in \mathcal{P}\}\}$$

will form a Grothendieck topology. Again, let's look at how this works in our running example.

Example 322 Suppose given the Lawvere-Tierney topology (on the poset of downsets) from example 318. Then, for instance, $J(z)$ will be computed as follows:

$$\begin{aligned} J(z) &= \{\downarrow z \cap \{x\}, \downarrow z \cap \{x,z\}, \downarrow z \cap \{x,y\}, \downarrow z \cap \{x,y,z\}, \downarrow z \cap \{w,x,y,z\}\} \\ &= \{\{x,z\} \cap \{x\}, \{x,z\} \cap \{x,z\}, \{x,z\} \cap \{x,y\}, \{x,z\} \cap \{x,y,z\}, \{x,z\} \cap \{w,x,y,z\}\} \\ &= \{\{x\}, \{x,z\}\}. \end{aligned}$$

We can similarly compute the rest of $J(p)$ for all $p \in \mathcal{P}$.

The following extended example gives us a chance to delve into some of these ideas in more depth, illustrating how Lawvere-Tierney topologies work more explicitly, in the context of graphs. This example will moreover help motivate the still more general definition of a sheaf with respect to a Lawvere-Tierney topology toward which we are building.

Example 323 Recall that if we take

$$\mathcal{G} := \boxed{V \underset{t}{\overset{s}{\rightrightarrows}} A}$$

for our indexing category, then presheaves on $\mathcal{G}$ recover directed graphs (or quivers), so that we have the equivalence of $\mathbf{Set}^{\mathcal{G}^{op}}$ and the category $\mathbf{Quiv}$. As a presheaf category (on $\mathcal{G}$, a small category), we know that $\mathbf{Set}^{\mathcal{G}^{op}}$ will have the requisite properties of a topos: (1) it has finite limits (terminal object 1 and pullbacks); (2) it has exponentials (i.e., the functor $- \times X$ has a right adjoint for each X); (3) it has a "truth-values" object Ω which classifies subobjects. While the first two properties of a topos are more or less standard constructions, the third can again use some explanation, and it will be instructive to see how this works in the particular case of graphs. This will further allow us to construct various explicit Lawvere-Tierney topologies.

The first thing to recall is the fact that since every object of a presheaf topos is built by gluing a suitable set of representables, and the representable functors provide us with a set of generators for the topos (since two arrows differ iff they differ over some representable), we can build the classifier Ω by considering subobjects of the representables. Moreover, we know that the subobject classifier in a presheaf topos $\mathbf{PreSh}(\mathbf{C})$ is the presheaf that sends

227. See Mac Lane and Moerdijk (1994, V.4) for a proof of this fact—which, together with proposition 319, lets us verify that $j \mapsto J$ and $J \mapsto j$ are inverse, proving the main result.

each object $c \in \mathbf{C}$ to the set of sieves on it,

$$\Omega : c \mapsto \{\text{sieves}(c)\},$$

and the sieves on c may equivalently be regarded as the set of subobjects of the representable presheaf Y_c. The classifier morphism $\top : 1 \to \Omega$ is then the natural transformation that selects, for each object $c \in \mathbf{C}$, the maximal sieve. Accordingly, the first step is to address the question: What are the representable functors?

In general, we know that there is one representable $Y_c = \mathrm{Hom}_{\mathbf{C}}(-, c)$ functor for each object c of $\mathbf{C}$, where this of course associates to c the set of all morphisms into c (and the definition on arrows supplied by composition). So in our present case, the representable functors giving us the "generic figures" that can be found in directed graphs will be

$$\mathrm{Hom}_{\mathcal{G}}(-, V) \text{ and } \mathrm{Hom}_{\mathcal{G}}(-, A),$$

which you might think of as supplying us with the "generic vertex" and "generic arrow" (and that is all we need to consider, as the only two objects of $\mathcal{G}$ are V and A). Now we need only range over the objects of $\mathcal{G}$, plugging them in to the above hom-sets, seeing what pops out. Let's do this first for $\mathrm{Hom}_{\mathcal{G}}(-, V)$:

$$\mathrm{Hom}_{\mathcal{G}}(V, V) = \{\mathrm{id}_V\} = \{V\}$$

$$\mathrm{Hom}_{\mathcal{G}}(A, V) = \emptyset$$

So the representable $\mathrm{Hom}_{\mathcal{G}}(-, V)$ will just consist of a single vertex or node (with no arrows)—where this can be thought of as the generic vertex. This will clearly have just two subobjects: V, for the vertex itself; and 0_V, for the empty vertex. These two subobjects will ultimately give us the two vertices of Ω, seen as a directed graph in its own right, that is, $\Omega(V) = \{V, 0_V\}$. These two subobjects of

$$\mathrm{Hom}_{\mathcal{G}}(-, V) = \boxed{\bullet}$$

are clearly ordered among themselves, in the sense that

$$0_V = \boxed{} \leq \boxed{\bullet_V} = V$$

Now for the representable $\mathrm{Hom}_{\mathcal{G}}(-, A)$, we get

$$\mathrm{Hom}_{\mathcal{G}}(V, A) = \{s, t\}$$

$$\mathrm{Hom}_{\mathcal{G}}(A, A) = \{\mathrm{id}_A\} = \{A\}.$$

This tells us that s and t furnish us with two vertices, and id_A (denoted A) for a single arrow. Altogether, the representable $\mathrm{Hom}_{\mathcal{G}}(-, A)$ can then be pictured as

$$\boxed{\bullet_s \xrightarrow{A} \bullet_t}$$

Conceptually, that the representable associated to V is given by a single vertex, while the representable associated to A is given by a single arrow, tells us that arbitrary figures built from such an indexing category will be a matter of building with directed arrows between vertices—and that's as it should be, since we are dealing with directed graphs!

The representable $\mathrm{Hom}_{\mathscr{G}}(-,A)$, for its part, will have five subobjects (natural names given on the right):

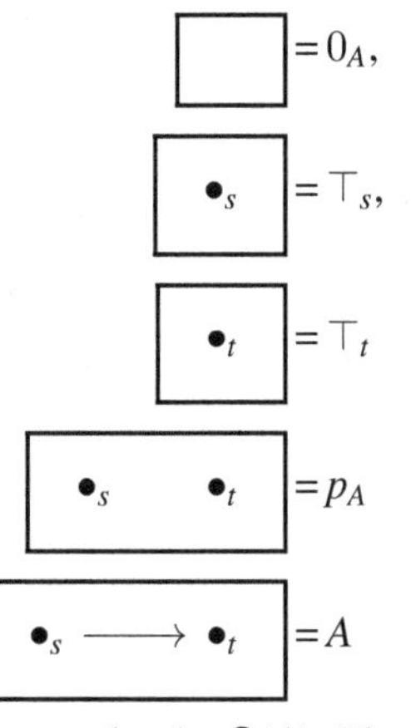

These will supply Ω with its arrow set, that is, $\Omega(A)$. These subgraphs of $\mathrm{Hom}_{\mathscr{G}}(-,A)$ are again clearly ordered by inclusion, an order that assembles into the Hasse diagram

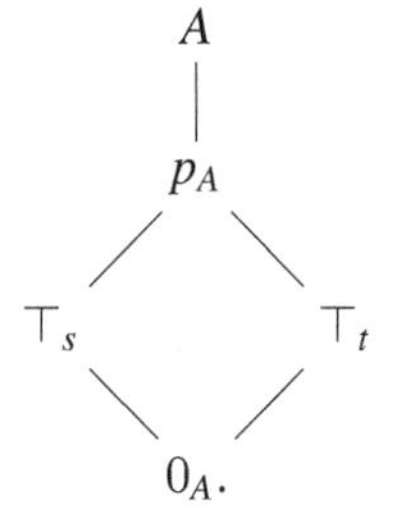

As Ω is itself a presheaf, the last thing to do, before assembling all this into Ω, is spell out how the right action works. As 0_A is the empty graph, we must have that $0_A s = 0_V$ and $0_A t = 0_V$. Moreover, $As = At = V$ and likewise $p_A t = p_A s = V$. Finally, $\top_s s = V$ and $\top_s t = 0_V$ (as t sends the only vertex of $\mathrm{Hom}_{\mathscr{G}}(-, V)$ into $\top_t$, not $\top_s$), and similarly $\top_t t = V$ and $\top_t s = 0_V$. Altogether, this data of Ω is supplied by the graph

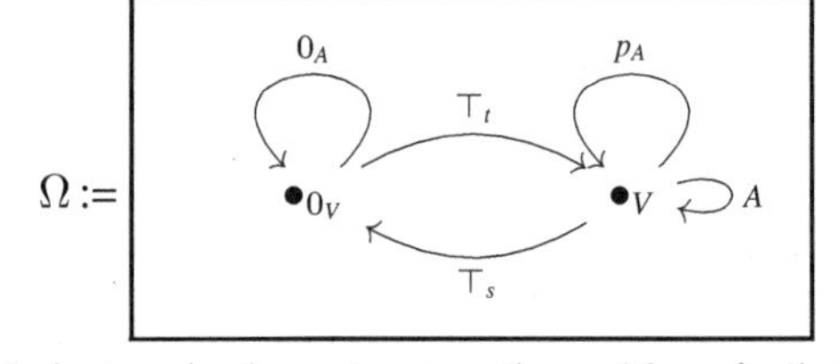

The terminal graph 1 is just a single vertex together with a single loop. Clearly the truth classifier arrow (graph homomorphism) $\top : 1 \to \Omega$ will then send the (only) vertex of 1 to V and the only arrow of 1 to A.

Now, suppose we are given a graph G and a subgraph $u : H \hookrightarrow G$. How does the classifying map $\chi_u : G \to \Omega$ work?

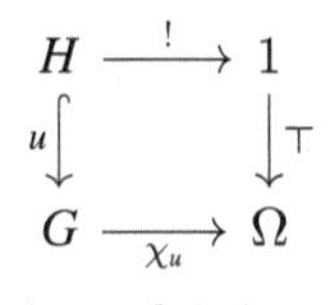

First off, on vertices: χ_u will map vertices of G that are not in H to 0_V, while vertices of G that are in H are mapped to $\bullet = V$. In other words, for vertices, the situation is simple: given a subgraph of a graph G, either a vertex of the graph G is or is not to be found in a certain subgraph. Ω has precisely two vertices to "represent" such a binary choice on vertices.

On arrows: if an arrow is in H, it will get mapped to A; if not, then the situation is more involved. In particular, there are a few alternatives:

- arrows whose source and target are not in H get mapped by χ_u to 0_A;
- arrows that have just their source (but not target) in H, get mapped by χ_u to $\top_s$;
- arrows whose target (but not source) is in H, get mapped by χ_u to $\top_t$; while
- arrows that have both their source and target in H get mapped by χ_u to p_A.

These four cases—plus the first taking an arrow that is indeed in the subgraph H to A—give us the five total possibilities, which five maps are represented by the five arrow-objects of Ω. In this way, the "classification" with arrows emerges as more refined or complicated than it is for vertices. In the end, the underlying idea here is that we are using Ω to tell us, for each part of a graph G, how much of it happens to be in a subgraph H. That there are more classifying maps for arrows than there are for vertices informs as that, as far as graphs are concerned, vertices are a comparatively simpler sort of structure than arrows.

In thinking of Ω as the "object of truth-values" for its topos, a natural question is how logical operators on such truth-values are to be defined, enabling us to reason within the topos. We can define the conjunction operator $\wedge : \Omega \times \Omega \to \Omega$ as the characteristic map of the subobject $\langle \top, \top \rangle : 1 \to \Omega \times \Omega$, and the negation operator $\neg : \Omega \to \Omega$ as the characteristic map of $\bot : 1 \to \Omega$. On vertices, since these have only two truth-values (0_V and V), $\wedge$ and $\neg$ behave just as you would expect—that is, just as they do in **Set**. However, on arrows, $\wedge$ acts as meet with respect to the order on the arrows of Ω, for instance, $p_A \wedge t = t$, $s \wedge t = 0_A$; while $\neg$ takes $\neg 0_A = A$ and $\neg A = 0_A$, but also $\neg p_A = 0_A$ (which means basically that this operator acts to ignore whether or not an arrow was in a subgraph, as long as both its source and target were in the subgraph).

Using these notions, we can take the *complement* of a subgraph $H \hookrightarrow G$ by taking the subobject $\neg H$ of G classified by $\neg \circ \chi_u$, and this will be the same as taking all the vertices of G that are not in H, together with all the arrows in G between those particular G-vertices. Note that this is the same as what we discussed back in example 197 (chapter 7), where $\neg X$ was treated as the largest subgraph disjoint from X. More explicitly, and in the general case of presheaf toposes,

Definition 324 For any subobject $A \rightarrowtail E$ in $\mathbf{Set}^{\mathbf{C}^{op}}$ and any object c in $\mathbf{C}$,

$$(\neg A)(c) = \{x \mid x \in E(c) \text{ and for all } f : b \to c, A(f)(x) \notin A(b)\}.$$

It is clear that, in general, $\neg A \vee A = E$ need not hold.

Returning back to the special case of graphs: suppose we apply $\neg$ again to the result, yielding $\neg\neg H$. This will be the same as adding to H all the arrows of G that have their source and target in H. More explicitly, and again in general,

$$(\neg\neg A)(c) = \{x \mid x \in E(c) \text{ and } \forall f : b \to c, \exists g : a \to b \text{ such that } A(g)(A(f)(x)) \in A(a)\}.$$

This composite map $\neg \circ \neg$ acting on subobjects in fact forms an example of a Lawvere-Tierney topology on a topos, specifically the *double negation topology* $\neg\neg : \Omega \to \Omega$. As such, the map $\neg\neg$ acts as a *closure operator*. In our particular case of subgraphs $H \hookrightarrow G$, this amounts to adding to the arrow set of H all the arrows of G between those vertices that are in H. Thus, $\neg\neg H$ will add to H all the arrows of G that have source and target in H.[228]

In terms of other Lawvere-Tierney topologies that exist for our graphs, in addition to the double negation topology just described, there are also always two trivial topologies available, namely that given by the identity on Ω and that given by

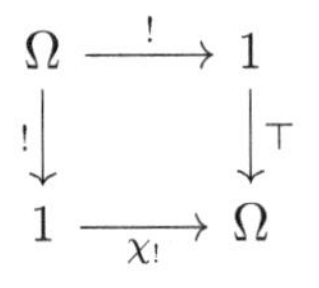

the composite diagonal map $\top \circ ! : \Omega \to \Omega$. Finally, there is one last, nontrivial, topology on **Quiv**, specified as follows. First, note that condition (1) on a topology, that is, preservation of truth, requires specifically that $j(V) = V$ and $j(A) = A$. Considering the other objects, if we then set $j(0_V)$ to 0_V, we would just recover either the identity topology or the double negation topology (after making suitably constrained choices for the rest of the object assignments). On the other hand, if $j(0_V) = V$, then setting $j(0_A) = A$ generates the other trivial topology $j = \top \circ !$. Therefore, we will instead set $j(0_A) = p_A$; then condition (3) on a Lawvere-Tierney topology implies that $j(\top_s) = j(\top_t) = p_A$, since $j(\top_s) \wedge j(\top_t) = j(\top_s \wedge \top_t) = j(0_A) = p_A$. Thus, altogether we have the following assignment:

$$j(V) = V = j(0_V),$$

$$j(A) = A, j(0_A) = p_A = j(\top_s) = j(\top_t) = j(p_A).$$

This assignment describes a nontrivial topology distinct from the $\neg\neg$ topology. It is sometimes called the *closed topology* on the global element p_A, defined as $(- \vee p_A)$. In total, there are four topologies or local operators for the Ω we have been describing for quivers.

Let us now take the opportunity to introduce some important terminology. This is entirely general, but we will continue to refer to quivers to keep things concrete.

Definition 325 Let $u : H \hookrightarrow G$ be a subobject (subgraph), with characteristic map $\chi_u : G \to \Omega$. Then, given a topology j, we call the subobject $\overline{H}$ classified by $j \circ \chi_u$ ("composition with j") the *closure* of H in G (with respect to the topology j).

We also say that H is *dense* provided its closure is equal to G, that is, $\overline{H} = G$.

For instance, the closure with respect to the closed topology above adds to a subgraph H of G all the vertices of the graph. Composition of a characteristic map with the closed topology j adds all the nodes of G to the subgraph H. The dense subobjects (with respect to the closed topology) of G are those subgraphs that include all the arrows of G (and

228. In this connection, it turns out that for a presheaf topos $\mathbf{Set}^{\mathbf{C}^{op}}$, the double negation (Lawvere-Tierney) topology coincides with the dense (Grothendieck) topology we explored in chapter 10. An illuminating proof of this general fact can be found in Mac Lane and Moerdijk (1994, 273). This $\neg\neg$ topology is rather important, and we return briefly to it below.

note that there is a *minimal* such dense subobject, namely that given by the arrow set of G itself). This should be evident since composition of the characteristic map with j does not add arrows to H—since besides A itself, no arrow of Ω is sent to A by j. In short: to be dense with respect to this closed topology, a subgraph H needs to already include precisely all the arrows of G.

What about the $\neg\neg$-topology? Here, closure amounts to adding to a subgraph H all the arrows of G that have source and target in H; this should also make some sense, since the characteristic map sends arrows to p_A and then $\neg\neg$ must take the arrow p_A to A. The dense subobjects of G are given by the subgraphs that include all the vertices of G (and there is a minimal such dense subobject, given by the vertex set of G itself). As before, to be dense, a subgraph H must include precisely all the vertices of G.

We have almost arrived at the point of being able to define a general sheaf for such topologies. One last set of definitions is needed.

Definition 326 An object X of a topos $\mathscr{E}$ with topology j is called *j-separated* (or just *separated*, when j is understood) provided for every object Y, every j-dense subobject $u : S \hookrightarrow Y$, and every morphism $f : S \to X$, there exists at most one $g : Y \to X$ "factoring" f through u:

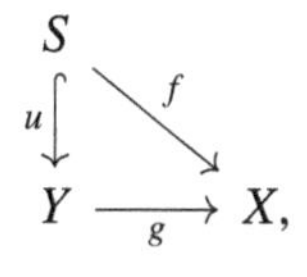

that is, making the above diagram commute.

We will then call an object *j-complete* (or just *complete*, j understood) provided such a unique factorization always exists.

Equipped with these notions, we can at last offer a rather succinct and general definition of a sheaf with respect to a (Lawvere-Tierney) topology, that is, for an elementary site.

Definition 327 (*Definition of j-sheaf*) A *j-sheaf* is an object F of a topos $\mathscr{E}$ that is both j-separated and j-complete.

To unpack this a little: we could have also said that F is a sheaf provided, given a map from a dense subobject of E into F, as in

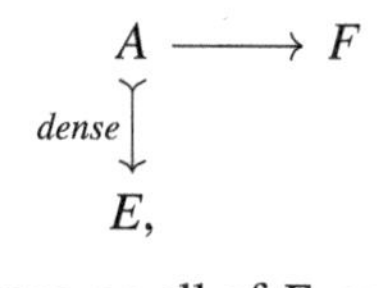

this can be uniquely extended to a map on all of E, as in

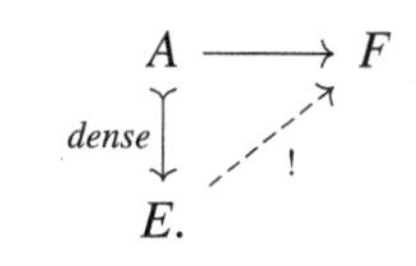

F is thus a j-sheaf provided for every dense monomorphism $m : A \rightarrowtail E$, composition with m induces the isomorphism $\mathrm{Hom}_{\mathscr{E}}(E, F) \xrightarrow{\cong} \mathrm{Hom}_{\mathscr{E}}(A, F)$.

In terms of this map, we could also have defined an object G as *separated* if for each dense $A \rightarrowtail E$, $\mathrm{Hom}_{\mathscr{E}}(E, G) \rightarrow \mathrm{Hom}_{\mathscr{E}}(A, G)$ is a monomorphism (so that the extension above need not exist, but will be unique if it does).

We use $\mathbf{Sh}_j(\mathscr{E})$ to denote the (full) subcategory (of $\mathscr{E}$) consisting of j-sheaves of $\mathscr{E}$, and $\mathbf{Sep}_j(\mathscr{E})$ for the (full) subcategory consisting of the separated objects.

Given the correspondence between Grothendieck topologies on a category $\mathbf{C}$ and Lawvere-Tierney topologies on the associated presheaf topos $\mathbf{Set}^{\mathbf{C}^{op}}$, presented in proposition 321, the following result illustrating how all this relates to the earlier notions of Grothendieck toposes may have been expected. Given $\mathbf{C}$ a small category and j a Lawvere-Tierney topology on the presheaf topos $\mathbf{Set}^{\mathbf{C}^{op}}$, we can form the corresponding Grothendieck topology J on $\mathbf{C}$ (as indicated by 321), and with such a setup, we have the following result:

Proposition 328 A presheaf $P \in \mathbf{Set}^{\mathbf{C}^{op}}$ will be a sheaf (in the sense of definition 327) for the Lawvere-Tierney topology j iff P is a J-sheaf (in the sense of definition 256).[229]

Thus, the definition of a sheaf for a Grothendieck topology J emerges as a special case of the elementary axiomatic approach to sheaves for a Lawvere-Tierney topology. While this sort of result helps reinforce the importance of the presheaf category construction, there are toposes that are not Grothendieck toposes (as we have seen), so the Lawvere-Tierney notions are accordingly a proper generalization of the Grothendieck notions.

Example 329 Before moving on, let us briefly look a little closer at this rather abstract definition of a j-sheaf via our particular case of **Quiv**. For the closed topology, j-sheaves will just be graphs with a single vertex, their topos being equivalent to **Set**. Here, the only j-separated object (not also j-complete) is the empty graph.

With the $\neg\neg$-topology, however, the story unfolds differently. The objects of **Quiv** that are $\neg\neg$-separated will be precisely the graphs without parallel arrows, that is, graphs with at most one arrow between each pair of vertices; the $\neg\neg$-complete graphs, for their part, have at least one arrow between each pair of vertices, that is, they are the complete graphs. Finally, then, $\neg\neg$-sheaves are precisely the complete graphs equipped with self-loops (which also forms a topos equivalent to **Set**), that is, a quiver or directed graph X is such a sheaf precisely when there exists exactly one arrow between any pair of vertices. If we denote the full subcategory of $\neg\neg$-separated quivers by $\mathbf{Sep}_{\neg\neg}(\mathbf{Quiv})$, and the further full subcategory consisting of $\neg\neg$-sheaves by $\mathbf{Sh}_{\neg\neg}(\mathbf{Quiv})$, the category of *sheaves* here forms a topos, and a *Boolean* one at that; however, note that $\mathbf{Sep}_{\neg\neg}(\mathbf{Quiv})$ is not quite a topos![230]

In general, for any topos $\mathscr{C}$, the topos of sheaves for the double-negation topology is Boolean, where this means that the law of the excluded middle $\models \phi \vee \neg\phi$ holds. The law of the excluded middle does not hold in general in a topos; it may indeed hold for certain individual formulas ϕ (and even for arbitrary individuals), without this allowing us to infer

229. See Mac Lane and Moerdijk (1994, V.4) for a proof.

230. In fact, it is what is called a *quasitopos*. Instead of having the ordinary subobject classifier of a topos, a quasitopos has a classifer only for *certain* subobjects. Moreover, as one might anticipate given that the separated objects are such a quasitopos, in general a quasitopos satisfies the uniqueness part of the sheaf axioms, but not the existence part. nLab Authors (2017) has a nice discussion of some of the topos features of quivers that we have been exploring, as well as the quasitopos subcategories mentioned above.

that the corresponding universal generalization holds. Those toposes for which it does hold are called "Boolean toposes."[231]

11.3.1 A Glimpse into Topos Logic

Lawvere and Tierney first found that to each topos, one can associate operations that are analogous to the usual logical operations of conjunction, disjunction, negation, implication, and (universal and existential) quantification, defined in a topos $\mathscr{E}$ on its object of truth-values Ω, where we think of the global elements $1 \to \Omega$ as the truth-values, and where the morphisms analogous to the usual logical operations satisfy categorical versions of the laws of a Heyting algebra. Those morphisms induce operations on the collection $\mathrm{Hom}_{\mathscr{E}}(1, \Omega)$ of truth-values, making it into a Heyting algebra. A topos can accordingly be regarded as a category that "models" intuitionistic higher-order logic (or set theory).

Lawvere and Tierney's elementary axiomatization, whereby we may reinterpret the specification of formal covers in terms of what is fundamentally a modal operator on an object of generalized truth-values, further suggests a number of fascinating connections to logic, just a few of which are indicated very sketchily in this section (some taken up more earnestly in the final section of this chapter). To better contextualize these remarks, notice that, in particular, the subobject classifier object in any presheaf topos will support a bi-Heyting algebra structure. This further implies that any such topos will support an infinite hierarchy of intermediate modalities.[232] Moreover, for any sheaf F on a site $(\mathbf{C}, J)$, the lattice formed by all the subsheaves of F also forms a complete Heyting algebra. Another way of thinking about this is that a **cHa** can be realized as a subobject lattice in a Grothendieck topos.[233] These sorts of facts, and many others that we have not explored, suggest that a number of strong connections to logic should be lurking here.

One particular feature of interest is that the laws satisfied by the logical operations admitted in a topos in general—and in particular the general logical laws satisfied by all sheaves—correspond not to classical logic, but to intuitionistic logic.[234] While we can provide a topos-theoretic interpretation of the logical connectives, there are a few subtleties.

231. A nice illustration of the interest in such a distinction can be found in the context of smooth infinitesimal analysis (SIA), mentioned at the beginning of the chapter. SIA is essentially a topos-theoretic approach to analysis based on a new construction of the continuum and infinitesimals via a "smooth real line" object R on which various axioms are placed. For such an R, equipped with a notion of location and a relation "=" of identity or coincidence of locations (i.e., $a \neq b$ is read as "a and b are distinguishable as locations or points"), it is not assumed that the identity relation is *decidable*, where this means that for any a, b, either $a = b$ or $a \neq b$. In some models of SIA, such as *basic smooth infinitesimal analysis* (BSIA), the law of excluded middle is positively refutable. However, in most models of SIA more generally, the law of excluded middle will be true in the restricted sense, namely whenever α is a *closed* sentence (having no free variables), then indeed $\alpha \vee \neg\alpha$ will hold. As Bell notes: "Thus, in smooth infinitesimal analysis, the law of excluded middle fails 'just enough' for variables so as to ensure that all maps on R are continuous, but not so much as to affect the propositional logic of closed sentences" (Bell 2008, 106). This is rather interesting, since the so-called well-adapted models of SIA still do not allow us to go on to infer from the fact that the law of excluded middle applies to arbitrary individuals/points that the corresponding universal generalization—that is, *for all* x and *for all* y in R either $x = y$ or $x \neq y$—holds. Certain elements of R simply cannot always be distinguished, though it does also contain points that can be distinguished; however, on account of the existence of the former, as Bell notes, R *cannot* be thought of as the sum total of its elements.

232. See Reyes and Zolfaghari (1996) and Zalamea (2012).

233. See Mac Lane and Moerdijk (1994, III.8) for details on this.

234. This does not mean that all toposes have an internal logic than is nonclassical, as evidenced by the rather conspicuous case of the topos $\mathscr{E} = \mathbf{Set}$; also, the topos of monoid actions is classical whenever M is a group.

For instance, in order to fully incorporate set-theoretic constructions into the formal language, we need to specify the topos semantics for symbols like $\in$. In this same vein, recall that in topos theory in general there will be a difference between $x \in_{\mathbf{1}} A$ and $x \in_c A$. One basic solution is to use intuitionistic type theory, where different types are provided by the distinct objects of a topos.[235] Complementing the Mitchell-Bénabou syntactic language, Kripke-Joyal provide a well-known semantics for toposes, translating the logician's "forcing" statements about a topos $\mathscr{E}$ into ordinary or naive assertions about $\mathscr{E}$. It turns out that via this semantics, it can be seen that the general logical laws satisfied by all sheaves coincide with those of intuitionistic logic—that is, where laws like the excluded middle, $\phi \vee \neg\phi$, and the law of double negation, $\neg\neg\phi \rightarrow \phi$, fail.

The basic idea of the Kripke-Joyal semantics is roughly this: (1) an individual element b (a constant) of a given type X is just understood as a morphism $b: 1 \rightarrow X$; (2) a k-argument property P of type $X_1 \times \cdots \times X_k$ is to be understood as a subobject P of $X_1 \times \cdots \times X_k$; (3) a k-argument function f from $X_1 \times \cdots \times X_k$ to Y is understood to be a morphism f from $X_1 \times \cdots \times X_k$ to Y; and (4) a term $t(x_1, \ldots, x_k)$ with free variables in the $x_1, \ldots, x_k$ (ranging over the sorts $X_1 \times \cdots \times X_k$) is to be understood as a morphism from $X_1 \times \cdots \times X_k$ to some codomain object, enabling the *interpretation* of this morphism to be defined. In the special case where the topos is the topos of presheaves $\mathbf{Set}^{\mathbf{C}^{op}}$—that is, the category of sheaves on $\mathbf{C}$ for the trivial (Grothendieck) topology—the semantics can be described more evocatively, and in this context it can be shown that the logic of sheaves includes classical logic as a special case. When applied to Grothendieck toposes, the Kripke-Joyal semantics emerges as none other than a sheaf semantics. Moreover, all the *forced formulas* in all the sheaves about a fixed space X will constitute a logic intermediate between intuitionistic and classical, which depends exclusively on the topology of X. The intermediate logic associated to a space X can be seen as a multivalued logic with values in the Heyting algebra on X. It is possible to enhance the logic of sheaves, and thus the intuitionistic logic, with new connectives (including modal operators) not expressible in terms of the usual ones.

There are many fascinating connections to logic that we might naturally further explore at this stage. But a proper development of the formal details would take further chapters of their own. Instead, we will just highlight certain high-level features of this story, leaving the reader to track down logical matters on their own.[236] Philosophically, that intuitionistic logic may be regarded as the "default" logic of sheaves in general can be seen as inviting perspectives that better attune with more continuous logics adequate to *extended things*, where whenever properties hold at a point of its domain of extension, they are expected to hold nearby that point. In this setting, what emerges is a strong emphasis on neighborhood relations over points or punctual properties, in which context classical logic can be seen as a limit or extreme case in a much vaster universe of more "relaxed" logics.[237]

Related to this is the appreciation of the fact that points are simply ideal limits in a more fundamental logic of neighborhoods. The intimate connection between intuitionistic logic and sheaves thus provides another important perspective on the subtle connections

235. For details on the Mitchell-Bénabou internal language of a topos, see Mac Lane and Moerdijk (1994).

236. There is already plenty to work with in Mac Lane and Moerdijk (1994) and Goldblatt (2006).

237. Caicedo (1995) takes up some of the ideas vaguely alluded to in this paragraph.

between continuity, generality, and the failure of the excluded middle, connections already suggested many years ago by Charles Peirce in other contexts:

> If we are to accept the common sense idea of continuity (after correcting its vagueness and fixing it to mean something) we must either say that a continuous line contains no points or we must say that the principle of excluded middle does not hold of these points. The principle of excluded middle only applies to an individual (for it is not true that "Any man is wise" nor that "Any man is not wise"). But places, being mere possibles without actual existence, are not individuals. Hence a point or indivisible place really does not exist unless there actually be something there to mark it, which, if there is, interrupts the continuity. (Peirce 1997, 6.168, marginal note)
>
> The *general* might be defined as that to which the principle of excluded middle does not apply. A triangle in general is not isosceles nor equilateral; nor is a triangle in general scalene. (Peirce 1997, 5.505)

Peirce's intuition that what makes something general is ultimately due to the failure of applicability of the law of the excluded middle, together with his repeated insistence that "continuity" (in some very broad but demanding philosophico-mathematical sense) is ultimately a form of generality, is sharpened in the close connection that emerges between sheaf logic (as the continuous logic of local validity) and intuitionistic logic (in which the law of the excluded middle is dropped).

11.4 Morphisms of Toposes

We have been exploring toposes—both elementary toposes and Grothendieck toposes—and we have begun to see some of the power of these concepts. However, we have not yet defined an appropriate notion of *morphisms of toposes*, so we have not really begun to tap the full potential of the topos construction. In these final two sections, we look closely at these notions and consider a few interesting examples.

11.4.1 Geometric Morphisms Defined

One might suspect that the notion of a morphism of toposes could be captured by a functor that preserves finite limits, exponentials, and the subobject classifier. This indeed defines a legitimate functor between toposes, namely a so-called *logical functor*. Such functors can play an important role in the theory—in particular, from the perspective of an elementary topos as the syntactic category of a higher-order (intuitionistic) type theory. Logical morphisms are those functors that preserve all the structure of the topos (that is, the finite limits and colimits, exponentials and subobject classifier; also "inherited" structure like the Heyting algebra structure of Ω, the validity of formulas, and so on). However, there is another natural type of morphism to consider between toposes: *geometric morphisms*. Viewed in a certain light, the notion of a geometric morphism can even be regarded as the more pertinent of the two sorts of morphisms. In the context of the overall perspective that regards toposes as "generalized spaces," geometric morphisms can be regarded as the corresponding "generalized continuous maps." As such, this sort of morphism can initially be thought of as preserving the geometric structure of toposes (compared to how logical morphisms can be thought of as preserving the elementary logical structure).

Formally, the definition of geometric morphisms between toposes just uses the concept of adjunctions and is rooted in the example of sheaves on topological spaces. We saw earlier how a continuous function $f: X \to Y$ between topological spaces induces a pair of functors—the inverse image functor f^* and the direct image functor f_*. These are such that, in fact, $f^* \dashv f_*$

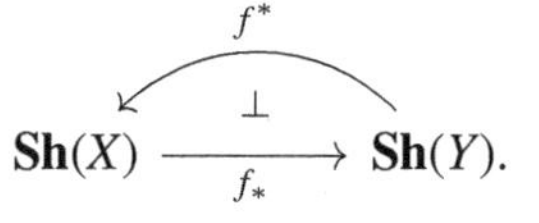

As a left adjoint, f^* clearly preserves (finite) colimits. But, actually, consideration of the definition of the inverse image functor f^* reveals that f^*, while not generally being a logical morphism of toposes (preserving *all* structure), does also preserve *finite limits* (pullbacks and the terminal object), that is, it is *left exact*.[238] This particular situation—where a morphism of toposes is not quite logical (preserving all structure), and yet, in addition to preserving colimits (by virtue of being a left adjoint), certain (finite) limits are preserved—motivates the following definition of a different sort of morphism.

Definition 330 A *geometric morphism* $f : \mathscr{F} \to \mathscr{E}$ between toposes consists of a pair of functors $f^* : \mathscr{E} \to \mathscr{F}$ and $f_* : \mathscr{F} \to \mathscr{E}$ such that $f^* \dashv f_*$ and also f^* is left exact. Modeled on the particular situation described above, f_* is generally referred to as the *direct image* part of f, and f^* the *inverse image* part of the geometric morphism.

If we have two geometric morphisms $f, g : \mathscr{F} \to \mathscr{E}$, a natural transformation

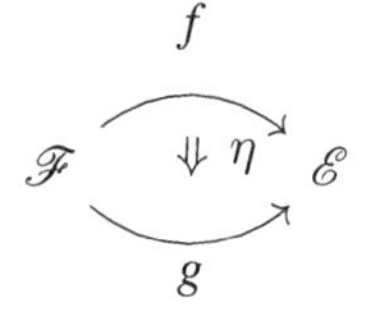

is given by a natural transformation $f^* \Rightarrow g^*$ between the inverse image parts (or equivalently, by adjunction, we could define this in terms of a map $g_* \Rightarrow f_*$ between direct image parts). Toposes together with geometric morphisms and the natural transformations between them forms a 2-category, where the objects are toposes, the 1-cells are the geometric morphisms, and the 2-cells are natural transformations as specified above; but we will not make much use of this formulation.

Finally, a geometric morphism $f : \mathscr{F} \to \mathscr{E}$ is said to be *essential* if the inverse image part f^* also has a left adjoint (in addition to having a right adjoint in f_*, which it does have by virtue of being a geometric morphism), usually denoted $f_!$.

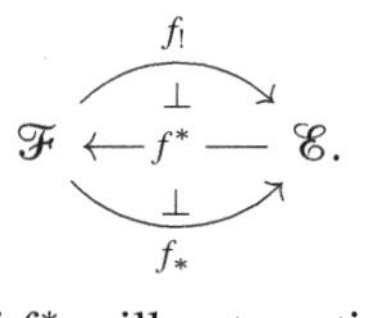

Note that the exactness property of f^* will automatically be satisfied whenever there is such a further left adjoint, since this makes f^* a right adjoint, and we know from RAPL (proposition 175) that right adjoints preserve limits.[239]

238. A proper proof of this fact can be found in Borceux (1994).

239. Following Grothendieck, the asterisk notation is meant to suggest functors that exist for every f, while the exclamation point is meant to suggest functors that exist only for special sorts of f. Moreover, for either of these, the subscript position is meant to indicate functors having the same direction as f, while the superscript position denotes functors going in the opposite direction of f.

We first look at a number of important but rather abstract examples of this notion. In the section that follows, we will look at some more concrete illustrations.

Example 331 We already saw, in motivating the definition of geometric morphisms, how any continuous map $f: X \to Y$ of topological spaces induces a geometric morphism $\mathbf{Sh}(f): \mathbf{Sh}(X) \to \mathbf{Sh}(Y)$ between the sheaves on the spaces. The direct image $\mathbf{Sh}(f)_*$ (or just f_*) is given, for each sheaf F on X, by $\mathbf{Sh}(f)_*(F)(V) = F(f^{-1}(V))$ for each $V \in \mathscr{O}(Y)$. The inverse image $\mathbf{Sh}(f)^*$ (or just f^*), for its part, will act on étalé bundles over Y, sending such a bundle $p: E \to Y$ to the étalé bundle over X (by "pulling back" p along $f: X \to Y$).[240]

Example 332 Generalizing the previous example, any morphism of sites $f: (\mathbf{C}, J) \to (\mathbf{D}, J')$ induces a geometric morphism $\mathbf{Sh}(f): \mathbf{Sh}(\mathbf{D}, J') \to \mathbf{Sh}(\mathbf{C}, J)$, with direct image the functor

$$- \circ f^{op}: \mathbf{Sh}(\mathbf{D}, J') \to \mathbf{Sh}(\mathbf{C}, J).$$

Example 333 Any functor $F: \mathbf{C} \to \mathbf{D}$ between small categories induces a geometric morphism (actually, an *essential* geometric morphism) between the associated presheaf categories

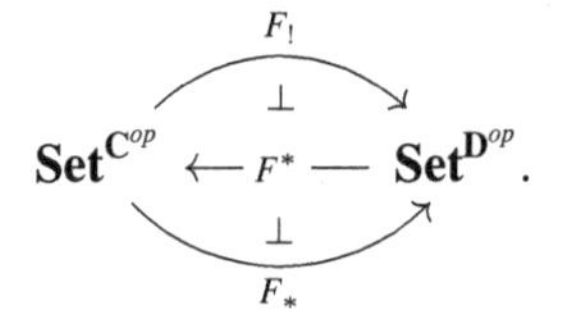

If $\mathbf{C}$ and $\mathbf{D}$ are both categories with finite limits, and $F: \mathbf{C} \to \mathbf{D}$ is left exact, then in the same diagram of adjoint functors, $F_!$ will also be left exact.

Example 334 For $(\mathbf{C}, J)$ a site, the inclusion functor ι of $\mathbf{Sh}(\mathbf{C}, J)$ into $\mathbf{Set}^{\mathbf{C}^{op}}$ is the direct image of a geometric morphism, with the inverse image given by the associated sheafification functor

$$\mathbf{Sh}(\mathbf{C}, J) \underset{\iota}{\overset{\mathbf{a}}{\rightleftarrows}} \mathbf{PreSh}(\mathbf{C}),$$

as described in theorem 291 (chapter 10).

Example 335 Another important example of a geometric morphism, one that we will explore in greater detail in the coming section, is given by the pair (Δ, Γ)

$$\mathbf{Set}^{\mathbf{C}^{op}} \underset{\Gamma}{\overset{\Delta}{\rightleftarrows}} \mathbf{Set},$$

where Δ is the *constant* (or *discrete*) functor, and Γ is the *points* (or *global sections*) functor.

In more detail: recall the *constant* presheaf functor $\Delta: \mathbf{Set} \to \mathbf{Set}^{\mathbf{C}^{op}}$ that, in the above case, will send each set $S \in \mathbf{Set}$ to the constant S itself, and each arrow to the identity morphism on that object, that is, $(\Delta S)(c) = S$ and $(\Delta S)(f) = \mathrm{id}_S$. We can then of course

240. Actually, with a fairly weak assumption on the space Y (namely, that the space is "sober"), via this construction we can actually show a *bijection* between the continuous maps from X to Y and the isomorphism classes of geometric morphisms from the topos $\mathbf{Sh}(X)$ to the topos $\mathbf{Sh}(Y)$. See Johnstone (1986).

consider morphisms between such a constant presheaf and some other presheaf, that is,

$$\mathrm{Hom}_{\mathbf{Set}^{\mathbf{C}^{op}}}(\Delta S, P).$$

But recall that a natural transformation $\Delta S \to P$ will just be a *cone* from the set S to the functor P. In more detail: if we consider an arbitrary presheaf $P : \mathbf{C}^{op} \to \mathbf{Set}$ and for $c \in \mathbf{C}$, the arrows (which are in fact natural transformations)

$$\Delta S \longrightarrow P,$$

we get that a typical arrow in $\mathbf{Set}^{\mathbf{C}^{op}}$ corresponding to these arrows is just a natural transformation, that is, a family of arrows of $\mathbf{C}$,

$$(\Delta S)(c) \xrightarrow{\xi(c)} P(c),$$

indexed by the various objects or nodes of $\mathbf{C}$ and such that

$$\begin{array}{ccccc} (\Delta S)(c) & \xrightarrow{\xi(c)} & P(c) & & c \\ {\scriptstyle (\Delta S)(e)}\downarrow & & \downarrow{\scriptstyle P(f)} & & \downarrow{\scriptstyle f} \\ (\Delta S)(d) & \xrightarrow[\xi(d)]{} & P(d) & & d \end{array}$$

commutes for each such edge $f : c \to d$ in $\mathbf{C}$. But when we apply the functor Δ, this commutative square collapses to the commutative triangle

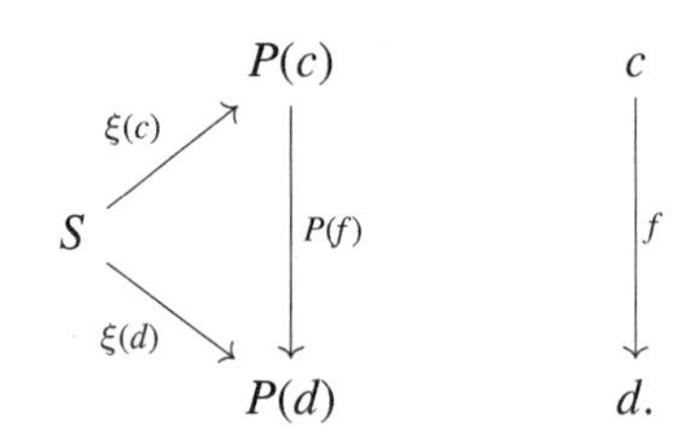

The definition guarantees that whenever the indexing category has composable edges, the corresponding composite triangles commute. The natural transformation $\xi : S \to P$ represented by the triangle gives a cone over P with summit vertex S. Recall also that the *limit* of P is then defined in terms of a universal cone, where a cone $\alpha : L \to P$ with vertex L is universal with respect to P when for every cone $S \to P$, there is a unique map $g : S \to L$ making

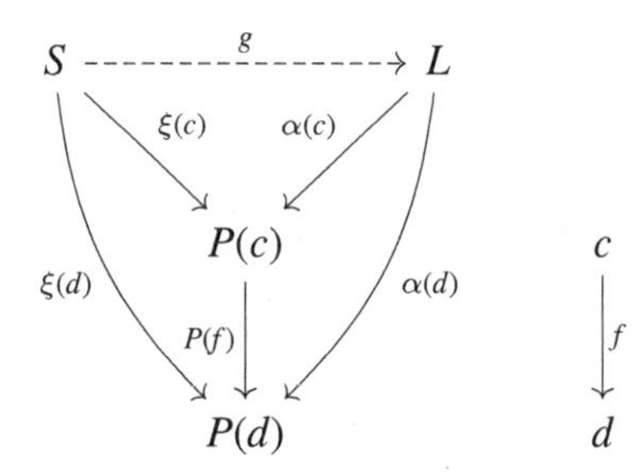

commute. In such a case, one usually refers to the universal cone by just the vertex $L = \varprojlim P$, and calls this the limit of P. Moreover, if every P has a limit in this sense, then the functor Δ has a right adjoint given by the limit.

Suppose, then, that there exists a functor Γ right adjoint to Δ, that is, such that $\Delta \dashv \Gamma$. Then, provided it exists, for each S and P, there will be a natural isomorphism

$$\mathrm{Hom}_{\mathbf{Set}^{\mathbf{C}^{op}}}(\Delta(S), P) \cong \mathrm{Hom}_{\mathbf{Set}}(S, \Gamma(P)).$$

In particular, for $S = \mathbf{1}$, the terminal presheaf (using bold font here to disambiguate from the terminal object 1 of **Set**), we would need

$$\mathrm{Hom}_{\mathbf{Set}^{\mathbf{C}^{op}}}(\mathbf{1}, P) \cong \mathrm{Hom}_{\mathbf{Set}}(S, \Gamma(P))$$

associating to each $\mathbf{1} \xrightarrow{\xi} P$ in $\mathbf{Set}^{\mathbf{C}^{op}}$ an element ξ of $\Gamma(P)$ in **Set**. But, on objects, this forces the definition of Γ to be: $\Gamma(P) = \mathrm{Hom}_{\mathbf{Set}}(\mathbf{1}, P)$; it is also forced on morphisms. Namely, it will be a function γ that assigns to each object c of $\mathbf{C}$ an element $\gamma_c \in P(c)$ in such a way that $\gamma_c|f = \gamma_d$ for $f : d \to c$ holds for every f in $\mathbf{C}$. This makes γ a natural transformation $\gamma : \mathbf{1} \to P$, where $\mathbf{1}$ is just the constant functor on $\mathbf{C}^{op}$! By taking the set $\Gamma(P)$ to consist of all such γ for P, we have constructed the functor $\Gamma : \mathbf{Set}^{\mathbf{C}^{op}} \to \mathbf{Set}$. In this case, for reasons that should be evident, such γ are called *global sections* or *points*, and Γ is the *global sections* or *points* functor. (Moreover, $\Gamma(P)$ is just the limit $\varprojlim P$.)

It is easy to verify that indeed, under such definitions, $\Delta \dashv \Gamma$. Let us sketch how we can appreciate that Δ is, moreover, left exact. To do this, we need to show that Δ preserves pullbacks and the terminal object. But given a pullback

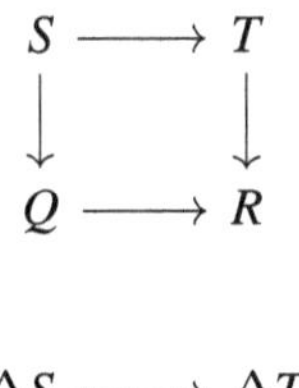

in **Set**, we need to show that

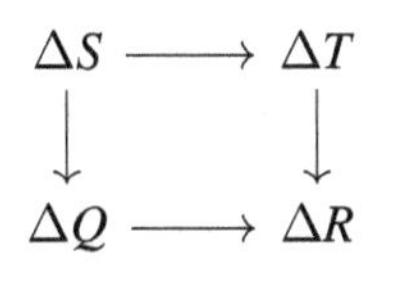

is a pullback in $\mathbf{Set}^{\mathbf{C}^{op}}$. But this is the same as showing that the diagram on objects is a pullback in **Set**; and such a diagram reduces back to the original diagram, on account of how Δ is defined. That Δ preserves the terminal object $\Delta 1 = \mathbf{1}$ is basically immediate.

This sketch suffices to show that the pair (Δ, Γ) forms a geometric morphism.

In fact, we can show that this is really the *only* geometric morphism from $\mathbf{Set}^{\mathbf{C}^{op}}$ to **Set**, revealing **Set** to be a sort of terminal object in the category of presheaf toposes with geometric morphisms between them. Suppose we have an arbitrary geometric morphism (p^*, p_*). Take S a set. Any set can be written $S = \coprod_{s \in S}\{s\}$. As a left adjoint, p^* preserves colimits, so

$$p^*(S) = p^* \coprod_{s \in S}\{s\}$$

and the entity on the right is isomorphic to $p^* \coprod_{s \in S} 1$. As a geometric morphism, p^* must in particular preserve the terminal object, so

$$\coprod_{s \in S} p^*(1) = \coprod_{s \in S} \mathbf{1} = \Delta(S).$$

Thus, we have shown that p^* is isomorphic to Δ. And since adjoints are unique, we must likewise have that $p_* \cong \Gamma$.

In the next section, we will see some particular examples of this geometric morphism, in the context of presheaf toposes that we are already familiar with. We will also see that there is actually a further functor Π left adjoint to Δ, making (Δ, Γ) an essential geometric morphism. In fact, (Δ, Γ) such that $\Pi \dashv \Delta \dashv \Gamma$ is a special case of example 333, taking $\mathbf{D} = \mathbf{1}$. In this connection, it is moreover straightforward to show how from a functor $F : \mathbf{C} \to \mathbf{D}$, the induced diagram of geometric morphisms

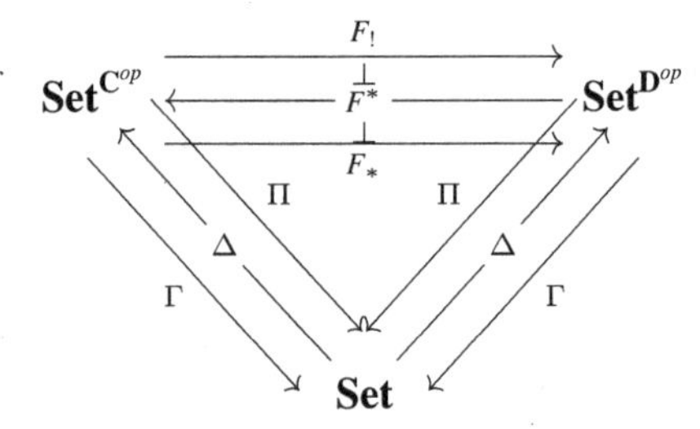

in fact commutes, where this means $F^* \circ \Delta = \Delta$, $\Gamma \circ F_* = \Gamma$, and $\Pi \circ F_! = \Pi$, in effect showing that the pair $(F^* \circ \Delta, \Gamma \circ F_*)$ is itself a geometric morphism.[241]

Example 336 Recall from example 333 how, given a functor $F : \mathbf{C} \to \mathbf{D}$ between two small categories, this functor gives rise to a diagram

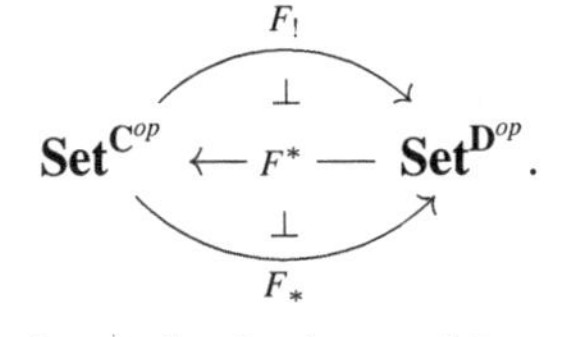

But as presheaf toposes, we can then ask whether and how the subobject objects Ω in these presheaf toposes compare. In fact, the functors $F_! \dashv F^* \dashv F_*$ will induce a further diagram of "internally adjoint" morphisms

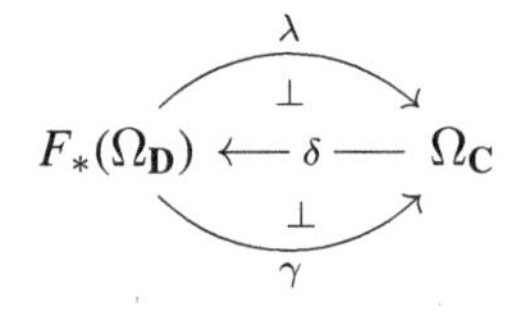

addressing this question.[242]

Moreover, we can then define two further operators $\Diamond, \Box : F_*(\Omega_{\mathbf{C}}) \to F_*(\Omega_{\mathbf{C}})$ by

$$\Diamond = \delta \circ \lambda, \Box = \delta \circ \gamma.$$

It can be shown that these satisfy that $\Diamond \dashv \Box$, and

1. $\Box \leq \text{id} \leq \Diamond$,
2. $\Box\Box = \Box, \Diamond\Diamond = \Diamond$,

241. Reyes, Reyes, and Zolfaghari (2008) contains a demonstration and discussion of this.

242. See Reyes, Reyes, and Zolfaghari (2008, chap.14) for details on the definitions of the adjoints in question.

revealing how we can construct modal operators within topos theory. Defined in this way, it has been suggested that to get the theory of modalities off the ground one might start by thinking of $\mathbf{Set}^{\mathbf{D}^{op}}$ as providing the background standard of constancy, taking $\mathbf{Set}^{\mathbf{C}^{op}}$ as the locus of change and modal qualifications.[243]

11.5 Toward Cohesive Toposes

In physics and other fields like materials science and even architecture, one will occasionally hear the tendency or force holding certain types of matter together referred to as their *cohesion*. The study of the various properties or qualitative features of material bodies, in their mutual interactions, reveals that bodies can have different sorts or amounts of resilience, elasticity, brittleness, bounciness, stiffness, flappiness, twistability, capability of creasing, "creep" (e.g., the distortion of a bow under prolonged loading; creep in textiles is why the knees of pants get baggy), and so on.

In a different context, the philosopher Hegel sought to understand, in a very general fashion, what makes determinate things *determinate*. In his "philosophy of nature," this amounted to inquiries into the main "general forms" describing the distinctive ways material bodies in particular can be seen to become determinate. Among such general forms was what Hegel called their *cohesion*. Hegel thought that, over and above the usual mechanisms of dynamic forces of attraction and repulsion, gravitational attraction, and chemical reactions, there are various ways that certain material bodies can be found to achieve determinacy essentially by "holding together" in a distinctive way. Hegel accordingly sought to advance a more refined concept of cohesion, as opposed to what he called "the common understanding of cohesion," which just refers to a particular snapshot of the "quantitative strength of connection between the parts of a body."[244] For Hegel, cohesion is a more fundamental phenomenon, even more fundamental than shape or density of a body, in that it does not even dictate any particular shape—and even if cohesiveness is fundamentally about the relations between the parts of a spatially-extended body, this is not something the analysis of which could be reduced to uniform geometrical considerations of the "mere outer shape" of a body. While in certain instances, cohesion will involve how a body retains a specific spatial configuration or shape, in general cohesiveness can be considered independently of shape. On this approach, cohesiveness is something like the most general form of a number of more particular processes and phenomena found throughout nature, all involving some "inward determination" of a material body that reveals itself only in physical interactions with other material bodies, in which interactions one can find a distinctive way of resisting being broken apart by forces of another body.

In exploring sheaves from various angles, and at various levels of generality, we have in a sense been exploring various sophisticated modifications of our notions of space, and making use of various "generalized spaces." In the (pre)sheaf context, the domain category from which we map our shapes of a certain figure has been seen to supply the form or blueprint according to which certain parts of a certain type are to hold together. How do the parts or points of any space hang together or cohere? Intuitively, certain toposes can

243. Details on this, and the perspective just mentioned, can be found in Reyes and Zawadowski (1993) and Reyes and Zolfaghari (1991).

244. Georg Wilhelm Fredrich Hegel (1991, 241).

appear to have a rather involved cohesiveness, while others (like **Set**) can appear to be almost devoid of cohesiveness. Informally, what different "spatial" categories seem to have in common is something like this feature of *cohesiveness*, which they may have in different ways and to differing degrees. Certain maps between toposes may then function to reveal comparisons in different degrees or types of cohesiveness.

The notion of a *cohesive topos* (or, sometimes, a *category of cohesion*)—first explored by Lawvere—basically emerged in an attempt to axiomatize those intuitively "cohesive" properties of a topos that would make it a setting in which some sort of generalized geometry (wherein familiar notions of geometry resurface in subsumed form) could take place. While this effort is still somewhat in its infancy, the reader may enjoy exploring some of these otherwise advanced notions through a slew of examples. In terms of the formalism, all one really needs to approach this is a good working sense of adjunctions and a basic understanding of toposes. With this, some accessible motivating examples should be enough to give the reader a small glimpse into this fascinating and admittedly somewhat mysterious new area of research.

11.5.1 Foray into Cohesive Toposes

We have been using the word "geometry" a lot recently, while suggesting that we intend this in a rather general way. But what sorts of things do we meet in our usual geometries? Among other things, a very primitive notion is that of a *point*. Euclid, for instance, thought it was so important that he included a definition of it as the very first sentence of the first part of his *Elements*. In category theory, we see points in terms of the trivial *terminal* category (single object and single morphism), and functors from such a category are used to pick out the objects or "elements" of any category.

Suppose you have a certain topos $\mathscr{M}$ that supports some degree of cohesion or activity (e.g., dynamical systems, i.e., a particular category of presheaves).[245] Suppose you have another topos $\mathscr{K}$ seemingly devoid of any internal cohesion and variation (e.g., the usual category of sets). We can similarly define a *point* of $\mathcal{M}$ as a geometric morphism

$$\mathcal{K} \xrightarrow{p} \mathcal{M}.$$

This terminology is entirely sensible, especially if we regard $\mathcal{M}$ as some presheaf topos, and $\mathcal{K}$ as **Set**, for **Set** acts as a sort of terminal object in the category of presheaves. Recall from the definition of a geometric morphism that such a p is then really a pair $p = (p^*, p_*)$

$$\mathbf{Set} \underset{p_*}{\overset{p^*}{\underset{\longrightarrow}{\overset{\longleftarrow}{\bot}}}} \mathbf{Set}^{\mathbf{C}^{op}}$$

with $p^* \dashv p_*$ and p^* left exact (i.e., preserving finite limits: preserving the terminal object $\mathbf{1}$ and pullbacks). Recall also that given two points p and q, a morphism from p to q will just be given by a natural transformation $p^* \Rightarrow q^*$ (and these will be bijective with those transformations between q_* and p_*). Thus, together with this notion of morphism, the points of $\mathbf{Set}^{\mathbf{C}^{op}}$ actually form a category $\mathbf{Pts}(\mathbf{Set}^{\mathbf{C}^{op}})$.

245. The ideas discussed here began with Lawvere. See, for instance, the engaging papers Lawvere (1994a, 1994b, 1996).

One can show that, for instance, **Pts(Set)** is just **1**, while **Pts(Grph)** is just the diagram with two vertices and the two nontrivial morphisms between them. One can also show that, for instance, $\mathbf{Pts}(\mathbf{Set}^{\mathbf{FinSet}^{op}})$ is isomorphic to **BoolAlg**, the category of Boolean algebras.

Example 337 Let's look closer at a rather simple example of this, namely **Pts(Set)** = **1**. Then

$$\mathbf{Set} \underset{p_*}{\overset{p^*}{\underset{\longrightarrow}{\overset{\longleftarrow}{\perp}}}} \mathbf{Set}^{\mathbf{1}^{op}} \simeq \mathbf{Set}.$$

As a geometric morphism, p^* must be exact. So we will have

$$p^*(A) = p^*\left(\coprod_{a\in A} 1\right) = \coprod_{a\in A} p^*(1) = \coprod_{a\in A} \mathbf{1} = A.$$

But then the adjoint relation $p^* \dashv p_*$ entails that $p^* = \mathrm{id} = p_*$, thus showing that the points of **Set** are just **1** itself.

Moreover, since we saw in the previous section that (Δ, Γ) was the only geometric morphism from **Set** to the presheaf category on **C**, the above just tells us that for $\mathbf{Set}^{\mathbf{1}^{op}} \simeq \mathbf{Set}$, $\Delta = \mathrm{id} = \Gamma$.

For concreteness, let us look at how Δ and Γ work in more particular cases.

Suppose you have the bouquet X

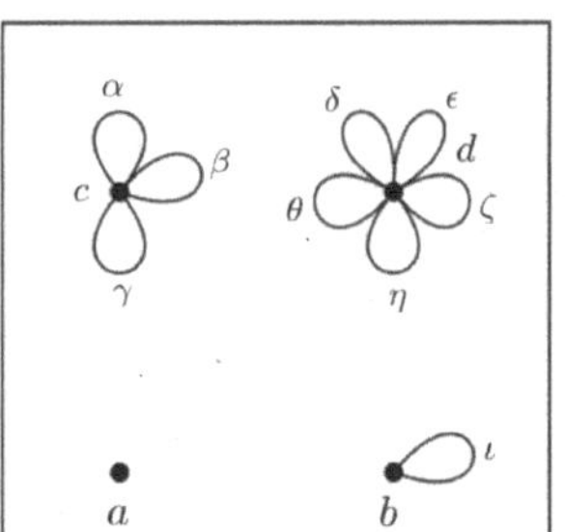

This is, of course, an object in $\mathbf{Set}^{\mathbb{B}^{op}}$ (which is the same as **Bouq**). What is ΓX? This will be the set whose elements are given by the loops of X, that is, $\Gamma(X) = \{\alpha, \beta, \gamma, \delta, \epsilon, \zeta, \eta, \theta, \iota\}$. If we had instead taken Y a trivial sort of bouquet, consisting of just vertices and no loops, then $\Gamma(Y)$ would just be the empty set. What about Δ? This functor goes from **Set** to $\mathbf{Set}^{\mathbb{B}^{op}} \simeq \mathbf{Bouq}$, so we have to consider how it acts on sets. Suppose S is a four-element set. Δ preserves the terminal object and colimits, so $\Delta(1+1+1+1) = \Delta(1) + \Delta(1) + \Delta(1) + \Delta(1) = \mathbf{1} + \mathbf{1} + \mathbf{1} + \mathbf{1}$, with $\mathbf{1}$ the terminal object of bouquets (the trivial bouquet). But this means that $\Delta(S)$ should just be the bouquet that consists of four vertices, each with a single loop. The adjunction (Δ, Γ) then tells us that it will be the same to look at the functions from a set S to ΓX as it will be to look at the bouquet homomorphisms between $\Delta(S)$ and a bouquet X.

If we instead took **Grph** or $\mathbf{Set}^{\mathscr{G}^{op}}$, Δ will take a set to the graph with as many directed loops (arrows with source and target the same) as there are elements in the set, while Γ will take a graph to the set having for elements the directed loops of the graph.

Example 338 Recall our earlier discussion of the *qua* category from examples 38 (chapter 2) and 196 (chapter 7). There, an interpretation X of $(\mathbf{A}, \mathscr{P})$ was just an object of the presheaf topos $\mathbf{Set}^{\mathbf{A}^{op}}$ together with a set of subobjects that correspond to the predicates in $\mathcal{P}$. One can see that the "global aspect" G is just the terminal object of **A** and we obtain the restriction of the functor X to this aspect G by applying the *global sections* functor Γ of the geometric morphism given by the pair (Δ, Γ):

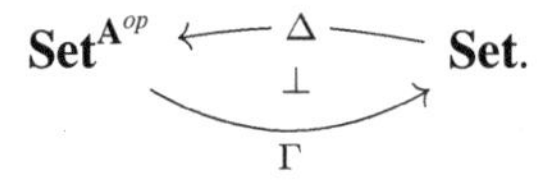

Because **A** has a terminal object, it follows that $\Gamma\Delta = \text{id}$, and that Δ is full and faithful. Moreover, since an interpretation assigns the same set to all the aspects, in the above light, this is just to say that the interpretation X is a constant presheaf. Putting these two things together, we can see that every morphism between two constant presheaves (different interpretations) will be a constant morphism. The functor Γ above basically *forgets* the aspects of X, returning just $\Gamma(X) = X(G)$ or the set of individuals (of the relevant kind). Moreover, for a predicate ϕ of X, the functor Γ, acting on ϕ, returns just a set-theoretic predicate, or subset, of $\Gamma(X)$, where this corresponds to the restriction of ϕ to the level of the global aspect. Altogether, interpretations X appear as constant objects in the presheaf topos $\mathbf{Set}^{\mathbf{A}^{op}}$ or as sets of individuals living in **Set**. These basically amount to two different points of view on X, where the constant objects of $\mathbf{Set}^{\mathbf{A}^{op}}$ have a rich structure, while the logic of **Set** is Boolean. The global judgment is carried out in **Set**, on the right of the above diagram, where Boolean logic is the rule. On the left, we have the richer logic supplied by the category that interprets the *qua* subcategory **A**.[246]

We have been exploring a little of the presheaf topos $\mathbf{Set}^{\mathbf{A}^{op}}$, for which every object X in this topos has a bi-Heyting structure. We also saw earlier how to interpret possibility and necessity operators in this context. But the logic of global judgments occurs in the Boolean setting of **Set**. Γ acts to take an X to **Set** where the predicates of $\Gamma(X)$ have the structure $(\mathcal{P}(\Gamma(X)), 0, 1, \wedge, \vee, c)$, where $c = \sim = \neg$ (c being the complement in **Set**), and where necessity and possibility operators reduce to the identity. While Γ preserves the action of $0, 1, \wedge, \vee, \sim$, and $\Box$, it does not preserve $\neg$ or $\Diamond$. In particular, $\Gamma\neg \neq c$, yielding a new operation $\Gamma\neg$ on $\mathcal{P}(\Gamma(X))$ which is not Boolean, and which acts as a kind of strong negation. As for $\Diamond$: a predicate may hold under some aspect, without being true at the global level. Thus, $\Gamma\Diamond$ also supplies yet another new operation on $\mathcal{P}(\Gamma(X))$. Both $\Gamma\neg$ and $\Gamma\Diamond$, mirroring their corresponding operations in $\mathbf{Set}^{\mathbf{A}^{op}}$, enrich the structure discussed, to give $(\mathcal{P}(\Gamma(X)), 0, 1, \wedge, \vee, c, \Gamma\neg, \Gamma\Diamond)$, a structure that, Reyes and Zolfaghari (1996) has suggested, provides a useful setting for global judgments in such situations. On the other hand, the rich general setting of $\mathbf{Set}^{\mathbf{A}^{op}}$ further suggests how we might model the patching together of local judgments regarding the applicability of certain predicates to a person *qua* the different hats they wear, or roles they play in life.

More generally, one can think of the result of applying the constant functor Δ as producing a subcategory of discrete spaces. In other words, Δ shows us, within the presheaf category (characterized by some sort of cohesiveness), what a space of *no cohesion* looks like. How do we be more formal about this?

246. The topos-theoretic features of this example are explored in further detail in Reyes and Zolfaghari (1996).

Recall that if there is a further left adjoint to p^*, then we call the geometric morphism p *essential*. The geometric morphism (Δ, Γ) will be essential provided there exists a further functor Π such that $\Pi \dashv \Delta \dashv \Gamma$. But for presheaf toposes, as we have been exploring, there is precisely such a functor Π, sometimes called the *connected components* functor. For any presheaf topos, we have

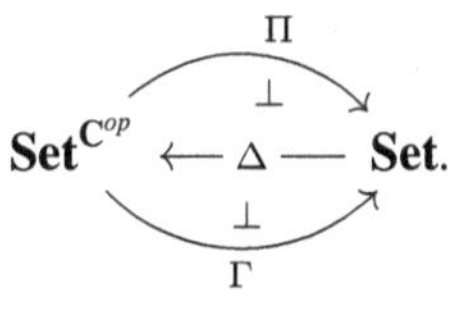

Recall that Γ was actually the *limit* of a functor. The above is in fact a restatement of a rather elementary result which has $\Pi = \text{colim}$ and $\Gamma = \lim$. The definition of Π is forced by the fact that, as a left adjoint, it must preserve colimits (the gluings). By considering particular categories, such as the category of graphs, it is easy to see how Π just picks out the connected components. In other contexts, this functor can also be thought of as assigning to each object in the cohesive topos $\mathscr{M}$ the cardinal representing the number of maps it supports into discrete sets.

We mentioned a moment ago that Δ in such a setup shows us what an extreme space of no cohesion looks like. This suggests that perhaps there exists a further functor from $\mathscr{K}$ back to $\mathscr{M}$ that—at the other extreme of the result of applying the discrete functor—would yield a space of total (or infinite or trivial) cohesion. As it turns out, in certain cases, there does indeed exist such a functor, where this is right adjoint to Γ, as in

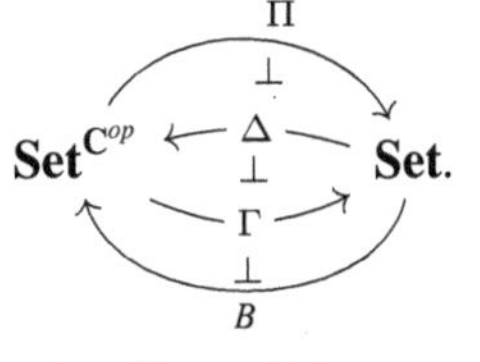

In these settings, a functor such as $B : \mathscr{K} \to \mathscr{M}$ is sometimes called the *chaotic* or *codiscrete* functor. The chaotic space produced by applying such a functor can be regarded as so "extremely cohesive" that, in moving a point to any other point, one need not concern oneself with the constraints put on the category by how the motion is parameterized (or the cohesion determined).[247] Following a suggestion of Lawvere, one might say that points in a discrete space are *distinct*, but points in a chaotic space are *indistinguishable* provided the chaotic spaces are connected.[248]

One can also show that if such a B exists, then the unit $\text{id} \to \Gamma\Delta$ of the adjunction $\Delta \dashv \Gamma$ is an isomorphism, and the counit $\Gamma B \to \text{id}$ of the adjunction $\Gamma \dashv B$ is also an isomorphism.[249]

For concreteness, when $B : \mathbf{Set} \to \mathbf{rGrph}$, landing in the category of reflexive graphs, $B(S)$ will yield a vertex for each of the elements of S and for each pair (x, y) of elements of

247. It is a fact that the functor Γ has a right adjoint B iff every representable (or generic figure) has a point.

248. See Lawvere (1994a).

249. A proof can be found in Reyes, Reyes, and Zolfaghari (2008). This reference also discusses further important consequences of this, namely the fact that the functor B will exist when the presheaf category satisfies a certain condition (that every nonempty presheaf P has a point $\mathbf{1} \xrightarrow{p} P$).

the set S, the functor will yield a unique arrow with source x and target y. In other words, $B(S)$ just describes the (directed) *complete graph* on the elements of the set S. In particular, suppose we have the set S consisting of four elements, which we can represent as a bag of four dots. Then $B(S)$ will be the graph with two arrows between each pair of vertices (one coming in and one going out), plus a single directed loop stationed at each of the four vertices.

Thus, more generally—that is, for certain special toposes $\mathscr{M}$ and $\mathscr{K}$—we can construct the adjoint quadruple diagram:

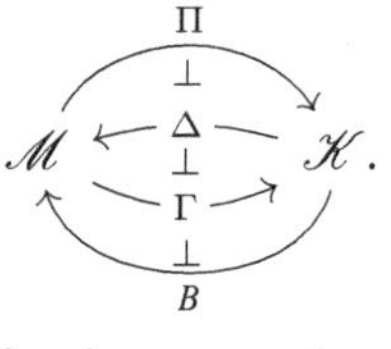

For an $\mathscr{M}$ where variation/dynamics is more relevant than cohesion, such as for the presheaf topos of dynamical systems (evolutive sets), the same adjunctions are sometimes called orbits $\dashv$ stationary points $\dashv$ equilibria $\dashv$ chaotic.

In the examination of such situations and what happens when we try to further compose such adjunctions, Lawvere noticed some curious things, which we look at more closely in the next example.

Example 339 Restricting our attention just to the adjoint functors $\Delta \dashv \Gamma \dashv B$ (defined as in the preceding example), we note that we can consider maps from the composite $\Delta\Gamma(M)$ to some M in $\mathscr{M}$ as well as maps from M to the composite $B\Gamma(M)$, that is,

$$\Delta\Gamma(M) \longrightarrow M \longrightarrow B\Gamma(M).$$

The space $\Delta\Gamma(M)$ on the left can be thought of as the closest approximation to M from the left given only its cardinality (points), while the space $B\Gamma(M)$ can similarly be thought of as its approximation from the right. For each object in the domain of the points functor, these maps provide an interval between which the object must lie, the endpoints being the two opposite subcategories, an interval that is in some sense relative to what the specific points functor does to the object on which it acts. Now, if we apply the points functor Γ again, we get a sequence of isomorphisms (in $\mathscr{K}$)

$$\Gamma\Delta\Gamma(M) \xrightarrow{\cong} \Gamma(M) \xrightarrow{\cong} \Gamma B\Gamma(M).$$

But this just says that even though the two composite maps are in general not isomorphisms in $\mathscr{M}$, applying the points functor yields an isomorphism of cardinals (in $\mathscr{K}$). The cardinal $\Gamma(M)$ or points(M) associated to a given M is at once isomorphic to the cardinal associated to the space $\Delta\Gamma(M)$ and to the cardinal associated to the space $B\Gamma(M)$. However, in the case of $\Delta\Gamma(M)$, all the points will be distinct, while in $B\Gamma(M)$ all points will be indistinguishable. In other words, we may have a definite number of points; however, via the unifying isomorphism such points will be *indistinguishable by any property*.

Lawvere (1994a) remarks how this curious situation appears to precisely capture the apparent paradox (first isolated by Cantor) that in an abstract set, all elements are distinct yet indistinguishable.[250]

The basic idea, then, is that $\mathscr{M}$ contains two opposed subcategories (the discrete and codiscrete objects) that, though inherently rather distinct, are rendered identical through the category $\mathscr{K}$. In more detail, the "unity of opposites" is precisely expressed through

$$\Gamma\Delta = \mathrm{id}_K = \Gamma B.$$

Lawvere interprets this "productive inconsistency" of having a definite number of points without these points being distinguishable by any property, and the underlying "unity of opposites," in terms of Hegelian dialectics. The basic idea is captured by the following diagram of natural transformations between the composite counit and unit functors:

$$\text{opposite}_1 \longrightarrow \text{unity} \longrightarrow \text{opposite}_2.$$

This situation is most clear in the particular case of the topos of reflexive graphs. In this context, Lawvere (1996) observed that while the notions of discrete and codiscrete are dual there, the full subcategories of discrete graphs and codiscrete graphs are each equivalent to the category of sets, and moreover, both the discrete and codiscrete graphs are identical when regarded in the category of all graphs. In such a situation, it seems entirely natural for Lawvere to have spoken of such adjunction pairs as embodying Hegel's idea of the unity and identity of opposites. In the case of such configurations in general, Lawvere speaks of *adjoint cylinders*, where the three functors involved are adjoint and the two composites are isomorphic to the identity in $\mathscr{K}$. Put otherwise: we have such an adjoint triple where there are two parallel functors that are adjointly opposite in that they are full and faithful, and moreover there exists a third functor that is left adjoint to one of them and right adjoint to the other functor; as subcategories included in the ambient category, they are opposite, but by neglecting the inclusions they are identical. In more detail, in the particular case of graphs, the category of sets gets embedded in the category of graphs through the action of the functor Δ (producing discrete graphs) and the action of the functor B (producing codiscrete graphs). These notions are dual; however, the resulting full subcategories of discrete graphs and codiscrete graphs are further equivalent to the category of sets, thereby yielding the relevant identity. From the perspective of $\mathscr{M}$, the discrete and the codiscrete are united; looked at from the other end, $\mathscr{K}$ is identified with $\mathscr{M}$ via inclusion of subcategories in two opposite ways. One might also think of this in terms of Hegel's discussion of *quantity* as the dynamic unity of the "moments" of discreteness and continuity.[251]

To make these ideas a little more concrete, consider the following.[252] If we set both $\mathscr{M}$ and $\mathscr{K}$ as the poset of natural numbers $\mathbb{N}$ (viewed as a category),[253] we can construct the two parallel functors $E, O : \mathbb{N} \to \mathbb{N}$, defined by $E(n) := 2n$, $O(n) := 2n + 1$, that is, the "even"

250. While there exist toposes with few connected objects such that for all $K <$ measurable cardinal we have that $\Delta(K) = B(K)$, in general these discrete and codiscrete maps are not equivalent.

251. See Georg Wilhelm Friedrich Hegel (2010, bk. 1, sec. II, chap. 1: "Quantity").

252. This very simple, but pedagogically useful, example of the adjoint cylinder construction is inspired by Lawvere (2000). The following exposition also follows nLab Authors (2018a, 2018b).

253. Technically such categories are not even toposes. The reader who cannot see why is invited to revisit the earlier section introducing toposes and try to see what makes them not qualify as toposes.

and "odd" functions. These functors obviously act to produce the two subcategories of $\mathbb{N}$, $\mathbb{N}_{even}$ and $\mathbb{N}_{odd}$, which is another way of saying that both functors are full and faithful. Such subcategories clearly are "opposed" to one another, at least in the sense that $\mathbb{N}_{even} \neq \mathbb{N}_{odd}$; however, performing the same sort of simple composite applications as above, we can produce the bijection $\mathbb{N}_{even} \xrightarrow{\cong} \mathbb{N}_{odd}$, through which they can be viewed as "identical." As subcategories, both can be seen to be united as the opposing parts of the containing category $\mathbb{N}$, in relation to which, by virtue of each being isomorphic to one another through their isomorphic maps to $\mathbb{N}$ itself, they are rendered identical. There indeed exists a third functor $T : \mathbb{N} \to \mathbb{N}$ that, together with E and O, will form the appropriate adjoint triple:

$$\mathbb{N} \xrightarrow{T} \mathbb{N}, \quad E \dashv T \dashv O \quad (E, O : \mathbb{N} \to \mathbb{N})$$

By definition of adjunctions, and given that we are working with posets, $E \dashv T$ and $T \dashv O$ just means that $E(n) \leq m$ iff $n \leq T(m)$ and $T(n) \leq m$ iff $n \leq O(m)$. Moreover, as long as T exists, we will further have that $TE = \text{id} = TO$, which just means in our particular case that $T(2n) = n$ and $T(2n+1) = n$, which forces the following piecewise definition of T as

$$T(k) = \begin{cases} \frac{k}{2} & \text{if } k \in \mathbb{N}_{even} \\ \frac{(k-1)}{2} & \text{if } k \in \mathbb{N}_{odd}. \end{cases}$$

The idea here is that the adjoint triple $E \dashv T \dashv O$ at the same time embraces the *identity* by the natural isomorphism $TE \xrightarrow{\cong} TO$, the *opposition* by the induced adjunction $ET \dashv OT$, and the *unity* by the idempotent relations $(E \circ T) \circ (E \circ T) = E \circ T$ and $(R \circ T) \circ (R \circ T) = R \circ T$. In this way, the middle functor T can be thought of as simultaneously identifying, opposing, and uniting E and O.

In the case of the category of presheaves on $\mathbf{C}$, Δ yields the discrete presheaves on $\mathbf{C}$ while B yields the codiscrete presheaves on $\mathbf{C}$, and Γ is their common projection. With such a setup, it is always the case that $\Gamma\Delta \cong \text{id} \cong \Gamma B$. In the particular context of presheaves, Lawvere takes the Hegelian notion of *Aufhebung* still further to yield a theory of dimension or "levels."[254] Roughly, a level is a functor from a given category into one that is "smaller," that moreover has both left and right adjoint sections which produce subcategories that in themselves are identical (in the smaller category) but that include themselves as subcategories in opposite ways (and which, moreover, give rise to the two composite idempotent functors on the given "larger" category). More specifically, given an adjoint cylinder situation between toposes, a *level* of a topos is defined as the inclusion of the right adjoint in this setup. In Lawvere's approach, again taking inspiration from Hegel, the *Aufhebung* of a level will be the smallest level that acts to resolve the component opposites (the opposing functors).[255] It is not the case that such an "*Aufhebung*-like" level always

254. See Lawvere (2002).

255. For more details on this theory of levels, see Lawvere (1989, 1996), and Kennett et al. (2011). In the toposes of simplicial sets, cubical sets, and reflexive globular sets—each definable in terms of presheaf categories—levels coincide with the notion of dimension.

exists for any given level, however in particular cases of special relevance to us (such as presheaf toposes over graphic categories), it does exist.[256]

This discussion could easily be extended in a number of directions, leading to a more thorough exposition of the notions and applications of cohesive toposes, in which context further adjoint cylinders arise, including situations relating to infinitesimally generated spaces; however, we leave curious readers to pursue these more advanced matters on their own.[257] For now, we content ourselves with observing that one sometimes finds further adjoint cylinders embedded in between $\mathscr{M}$ and $\mathscr{K}$ via intermediate categories $\mathscr{L}$ that are less "abstract" than **Set** but with a simpler sort of cohesion than $\mathscr{M}$:

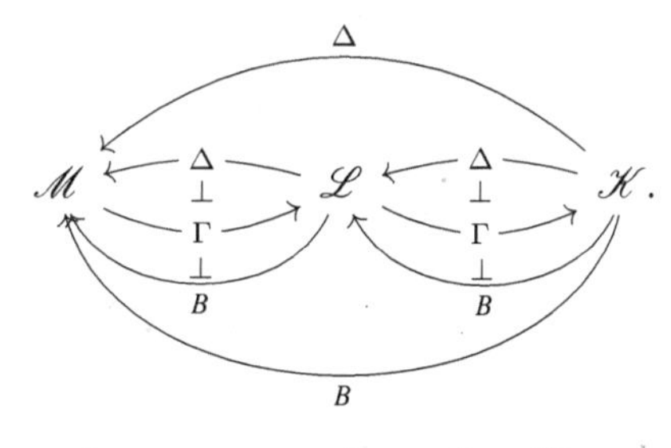

The basic idea here is that various toposes such as $\mathscr{M}$ or $\mathscr{L}$ are contrasted with the extreme case of $\mathscr{K}$ via geometric morphisms; but the diagram above suggests that we extend this to consider intermediate toposes and chains of maps between such adjoint triples, with the effect that the various determinations of cohesivity or variation in toposes themselves can be compared.

11.5.2 Philosophical Pass: A New Dialectical Science?

Box 11.1
A New Dialectical Science?

An object that arises in a "spatial" category (a category with some cohesion or variation) can be examined via *levels*, constructions that provide a precise formulation of the unity and identity of opposites so characteristic of the (originally vaguely formulated) philosophical concept of dialectics, making higher-order relations between general forms of cohesiveness amenable to more exact solution. In the above exposition, we even saw how, from one perspective, the moments of discreteness (in the form of zero cohesion) and continuity (in the form of total cohesion) could be unified. However, this was a rather extreme case. By considering intermediate cylinders and passages between adjoint triples (or quadruples) of various toposes, we move beyond the case of relating a single category to the extreme case of discreteness or constancy (as in $\mathscr{K}$). We can now examine sequences of intermediate categories that are interlocked via cylinder maps of their own, opening onto a more refined dialectical science of cohesion, by which ultimately one could systematically characterize and compare the differing properties of cohesion and variation that emerge in certain universes or models

256. See Lawvere (2002) for details. Informally, "graphic categories" can just be thought of as certain simple enough categories that allow for finite graphic display or presentation once one constructs their corresponding presheaf category.
257. See, for instance, Lawvere (2007).

for mathematical theories that treat of objects with some dynamics. Using such intermediate adjoint cylinders, the quality of dimension or level in spaces can be compared. (See Lawvere [1994b] for more details.)

At a fundamental or more philosophical level, it could be argued that what is going on here is that the question of what is more variable, more cohesive—or how various models for certain mathematical theories dealing with dynamical phenomena or settings of cohesiveness are differently variable or cohesive—should involve functorial comparisons. The further suggestion could be made that, in addition to examining such sequences of adjoint cylinders as shown above, through the discovery and study of left-exact functors between two toposes (which may not have adjoints) we should be able to make even more precise the notion of greater or lesser discreteness (noncohesion or constancy) versus continuity (cohesion or variability). In a sense, this latter suggestion is a natural extension of the perspective of the categorical notion of continuity via (co)continuous functors (defined as preserving limits, and not just those that are finite). Going somewhat further, we could perhaps make use of such functors to begin to construct general metrics measuring something like "what it would cost" to make an "almost-adjunction" an honest adjunction; another use of the resulting metric might be to begin to more precisely analyze how far we are from the more extreme or trivial settings of infinite cohesion (continuity) on the one hand and zero cohesion (discreteness) on the other. This sort of approach should open onto a much richer terrain of dialectical subtleties, and it has the potential to provide a powerful weapon in the ever-frustrating yet beguiling dialectic of the continuous and the discrete at the heart of so much of mathematics and human thought.

Appendix (Revisiting Topology)

In which we continue the thread from chapter 4, introducing the reader to certain connections between (modal) logic and general topology, using this to tell one possible story that can address the three questions raised at the end of chapter 4, and finally where we briefly consider how this story might relate to the bigger story we have been telling, throughout the book, about the rethinking of space.

Returning to the realm of general (point-set) topology: recall how, as we ascended in generality, leaving the notion of distance behind, the resulting notions of the open (closed) sets of a topology really just capture requirements on the interrelations of the subsets of a given set, requirements essentially governing how new members of a collection of subsets can be built from old ones (through intersection and union). In the usual treatment of these matters in general topology, the defining axioms—stability under finite intersection and stability under arbitrary union—are often left somewhat inscrutable. The aim of this appendix is to begin to lift the veil on such matters by introducing the reader to some notable connections with logic, and using this to tell a story that helps us get a better handle on the "meaning" of the axioms.

To get a better sense of what we are doing here, briefly consider the history of aviation. Consider how, for centuries, early aeronautical engineers attempted to build airplanes by looking to birds and bats, mimicking their flapping wings, general shape, and way of flying. By closely observing such creatures, they came to believe that the secret to manned flight was going to be found in devising complex machines that closely mimicked the flapping-wing techniques of birds. On the whole, these attempts failed, and airplanes achieving sustained flight were long deemed impossible. As an improved understanding of the general principles of aerodynamics was gradually attained, it would occur to people that the apparently essential feature of a bird's wings—the flapping of its wings—could be abandoned. Of course, by dropping this feature, and constructing fixed-wing aircraft, sustained long-range load-carrying flight became possible. The rest is history.

Similarly, it was realized by early "explorers" in analysis that we can drop the apparently essential notion of distance and retain a workable notion of open sets—in fact, achieving an even more powerful notion, capable of accommodating new constructions and broadening our treatment of continuity. But just as a fixed wing is still a wing despite not flapping, the introduction of open sets that no longer deploy a notion of distance—and the attendant general definition of continuity—raises a number of questions concerning such a construction and its relation to the old approach. Following this idea, the three questions we introduced in chapter 4 present themselves. To continue the analogy, such questions remind us of how, despite all the advances in aerodynamics, there apparently remains little agreement on what generates the aerodynamic force known as lift, that is, what keeps things in the air![258] Likewise, there is a surprising lack of explicit awareness or agreement among mathematicians regarding why the defining features of a (general) topology are what they are. Most mathematicians will not hesitate to point to the metric space setting as motivation for the more general notion of a

258. See, for instance, Regis (2021).

topology; and they will be able to point to instances exhibiting the power of the more general notion and show off its utility in capturing nonmetrizable spaces; but it is rare to find any general story from topologists about what is so special about the particular axioms used to define a topology: namely, that the members of the collection should be stable under arbitrary union and under finite intersection.

The aim of what follows is to explore some aspects of an account, and connections to matters beyond topology, that might help clarify why the decisive defining properties of a topology are what they are. Exploration of this question will get at a number of related issues, such as why there is this size asymmetry, what this may reveal about what general topology is about, and why open sets typically seem to be preferred to closed sets (and whether this is even conceptually justifiable or just a historical accident). Answers to these questions appear to have something to do with characterizations and assumptions about the nature of continuity, but one can also argue that they extend to deep connections with logic, models of observations, and the structure of verification, and what these connections mean for the treatment of points and boundaries.

The appendix includes five sections. Section A.1 informally motivates the interpretation of general topological notions in terms of certain logical notions and the nature of verification. Section A.2 makes precise the ideas of the preceding section, opening onto a presentation of modal logic and grounding a deep connection between a certain modal logic and topology. Section A.3 steps back and discusses the ideas behind such a connection, fleshing out some of the philosophical implications both for how we see topology and how we think about certain modalities. In this section, we will also unearth a potential response to the first of the three questions raised in the present chapter. Section A.4 takes up the question of "why opens?" and offers some observations and conjectures concerning the precedence given to them in the modern treatment of topology; we also take the opportunity to discuss some potential (generally ignored) advantages of working with closed sets, especially in relation to sheaves. The appendix concludes, in section A.5, with a return to the third question first raised in the Philosophical Pass of chapter 4—*What is general topology really about?*—and briefly relates such ideas to the broader conceptual advances (regarding space and "topologies") we have seen throughout this book.

A.1 Conceptual Motivation: Topology as Logic of Finite Observations

While deceptively obvious at first glance, it turns out to be a surprisingly deep observation that

> topological reasoning is intimately bound up with reasoning with approximations (or under conditions of error tolerance).

Let us spell this out a bit, using a common illustration.[259] Suppose you are a traffic officer charged with observing cases of speeding vehicles, and that you have radar guns—always with some fixed error tolerance—for precisely this purpose. The radars clock speeds, so as long as a car is moving, it is observable by a radar. In other words, we are imagining that there is some overall field of observation,

10 20 30 40 50 60

the nonnegative real line, while various subsets of this represent possible, more specific observations, which can be captured by our measuring devices. Your department has reserved a special name for the portion of this field of observation involving speeds greater than 50 mph—these have the observable property of *speeding*.

10 20 30 40 50 60

In other words, the property *speeding* can be thought of as being identified with $(50, \infty)$.

Now suppose a car is going 51 mph.

259. For instance, Moss and Parikh (1992) make use of this same metaphor of radars.

"Clearly," the car going 51 mph is speeding, as 51 "clearly" is in the interval $(50, \infty)$. But, as we already insisted in the dialogue in chapter 4, everything depends on our account of what this "clearly" actually involves. Let us first make a relevant observation. Surely, if 51 is indeed a case of speeding, since such events are observable, we should be able to produce a *verification* (or *piece of evidence*) that tells us unequivocally that what we are observing is indeed a case of speeding. As our radars supply us with our windows of observation, perhaps certain of these can give us the verification we are looking for. Let us consider various cases of such a possible verification.

Case 1: Suppose we witness 51 mph with a measuring device or radar with an error of ± 2, which is equivalent to adopting a somewhat restricted observational window of $(49, 53)$.

Using such a radar as our measuring device meant to serve as verification of the property in question, we get back only an equivocal answer about whether or not 51 ought to count as having that property: it can really only tell you that there is both speeding and not speeding going on. As potential evidence, the radar just tells us that the reading of the radar is 51. But we want to assert something stronger, namely that "there is speeding." After all, it may be that this radar shows 51 but the car is really going 49.5 mph. Such a measurement device cannot serve as evidence verifying that 51 is a case of speeding.

Case 2: Suppose our department gets more resources and allocates some of those to acquiring more refined radars, with an error of just ± 0.5 mph. Now, as we use such a radar and it shows 51, we have that for the resulting observational window $(50.5, 51.5)$,

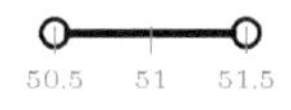

there will only be speeding *as far as the radar can see*! Giving us a read-out of 51, such a radar indeed serves as verification of speeding: for, directed at 51, it does not witness anything but cases of speeding. After all, unlike in the previous case, when pointed at 51, there is nothing that this radar can even see that is not speeding, so the entire radar cannot but verify that 51 really is speeding. In a moment, we will more closely consider the general features of this device; but before doing so, let us consider one last case.

Case 3: But why deal with such approximate regions? Why not just zoom in until we were *at the point itself*, on the nose? Provided we allow that this is even something we could really do (instead of an impossibility that would require our department had infinite resources), our resulting observational windows would then collapse down to the point itself, and by construction, operating with this degenerate window of observation, there would be nothing around, nothing *in the vicinity*, that could inform about how it related to its environs—in particular, whether or not 51 was in $(50, \infty)$. Informally, to verify that you are in (or not in) some interval, such as $(50, \infty)$, you need to be able to "look around"—that is, you need the presence of others (in or not in the interval in question) to help inform you about where you are. If we really could find such an exact measuring device, it could not tell us that 51 was speeding (or not), since there is nothing for it to see except the point itself—and, remember, we are asking about evidence for 51 *being in some interval* (or having some observable property). The basic idea here, in short, is that

> *verifiable knowledge* of an actual location/speed belonging in a region of observation is determined by observing the relationship of the region to environments/neighborhoods of the actual location/speed.

Moreover, as our discussion of the previous cases already suggested, if an observation made by a measuring device of some s is to serve as evidence, or verify a property, then anything that can be said of s must hold not just of s, but also of any other s' within the observational window of that device. The present point is: a hypothetical exact radar would not constitute a verification of anything—for it cannot speak to the relationship of what it purports to observe (the actual speed/point) with any other observable.

Of the three cases presented above, only case 2 describes something that could verify that 51 constituted speeding (i.e., was *in* $(50, \infty)$). Let us thus look at case 2 more closely. First, observe that, while the radar with an error of ± 0.5 mph worked just fine—in a way that the radar from case 1 (with an error of ± 2 mph) did not—there is in fact nothing special about the number 0.5 mph. A number of other radars would have served the same purpose: for instance, any radar of error smaller than ± 0.5 mph; in fact, so too would any radar with error $\pm 1 - \epsilon$ for any $0 \leq \epsilon < 1$.

Now suppose you have an old radar R_1 that can approximate the speed up to an error of ± 1, so that we can think of this radar as providing evidence of the actual case of 51 mph being a case of speeding, on account of its observation of $R_1 := (50, 52)$. But suppose you have another newer more expensive radar R_2 that observes cars with an error of only ± 0.5 mph, so that you can also supply evidence of speeding in $R_2 := (50.5, 51.5)$. This refined instrument R_2 will be able to prove its value in being able to observe instances of speeding that the first radar R_1 could not, for example, when a car is clocked at 50.75 mph. In other words, the two radars do not verify exactly the same things. However, there is an important asymmetrical relationship between what they can verify:

if R_1 can verify an instance of speeding, then R_2 can as well;

but

if R_2 can verify an instance of speeding, R_1 need not be capable of verifying it.

The latter fact describes how R_2 is a *better approximation* (and so represents a possible increase in knowledge). On the general way of looking at things that we have started to motivate, subsets are regarded as the *possible observations*, so that “open sets” containing a point are effectively just pieces of evidence *qua* observable properties concerning actual states (represented by a point or definite speed). We can further regard the key relation between the various subsets of a space as follows:

$x \in V \subseteq U$ says that V is a better approximation to x.

If you think of the open sets in terms of generalized rulers or measuring devices, the idea is that smaller rulers $V \subseteq U$ give you more refined measurements. One might also begin to think of such better approximation sets in terms of *effort*—the more refined the set, the more effort. The interpretation in terms of effort is meant to represent an action—such as a measurement, computation, or approximation—that in general may result in an increase in knowledge or what is knowable. If one thinks of the open sets as involving possible observations, the idea is that for $V \subseteq U$, V is a more sophisticated or refined means of observing properties. As such, it will in principle “cost more effort” compared to any cruder observation. If a radar is attempting to determine whether or not a car is speeding in a 50 mph speed zone, but the accuracy of the radar is ± 2, then a car going 51 mph cannot be verified as speeding; on the other hand, supposing a more refined traffic radar has an accuracy of ± 0.5 mph, then a car going 51 mph can be verified as speeding. The second, more refined radar—and so the one that surely costs more effort to make—can detect a property, namely that of “speeding,” which the first radar cannot. That it can detect properties that “fly under the radar” (pun intended!) of cruder radars justifies our thinking of smaller sets in terms of “better approximations.”

However, note that whenever a given radar detects a property, any more refined or costly radar must also detect that property. This last sentence gives a way of thinking about the key feature of “openness,” and appeals to the key idea of the

Paradigm of Truth Continuity: an observable property will be true of a point whenever a verification (approximation) can be supplied, such that this continues to be true in all better approximations.

Such truth would of course break down provided there was at least one more refined approximation for which the property ceased to be true. On this interpretation that we are starting to suggest, there is a very natural way of understanding the full force of the finite intersection and arbitrary union conditions that define a topology. For reasons that will become apparent in a moment, we will describe the paradigm above by saying

$\Box\phi$ is true at $p \in U$ provided for all refinements $V \subseteq U$ of the observation U, ϕ is still true at $p \in V$.

For now, you might think of $\Box$ in terms of supplying a sort of radar for some observable property—the old radar guns used to have rectangular windows! The reader might want to contemplate on their own some other sensible conditions one might expect to obtain for this $\Box$.

A.2 Explicit Connections to Modal Logic

Recall the discussion of modal logic from chapter 7.2. The first modern formal analysis of modalities is typically attributed to the work of C. I. Lewis, beginning in 1912 and culminating in a paper in 1932.[260] Lewis constructed a series of axiomatic systems of modal logic—called $S1$ through $S5$—based on which of the axioms were held to legislate over the modalities. Lewis was first motivated by concerns over what he took to be an eliding of an important difference in how implication was understood by the algebraists and logicians of his time, on the one hand, and the ordinary meanings of implication, on the other. He thought that these concerns could be met by using an *impossibility* ($\neg\Diamond$) operator to define a more appropriate notion of implication, and accordingly he came to introduce systems taking as primitive the connectives negation, conjunction, and possibility ($\Diamond$).

Gödel would later[261] advance this by initiating a simplified presentation of modal logics: here, modal logic emerges as just an extension of propositional logic—the modal operator $\Box$, or $\Diamond$, is added as an additional connective to propositional logic, together with additional axioms and rules governing its behavior. In that very brief one-page article, Gödel happened to be concerned with formalizing assertions of *provability*, which he sought to capture by means of a propositional connective B (for *beweisbar*), so that $B\alpha$ was to be read as "α is provable" or "it is provable that α." In that paper, he revealed how to define a system given by the axioms and rules of the usual propositional logic together with the following axioms governing his new operator B:

1. $B\alpha \to \alpha$
2. $B(\alpha \to \beta) \to (B\alpha \to B\beta)$
3. $B\alpha \to BB\alpha$

together with the inference rule

from α, infer $B\alpha$.

Gödel then stated that this system is none other than what Lewis had described as the system $S4$, after translating B to $\Box$. He also stated that the appropriate translation of the law of the excluded middle is not derivable, making the further suggestion that the translation of any theorem of the intuitionistic propositional calculus of Heyting is derivable in his system, and conjectured that the converse was true as well. This is of some significance, as we shall see.

These days, it is common to follow Gödel's general approach of presenting modal logic as an extension of propositional logic. While the particular axioms and inference rules adopted do depend on the context of use, or on which properties we expect our modal connectives to have, the main axioms used in modal logic, built on top of a complete axiomatization of propositional logic together with the rules of inference modus ponens, start by adjoining the basic necessitation inference rule

(Necessitation rule **N**) $\frac{\phi}{\Box\phi}$,

260. See, for instance, Lewis (1912).
261. Gödel (1933).

or $\phi \vdash \Box\phi$, where the turnstile symbol $\vdash$ here just tells us that whatever follows it represents either an axiom or a formula that can be derived as a theorem. Thus, this just says that if ϕ can be inferred (is provable) from no assumptions (beyond the axioms of logic), then ϕ is necessary in the modal logic—that is, if ϕ is a theorem (just derivable from nonmodal good old-fashioned logic), then $\Box\phi$ is also a theorem. This can also be construed as an *axiom*, in which case it reads

(**N** as an axiom) $\vdash \Box\top$.

The weakest (or most minimal) modal logic that has proven to be useful is called K, which involves an augmentation of propositional logic with $\Box$ and the rule **N**, together with the further axiom

(Axiom **K**) $\Box(\phi \to \psi) \to (\Box\phi \to \Box\psi)$.

The axiom systems of many other commonly used modal logics are extensions of K: by adding further axioms, we get other well-known modal systems. K on its own assumes very little, and so within K we cannot even prove that "if $\Box\phi$ is true, then ϕ is true." To address this, we could add the further axiom

(Axiom **T**) $\Box\phi \to \phi$,

an axiom that indeed holds in most modal logics.[262] Other elementary axioms are

(Axiom **4**) $\Box\phi \to \Box\Box\phi$,

sometimes called the "Positive Introspection" axiom, and

(Axiom **B**) $\phi \to \Box\Diamond\phi$

(Axiom **D**) $\Box\phi \to \Diamond\phi$

(Axiom **5**) $\Diamond\phi \to \Box\Diamond\phi$.

Various axioms, when conjoined, allow us to formulate other axioms, which are occasionally useful enough to be given their own name. For instance, **K** with the rule **N** can be shown to entail

(Axiom **R**) $\Box(\phi \wedge \psi) \leftrightarrow (\Box\phi \wedge \Box\psi)$.

Variously combining certain of these axioms yields distinct axiom systems corresponding to some notable modal logics (the usual names are given on the left):

- $K := \mathbf{N} + \mathbf{K}$
- $T := \mathbf{N} + \mathbf{K} + \mathbf{T}$
- $S4 := \mathbf{N} + \mathbf{K} + \mathbf{T} + \mathbf{4}$
- $S5 := \mathbf{N} + \mathbf{K} + \mathbf{T} + \mathbf{5}$ (or $\mathbf{N} + \mathbf{K} + \mathbf{T} + \mathbf{4} + \mathbf{B}$).

Observe that K through $S5$ form a nested hierarchy of systems, ultimately built on top of the minimal necessitation rule. Modal logics for which the necessitation rule holds—or which assume axiom **N**—are called *normal* modal logics. This nested hierarchy thus describes how the main normal modal logics relate. There are further, less commonly used normal modal logics, intermediate to those above, such as

- $D := \mathbf{N} + \mathbf{K} + \mathbf{D}$
- $K4 := \mathbf{N} + \mathbf{K} + \mathbf{4}$
- $B := \mathbf{N} + \mathbf{K} + \mathbf{T} + \mathbf{B}$
- $D4 := \mathbf{N} + \mathbf{K} + \mathbf{D} + \mathbf{4}$.

Calling this a "hierarchy" makes sense, as some of these logics are sublogics of others; for instance, $K4$ is a sublogic of $S4$ in the sense that all formulas valid in $K4$ are also valid in $S4$.

262. Anticipating, one could think of this as the "truthfulness of verification" principle.

Both for reasons internal to modal logic and for reasons we shall appreciate in a moment, the logic *S4* is of special interest. An equivalent axiomatization—one that will prove to be even more revealing—of *S4* can be given as follows:

Definition 340 The modal logic *S4* is defined as the logic for which the following axioms hold:

- $\Box\top$ (Axiom **N**)
- $\Box\phi \rightarrow \phi$ (Axiom **T**)
- $\Box\phi \rightarrow \Box\Box\phi$ (Axiom **4**)
- $\Box(\phi \wedge \psi) \leftrightarrow (\Box\phi \wedge \Box\psi)$ (Axiom **R**)

and where modus ponens

$$\frac{\phi \rightarrow \psi \qquad \phi}{\psi}$$

and monotonicity

$$\frac{\phi \rightarrow \psi}{\Box\phi \rightarrow \Box\psi}$$

are the only rules of inference.

There is also the following derivable theorem, which is often of use:

$$\Box\phi \vee \Box\psi \leftrightarrow \Box(\Box\phi \vee \Box\psi) \qquad \textbf{(or)}.$$

While *S4* is typically presented via **K, N, T** and **4**, it is easy to show that **Monotonicity** and **R** entail **K**; likewise, **N** and **K** can be shown to entail **Monotonicity** and **R**. Observe also that **4** together with **T** in fact gives us $\Box\phi \leftrightarrow \Box\Box\phi$.

While these axioms are of some interest in their own right, we can already anticipate where we are going with all this by making a purely formal observation, involving nothing more than pattern matching. Recall that we can express a topology, and its open sets, in terms of an interior operator. Have another look at the axioms of *S4* given in definition 340. Now recall the Kuratowski axioms governing the interior operator, reproduced here for convenience:

- **(i1)** $\textbf{int}(X) = X$ (it preserves the total space);
- **(i2)** $\textbf{int}(A) \subseteq A$ (it is intensive);
- **(i3)** $\textbf{int}(\textbf{int}(A)) = \textbf{int}(A)$ (it is idempotent);
- **(i4)** $\textbf{int}(A \cap B) = \textbf{int}(A) \cap \textbf{int}(B)$ (it preserves binary intersections).

This striking, apparently purely formal, similarity suggests that the logical connective $\Box$ might be some sort of interior operator, and that there may be some close connections between the modal logic *S4* and topology. Indeed, *S4*, while having evolved for rather distinct purposes and in different contexts, turns out to be ordinary topology in disguise! Viewed from the other direction, this intimate connection with modal logic and features of certain logics and their interpretations can help us shed light on the defining axioms of a topology.

S4 as the Logic of Topological Spaces

The modern study of propositional logic largely got its start in the nineteenth century as algebra—in the tradition of Boole, who initiated a revolutionary conceptual shift, by combining algebra with logic. Boole had revolutionized logic, and began the project of the mathematization of logic, by applying notions and methods from symbolic algebra to the treatment of logical arguments, while simultaneously revolutionizing algebra by freeing it from its narrow application to arithmetic. By the time Lewis's work on modal logic systems appeared, many advances in the study of algebras had been made, and it wasn't long before modal systems were seen in an algebraic light. One of the overriding morals of the work of the logician Tarski—who proved a major result concerning the relation between *S4* and topology—involves the insight that the conditions defining a topology (in terms of conditions on open sets, or in terms of an interior operator) are at bottom *algebraic*.

A prominent class of examples of Boolean algebras can be obtained, in general, as follows: let X be any nonempty set and $\mathbb{P}(X)$ the collection of all subsets of X. Then, for subsets A and B of X, if we define $\neg A$ as the set-theoretic complement of A (in X), $A \vee B$ as the union of A and B, and $A \wedge B$ as the intersection of A and B, the set $\mathbb{P}(X)$ becomes a Boolean algebra—an algebra of sets. In this spirit, after some maneuvering involving taking equivalence classes (identifying expressions that are logically equivalent), the set of all propositions of the propositional logic forms a nontrivial Boolean algebra.

Actually, a topological space itself can be represented as the Boolean algebra of all subsets of the space—in this setting, the interior operator is construed as an operator on the algebra, where it takes elements (subsets) of the algebra to other elements of the algebra. As such, a topological space can be represented as a particular algebra equipped with a particular operator—and, in this light, the Kuratowski axioms emerge as nothing other than algebraic equations. More explicitly, the equations will stipulate that the interior of the top element of the algebra is equal to the top element; the interior of any element is less than or equal to that element; the interior of the meet of two elements is equal to the meet of the interiors; and the interior of the interior of any element is equal to the interior of that element. In symbols,

- **(i1*)** $I(\top) = \top$;
- **(i2*)** $I(a) \leq a$;
- **(i3*)** $I(I(a)) = I(a)$;
- **(i4*)** $I(a \wedge b) = I(a) \wedge I(b)$.

If we substitute $\Box$ for I in the above equations, we can immediately see that we will just recover the logic $S4$; alternatively, substituting I for $\Box$ in the $S4$ axioms, we should recover the algebraic version of the topological interior axioms governing a topological space. It is intriguing that axioms first formulated to model reasoning with propositions under situations of governed by dual modalities like "possibility" and "necessity," after a very natural translation, end up recapitulating the very same axioms that mathematicians had been using to describe a space.

This connection further motivates the development of a topological semantics for modal logic. Here is one way of starting to see how this goes. Ordinary propositional calculus can be regarded as basically set theory in disguise. If we assume we start with some set X, we might attempt to translate a logical proposition of our propositional logic into a statement about sets by assigning atomic (noncompound) propositions to subsets of X, regarding $\wedge$ as $\cap$, $\vee$ as $\cup$, $\top$ as X, $\bot$ as $\emptyset$, and $\neg$ as the complement (in X). A proposition would then be regarded as true precisely when its translation always equals X, the entire set. Typically, when we think of a way of assigning truth to the various propositions of our logic, we think of interpretation (or valuation) of a formula in terms of extensions of a function from the propositional values into a "truth value" set $\{0, 1\}$. But really, an interpretation could just be seen as an assignment of meaning to the symbols that make up a formula, and the truth-values need not be confined to 0 (false) and 1 (true), but can just be some sets, where these convey information concerning the proposition. Following this lead, we can instead interpret propositions as being assigned to subsets of $\mathbb{P}(X)$, where these subsets effectively act to inform us about *where* the proposition is true or holds. We then make the appropriate substitutions for Boolean connectives as indicated, from which we can also express the implication $\rightarrow$ as $\subseteq$, and $\leftrightarrow$ as $=$.

Suppose we have a mapping $v : \Phi \rightarrow \mathbb{P}(X)$ that sends propositional variables to subsets of X, where we think of this as a general "valuation" in the algebra of subsets of X, as indicated above. In this way, for ϕ an arbitrary formula, we can let

$$v(\phi) = \{x \in X \mid \phi \text{ is true at } x\}.$$

Then, if ϕ and ψ are propositional variables, with $v(\phi) = A$ and $f(\psi) = B$, where A, B are subsets of X, by interpreting logical connectives as set operations, as indicated above, we can extend the function v just as you might expect, so that every formula is mapped to a subset of X:

- $v(\neg\phi) = X \setminus v(\phi)$;
- $v(\phi \vee \psi) = v(\phi) \cup v(\psi)$;

- $v(\phi \wedge \psi) = v(\phi) \cap v(\psi)$;
- $v(\phi \rightarrow \psi) = (X \setminus v(\phi)) \cup v(\psi)$.

Observe also that we must have $v(\top) = X$. If a formula ϕ is mapped to the entire set X by all mappings v, then we say that ϕ is *valid* in X. Moreover, observe that $\phi \rightarrow \psi$ will be valid with respect to any interpretation in X if and only if $v(\phi) \subseteq v(\psi)$ for any mapping v that sends propositional variables to subsets of X. This can be seen by noting that $\phi \rightarrow \psi$ will be valid with respect to any interpretation v in X if and only if $v(\phi \rightarrow \psi) = v(\neg\phi \vee \psi) = v(\neg\phi) \cup v(\psi) = (v(\phi))^c \cup v(\psi) = X$—and this will be true precisely when $v(\phi) \subseteq v(\psi)$, for any v.

So far, none of this is that interesting. The important bit comes with the treatment of the modal operator, which is where the topological semantics really comes into force (and derives its name). We have already anticipated how the logical connective $\Box$ might plausibly be interpreted as follows: for any mapping v that maps propositional variables to subsets of X and for any propositional variable ϕ, we interpret $\Box$ as stipulating

$$v(\Box\phi) = \mathbf{int}(v(\phi)).$$

Supposing we have a space, then, the topological semantics proceeds as follows. We define a *topological model* on a topological space X. This is a pair (X, v), with $v : \Phi \rightarrow \mathbb{P}(X)$ a valuation that takes values in the set of all subsets of X, as before. Here, we can define the truth of any formula ϕ at a point x of a topological model $\mathcal{M} = (X, v)$ by induction—where, we use the notation $\mathcal{M}, x \Vdash \phi$ to mean "x semantically entails ϕ with respect to v" (or "ϕ is satisfied by v at $x \in X$"), and this is the case provided $x \in v(\phi)$. Where $\mathcal{O}$ is the set of open neighborhoods of x, we can then show, just as you might expect, that:

- $\mathcal{M}, x \nVdash \bot$;
- $\mathcal{M}, x \Vdash \phi \wedge \psi$ iff $x \Vdash \phi$ and $x \Vdash \psi$;
- $\mathcal{M}, x \Vdash \phi \vee \psi$ iff $x \Vdash \phi$ or $x \Vdash \psi$;
- $\mathcal{M}, x \Vdash (\phi \rightarrow \psi)$ iff $\mathcal{M} \nVdash \phi$ or $\mathcal{M}, x \Vdash \psi$;
- $\mathcal{M}, x \Vdash \neg\phi$ iff $\exists U \in \mathcal{O}$ s.t. $\forall y \in U\ y \nVdash \phi$;
- $\mathcal{M}, x \Vdash \Box\phi$ iff $\exists U \in \mathcal{O}$ s.t. $(x \in U$ and $\forall y \in U(\mathcal{M}, y \Vdash \phi))$.

In this way, we can say how

$$v(\Box\phi) = \mathbf{int}(v(\phi))$$

by noting that

$$\begin{aligned} x \in v(\Box\phi) &\text{ iff} \\ \mathcal{M}, x \Vdash \Box\phi &\text{ iff} \\ \exists U \in \tau \text{ s.t.}(x \in U \text{ and } \forall y \in U(\mathcal{M}, y \Vdash \phi)) &\text{ iff} \\ \exists U \in \tau \text{ s.t.}(x \in U \text{ and } \forall y \in U(y \in v(\phi))) &\text{ iff} \\ \exists U \in \tau \text{ s.t. } (x \in U \text{ and } U \subseteq v(\phi)) &\text{ iff} \\ & x \in \mathbf{int}(v(\phi)). \end{aligned}$$

In other words, the formula $\Box\phi$ is true throughout the interior of the set of points where ϕ is true. In the presence of such a model and associated semantics, we then say that a formula ϕ is *satisfied* by (or in) the model provided it is true throughout the entire space (i.e., $v(\phi) = X$), and that ϕ is *valid* in the space X provided it is satisfied in *every* model defined over X.

Moreover, given the modal translation, we can in fact say that

- $v(\neg\phi) = \mathbf{int}(v(\phi)^c)$;
- $v(\phi \rightarrow \psi) = \mathbf{int}(v(\phi)^c \cup v(\psi))$.

Building on this, the axioms of *S*4, once interpreted topologically, simply restate the Kuratowski conditions that a topological interior is expected to satisfy. Of course, the rule **N** just corresponds to the condition

$$\mathbf{int}(X) = X.$$

And recalling that combining axiom **K** with the rule **N** gives us our alternative axiom **R**, it is easy to see how this corresponds to the condition

$$\mathbf{int}(A) \cap \mathbf{int}(B) = \mathbf{int}(A \cap B).$$

Finally, axiom **T** just amounts to asserting that the interior of a region is a subset of that region,

$$\mathbf{int}(A) \subseteq A,$$

while axiom **4** (added to axiom **T**) states that the interior of the interior of a region is just the interior of that region,

$$\mathbf{int}(A) \subseteq \mathbf{int}(\mathbf{int}(A)).$$

That **int** is monotone in the sense that

$$A \subseteq B \Longrightarrow \mathbf{int}(A) \subseteq \mathbf{int}(B)$$

is represented by the fundamental inference rule

$$\frac{\phi \to \psi}{\Box\phi \to \Box\psi}.$$

Exploiting the striking formal similarity between Gödel's axioms for provability logic and Kuratowski's axioms for a topological space, McKinsey and Tarski proved that, under the above interior-based interpretation, the topologically valid statements are exactly those provable in the the modal system *S4*—that is, that *S4* is the logic of topological spaces.[263] More explicitly, we can not only show that *S4* is sound with respect to topological spaces in general, via the topological semantics, but we can also show that it is complete. To begin to appreciate this, recall how both propositional modal logic and topological spaces can be seen as algebras. In describing a valuation, then, we could have described these valuations without assuming anything about being given a topological space. Instead, given a set X, v is just a map from propositions to $\mathbb{P}(X)$. Now, we can consider some mapping

$$i : \mathbb{P}(X) \to C,$$

where $C = i(\mathbb{P}(X)) \subseteq \mathbb{P}(X)$. In other words, with the map i we are considering a choice of some collection of subsets of X.

In relation to our valuation and what to do about the valuation of $\Box\phi$, we can then say that $v(\Box\phi) = i(v(\phi))$. But what should i be? Here is the fascinating result. Given any set X, we can show that *any* interpretation of $\Box$ in X that happens to satisfy precisely the axioms of *S4* will be such that the image of this interpretation forms a topology on X. *S4* is sound with respect to any interpretation in any topological space; moreover, *S4* can be shown to be complete over all topological spaces—in other words, if a formula is valid in every topological space, then it will be derivable from the logic *S4*. Supposing the map i makes *S4* sound, we can show that the set $C = i(\mathbb{P}(X))$ is a topology on X and that it is one only on the condition that it satisfies the axioms and rules of *S4*!

In short, let us now ask the following question: Is the set $C = i(\mathbb{P}(X))$ a topology for X? What conditions—axioms and rules—must i satisfy to guarantee that $i(\mathbb{P}(X))$ is a topology on X? As it turns out, the set $C = i(\mathbb{P}(X))$ is indeed a topology, and is one precisely when it satisfies all the axioms of the modal logic *S4* (where no proper sublogic of *S4* can guarantee that C is a topology). The following table shows how the particular modal axioms transfer over to topological conditions – in particular, a checkmark means that the collection C does have the property in question, while an × means that C does not have that property, when governed by the particular modal axioms of that row.

263. See McKinsey and Tarski (1944).

If we drop any of the axioms of $S4$, we can see that C will not necessarily be a topology—in other

Modal axioms	Topological axioms			
	$X \in C$	$\emptyset \in C$	$X_1, X_2 \in C \Rightarrow (X_1 \cap X_2) \in C$ (finite intersection)	$\{X_i\}_{i \in I} \subseteq C \Rightarrow \bigcup_{i \in I} X_i \in C$ (arbitrary union)
$\mathscr{L}_\Box + \mathbf{N}$	✓	×	×	×
$\mathscr{L}_\Box + \mathbf{N}, \mathbf{T}, \mathbf{4}$	✓	✓	×	×
$\mathscr{L}_\Box + \mathbf{N}, \mathbf{K}$	✓	×	✓	×
$\mathscr{L}_\Box + \mathbf{N}, \mathbf{K}, \mathbf{4}$	✓	×	✓	×
$\mathscr{L}_\Box + \mathbf{N}, \mathbf{K}, \mathbf{T}$	✓	✓	✓	×
$\mathscr{L}_\Box + \mathbf{N}, \mathbf{K}, \mathbf{4}, \mathbf{T}$	✓	✓	✓	✓

words, no proper sublogic of **S4** can guarantee that C is a topology.[264] By observing the table, we can even decipher which axioms of modal logic must correspond to, or are responsible for, which of the axioms of topological spaces.

- **N** just stipulates that the whole space is open, that is, that $X \in C$;
- **T** with **N** gives that $\emptyset \in C$;
- **R** (equivalently, **N** and **K**) is just the stability under finite intersection condition; and
- **or** (derived from **N**, **K**, **4**, and **T** together) is the condition that open sets are stable under finite unions (which can be extended to arbitrary unions by using an infinitary extension of the modal language).

McKinsey and Tarski further showed that $S4$ is sound and complete with respect to particular spaces—namely, the logic of any dense-in-itself metrizable space.[265] This result implies, in particular, that $S4$ is the logic of the real line $\mathbb{R}$ (endowed with the usual topology), the rationals $\mathbb{Q}$, the Cantor space, or any Euclidean space. This basically shows how $S4$ fundamentally characterizes any dense-in-itself metric space, in particular the spaces we are most familiar with—thereby solidifying the spatial core of modal logic.

Altogether, *soundness* of $S4$ for a given topological space is more or less had for free, where soundness just means that if something is provable, then it is valid, that is,

$$\vdash_{S4} \phi \text{ implies } \Vdash_X \phi.$$

For its part, the completeness of $S4$ for a given topological space X is trickier. Completeness of $S4$ for a space X asserts that every validity in X is provable within $S4$, that is,

$$\Vdash_X \phi \text{ implies } \vdash_{S4} \phi.$$

The equivalent contrapositive of this claim of completeness can be more helpful:

$$\nvdash_{S4} \phi \text{ implies } \nVdash_X \phi;$$

that is, if ϕ is *not* a theorem of $S4$, then it is *not* valid in X—which formulation effectively informs us that checking for completeness, in this context, amounts to checking that space for a class of refuting models. McKinsey and Tarski's result concerning how $S4$ characterizes any dense-in-itself metric space basically informs us that in any dense-in-itself metric space there will always be the resources to refute all nontheorems of $S4$.

264. Xu (2016) has a nice discussion of these matters; the idea for the above table came from that work.

265. A space is *dense-in-itself* when every point is the limit of other points in the space. The most familiar spaces of all—including any finite-dimensional Euclidean space—are such a space.

Stepping back, observe that we can also define

$$v(\Diamond\phi) = v(\neg\Box\neg\phi) = (v(\Box\neg\phi))^c = (\mathbf{int}(v(\neg\phi)))^c = (\mathbf{int}((v(\phi))^c))^c,$$

where this is just the closure of the set $v(\phi)$. Moreover, in general, we can construe the boundary as

$$\partial(\phi) = \Diamond\phi \wedge \Diamond\neg\phi.$$

In the context of modal logics, the notion of boundary generally captures something like the *contingency* of ϕ.

In connection with the topological semantics, we can also observe that for the law of excluded middle $\phi \vee \neg\phi$ to be valid in a topological space X, each closed set must also be open. A concrete counterexample to the excluded middle being valid can be given by the Sierpiński space described earlier, where $\{a\}$ is closed and $\{b\}$ is open, and where we define $v(\phi) = \{b\}$. Then, $a \nVdash_v \phi$. But since the only open neighborhood of a is the entire space $\{a, b\}$, which does indeed contain a point that satisfies ϕ, we must have $a \nVdash_v \neg\phi$. Thus, $a \nVdash_v \phi \vee \neg\phi$. This fundamentally gets at the deep connections with Heyting algebras as models of intuitionistic propositional logic.

Altogether, by interpreting each ϕ as a subset A of a topological space X (so that each propositional variable represents a region of the space, and so too does any formula), and making the appropriate substitutions, we can see that the modal logic $S4$ axioms governing $\Box$ reduplicate those of the topological interior operator. Going the other way, with the axioms of $S4$, we appear to have rediscovered the algebraic version of Kuratowski's axioms for a topological interior. This connection is not merely formal, but lets us attempt a justification of the defining features of a topology.

One of the merits for the topologist of seeing $S4$ as the logic of topological spaces—or topology in disguise—is to help ground an important verificationist interpretation, in which setting the axioms of a topology appear to achieve some clarification. This is explored in the next section.

One of the merits of the topological semantics for the modal logician, for their part, is that it does not validate

$$\neg\Box\phi \rightarrow \Box\neg\Box\phi,$$

known by philosophers under the name "negative introspection" principle, which seems to require that if one cannot verify something as true, then one must be able to verify that one cannot verify it as true. Some have argued that a good notion of *belief* is captured by *not knowing that you do not know*—so that one can define a belief operator

$$B\phi := \Diamond\Box\phi.$$

In those terms, the principle of negative introspection effectively makes it impossible to have incorrect beliefs, for one always knows whatever one believes that one knows. The principle is generally regarded as undesirable—especially applied in epistemic contexts, involving verifications and knowledge—since it seems to say that one cannot believe one knows things that one does not actually know. Rational agents are constantly believing they know things that they do not in fact know, so there are reasons for not wanting to enforce this principle. That the topological semantics does *not* validate it is encouraging.

Finally, there are merits to this connection that go beyond any benefit to one side or the other. For instance, in example 198 (chapter 7), we mentioned Kuratowski's 14-set theorem, which says that in a topological space, there can be no more than fourteen distinct sets that can be generated from a fixed set by taking closures and complements (or using the interior, closure, and complements). Why this should be true seems never to be explained—and one is left with the feeling that such a result is extremely mysterious, if not entirely arbitrary. In modal logic textbooks, one finds sequences of boxes and diamonds formed to get finite modalities—but, on account of certain logical equivalences, sequences of modal operators can often be reduced. One is sometimes asked to show that in $S4$ in particular, up to equivalence, there are only fourteen modalities. Again, there is not usually an explanation of why this should be the case. Authors in either camp rarely seem to acknowledge that these results are two ways of looking at the same result.

A.2.1 Models of Information

The results of McKinsey and Tarski laid the groundwork for further work into spatial logics. Moreover, the completeness results in particular drew the attention of epistemic logicians, leading certain logicians to use the deep connection between modal logic and topology to reflect features of $S4$ as an epistemic system back onto topology—where this led to an epistemic reevaluation of the interior semantics, in the context of which topologies came to be seen as *models for information*. The idea of regarding topological spaces as codifying information structures had itself been considered since the 1930s, leading to the use of topological spaces as models for intuitionistic languages.[266] Building on this use of topological models for broad classes of intuitionistic languages, the interior semantics sketched above has been seen as providing insight into evidential approaches to knowledge.

In the topological models for intuitionistic languages, as used in epistemic settings, elements of a given open basis are taken to be "pieces of observable evidence," while open sets of the topology are generally treated as "observable properties" that can be verified based on the observable evidence. Evidence as open sets can be grounded in thinking of a collection of observable evidence *directly* obtainable by an agent—through things such as computation, approximation, measurement—as a subbasis, so that the collection of observable evidence forms a basis for the topology. Furthermore, taking open sets as pieces of evidence is further supported by how measuring devices used to compute things like the speed of a car or height of a person are always *approximating devices*—in the sense that they do not give exact values, but are defined by a definite range of error tolerance.[267]

On this view of things, points in a topological space X are seen as (partial) states of information (about how things actually are).[268] Taking increasingly accurate measurements with less error-tolerant approximating devices gives rise to better and better approximations to the actual state. Building on this, the meaning of the fundamental logical connective as a kind of interior can be developed in terms of the notion of *verifiability*.[269] Open sets correspond to properties that are *in principle verifiable* by the agent—in short, open sets are being construed in terms of meaningful propositions, where these are just propositions whose truth is equivalent to their verifiability.

The general idea is that there are properties that are observable (directly) by an agent. These are regarded as forming an open basis for a topology. Because open sets can be expressed as those sets equal to their interior, we can see **int** of those properties as the set of states in which the properties are verifiable, suggesting that we read $\Box$ as follows:

$\Box\phi$: = "it is verifiably true that ϕ."

Another way to see $\Box$ is as expressesing the notion of "continuous truth."

Fundamentally, we are understanding "verification" here in such a way that

> to verify something is to observe something that entails it,

where this is entirely in keeping with the earlier view of a neighborhood as a kind of proof or evidence. Open neighborhoods U of an "actual state" x act as sound or truthful evidence. An actual state x is then in the interior of ϕ if and only if there exists a sound piece of evidence U that justifies ϕ. If an open U is included in a set representing a proposition ϕ, then we can say the proposition ϕ is entailed or justified by the evidence U. That agent knows ϕ (to be true) if they have a correct justification for it, where this is based on a sound piece of evidence justifying it.

As Bezhanishvili and Holliday (2019) puts it, given any set U of states of information,

> U is *verifiable*, relative to one's current information state, iff it is possible to achieve such verification of U after a finite amount of time, starting from the current information state.

266. See, for instance, Troelstra and Dalen (1988) for discussion.

267. See Baltag et al. (2019) for further elaboration on this.

268. See Dalen (2002) and Scott (1968).

269. This way of looking at things is nicely summarized in Bezhanishvili and Holliday (2019). One can find further variations of this overall idea in developments by computer scientists, for instance in domain theory (see Abramsky 1987, Vickers 1996) and work in formal epistemology.

Seen in this light, the features of the logic of verifiability, or the process of finite observation, can begin to shed some light on why the conditions of a topology are what they are. By taking the family of observable propositions as a basis, the features of the modal operator generate what we can think of as a "structure of verifiability"—and these just recapitulate, in another guise, the core defining properties of a topology. Moreover, this interpretation comes with a natural dual notion whereby closed sets can be expressed in terms of *falsifiable properties*: whenever a property is false, it is falsified by a sound piece of evidence. Finally, on this picture, the set of *boundary points* will correspond to those properties that are neither verifiable nor falsifiable.[270]

A.3 The Idea of All This

Making use of the approach sketched above, and taking advantage of the intimate connection between the interior and the modal operator $\Box$, we can attempt something like a *justification* of the axioms of the interior operator, that is, of the defining features of a topology.[271]

First of all, by taking $\mathbf{int}(U)$ to be the set of states in which U is verifiable, it is evident that $\mathbf{int}(U) \subseteq U$. Next, if it is possible to perform any finite sequence of possible verifications in a finite amount of time, then the **int** operator ought to distribute over finite intersections—where this effectively just means that observations can be refined in a finitary way. This is something like a continuity condition on our verifications. By contrast, we are *not* assuming that it is possible to perform an infinite sequence of verifications in a finite amount of time. Put otherwise, while observations can indeed be refined in a finitary manner—such that *each* observation can be refined by another—we do not assume that there generally exists a single and universal observation that could embrace *all* the observations in one go (which corresponds to not assuming that **int** distributes over arbitrary intersections). The "stability under *finite* intersections" condition then captures the idea of that agent's ability to conjoin only finitely many pieces of evidence into a single piece of evidence. An infinite conjunction of such properties would require confirming or verifying all of them, something it is not reasonable to assume is available to an agent capable of producing verifications. For the justification of the last condition on the interior operator, we can assume that we are working with a model of verification

> according to which by verifying U, we are also verifying that one has verified U, which implies that if it is possible to verify U, then it is also possible to verify that it is possible to verify U. (Bezhanishvili and Holliday 2019)

270. In section A.1, we sometimes used the language of "effort." Technically, modeling effort involves some extension of the semantics just sketched. If, following Moss and Parikh (1992), we let $(X, \mathcal{O})$ be a "subset space," where X is a nonempty set of states and $\mathcal{O}$ is a collection of subsets of X (not necessarily forming a topology, though topologies provide a particular case of this construction), then again elements of $\mathcal{O}$ are = *possible observations* or *possible observation sets*. Formulas are interpreted not just with respect to the actual state, but with respect to pairs of the form (x, U), where $x \in U \in \mathcal{O}$, and x represents the way the actual state of affairs happens to be. The *neighborhood* U with $x \in U \in \mathcal{O}$ is interpreted as a truthful observation that can be made about the actual state x. Moss and Parikh give a subset space semantics as follows: given a pair (x, U), the modality $\Box$ quantifies over all subsets of U in $\mathcal{O}(X)$ that include the actual state x, while another modality K (acting as a "knowledge" modality) quantifies over the elements of U. In other words, we have a logic that formalizes reasoning about sets and points: $\Box$ quantifies *over* the sets, while K quantifies *in* the sets. In this setup, $(x, U))$ is called a *neighborhood situation* if U is a neighborhood of x, that is, $x \in U \in \mathcal{O}$. If at (x, U) ϕ is known, this is to be interpreted as saying that we can move from the given reference point x to any other point y in the given neighborhood situation (x, U). Similarly, using the $\Box$ modality, we can shrink the neighborhood around a given reference point. In this manner, knowledge (via K) is interpreted locally in a given truthful observation set U. Effort (via $\Box$), for its part, is interpreted as neighborhood-refinement, where "more effort" corresponds to a smaller neighborhood, and a smaller neighborhood corresponds to a more refined truthful observation—and so, a possible increase in knowledge. The smaller the observation set is, the more informative whatever information we have, and the more effort we will have spent to obtain this. See Moss and Parikh (1992) for more details.

271. The main ideas of the next paragraph are covered in Bezhanishvili and Holliday (2019), which has a valuable discussion of these, and related, matters.

In other words, $\mathbf{int}(U) \subseteq \mathbf{int}(\mathbf{int}(U))$. These conditions taken together further imply things about disjunctions. Observe that an arbitrary disjunction (a union) of opens-as-properties gets verified by confirming *any one of them* and thus also requires only a finite amount of evidence.

Regarding the asymmetry in the finiteness conditions on how verification distributes over union and intersection, there is a closely related tradition of computer scientists who have taken to describing the key properties of a topology in terms of a key computational or logical idea, namely that

> open sets are analogous to *semidecidable* properties.[272]

The idea here is that the notion of *(semi)decidability* may help understand the axiomatic stability properties that characterize a topology—where, computationally speaking, an observable property of a data type is taken to correspond to a semidecision procedure.

In its native setting, one attributes the property of semidecidability to a theory or logical system. A theory is said to be semidecidable provided there is an effective method which, given an arbitrary formula, will always tell correctly when the formula is in the theory, but acts differently when the formula is not in the theory: in that case, it may give either a negative answer or no answer at all. A logical system, for its part, is regarded as semidecidable if there is an effective method for generating theorems (and only theorems) such that every theorem will eventually be generated—yet, in a semidecidable system there may be no effective procedure for checking that a formula is *not* a theorem. Applied to our present setting: this concept would help describe how, if something is observably true of a set, this can be *decided* by some evidence or verification; but if it is false, the same procedure may "run forever" and not have the means of verifying that it is not true of the set.

To better see what is going on here, referring back to our informal discussion in A.1, consider the following proposition or assertion: "The car is speeding." Affirmations and refutations of such propositions can be carried out by means of what can actually be observed, and an observation must be made in finite time (i.e., after a finite number of steps), in particular by supplying a piece of evidence. Whenever the car is unmistakably speeding, so that it is definitely true, such a proposition can be affirmed. Whenever the car is unmistakably not speeding, so that it is definitely false, it can be refuted. But in an important way, the real crux of the matter boils down to how all the other borderline cases are handled. The two extremes can help reveal what is going on here. Vickers (1996) describes how

> An assertion is *affirmative* (or *affirmably true*) iff it is true precisely in the circumstances when it can be affirmed.

Then,

> if we declare that for all border cases the assertion is *false*, then by "true" we will mean "affirmably true."

In other words—on the affirmative interpretation—any sets corresponding to true are treated as *open sets*.

We can also speak of refutative assertions as follows:

> An assertion is *refutative* iff it is false precisely in the circumstances when it can be refuted.

Then,

> if we declare that for all border cases the assertion is *true*, then by "true" we will mean "irrefutable" (in the sense of being not falsifiable).

In other words – on the refutative interpretation – sets corresponding to true are treated as *closed sets*.

272. This suggestion apparently first appeared in Smyth (1983); it is discussed in Vickers (1996), which inspired the next few pages.

Let us declare the sets for which we can make affirmations or refute the assertion *open* – since they do not include any boundary. Asserting "The car is speeding" means the car is definitely speeding, while refuting it means it is definitely not speeding.

Suppose, for instance, we can agree that the car really is speeding, where speeding means any speed strictly greater than 50 mph. As suggested in A.1, this affirmation might be tested by using finer and finer "radar guns." Then we can *affirm* the assertion—such an observation can be made—since if the speed is indeed greater than 50 mph, you will eventually discover this, provided only that you use a radar or measuring device with a granularity finer than ± 1.

As Vickers stresses, working with the affirmative interpretation, we can surely affirm $\phi \vee \psi$ either by affirming ϕ or by affirming ψ. This extends to further disjunctions, even to the point where we have an entire family of infinitely many assertions $\{\phi_i\}$—the infinite disjunction $\bigvee_i \phi_i$ can be affirmed by just affirming *any* individual ϕ_i. In short, any disjunction, even an infinite one, of affirmative assertions is still affirmative—we need affirm only one!

We can affirm $\phi \wedge \psi$ by affirming both ϕ and ψ. This extends to any finite conjunction. But to affirm an infinite conjunction $\bigwedge_{i \in I} \phi_i$, we must affirm every single one and that will require an infinite amount of work. Thus, we can only say: any *finite* conjunction of affirmative assertions is still affirmative.

Under different circumstances, it may be that the car is decidedly not speeding – in that instance, the assertion is definitely false, and so can be refuted. To affirm "not ϕ" ($\neg\phi$), we will have to make a finite observation evidencing that ϕ is definitely false (i.e., we refute ϕ).

Under the refutative interpretation, for its part, assertions correspond to falsifiable ones. And for these, the following holds: For conjunctions, if ϕ is falsifiable and ψ is any statement (falsifiable or not), then ϕ and ψ is falsifiable—it is falsifiable by *any* observation that falsifies ϕ. For disjunctions, if you have two falsifiable statements ϕ, ψ, then $\phi \vee \psi$ is falsifiable since you can demonstrate the disjunction to be false merely by supplying two observations, the first falsifying ϕ and the second falsifying ψ. But this does not extend to arbitrary disjunctions. This gives another way of seeing the conditions on closed sets.

The logic of refutative assertions will be such that it is stable under arbitrary conjunctions and finite disjunctions, while the logic of affirmative assertions will be stable under finite conjunctions and arbitrary disjunctions. Whether we take up the logic of affirmative assertions or the logic of refutative assertions, we capture the essence of a topology.

Finally—and this is the point!—suppose "The car is speeding" (ϕ) is false and, as it turns out, the car is going *exactly* 50 mph. Then, on the affirmative interpretation (i.e., using opens), you will never discover whether or not ϕ is true, regardless of how refined the radar used. We will never be able to affirm that it is false. This borderline case helps reveal the key idea of thinking of all this in terms of semidecidability: there is some test that you can perform such that, if ϕ is true, you will eventually discover this, but if it is false, you may never discover this. On the refutative side of things (i.e., using closed sets): if ϕ is false, you will eventually be able to falsify it, but if it is true, you may never discover this.

Altogether, this accords with the reading

> $\Diamond\phi :=$ "It is not verifiably false that ϕ" (or, equivalently, "The hypothesis that ϕ cannot be falsified"),

according to which $\Diamond$ is a closure operator, and $S \subseteq X$ is falsifiable if and only if $\mathbf{cl}(S) \subseteq S$—another way of seeing that we can falsify a falsifiable proposition unless it happens to be true. Furthermore, a boundary of a set $S \subseteq X$ is just $\partial(S) = \mathbf{cl}(S) \setminus \mathbf{int}(S)$, so that x is a boundary point of S precisely when there is no true piece of evidence that supports neither S nor $\neg S$.

A.4 Why Opens?

In chapter 4, I mentioned that I would attempt to address the question of why, in spite of the historical precedence given to closed sets, present-day topologists often seem to regard open sets as more primitive. Building on previous discussions, here are some plausible reasons one could give for such a preference:

1. The epistemic-topological framework sketched above helps reveal how key axioms of a topology appear to encode the notion of semidecidability, that is, the notion of a condition whose truth can be verified in finite time (but whose falsehood cannot necessarily be verified in finite time). It is sometimes suggested that this interpretation encodes the intuition from metric space settings where verifying that a point is in an open ball can be done with finite-precision computations by just producing an even smaller open ball in which it lies, yet where it is not necessarily possible to do the same sort of thing for a closed ball (if the point lies on the boundary, that we cannot make arbitrarily precise measurements in finite time becomes an issue).

   In this same epistemic-topological framework, another more general reason could have to do with the following. We can technically freely consider things either in terms of affirmative propositions or refutative propositions and convert between them. Affirmative assertions, though, for their part, do seem more primitive—or less derivative or parasitic—than their refutative counterparts. And affirmative assertions are the ones that correspond—in both intuitive and precise ways—to open sets. This priority of the affirmably true might be further developed into something like a proper defense of the priority of open sets.
2. Another reason I have heard for preferring open sets over closed sets is the apparent ease with which certain formulations of continuity, and related results, can be given using open sets and potentially infinite unions thereof—compared to using closed sets, where one must make restrictions to the finite case. Indeed, this can be related to sheaves as well. Over and above the important historical connection between sheaves and local homeomorphisms, there is a viable explanation for the general preference for working with open sets in that what has been called the *open pasting lemma*—or the "local formulation of continuity" (as in Munkres 2015)—stipulates that if $X = \cup_i U_i$ is a possibly infinite union of open sets with continuous sections $s_i : U_i \to E$ that agree on their overlaps, then it follows that the (set-theoretically defined) section $s : X \to E$ will be continuous as well. However, if closed sets are used here, then the gluing argument works only for those covers that have *finitely* many closed sets.[273] More generally, if it is really the case that having arbitrary unions of open sets at one's disposal (such that this is open) is somehow more useful than having arbitrary intersections of closed sets (such that this is closed), then I think that this sort of justification has some merits and deserves to be probed further. On the other hand, to the extent that the results and formulations adduced as defense in fact really only highlight a lack of human ingenuity, or some feature of the current practice of mathematics, preventing us from finding the right formulation and obtaining the same results while using closed sets, then this reason is less compelling.
3. A last reason I will list—which I have not seen developed before—is a little more complicated and difficult to parse. Fundamentally, the idea is that the preference may be influenced by differences in the underlying algebraic structures of open sets versus closed sets.

   The lattice of open subsets of a topological space supplies us with an important and canonical example of a Heyting algebra, while its dual—the lattice of closed subsets—supplies a canonical example of a co-Heyting algebra. For a co-Heyting algebra (like that of closed subsets of a space), we have seen that there is an induced operator $\sim$ that could be called *paraconsistent negation*, some of whose properties were explored in chapter 7. The impact of this alternative negation operator can be initially appreciated by considering that $\mathbf{cl}(U) \cap \mathbf{cl}(U^c)$, where $\mathbf{cl}(U^c) := \sim U$, will in general be nonempty. By defining such an alternative negation with $\sim$ ("non"), as the closure of the complement, we get a co-Heyting algebra of closed sets. In the general setting of co-Heyting algebras, which model paraconsistent logics, inconsistent theories are modeled as the ones that include the formulas that are true at the boundaries. As such, the logic of closed set topologies can be seen as being *paraconsistent* or forming a *paraconsistent algebra*, as a formula ϕ and its paraconsistent negation $\sim\phi$ may intersect at the boundary of their extensions.

273. Curry (2014) makes this observation.

Recall that paraconsistency is the blanket term for a logical system where the principle of noncontradiction may fail, that is, we can have $\alpha\wedge$ not-α. In general, paraconsistent structures are those for which theories that are true at boundary points include formulas and their negation; a boundary of a theory T can be construed as those sequents that neither "fully follow from" T nor "fully contradict" T. Note that, with a paraconsistent logic, this does not mean that *all* contradictions are true—rather, that there are some contradictions that do not entail a trivial theory. In general, the idea here is that the extension of the conjunction of some formulas and their negations may not be the empty set. As such, viewed in semantic terms, there may exist some states for which a formula and its negation are true. There are many different logical structures that are used to represent paraconsistent logics, but co-Heyting algebras have been seen as an especially natural choice.

The reason why closed set topologies form a paraconsistent structure has to do with the fact that theories that are true at boundaries include formulas and their negation—in other words, a formula α and its paraconsistent negation $\sim\alpha$ intersect at the boundary of their extensions. In such settings, the boundary operator also plays a decisive and active role. In general, boundaries play an important role in supplying topological semantics for paraconsistent logics.[274] As Lawvere writes,

> That the notion of boundary is just that of "logical contradiction" (within the realm of closed sets) follows at once from the intuitive notion of motion: indeed, since the unit interval is connected, any continuous path which is in A at time 0 and in $\sim A$ at time 1 must at some intermediate time be in both A and $\sim A$, i.e., must pass through the boundary of A. (Lawvere 1986, 10)

We have seen that sheaves fundamentally involve assignments of data to a base topology. In the usual presentations of sheaves on topological spaces, it is standard practice to almost exclusively work with open set topologies. But in the sheaf construction, the base topology is of special importance. For one thing, the algebra of the base space topology can be seen in terms of the algebra of the sheaf section structure. And historically at least, sheaves were defined over closed sets before they were defined over open sets.[275] By considering sheaves with respect to closed set topologies, then, the sheaf morphisms ought to reflect this algebra—so, using closed sets, we could have morphism algebras equipped with paraconsistent negations. Moreover, by taking closed sets for our base topology, we could introduce into the sheaf construction the valuable notion of boundary—something that will not exist for the corresponding open set sheaf construction. As William James has suggested, building on suggestions of Lawvere:[276]

> One area in which this may work for us is the mathematics of physics where the boundaries of a body are as important as the parts of a body inasmuch as physics concerns itself with the interactions of bodies in a system. Lawvere in the introduction to *Categories in Continuum Physics*, mentions the speculation that there is a role for a closed set sheaf in thermodynamics as a functor from a category of parts of a body to a category of "abstract thermodynamical state-and-process systems" (Lawvere 1986, 9). Lawvere recognises the particular properties of closed set topologies that make them interesting to us, namely that as algebras they provide us with a formalisation of what we call a paraconsistent negation. Sheaves are then of interest to us in our project of developing paraconsistent logic in categories for the way in which they transport algebras of a topology into the structure of a category of sheaves over that topology. (James 1996, 127–128)

274. See Başkent (2013), Goodman (1981), and Mortensen (2000).

275. Historically, in 1946 Jean Leray first defined a sheaf as a way of assigning modules to closed sets in an inclusion-reversing manner. For a historical account of the early development of the sheaf concept, the reader can consult Fasanelli (1981).

276. The interested reader should also consult James (1992, 1995).

There may indeed be considerable virtues to be had by giving more attention to closed sets, especially in relation to sheaves, and the reader is urged to explore these further. For instance, by looking at sheaves over closed sets, ultimately paraconsistency could be integrated into topos theory. Moreover, we can mention that in any presheaf topos, the lattice of all subobjects of any given object is an example of a co-Heyting algebra (as well as being a Heyting algebra).

All this is to make the following point: The co-Heyting operations are in general not preserved by substitution (inverse image) along maps—and this is as opposed to the Heyting "not" ($\neg$) and even the "possibility"-like operators given by Grothendieck topologies. I believe that this—and subtleties having to do with how boundaries and continuity interact—may contribute (albeit in the background, or beneath awareness) to the preference for working with open set topologies. By encoding the notion of "having no boundary," open sets are able to avoid some of the subtleties associated with explicitly working with boundaries and the induced paraconsistent structure supporting some inconsistency tolerance (which generally seems to require more careful treatment). By comparison, there are close connections between Heyting algebras and intuitionistic logic (which logic is particularly useful for representing a number of situations relevant to the mathematical treatment of continuity). On account of these features of the algebraic structures of open versus closed sets, we might propose the following:

> topology in terms of open sets $\approx$ algebra of observables where the law of excluded middle can fail but the law of noncontradiction holds in general;

> topology in terms of closed sets $\approx$ algebra of observables where the law of noncontradiction can fail but the law of excluded middle holds in general.

Finally, related to this issue of boundaries: a great deal of mathematics is concerned with *continuity* in its many different forms. Inquiring into the continuity of something, say some function f defined on some region, at a point p, is to invite comparison of the behavior of f *all around* p—yet, at boundaries, we cannot necessarily go around the point without leaving our area of concern. This merely suggests one of the related ways in which one might expect to find, on account of the general features of boundaries, that there is a more natural alignment of continuity with open, rather than closed, sets—as the former exclude all boundaries, while the latter include them all.

A.5 What Is Topology Really About?

At least as far as general (point-set) topology goes, the story just told suggests that the essence of topology can be seen as being closely connected to certain features of logic and the logic of verification, specifically as this relates to the implications of how we formulate notions of *negation* and our dealings with *boundaries*. Guided by such a story, a think a reasonable case can be made for wanting to view

> topology as a formal framework for studying reasoning about systems of parts of a whole in such a way that the nondegeneracy of *boundaries* is respected.

In particular,

> open set topology $\approx$ the structure of reasoning in the absence of boundaries (or with approximations),

while

> closed set topology $\approx$ the structure of reasoning in the presence of boundaries.

Depending on whether we choose to admit boundaries or not, certain prices will have to be paid—involving the tolerance of violations of a generalized "law of noncontradiction," and how negations must be treated. And it is arguably these prices that are truly determinative of the structure we call a topology.

However, what does all this have to do with the wider notions we have explored in this book—from the familiar topologies to Grothendieck topologies to Lawvere-Tierney topologies? Already in proposition 132 (chapter 5) we saw that we can effectively drop the open sets of a topological space X and just look at the category of sheaves on X—this later category can then be used, by looking at the subsheaves of the terminal sheaf, to recover the original data of the open sets. This suggests that, in some sense, sheaves are the more essential of the two, even if at first it may appear as if the space on which sheaves are built needs to be assumed by the sheaf construction. If that is indeed the case, how can we reconcile everything we said about sheaves—their power, how to think about them, and how to use them to think more deeply about space—with what this particular logical story seems to suggest about the nature of topology and space? I leave that to the reader to ponder!

References

Abramsky, Samson. 1987. "Domain Theory and the Logic of Observable Properties." PhD diss., Queen Mary College, University of London.

Abramsky, Samson. 2014. "Contextual Semantics: From Quantum Mechanics to Logic, Databases, Constraints, and Complexity." *Bulletin of the European Association for Theoretical Computer Science* 113:137–163. arXiv: 1406.7386.

Artin, Michael, Alexander Grothendieck, and Jean-Louis Verdier, eds. 1972. *Théorie des Topos et Cohomologie Etale des Schémas. Seminaire de Géometrie Algébrique du Bois Marie, 1963–1964.* Vol. 1. Lecture Notes in Mathematics (French) 269. Berlin: Springer-Verlag.

Awodey, Steve. 2010. *Category Theory.* 2nd ed. Oxford: Oxford University Press.

Baltag, Alexandru, Nick Bezhanishvili, Aybüke Özgün, and Sonja Smets. 2019. "A Topological Approach to Full Belief." *Journal of Philosophical Logic* 48:205–244.

Başkent, Can. 2013. "Some Topological Properties of Paraconsistent Models." *Synthese* 190:4023–4040.

Bell, John L. 2008. *A Primer of Infinitesimal Analysis.* 2nd ed. Cambridge: Cambridge University Press.

Belnap, Nuel. 1992. "How a Computer Should Think." In *Entailment: The Logic of Relevance and Necessity,* edited by Alan Ross Anderson, Nuel Belnap, and J. Michael Dunn, vol. 2. Princeton, NJ: Princeton University Press.

Berlekamp, Elwyn R. 1982. *Winning Ways for Your Mathematical Plays. Vol. 1, Games in General.* New York: Academic Press.

Berlekamp, Elwyn, and David Wolfe. 2012. *Mathematical Go Endgames: Nightmares for Professional Go Players.* Wellesley, MA: Ishi Press.

Bezhanishvili, Guram, and Wesley H. Holliday. 2019. "A Semantic Hierarchy for Intuitionistic Logic." *Indagationes Mathematicae* 30 (3): 403–469.

Bobick, Aaron. 2014. "Binary images and Morphology." Presentation, GA Tech, School of Interactive Computing. http : //www.cc.gatech.edu/~afb/classes/CS4495-Fall2014/slides/CS4495-Morphology.pdf.

Borceux, Francis. 1994. *Handbook of Categorical Algebra. Vol 3, Categories of Sheaves.* Cambridge: Cambridge University Press.

Caicedo, Xavier F. 1995. "Logica de los haces de estructuras." *Revista de la Academia Colombiana de Ciencias Exactas, Fisicas y Naturales* 19, no. 74:569–586.

Cannizzo, Jan. 2014. "Schreier Graphs and Ergodic Properties of Boundary Actions." PhD diss., University of Ottawa.

Caramello, Olivia. 2012. "Atomic Toposes and Countable Categoricity." *Applied Categorical Structures* 20, no. 4:379–391. http://arxiv.org/pdf/0811.3547.pdf.

Ceccherini-Silberstein, Tullio, and Michel Coornaert. 2010. *Cellular Automata and Groups.* Heidelberg: Springer.

Cockett, J. R. B., G. S. H. Cruttwell, and K. Saff. 2010. "Combinatorial Game Categories." http://www.reluctantm.com/gcruttw/publications/CGC.pdf.

Connes, Alain. 1983. "Cohomologie cyclique et foncteurs Extn." *C. R. Acad. Sci. Paris Sér. I Math.* 296 (23): 953–958.

Connes, Alain, and Caterina Consani. 2015. "Cyclic Structures and the Topos of Simplicial Sets." *Journal of Pure and Applied Algebra* 219 (4): 1211–1235.

Conway, John H. 2000. *On Numbers and Games.* 2nd edition. Natick, Mass: A K Peters/CRC Press, December 11. ISBN: 978-1-56881-127-7.

Curry, Justin Michael. 2014. "Sheaves, Cosheaves and Applications." PhD dissertation, University of Pennsylvania. http://arxiv.org/pdf/1303.3255.pdf.

Curry, Justin Michael. 2019. "Functors on Posets Left Kan Extend to Cosheaves: an Erratum." http://arxiv.org/pdf/1907.09416.pdf.

Curry, Justin Michael, Robert Ghrist, and Vidit Nanda. 2015. "Discrete Morse Theory for Computing Cellular Sheaf Cohomology." http://arxiv.org/pdf/1312.6454.pdf.

Dalen, Dirk van. 2002. "Intuitionistic Logic." In *Handbook of Philosophical Logic,* 2nd ed, Vol. 5, edited by Dov M. Gabbay and Franz Guenthner, 1–114. Dordrecht: Springer.

Eilenberg, Samuel, and Saunders Mac Lane. 1945. "General Theory of Natural Equivalences." *Transactions of the American Mathematical Society* 58, no. 2:231–294.

Fasanelli, Florence Dowdell. 1981. "The Creation of Sheaf Theory." PhD dissertation, American University, Washington, DC.

Fong, Brendan, and David I. Spivak. 2020. "Seven Sketches in Compositionality: An Invitation to Applied Category Theory." http://arxiv.org/pdf/1803.05316.pdf.

Gelbaum, Bernard R., and John M. H. Olmsted. 1962. *Counterexamples in Analysis.* Mineola, NY: Dover.

Ghrist, Robert. 2014. *Elementary Applied Topology.* CreateSpace Independent Publishing Platform.

Gödel, Kurt. 1933. "Eine Interpretation des intuitionistischen Aussagenkalküls." *Ergebnisse eines Mathematischen Kolloquiums* 4:39–40.

Goguen, Joseph A. 1992. "Sheaf Semantics for Concurrent Interacting Objects." *Mathematical Structures in Computer Science* 2 (2): 159–191.

Goldblatt, Robert. 1980. "Diodorean Modality in Minkowski Spacetime." *Studia Logica* 39, no. 2:219–236.

Goldblatt, Robert. 1992. *Logics of Time and Computation.* Stanford, CA: Center for the Study of Language.

Goldblatt, Robert. 2003. "Mathematical Modal Logic: A View of Its Evolution." *Journal of Applied Logic* 1 (5): 309–392.

Goldblatt, Robert. 2006. *Topoi: The Categorial Analysis of Logic.* Revised ed. Mineola, NY: Dover.

Goodman, Nicolas D. 1981. "The Logic of Contradiction." *Mathematical Logic Quarterly* 27, no. 8:119–126.

Gould, Glenn. 1990. "The Grass Is always Greener in the Outtakes." In *The Glenn Gould Reader,* edited by Tim Page, 357–367. New York: Vintage.

Grothendieck, Alexander. 1957. "Sur quelques points d'algèbre homologique." *Tohoku Mathematical Journal* 9 (2): 119–221.

Grothendieck, Alexander. 1986. "Récoltes et Semailles: Réflexions et témoignage sur un passé de mathématicien."

Guy, Richard, John Horton Conway, and Elwyn Berlekamp. 1982. *Winning Ways for Your Mathematical Plays: Games in Particular.* London: Academic Press.

Hansen, Jakob, and Robert Ghrist. 2020. "Opinion Dynamics on Discourse Sheaves." http://arxiv.org/pdf/2005.12798.pdf.

Hatcher, Allen. 2002. *Algebraic Topology.* Cambridge: Cambridge University Press.

Hegel, Georg Wilhelm Fredrich. 1991. *Encyclopedia of the Philosophical Sciences in Outline and Critical Writings.* 1st ed. Edited by Ernst Behler. Translated by Taubeneck. New York: Continuum.

Hegel, Georg Wilhelm Friedrich. 2010. *The Science of Logic.* Edited and translated by George di Giovanni. Cambridge: Cambridge University Press.

Hlavac, Vaclav. 2020. "Grayscale Mathematical Morphology." Presentation, Czech Technical University in Prague. http://people.ciirc.cvut.cz/~hlavac/TeachPresEn/11ImageProc/71-06MatMorfolGrayEn.pdf.

Hubbard, John H., ed. 2006. *Teichmüller Theory and Applications to Geometry, Topology, and Dynamics.* Ithaca, NY: Matrix.

James, William. 1992. "Sheaf Spaces On Finite Closed Sets." *Logique et Analyse* 35, no. 137:175–188.

James, William. 1995. "Closed Set Sheaves and Their Categories." In *Inconsistent Mathematics,* edited by Chris Mortensen, 115–124. Dordrecht, Springer.

James, William. 1996. "Closed Set Logic in Categories." PhD dissertation, University of Adelaide. https://digital.library.adelaide.edu.au/dspace/handle/2440/18746.

Johnstone, Peter T. 1986. *Stone Spaces.* Cambridge: Cambridge University Press.

Johnstone, Peter T. 2002. *Sketches of an Elephant: A Topos Theory Compendiumm vol. 1.* 1st edition. Oxford; New York: Clarendon Press, November 21. ISBN: 978-0-19-853425-9.

Johnstone, Peter T. 2014. *Topos Theory.* Reprint edition. Mineola, NY: Dover.

Joyal, A. 1977. "Remarques sur la théorie des jeux à deux personnes." *Gazette des Sciences Mathematiques du Québec* 1 (4): 46–52.

Kelly, G. M. 2005. "Basic Concepts of Enriched Category Theory." *Reprints in Theory and Applications of Categories,* no. 10: 143. http://www.tac.mta.ca/tac/reprints/articles/10/tr10.pdf.

Kennett, Carolyn, Emily Riehl, Michael Roy, and Michael Zaks. 2011. "Levels in the Toposes of Simplicial Sets and Cubical Sets." *Journal of Pure and Applied Algebra* 215 (5): 949–961.

Kock, Anders, ed. 2006. *Synthetic Differential Geometry.* 2nd ed. Cambridge: Cambridge University Press.

Lawvere, F. William. 1986. "Categories in Continuum Physics: Introduction." In *Categories in Continuum Physics,* edited by F. William Lawvere and Stephen H. Schanuel, 1–16. Berlin: Springer.

Lawvere, F. William. 1989. "Display of Graphics and Their Applications Exemplified by 2-Categories and the Hegelian 'Taco'." In *Proceedings of the First International Conference on Algebraic Methodology and Software Technology,* 51–75. University of Iowa.

Lawvere, F. William. 1994a. "Cohesive Toposes and Cantor's Lauter Einsen." *Philosophia Mathematica* 2, no. 1:5–15.

Lawvere, F. William. 1994b. "Tools for the Advancement of Objective Logic: Closed Categories, and Toposes." In *The Logical Foundations of Cognition,* edited by John Macnamara and Gonzalo E. Reyes, 43–56. New York: Oxford University Press.

Lawvere, F. William. 1996. "Unity and Identity of Opposites in Calculus and Physics." *Applied Categorical Structures* 4:167–174.

Lawvere, F. William. 2000. "Adjoint Cylinders." Message to catlist. November. https://www.mta.ca/~cat-dist/archive/2000/00-11.

Lawvere, F. William. 2002. "Linearization of Graphic Toposes via Coxeter Groups." *Journal of Pure and Applied Algebra* 168, no. 2–3:425–436.

Lawvere, F. William. 2005. "Taking Categories Seriously." *Reprints in Theory and Applications of Categories* 8:1–24. https://github.com/mattearnshaw/lawvere/blob/master/pdfs/1986-taking-categories-seriously.pdf.

Lawvere, F. William. 2006. "Diagonal Arguments and Cartesian Closed Categories." *Reprints in Theory and Applications of Categories* 15:1–13. http://emis.matem.unam.mx/journals/TAC/reprints/articles/15/tr15.pdf.

Lawvere, F. William. 2007. "Axiomatic Cohesion." *Theory and Applications of Categories* 19, no. 3:41–49.

Lawvere, F. William. 2011. "Euler's Continuum Functorially Vindicated." In *Logic, Mathematics, Philosophy, Vintage Enthusiasms,* vol. 75. The Western Ontario Series in Philosophy of Science. Springer, Dordrecht.

Lawvere, F. William, and Robert Rosebrugh. 2003. *Sets for Mathematics.* Cambridge: Cambridge University Press.

Lawvere, F. William, and Stephen H. Schanuel. 2009. *Conceptual Mathematics: A First Introduction to Categories.* 2nd ed. Cambridge: Cambridge University Press.

Leinster, Tom. 2014. *Basic Category Theory.* Cambridge: Cambridge University Press.

Lewin, David. 2010. *Generalized Musical Intervals and Transformations.* 1st edition. Oxford: Oxford University Press, November 16. ISBN: 978-0-19-975994-1.

Lewis, C. I. 1912. "Implication and the Algebra of Logic (IV)." *Mind* 21 (84): 522–531.

Lindenhovius, Bert. 2014. "Grothendieck Topologies on a Poset." http://arxiv.org/pdf/1405.4408.pdf.

Mac Lane, Saunders. 1998. *Categories for the Working Mathematician.* 2nd edition. New York: Springer, September 25. ISBN: 978-0-387-98403-2.

Mac Lane, Saunders, and Ieke Moerdijk. 1994. *Sheaves in Geometry and Logic: A First Introduction to Topos Theory.* Corrected edition. New York: Springer.

Mazzola, G. 1985. *Gruppen und Kategorien in der Musik: Entwurf einer mathematischen Musiktheorie.* OCLC: 12800630. Berlin: Heldermann. ISBN: 978-3-88538-210-2.

Mazzola, Guerino, and Moreno Andreatta. 2006. "From a Categorical Point of View: K-Nets as Limit Denotators." *Perspectives of New Music* 44 (2): 88–113. ISSN: 0031-6016. https://www.jstor.org/stable/25164629.

McKinsey, J. C. C., and Alfred Tarski. 1944. "The Algebra of Topology." *Annals of Mathematics* 45, no. 1 (1): 141–191.

Moerdijk, Ieke, and Gonzalo E. Reyes. 1991. *Models for Smooth Infinitesimal Analysis.* New York: Springer.

Moore, Gregory H. 2008. "The Emergence of Open Sets, Closed Sets, and Limit Points in Analysis and Topology." *Historia Mathematica* 35 (3): 220–241.

Mortensen, Chris. 2000. “Topological Separation Principles and Logical Theories.” *Synthese* 125, no. 1:169–178.

Moss, Lawrence S., and Rohit Parikh. 1992. “Topological Reasoning and the Logic of Knowledge: Preliminary Report.” In *TARK ’92: Proceedings of the 4th Conference on Theoretical Aspects of Reasoning about Knowledge,* 95–105. San Francisco, CA: Morgan Kaufmann.

Munkres, James R. 2015. *Topology.* 2nd ed. Hoboken, NJ: Pearson.

nLab Authors. 2017. “nLab: Quiv.” Accessed October 2017. https://ncatlab.org/nlab/show/Quiv.

nLab Authors. 2018a. “nLab: adjoint modality.” Accessed May 2018. https://ncatlab.org/nlab/show/adjoint+modality.

nLab Authors. 2018b. “nLab: Aufhebung.” Accessed May 2018. https://ncatlab.org/nlab/show/Aufhebung.

nLab Authors. 2019. “motivation for sheaves, cohomology and higher stacks in nLab.” Accessed March 2, 2019. https://ncatlab.org/nlab/show/motivation+for+sheaves%2C+cohomology+and+higher+stacks.

Noll, Thomas. 2005. “The Topos of Triads.” Edited by H. Fripertinger and L. Reich. *The Colloquium on Mathematical Music Theory* (Nr. 347): 1–26. ISSN: 1016-7692.

Payzant, Geoffrey. 2008. *Glenn Gould, Music and Mind.* 6th Updated & Revised edition. Toronto: Prospero Books.

Peirce, Charles Sanders. 1997. *Collected Papers of Charles Sanders Peirce.* Edited by Charles Hartshorne, Paul Weiss, and Arthur W. Burks. 8 vols. Bristol: Thoemmes Continuum.

Penrose, Roger. 1992. “On the Cohomology of Impossible Figures.” *Leonardo* 25, no. 3:245–247. doi:10.2307/1575844.

Phillips, Tony. 2021. “The Topology of Impossible Spaces.” American Mathematical Society. http://www.ams.org/publicoutreach/feature-column/fc-2014-10#comments.

Popoff, Alexandre, Moreno Andreatta, and Andree Ehresmann. 2018. “Groupoids and Wreath Products of Musical Transformations: A Categorical Approach from poly-Klumpenhouwer Networks.” http://arxiv.org/pdf/1801.02922.pdf.

Praggastis, Brenda. 2016. “Maximal Sections of Sheaves of Data over an Abstract Simplicial Complex.” https://arxiv.org/pdf/1612.00397.pdf.

Regis, Ed. 2021. “No One Can Explain Why Planes Stay in the Air.” doi:10.1038/scientificamerican0220-44. https://www.scientificamerican.com/article/no-one-can-explain-why-planes-stay-in-the-air/.

Reyes, Gonzalo E., and Marek W. Zawadowski. 1993. “Formal Systems for Modal Operators on Locales.” *Studia Logica: An International Journal for Symbolic Logic* 52, no. 4 (4): 595–613.

Reyes, Gonzalo E., and Houman Zolfaghari. 1991. “Topos-Theoretic Approaches to Modality.” In *Category Theory,* edited by Aurelio Carboni, Maria Cristina Pedicchio, and Guiseppe Rosolini, 359–378. Berlin: Springer.

Reyes, Gonzalo E., and Houman Zolfaghari. 1996. “Bi-Heyting Algebras, Toposes and Modalities.” *Journal of Philosophical Logic* 25 (1): 25–43.

Reyes, Marie La Palme, John Macnamara, Gonzalo E. Reyes, and Houman Zolfaghari. 1999. “Models for Non-Boolean Negations in Natural Languages Based on Aspect Analysis.” In *What Is Negation?,* edited by Dov M. Gabbay and Heinrich Wansing, 241–260. Dordrecht: Kluwer Academic: Springer Netherlands.

Reyes, Marie La Palme, Gonzalo E. Reyes, and Houman Zolfaghari. 2008. *Generic Figures and Their Glueings: A Constructive Approach to Functor Categories.* Monza: Polimetrica.

Riehl, Emily. 2016. *Category Theory in Context.* Mineola, NY: Dover.

Robinson, Michael. 2014. *Topological Signal Processing.* New York: Springer.

Robinson, Michael. 2015. "Pseudosections of Sheaves with Consistency Structures." Washington, DC: No. 2015–2.

Robinson, Michael. 2016a. "Sheaf and Cosheaf Methods for Analyzing Multi-Model Systems." http://arxiv.org/pdf/1604.04647.pdf.

Robinson, Michael. 2016b. "Sheaves Are the Canonical Datastructure for Sensor Integration." http://arxiv.org/pdf/1603.01446.pdf.

Robinson, Michael. 2018. "Assignments to Sheaves of Pseudometric Spaces." *Compositionality* 2, no. 2. http://arxiv.org/pdf/1805.08927.pdf.

Schultz, Patrick, and David I. Spivak. 2017. "Temporal Type Theory: A Topos-Theoretic Approach to Systems and Behavior." http://arxiv.org/pdf/1710.10258.pdf.

Schultz, Patrick, David I. Spivak, and Christina Vasilakopoulou. 2016. "Dynamical Systems and Sheaves." http://arxiv.org/pdf/1609.08086.pdf.

Scott, Dana. 1968. "Extending the Topological Interpretation to Intuitionistic Analysis." *Compositio Mathematica* 20:194–210. http://www.numdam.org/article/CM_1968__20__194_0.pdf.

Serra, Jean. 1986. "Introduction to Mathematical Morphology." *Computer Vision, Graphics, and Image Processing* 35, no. 3:283–305.

Serra, Jean, and Luc Vincent. 1992. "An Overview of Morphological Filtering." *Circuits, Systems and Signal Processing* 11 (1): 47–108.

Shepard, Allen Dudley. 1985. "A Cellular Description of the Derived Category of a Stratified Space." PhD diss., Brown University.

Sherman, David. 2004. "Variations on Kuratowski's 14-Set Theorem." http://arxiv.org/pdf/math/0405401.pdf.

Shulman, Michael A. 2008. "Set Theory for Category Theory." http://arxiv.org/pdf/0810.1279.pdf.

Siegel, Aaron N. 2013. *Combinatorial Game Theory.* Providence, RI: American Mathematical Society.

Smith, Justin R. 2014. *Introduction to Algebraic Geometry.* CreateSpace Independent Publishing Platform.

Smyth, M. B. 1983. "Power Domains and Predicate Transformers: A Topological View." In *Automata, Languages and Programming,* edited by Josep Díaz, 662–675. Berlin: Springer.

Spivak, David I. 2014. *Category Theory for the Sciences.* Cambridge, MA: MIT Press.

Srinivas, Yellamraju V. 1991. "Pattern Matching: A Sheaf-Theoretic Approach." Dissertation, University of California, Irvine.

Srinivas, Yellamraju V. 1993. "A Sheaf-Theoretic Approach to Pattern Matching and Related Problems." *Theoretical Computer Science* 112 (1): 53–97.

Stubbe, Isar. 2005. "The Canonical Topology on a Meet-Semilattice." *International Journal of Theoretical Physics* 44 (12): 2283–2293.

Troelstra, A. S., and D. van Dalen. 1988. *Constructivism in Mathematics: An Introduction.* Vol. 1. Amsterdam: North-Holland.

Tymoczko, Dmitri. 2011. *A Geometry of Music: Harmony and Counterpoint in the Extended Common Practice.* 1st edition. New York: Oxford University Press, March 21. ISBN: 978-0-19-533667-2.

Vickers, Steven. 1996. *Topology via Logic.* Cambridge: Cambridge University Press.

Xu, Bing. 2016. "Investigation of the Topological Interpretation of Modal Logics." PhD diss., Fresno State University.

Yanofsky, Noson S. 2003. "A Universal Approach to Self-Referential Paradoxes, Incompleteness and Fixed Points." http://arxiv.org/pdf/math/0305282.pdf.

Zalamea, Fernando. 2012. *Peirce's Logic of Continuity: A Conceptual and Mathematical Approach.* Boston, MA: Docent.

Zalamea, Fernando. 2013. *Synthetic Philosophy of Contemporary Mathematics.* Falmouth, UK: New York: Urbanomic/Sequence Press, January 15. ISBN: 978-0-9567750-1-6.

Index